NATO ASI Series

Advanced Science Institutes Series

A series presenting the results of activities sponsored by the NATO Science Committee, which aims at the dissemination of advanced scientific and technological knowledge, with a view to strengthening links between scientific communities.

The Series is published by an international board of publishers in conjunction with the NATO Scientific Affairs Division

A Life Sciences **B Physics**	Plenum Publishing Corporation London and New York
C Mathematical and Physical Sciences **D Behavioural and Social Sciences** **E Applied Sciences**	Kluwer Academic Publishers Dordrecht, Boston and London
F Computer and Systems Sciences **G Ecological Sciences** **H Cell Biology**	Springer-Verlag Berlin Heidelberg New York London Paris Tokyo

The ASI Series Books Published as a Result of Activities of the Special Programme on CELL TO CELL SIGNALS IN PLANTS AND ANIMALS

This book contains the proceedings of a NATO Advanced Research Workshop held within the activities of the NATO Special Programme on Cell to Cell Signals in Plants and Animals, running from 1984 to 1989 under the auspices of the NATO Science Committee.

The books published as a result of the activities of the Special Programme are:

Vol. 1: **Biology and Molecular Biology of Plant-Pathogen Interactions.** Edited by J. A. Bailey. 1986.

Vol. 2: **Glial-Neuronal Communication in Development and Regeneration.** Edited by H. H. Althaus and W. Seifert. 1987.

Vol. 3: **Nicotinic Acetylcholine Receptor: Structure and Function.** Edited by A. Maelicke. 1986.

Vol. 4: **Recognition in Microbe-Plant Symbiotic and Pathogenic Interactions.** Edited by B. Lugtenberg. 1986.

Vol. 5: **Mesenchymal-Epithelial Interactions in Neural Development.** Edited by J. R. Wolff, J. Sievers, and M. Berry. 1987.

Vol. 6: **Molecular Mechanisms of Desensitization to Signal Molecules.** Edited by T. M. Konijn, P. J. M. Van Haastert, H. Van der Starre, H. Van der Wel, and M. D. Houslay. 1987.

Vol. 7: **Gangliosides and Modulation of Neuronal Functions.** Edited by H. Rahmann. 1987.

Vol. 9: **Modification of Cell to Cell Signals During Normal and Pathological Aging.** Edited by S. Govoni and F. Battaini. 1987.

Vol. 10: **Plant Hormone Receptors.** Edited by D. Klämbt. 1987.

Vol. 11: **Host-Parasite Cellular and Molecular Interactions in Protozoal Infections.** Edited by K.-P. Chang and D. Snary. 1987.

Vol. 12: **The Cell Surface in Signal Transduction.** Edited by E. Wagner, H. Greppin, and B. Millet. 1987.

Vol. 19: **Modulation of Synaptic Transmission and Plasticity in Nervous Systems.** Edited by G. Hertting and H.-C. Spatz. 1988.

Vol. 20: **Amino Acid Availability and Brain Function in Health and Disease.** Edited by G. Huether. 1988.

Vol. 21: **Cellular and Molecular Basis of Synaptic Transmission.** Edited by H. Zimmermann. 1988.

Series H: Cell Biology Vol. 21

Cellular and Molecular Basis of Synaptic Transmission

Edited by

Herbert Zimmermann

AK Neurochemie
Zoologisches Institut der J. W. Goethe-Universität
Siesmayerstraße 70, 6000 Frankfurt am Main, FRG

Springer-Verlag
Berlin Heidelberg New York London Paris Tokyo
Published in cooperation with NATO Scientific Affairs Division

Proceedings of the NATO Advanced Research Workshop on Cellular and Molecular Basis of Neuronal Signalling held in Göttingen, FRG, September 9–13, 1987

ISBN-13: 978-3-642-73174-7 e-ISBN-13: 978-3-642-73172-3
DOI: 10.1007/978-3-642-73172-3

Library of Congress Cataloging-in-Publication Data. NATO Advanced Research Workshop on Cellular and Molecular Basis of Neuronal Signalling (1987 : Göttingen, Germany) Cellular and molecular basis of synaptic transmission / edited by Herbert Zimmermann. p. cm—(NATO ASI series. Series H, Cell biology ; vol. 21) "Proceedings of the NATO Advanced Research Workshop on Cellular and Molecular Basis of Neuronal Signalling, held in Göttingen, FRG, 9–13 September 1987"—Verso t.p.
ISBN-13: 978-3-642-73174-7 (U.S.)
1. Neural transmission—Congresses. I. Zimmermann, Herbert, 1944- . II. Title. III. Series.
QP364.5.N35 1988 599'.0188—dc 19 88-12355

Softcover reprint of the hardcover 1st edition 1988

2131/3140-543210

The Honorary Chairman, Victor P. Whittaker

In more than 40 years of a highly productive scientific career he has had numerous pupils many of whom have followed him as successful neuroscientists in their own right. So it is not surprising that many of the participants had previously worked with him at certain stages of their careers. The contribution to the Workshop exemplified developments which originated from his pioneer work in neurobiochemistry. The study of the synaptosome (he coined the word in 1964) has now aquired a molecular biological dimension. On the occasion of his recent (but far from complete) retirement the spirit of the lectures expressed the gratitude of those who have had the opportunity of working with and learning from him.

PREFACE

Synaptic transmission has a key role to play in the function of the nervous system.The synapse is not simply the site of signal transfer from one cell to another; it displays a great deal of plasticity. It modulates and is subject to modulation, at both the presynaptic and postsynaptic sides. Learning and memory are likely to manifest themselves in the molecular properties of synapses. Almost a century since Sherrington coined the word "synapse" we are at the verge of understanding its function at the molecular level. Major steps towards this goal were the discovery of the physiological properties of the synapse by electrophysiologists at the beginning of the 1950s, the advent of electron-microscopical techniques for the analysis of neuronal fine structure at about the same time and the application of the ultracentrifuge for the study of neuronal biochemistry at the subcellular level.

The isolation of synaptosomes from guinea-pig brain by Whittaker and his colleagues at the beginning of the 1960s was a milestone on our way to disclosing the molecular processes underlying synaptic signalling. It provided the biochemist with an in vitro preparation with which to study transmitter uptake, synthesis, storage and release in the various transmitter systems. It has led to the identification and characterization of individual molecules involved in these processes. But it also allowed the testing of pharmacologically active substances on these mechanisms and thus contributed to their development and therapeutical application in various branches of neuropathology. Finally this study paved the way for the isolation not only of synaptic vesicles by the same group but also of other cellular and subcellular fractions derived from neural tissue and the identification of their molecular properties.

The understanding of a principal cell biological mechanism is almost always the result of an integrative approach, in both thinking and technique. Morphological techniques now allow the documentation of synaptic vesicle exocytosis in the millisecond range. Immunocytochemical studies at the light-microscopical and electron-microscopical level reveal the colocalization of synaptically active substances in many types of neurones which could not have been imagined only a few years ago. The concept of corelease not only of two but presumably of several physiologically active substances from the same nerve terminal demands a substantial revision of the classical model of synaptic signalling. The application of the principal techniques of molecular genetics has greatly accelerated the identification of individual synaptic proteins. The leading edge is still with the characterization of the transmitter receptors and the ion channels which profited in its early phase from the occurrence of natural ligands of high affinity. The present techniques no longer require this advantage and we now have the molecular description of many protein species including synaptic vesicle and synaptic vesicle-associated proteins.

While some of the functional aspects are already well understood at the molecular level others are just being discovered. This includes the functional characterization of presynaptic ion channels - of great importance for the function and functional modulation of the nerve terminal - and of the enzymes involved in the extracellular degradation of substances coreleased with classical neurotransmitters.

The NATO workshop held in Göttingen from September 9-13, 1987 brought together scientists from the various disciplines which are rapidly increasing our knowledge of synaptic transmission. It focused on the integrative approach which allows scientists with diverse and technically highly specialized approaches to learn from each other and so jointly promote the understarding of synaptic function. Of course such a meeting only permits a limited number of contributions from individual fields which by no means can be completely covered. The aim of the workshop was rather to highlight major developments in the study of the cellular and molecular basis of synaptic transmission. The participants undoubtedly learned from each other and gained ideas for the further development of their subject both from the conceptual and technical aspects.

We are indebted to NATO whose financial support made the workshop and the publication of this volume possible.

Herbert Zimmermann

ORGANIZING COMMITTEE

DOWDALL, M.J. — Department of Zoology, School of Biological Sciences, Nottingham University, University Park, Nottingham NG7 2RD, UK

FONNUM, F. — Norwegian Defence Research Establishment, Division for Environmental Toxicology, PO Box 25, N-2007 Kjeller, Norway

STADLER, H. — Abteilung Neurochemie, Max-Planck-Institut für biophysikalische Chemie, Am Fassberg, D-3400 Göttingen, FRG

ZIMMERMANN, H. — AK Neurochemie, Zoologisches Institut der J.W. Goethe-Universität, Siesmayerstr. 70, D-6000 Frankfurt am Main, FRG

CONTRIBUTORS

BETZ, H. — Zentrum für Molekulare Biologie, Universität Heidelberg, Im Neuenheimer Feld 282, D-6900 Heidelberg, FRG

BLAUSTEIN, M.P. — University of Maryland School of Medicine, Department of Physiology, 655 West Baltimore Street, Baltimore, Maryland 21201, USA

BRADFORD, H.F. — Department of Biochemistry, Imperial College of Science and Technology, London SW7 2AZ, UK

CARVALHO, A.P. — Centro de Biologia Celular, Departamento de Zoologia, Universidade de Coimbra, 3049 Coimbra Codex, Portugal

CECCARELLI, B. — Department of Medical Pharmacology, Center for the Study of Peripheral Neuropathies and Neuromuscular Diseases, CNR Center of Cytopharmacology, University of Milan, Via Vanvitelli 32, 20129 Milano, Italy

CLEMENTI, F. — Department of Medical Pharmacology, CNR Center of Cytopharmacology, University of Milan, Via Vanvitelli 32, 20129 Milano, Italy

COUTEAUX, R. — Institut des Neurosciences du C.N.R.S., Départment de Cytologie, Université Pierre-et-Marie-Curie, 75005 Paris, France

DARLISON, M.G.	MRC Molecular Neurobiology Unit, University of Cambridge Medical School, Hills Road, Cambridge CB2 2QH, UK
FOX, G.Q.	Abteilung Neurochemie, Max-Planck-Institut für biophysikalische Chemie, Am Fassberg, D-3400 Göttingen, FRG
HEILBRONN, E.	University of Stockholm, Unit of Neurochemistry and Neurotoxicology, S-106 91 Stockholm, Sweden
ISRAËL, M.	Centre National de la Recherche Scientifique, Laboratoire de Neurobiologie cellulaire et moléculaire, 1, avenue de la Terrasse, 91190 Gif-sur-Yvette, France
KADOTA, K.	Department of Neurochemistry, Psychiatric Research Institute of Tokyo, Setagaya-ku, Tokyo 156, Japan
KLEIN, R.L.	University of Mississippi Medical Center, Department of Pharmacology, 2500 North State St., Jackson, MS 39216-4505, USA
KRIEBEL, M.E.	Department of Physiology, SUNY Health Science Center, 750 E. Adams St., Syracuse, New York 13210, USA
KRNJEVIĆ, K.	Anaesthesia Research Department, McGill University, 3655 Drummond Street, Montréal Québec Canada H3G 1Y6
MAELICKE, A.	Max-Planck-Institut für Ernährungsphysiologie, Rheinlanddamm 201, D-4600 Dortmund, FRG
MICHAELSON, D.	Department of Biochemistry, Tel Aviv University, Ramat Aviv, Israel 69978
NICHOLLS, D.	Department of Biochemistry, University of Dundee, Dundee DD1 4HN, Scotland
PAPPAS, G.D.	Department of Anatomy and Cell Biology, University of Illinois at Chicago, P.O. Box 6998, Chicago, Illinois 60680, USA
PARSONS, S.M.	Department of Chemistry and the Neurosciences Research Program, Institute of Environmental Stress, University of California, Santa Barbara, CA 93106, USA
RAITERI, M.	Istituto di Farmacologia e Farmacognosia, Università di Genova, Viale Cembrano 4, I-16148 Genova, Italy

RICHARDSON, P.J.	Department of Clinical Biochemistry, University of Cambridge Clinical School, Hills Road, Cambridge CB2 2QR, UK
SIMON, E.J.	Departments of Psychiatry and Pharmacology, New York University Medical Center, New York, N.Y. 10016, USA
STJÄRNE, L.	Department of Physiology, Karolinska Institutet, S-10401 Stockholm, Sweden
SUSZKIW, J.B.	Department of Physiology and Biophysics, University of Cincinnati School of Medicine, Cincinnati, Ohio 45267-0576, USA
THURESON-KLEIN,Å.K.	University of Mississippi Medical Center, Department of Pharmacology, 2500 North State St., Jackson, MS 39216-4505, USA
VOLKNANDT, W.	AK Neurochemie, Zoologisches Institut der J.W. Goethe-Universität, Siesmayerstr. 70, D-6000 Frankfurt am Main, FRG
WHITTAKER, V.P.	Abteilung Neurochemie, Max-Planck-Institut für biophysikalische Chemie, Am Fassberg, D-3400 Göttingen, FRG
WIEDENMANN, B.	Medizinische Universitätsklinik, Department of Medicine, University of Heidelberg, Bergheimer Str. 58, 6900 Heidelberg, FRG
WINKLER, H.	University of Innsbruck, Department of Pharmacology, Peter-Mayr-Straße 1, A-6020 Innsbruck, Austria
WITZEMANN, V.	Abteilung Neurochemie, Max-Planck-Institut für biophysikalische Chemie, Am Fassberg, D-3400 Göttingen, FRG

CONTENTS

DEVELOPMENT OF THE SUBJECT AND STATE OF THE ART

STRUCTURAL ORGANIZATION OF THE SYNAPSE AND DYNAMICS OF SYNAPTIC ORGANELLES

PRESYNAPTIC REGULATION OF TRANSMITTER SYNTHESIS AND RELEASE

THE CELLULAR BASIS OF SYNAPTIC TRANSMISSION: AN OVERVIEW

V.P. Whittaker
Arbeitsgruppe Neurochemie, Max-Planck-Institut für biophysikalische Chemie, Göttingen, FRG

INTRODUCTION

The beginning of cell biology and the role of instrumentation in its development The recent growth of interest of biologists of all complexions in the techniques of molecular genetics has tended to thrust into the background an earlier revolution in our understanding of cellular function which has by no means yet run its full course and which led to the emergence of a new subject, that of *cell biology*. One of the two instruments of that revolution was the electron microscope: or, rather, the development of reliable, user-friendly commercial electron microscopes and ancillary equipment such as knife-makers and ultratomes as well as improvements in embedding plastics.

Of course 'user-friendly' is a relative term: I well remember the Siemens ÜM 11 which required the evolution of a new type of human being, *Homo sapiens Siemensiensis*, who had very long arms to reach the adjustment knobs at the top of the instrument while still remaining seated in front of the fluorescent screen! The replacement of methyl methacrylate - a very destructive embedding plastic - by epoxy resin, by Audrey Glauert in the early fifties, was due to the fortunate accident that her brother worked for the firm in Duxford, near Cambridge (UK) that developed Araldite - the magic glue that held together the wooden second-world-war photographic reconnaisance aircraft, the Mosquito, wich could penetrate the German radar screen and bring back photographs of the V1 and V2 launch sites. On such coincidences is progress in science very often dependent.

The other instrument of this revolution was the Model L preparative ultracentrifuge developed and marketed by Spinco, later bought up by Beckman. It was at last possible to sediment and concentrate the smallest particles generated by homogenization and the introduction of the swing-bucket rotor, the old SW 39 and its three 5 ml cups and later the larger capacity SW 25 with its three 38 ml cups enabled density gradient separations to be performed on what was for then a relatively large scale.

NATO ASI Series, Vol. H21
Cellular and Molecular Basis of Synaptic Transmission
Edited by H. Zimmermann

Subcellular fractionation - a process whereby cells are broken open by the application of liquid shear and the resultant homogenate fractionated by a combination of moving boundary ('differential') and density-gradient centrifuging - is a much older technique than generally realized. Thus Benson and Hoerr succeeded in isolating mitochondria - recognized as Janus-Green-positive organelles - in the thirties and so confirmed that these structures were true organelles and not staining artefacts, as many believed at the time. Also, ahead of full morphological characterization, Claude developed his four-fraction scheme of subcellular fractionation, whereby what we would now call nuclear, mitochondrial, microsomal and cell-sap fractions were separated in increasingly intense centrifugal fields; however, only one of these, the nuclear, could be characterized by the current light-microscopic methods. Hogeboom, Schneider, Lehninger, Siekevitz and others characterized these fractions enzymically and this led to the main generalization of cell biology - the concept of the specific metabolic roles of the various cell organelles within the total economy of the cell, a practical outcome of which is the use of enzymes or other organelle-specific components as 'markers', the distribution of which throughout the fractionation scheme is such a valuable guide to the purification of a particular organelle.

Identification of subcellular fractions The pioneer work which correlated the biochemical studies of subcellular fractions with the new knowledge of cell structure provided by the electron microscope was largely initiated by Palade and Siekevitz at the Rockefeller Institute. Thus the 'large granule fraction' was identified as consisting mainly of mitochondria and the 'small granule fraction' as small, vesiculated membrane fragments mainly derived from the rough and smooth endoplasmic reticulum discovered by Porter and much studied morphologically by Palade and by Sjöstrand. From this union of electron microscopy and subcellular fractionation the new science of cell biology was born and this was soon institutionalized by the Rockefeller-sponsored Journal of Cell Biology and similar journals in other countries.

Most of the pioneer work was done with liver: brain tissue was probably regarded - and rightly - as too complex to lend itself to subcellular fractionation. There was, indeed, throughout the fifties, much discussion of the true enzymic makeup of mitochondria and other subcellular particles. By inserting an extra step into the four-fraction scheme, DeDuve and associates succeeded in obtaining a fraction enriched in particles - which they named lysosomes - which sequestered the acid hydrolases of the liver cell, a group of enzymes collectively capable of breaking down most cell constituents. Subsequently these were identified as the peribiliary dense bodies. A somewhat similar sequence of events later enabled this group to identify yet another liver organelle, the peroxysome, as the cellular site of a group of oxidizing enzymes, among them uricase. This established mitochondria as the organelles responsible for cell respiration and as the site of oxidative phosporylation and the electron transport chain.

APPLICATION OF SUBCELLULAR FRACTIONATION TO BRAIN TISSUE

The problem of bound acetylcholine My own interest in applying subcellular fractionation techniques to brain tissue surfaced when I moved from Cincinnati to Babraham and was sparked off by wondering what 'bound acetylcholine' was and what structures it was associated with. It was already known that if brain tissue was minced in iso-osmotic saline, very little acetylcholine was extracted: extraction required trichloroacetic acid or other protein denaturing agents. I soon established that homogenization in iso-osmotic sucrose also conserved bound acetylcholine; this opened the way for following the distribution of bound acetylcholine in the various subcellular fractions and for devising methods of concentrating the relevant subcellular particles in one fraction.

My inspiration for this project was derived from two sources. The first of these was the work of DeDuve and colleagues on the lysosome; here it had been demonstrated that the lysosomal contents were sequestered within an osmotically sensitive lipoprotein membrane and so prevented from acting indiscriminately on cell constituents. The second was the recent isolation of adrenaline-containing chromaffin granules from the adrenal medulla by Blaschko, Hagen and Welsh in Oxford and Hillarp in Lund. Here a pharmacologically active hormone and neurotransmitter was likewise sequestered within an osmotically labile organelle. Might not acetylcholine be similarly sequestered?

Fig. 1 shows the results I obtained. The primary fraction showing the highest bound acetylcholine content was the crude mitochondrial (P_2) fraction; on subfractionating this through a simple step-gradient there was a clear separation of most of the mitochondrial markers from bound acetylcholine, which remained associated with particles of intermediate density (fraction B).

<u>The discovery of the synaptosome and the isolation of synaptic vesicles</u> The relatively crude fixation, dehydration and embedding technique I initially used for identifying the acetylcholine-rich fraction led me to believe (Whittaker 1959) that I had isolated synaptic vesicles, then recently discovered and postulated to be responsible for the then also recently discovered quantized release of acetylcholine from nerve terminals. Further consideration convinced me that particles of the size range of synaptic vesicles could not possibly have sedimented in the way my 'acetylcholine-particles' had done, if they had a density of a reasonable magnitude. I needed to reevaluate my morpholgical data.

Here I had the good fortune to meet George Gray through mutual friends. George advised me on better preparation techniques and his pioneer work on the fine structure of cortical synapses enabled him instantly to identify the main component particles of the 'B fraction' as detached or 'pinched-off' nerve terminals (Whittaker 1960; Gray and Whittaker 1960,1962). The synaptic vesicles were concentrated in the fraction but enclosed within the terminal bags. The story has been told in a recent 'Citation Classic' (Gray and Whittaker 1981).

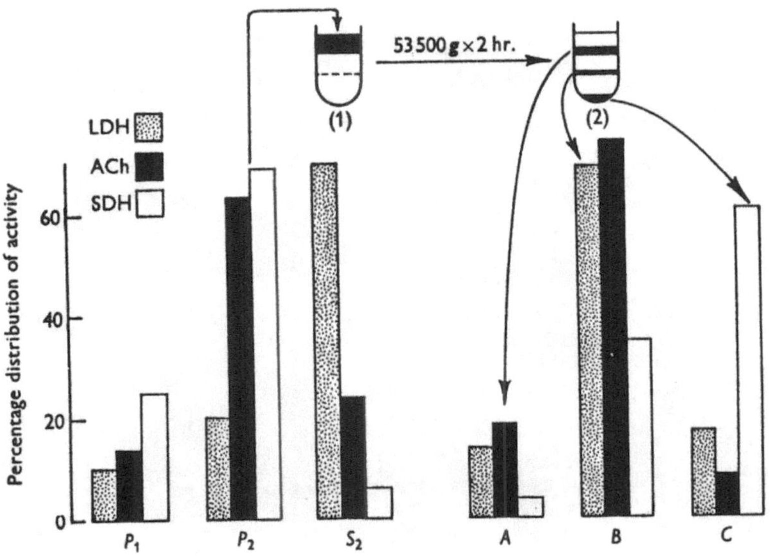

Fig. 1 *The subcellular distribution of bound acetylcholine. The blocks show the distribution (as a percentage of the activity of the parent fraction recovered in its subfractions) of bound acetylcholine* (ACh, black blocks) *and markers for the cytosol* (lactate dehydrogenase, LDH, stippled blocks) *and mitochondria* (succinate dehydrogenase, SDH, white blocks) *in the primary subcellular fractions P_1 (nuclear), P_2 (crude mitochondrial) and S_2 (microsomes and cytosol), and in the subfractions A (myelin), B (synaptosomes) and C (mitochondria) of fraction P_2. The parent fraction for P_1, P_2 and S_2 is the homogenate, that of the A, B and C, fraction P_2.*

Synaptosomes - as we were later (Whittaker et al 1964) to call the pinched-off nerve terminals - retain all the morphological features of the presynaptic nerve terminals *in situ:* a bulbous package 0.5 to 0.6 µm in diameter contains numerous synaptic vesicles 0.05 µm in diameter and one or more small mitochondria and often has a length of postsynaptic thickening adhering to it (Fig. 2). Presynaptic dense projections, endoplasmic reticulum and microtubules can also be identified by appropriate staining methods. Clearly, under my original fixation conditions, the package had burst, releasing the vesicles. Since mitochondrial morphology had been reasonably well preserved by the original preparative technique, we could no longer assume that synaptosomes were as rugged as mitochondria.

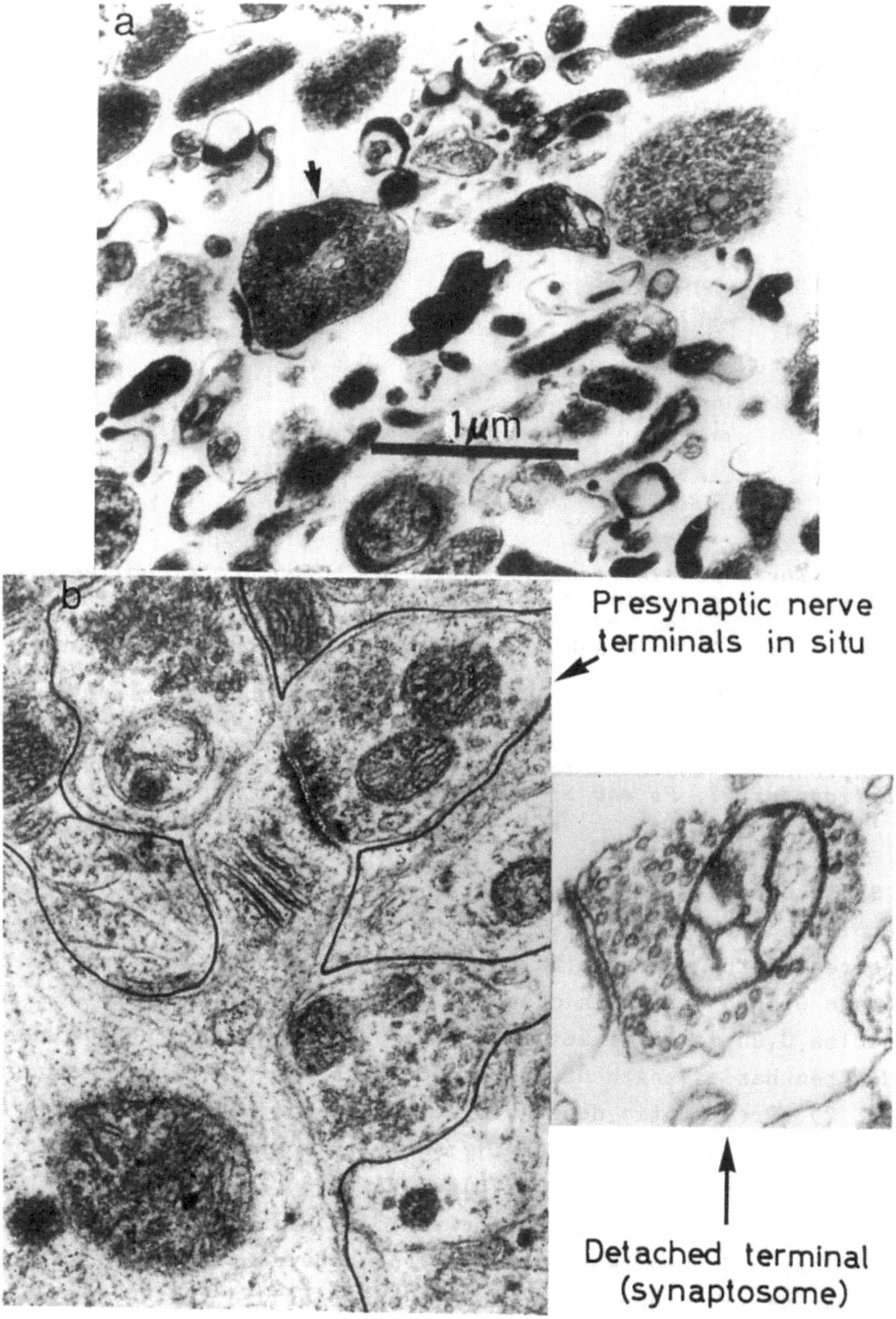
a
1 μm
b
Presynaptic nerve terminals in situ
Detached terminal (synaptosome)

Fig. 2 (left) *(a) Electron micrograph of fraction B of the fractionation scheme shown in Fig. 1 (from Gray and Whittaker 1962). The identification of the main structures present as pinched-off nerve endings was particularly evident from particles like that indicated by the arrow which shows all the main features of the nerve terminal, including synaptic vesicles, a small mitochondrion and an attached piece of post-synaptic membrane. This identification is further exemplified in (b) where a synaptosome is compared with a presynaptic nerve terminal* in situ.

Fig. 3 (above) *Negative staining (Horne and Whittaker 1962) of (a) synaptosome (b) synaptic vesicles.*

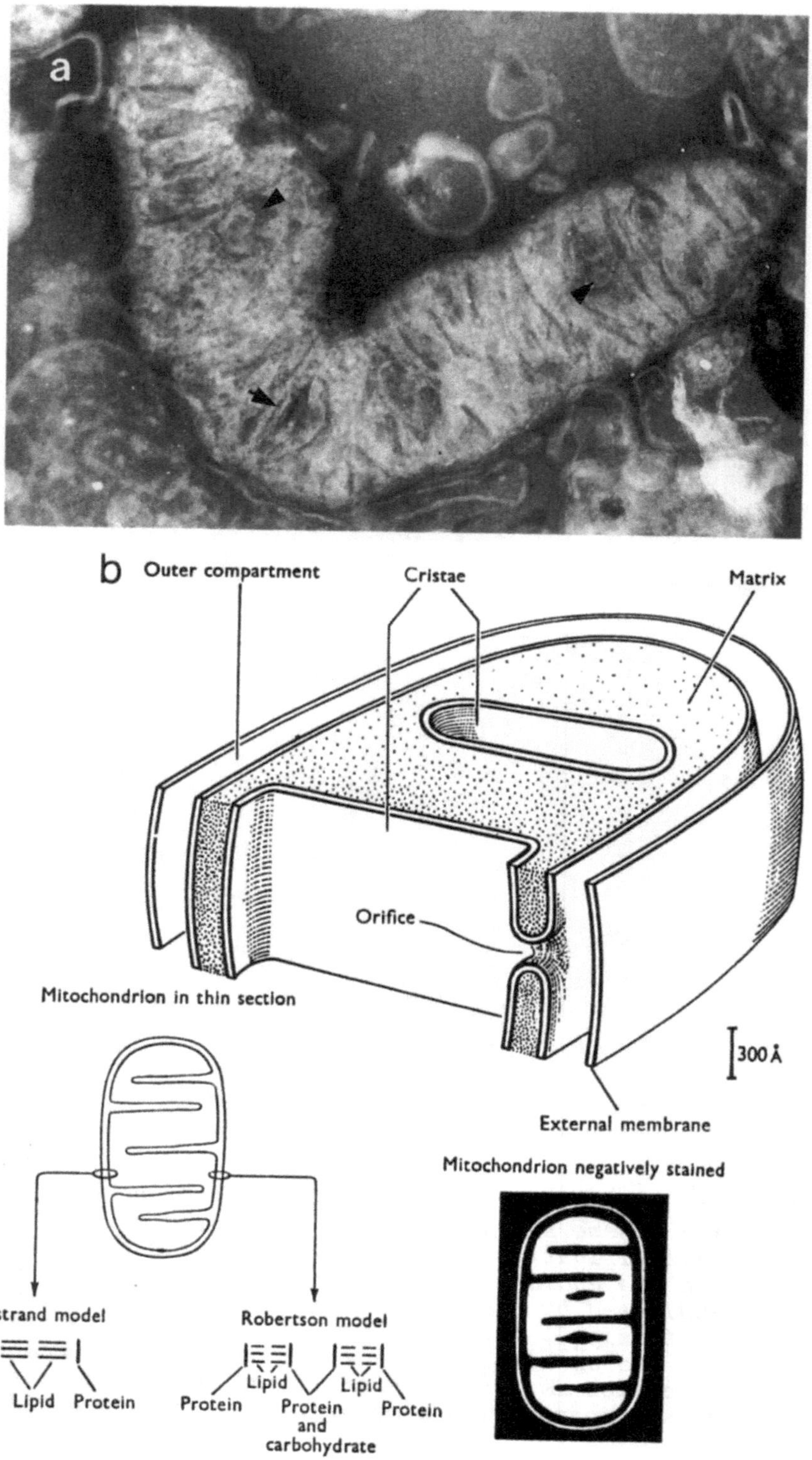
a
b
Outer compartment
Cristae
Matrix
Orifice
Mitochondrion in thin section
300 Å
External membrane
Mitochondrion negatively stained
Sjöstrand model
Protein
Lipid
Protein
Robertson model
Lipid
Lipid
Protein
Protein
and
carbohydrate
Protein

Gray's collaboration lasted only two months, but after that it was fortunately possible to set up, with the help of the Wellcome Foundation, an electron microscope unit in my Institute (the Institute of Animal Physiology at Babraham, near Cambridge UK) and to secure the appointment of Bob Horne as its Director. As we continued the research on synaptosomes and, more especially, concentrated on isolating the synaptic vesicles from them, we badly needed a rapid method of characterizing the fractions morphologically. We found this, at Bob Horne's suggestion, in the use of negative staining, a technique in which the fixed particles are, so to speak, embedded in a film of dried sodium phosphotungstate which reveals their fine structure in great detail in negative contrast (Horne and Whittaker 1962; Figs 3 and 4). Mitochondria are particularly good subjects for this method (Fig. 4a) and the distribution of stain showed that the intramembranous and cristal spaces communicate with each other and are hydrophilic thus enabling us to decide between two rival theories (Whittaker 1963; Fig. 4b).

The discovery that the particles of the B band were pinched-off nerve terminals explained something which had hitherto been ignored: the presence in the fraction of small, but significant amounts of the cytosolic marker lactate dehydrogenase and the mitochondrial marker succinate dehydrogenase (Fig. 1). These had originally been attributed to contamination, but they were now seen to be essential components of the fraction, the former expressing the entrapped cytosol of the synaptosome, the latter, its terminal mitochondria.

I now concentrated on obtaining a pure preparation of synaptic vesicles uncontaminated by the other organelles of the synaptosome (Whittaker et al 1964). The team (Fig. 5) consisted of Arthur Michaelson, Jeanette Kirkland and myself, assisted by

Fig. 4 (left) *(a) Negatively stained mitochondrion. (b) The structure of the mitochondrion deduced from its negatively stained image (Whittaker 1963). The constricted openings of the cristal spaces are imaged in (a) (arrows) and were interpreted by means of stereo-pairs. The observations favour the Palade-Robertson structural model by showing that the cristal spaces are hydrophilic.*

Fig. 5 *The team that isolated pure monodisperse synaptic vesicles from synaptosomes. Right to left: Victor Whittaker, Nick Fincham, Gordon Dowe, Jeanette Kirkland, Arthur Michaelson and Flo Coad (cleaner).*

Gordon Dowe and Nick Fincham; Flo Coad kept us well supplied with clean glassware! Lysed synaptosomes were placed on a sucrose step gradient and separated into five fractions designated O, D, E, F, G, H and I. The frog rectus abdominis muscle, used in our earlier work, proved too insensitive and we switched to the leech dorsal muscle micro-assay originally devised by Gaddum. We improved this, turning it into a reliable micro-assay (threshold, 1-2 $pmol.ml^{-1}$) which we still routinely use. I shall never forget the thrill I experienced when I saw for the first time a morphologically pure fraction of synaptic vesicles covering the grid with that crisp but soft contrast characteristic of particles in negative stain bound by a limiting lipoprotein membrane (Fig. 3b). Although very late, Arthur and I immediately returned to the lab and assayed the fraction, first after acid treatment to release bound acetylcholine, then without treatment to confirm that the transmitter was in a bound state and unable to produce a response when the vesicles were intact. Not satisfied, as Arthur recalls, I then asked for some cholinesterase to check that the liberated activity could be destroyed by this enzyme and was thus authentic acetylcholine. There was none available so I pricked my finger and put a drop of blood into the test-tube. The activity disappeared!

Bound acetylcholine was associated, in our fractionation scheme, with two fractions, that of monodisperse vesicles (fraction D) and another (fraction H) of vesicles attached to fragments of presynaptic plasma membrane. With Larry Chakrin (Chakrin et al 1972) and later with Alan Barker and Mike Dowdall (Barker et al 1972) in work confirmed by Irene von Schwarzenfeld (1979) we found that acetylcholine, newly synthesized from a radioactive precursor, preferentially entered the fraction H vesicles. Von Schwarzenfeld was later to show (see below) with false transmitters that the fraction H vesicles were the preferred source of evoked transmitter release. Thus was obtained the first indication of synaptic vesicle metabolic heterogeneity which we now recognize to be a fundamental phenomenon which resolves some of the earlier paradoxes of vesicular transmitter release.

It is often stated in the literature that De Robertis, an Argentinian electron microscopist, independently discovered synaptosomes. In fact, it is impossible to find any publication of his - abstract, review or paper - which predates my oral presentations in Philadelphia in April 1960 and Varenna in June 1960 (the latter published as a chapter in a book; Whittaker 1960) or my abstract with Gray in July 1960 (Gray and Whittaker 1960). Following my 1958-59 publications, which he had read, De Robertis's aim was to isolate synaptic vesicles; he expected them to be released by homogenization and to be recovered in the microsome fraction. As he told me in Varenna, and as was apparent from his presentation, his only 'marker' was vesicle morphology. Had he assayed acetylcholine, as we did, he would have concentrated on the crude mitochondrial fraction and subfractionated this. He had electron micrographs of this fraction and I, as a young and naive scientist was able to point out the synaptosome profiles in these during our conversation before his lecture. None of this is apparent in the published version submitted after the meeting.

His results puzzled us until we realized that he had included cerebellum in his homogenates, whereas we had excluded it. The large mossy-fibre endings in cerebellum (Israel and Whittaker 1965) tend to break up on homogenization and release their vesicles which are recovered in small numbers in the microsomal fraction, where De Robertis, in his Varenna lecture, was able to point to them.

De Robertis has a better claim to have isolated vesicles from synaptosomes before we did, but we already knew, before the appearance of his paper in Nature (De Robertis et al 1962), that pure vesicles could not be obtained from synaptosomes by moving boundary sedimentation; such preparations are contaminated with, among other things, significant amounts of plasma membranes. This led to considerable controversy about the composition of synaptic vesicles (see Whittaker 1964). This served a useful purpose, as Frode Fonnum has pointed out, since, when a new technique is introduced, it is important to get the facts right otherwise the literature may be muddied for years. De Robertis (Lapetina et al 1968) eventually tacitly accepted our point of view when he subfractionated his vesicle fraction on a step gradient, a simplified form of our 'O-I gradient'.

Synaptosomes retain, to a degree that still astonishes me, almost all the properties - morphological, biochemical and physiological - of the nerve terminal *in situ*. These have been extensively reviewed (Whittaker 1984a) and many aspects of them will be covered in subsequent presentations in this Workshop. They have also been used extensively in industrial pharmacological research as a test system for drugs which act on synaptic transmission. I often wish I had patented the synaptosome, but at the time the emergence of the Yuppy-scientist still lay in the future! The ability of synaptosomes to take up, in carrier-mediated fashion, a variety of transmitters, transmitter precursors and metabolites such as glucose and adenosine is well known and allows the testing of drugs and toxins on such processes. Synaptosomes are in effect miniature cells and allow the properties of mitochondria to be studied *in situ*. Their membranes are rich in channels and in receptors for substances - some endogenous - which regulate transmitter release. They contain not only the classical neurotransmitters but also - as Cleugh, Gaddum and I were the first to show (Cleugh et al 1964) many years ago with substance P in an almost forgotten paper in the Journal of Physiology - they contain, in highly localized fashion, many neuropeptides. We shall be hearing from Blaustein, Carvalho, Nicholls, Raiteri, Michaelson and others more about these aspects later.

Impact on subcellular fractionation Our work on the synaptosomes, it is fair to say, greatly enlarged biologists' appreciation of what could be achieved by subcellular fractionation techniques and stimulated their ingenuity. It became possible, using milder methods of tissue disruption, to extend the techniques to the isolation of cells and even of multicellular structures (Whittaker 1976). It is inappropriate to call such methods 'subcellular fractionation', so we named our Advanced Study Courses in 1973-76 'Tissue fractionation techniques in neurochemistry'. Among the structures of this type that have been isolated include cerebral capillaries, neuronal cell bodies, oligodendrocytes, astrocytes and, to give two non-nervous examples, adrenal medullary cells and whole kidney glomerulae. Almost any structure can now be isolated from nervous tissue by tissue fractionation techniques: growth-cones, dendrites and axons must be added to the list already given. Particularly when embryonic material is used, the isolated cells are often viable when held under the conditions of tissue culture. An important task for the future is fully to characterize the functional membrane proteins in these diverse membrane preparations.

At the other end of the scale, tissue fractionation techniques merge into the methods of membrane protein isolation. Synaptic plasma membrane fragments, prepared by the fractionation of lysed synaptosomes may for example (Ducis and Whittaker 1985) be used in the study of membrane transporters after energizing them with imposed ionic gradients, the specificity of the ion-requirements of the transporter studied, the system reconstituted in artificial membranes, and finally the transporter isolated and its sequence determined by the cDNA cloning technique.

Newer tissue fractionation methods In recent years, the classical centrifugal moving boundary and density gradient separation methods have been supplemented in various ways. The invention of the zonal rotor, loaded and unloaded dynamically (i.e. while spinning) has greatly increased resolution and capacity (Whittaker et al 1972); of other methods, exclusion chromatography on porous glass (Morris 1973; Nagy et al 1976; Giompres et al 1981a) or plastic beads (Stadler and Tsukita 1984), has been the most successful. Immunoabsorption methods, discussed below, have a great potential for future development.

<u>The problem of neuronal heterogeneity</u> Neurones and their processes may be classified in various ways and it is entirely conceivable that each neurone in the mammalian brain is a unique individual, different from all its neighbours in its pattern of surface markers and connexions.

A simpler classification is by transmitter content, though, as a result of the discovery of the neuropeptides, this may be complex enough! It would be extremely useful to separate out neurones - and synaptosomes - of single transmitter types. This is not possible by conventional subcellular fractionation techniques: there is not enough difference in biophysical properties to permit this.

The best approach is to utilize tissues containing one main type of innervation. In the adrenergic system one can use adrenal medullary cells whose catecholamine-storing granules are in effect hypertrophied synaptic vesicles. Though not a perfect model, its relevance to adrenergic transmission is clear from the ability of these cells to assume a neuronal morphology in tissue culture.

Though so far little exploited, crustacean muscle should be a potential source of glutamergic or 'gabaergic' terminals, though the abundance of nerve terminals in such tissues is low. The electric lobes of *Torpedo*, which contain fair numbers of synapses of only one type - thought to be glutamergic - should be a better source, since synaptosomes can be obtained from this tissue (V.P. Whittaker, unpublished observations).

THE ELECTRIC ORGAN OF *TORPEDO* AS A MODEL CHOLINERGIC SYSTEM

<u>Cholinergic synaptic vesicles: isolation and functional states</u> For cholinergic terminals, the electric organs of torpedine electric rays have proved an excellent source: it contains 500 to 1000 times more synaptic material than muscle. We introduced this model system into cholinergic neurochemistry in 1964 (Sheridan and Whittaker 1964; Sheridan et al 1966). Our first attempts to isolate synaptic vesicles were quite successful; however purification on a density gradient was hampered by the difficulty of reproducibly redispersing the sedimented vesicles. Later work included a small but significant imporvement: the application to the gradient of the original cytoplasmic extract without prior sedimentation of the vesicles (Israël et al 1970). Other improvements have been: the rapid comminution of large amounts of electric organ by freezing it in liquid nitrogen followed by the mechanical crushing of the brittle tissue, and the use of large capacity high resolution density-gradient centrifuging in a zonal

rotor (Whittaker et al 1972; Ohsawa et al 1979). For highest purity, a preliminary discontinuous gradient separation to remove soluble cytoplasmic protein is a useful step (Tashiro and Stadler 1978).

Such techniques have enabled my group in recent years to prepare biochemically and morphologically pure cholinergic synaptic vesicles, to characterize a number of their functional components (for review see Whittaker 1984c) and to define, in considerable detail, their biophysical properties especially their response to changes in osmotic loading and external osmotic pressure (Giompres and Whittaker 1986). Vesicles may be isolated from unstimulated tissue or from tissues stimulated for varying lengths of time after varying periods of recovery.

Such studies have led to the recognition of three main subpopulations of vesicles in different functional states: fully charged 'reserve vesicles', with contents of acetylcholine in excess of 200 000 molecules per vesicle (Ohsawa et al 1979); recycling vesicles with a reduced load, thicker membrane and smaller water content (Zimmermann and Denston 1977; Giompres et al 1981b); and immature, newly-arrived vesicles transported to the terminal from their site of formation in the cell body by fast axonal transport (Kiene and Stadler 1987). The three types differ in density and may be separated by high-resolution density-gradient separation in a zonal rotor. The proportion of the total vesicle population in these functional states - and especially their distribution between the reserve and recycling pools - is a function of previous activity. All three types of vesicle are present in freshly excised electric tissue but after perfusion for only 2 h, during which time both axonal transport and impulse traffic are cut off, most of the immature and recycled vesicles have time to fill up and join the reserve pool which now contains nearly 80% of the tissue acetylcholine (Table 1). This redistribution is indicated by the sharper and more symmetrical acetylcholine peak in the zonal density gradient (Fig. 5).

The development of a perfused electric organ preparation (Zimmermann and Denston 1977) has made it possible to introduce isotopically labelled acetylcholine precursors into the terminals and thus to explore the kinetics of acetylcholine turnover in the cytoplasmic and vesicular pools and their interrelation (Agoston et al 1986). From this it has become clear that the cytoplasmic pool comprises only about 20% of the total in resting tissue (Weiler et al 1982), even less in stimulated (Agoston et al 1986) and is subject to futile recycling - that is, continuous turnover characteristic of a metabolic pool with regulatory properties. When a recycling vesicle pool is generated by stimulation this fills up from the cytoplasmic pool and comes into isotopic equilibrium with it. By contrast, the fully charged reserve vesicles fail to exchange their unlabelled acetylcholine with the cytoplasmic pool (Kosh and Whittaker 1985). Acetylcholine released on stimulation has the same specific radioactivity as that of the recycling vesicular (and cytoplasmic) pool (Suszkiw et al 1978).

False transmitters We can however distinguish between the cytoplasmic and recycling vesicular pool by introducing false transmitter precursors into the system (reviewed by Whittaker and Luqmani 1980). These are choline analogues differing from choline itself by 2 or more carbon atoms. These compounds (and their acetate esters) in general distribute themselves differently between cytoplasm and recycling vesicles from newly-synthesized acetylcholine. Thus by measuring the ratio in which they are released on stimulation and matching this with that of the labelled pools we can identify the pool from which release has taken place (von Schwarzenfeld 1979; Luqmani et al 1980). The result is unequivocal: resting release (or leakage) takes place from the cytoplasmic pool (and is part of the futile recycling mechanism): stimulated release takes place from the recycling vesicle pool.

The presynaptic plasma membrane Our early subcellular fractionation studies on electric organ (Sheridan et al 1966) showed that synaptosomes were formed on homogenization, albeit in very low yield, probably because the mechanical conditions favouring synaptosome formation in brain do not prevail in electric tissue. However, by using young fish, the yield can be improved. The size of the synaptosomes varies according to the procedure; those originally prepared in my group (referred to as T-sacs; Dowdall and Zimmermann 1977) were rather small; later, the procedure was modified to produce larger ones (Israël et al 1976).

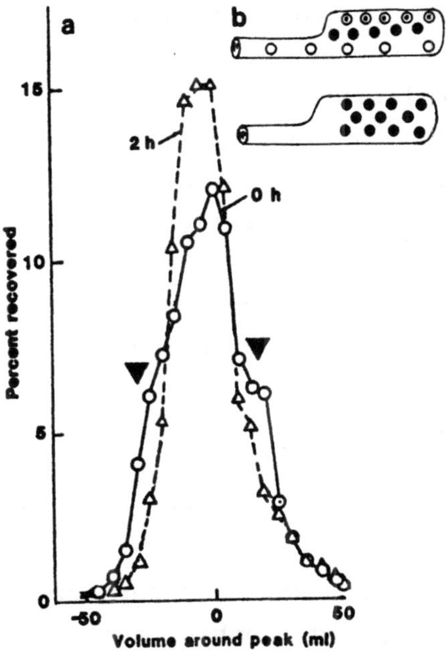

Fig. 6 (a) Effect of perfusion for two hours (2 h) on the density distribution of vesicular acetylcholine in a zonal gradient. Note that the shoulders (arrows) indicative of vesicle heterogeneity in freshly excised (0 h) tissue have largely disappeared at 2 h. The inserts (b) show how this could come about by the cutting off of the supply of immature vesicles from the axons and the cessation of impulse traffic or adventitious stimulation giving immature (o) and recycling (⊙) vesicles the chance to fill up with acetylcholine and join the reserve pool (●). The acetylcholine contents and relative proportions of the three vesicle populations in this experiment at 0 and 2 h are given in Table 1.

Table 1 Distribution of acetylcholine in vesicle pools

Recovery time (h)	% recovered	VP_0 molecules x 10^{-3} per vesicle	% recovered	VP_1 molecules x 10^{-3} per vesicle	% recovered	VP_2 molecules x 10^{-3} per vesicle
0	8	67 ± 28 (7)	55	293 ± 37 (6)	37	140 ± 37 (3)
2	3	11 ± 2 (2)	79	246 ± 28 (7)	18	87 ± 9 (4)

However there is very little difference in properties between the two preparations.

Electromotor synaptosomes in bulk lack the metabolic competence of mammalian brain synaptosomes (Richardson and Whittaker 1981) though some give signals with fluorescent probes indicative of the presence of a membrane potential (Meunier 1984). The most likely explanation is that a few percent (possibly those containing mitochondria) are well-sealed and capable of ion-pumping, and the rest are poorly sealed and deficient in metabolism, acetylcholine and choline acetyltransferase content. The latter would explain why the increase in the concentration of cholinergic markers in the preparation relative to that of the original tissue is so much lower than would be expected when a structure in which the markers are specifically located but accounting for 3% or less of a tissue is isolated and purified.

One of the most useful features of the synaptosome preparation is that it is a good source of presynaptic plasma membranes. Such membranes possess a variety of functions: they contain the transporters for energy-yielding substrates like glucose, for transmitters or their precursors and for adenosine; they possess receptors for regulatory and trophic factors; they are the partner of the synaptic vesicles in exocytosis; and they may present surface markers of importance in cell-cell recognition. Vesiculated fragments of these membranes from electric organ avidly take up choline and its near analogues under the influence of imposed, inwardly directed sodium and outwardly directed potassium gradients; the transporter has been sucessfully transferred in functional form to artificial lipid membranes in reconstitution experiments (Ducis and Whittaker 1985) and considerable progress has been made in isolating it in pure form. The choline binding site appears to be on a protein of molecular mass 40-50 kDa and isoelectric point 5.1 and its function is sodium-dependent and hemicholinium-sensitive (Rylett and Whittaker 1987).

Another potentially valuable feature of this preparation is that it should be a good source of putative cholinergic-specific surface markers. To investigate this we raised antisera to electromotor presynaptic plasma membranes by injecting them into sheep. The presence of antibodies which recognized cholinergic-specific surface antigens common to *Torpedo* and mammals was then tested for by seeing whether they could activate the complement-induced lysis of the cholinergic subpopulation of guinea-pig cortical synaptosomes (Jones et al 1981). The test was positive: about 85% of the soluble cholinergic marker choline acetyltransferase was released, but no soluble markers for other transmitter types and about 6% of soluble general cytoplasmic markers (e.g. lactate dehydrogenase) was released, which is approximately the proportion of cholinergic synaptosomes in the total population.

Subsequent work (Richardson et al 1982; Ferretti and Borroni 1986) has shown that the antigen (designated Chol-1) is a family of minor gangliosides of varying sialic acid content with, probably, a common epitope. The main component in pig brain is a trisialoganglioside containing glucose, galactose and N-acetylgalactosamine (S. Sonnino, personal communication). Its function is not yet known, but embryological studies suggest that its appearance coincides with functional cholinergic synaptogenesis (Fiedler et al 1986). Possibly it is necessary to consolidate synapse formation, perhaps by providing the optimum lipid environment for a presynaptic receptor for components of the postsynaptic membrane, thereby strengthening their interaction.

By use of a column of an immobilized high-titre anti-sheep IgG monoclonal antibody, cholinergic synaptosomes pretreated with anti-Chol-1 antiserum may be separated from synaptosomes of non-cholinergic origin (Richardson et al 1984). These are proving extremely useful for investigating the structure and function of mammalian central cholinergic nerve terminals, and the technique represents a new development in subcellular fractionation of considerably wider potential application. We shall be hearing more about this from Richardson and Bradford.

FUTURE PROBLEMS IN THE CELL BIOLOGY OF SYNAPTIC FUNCTION

Exo- and endocytosis Although much detailed work still needs to be done, we can claim that the functional components of the cholinergic system in cytoplasm, synaptic vesicles and presynaptic plasma membranes have been fairly well characterized, further, that the role of the synaptic vesicles, in storing transmitter and releasing it on stimulation, is also well established. Recent work has also given important insights into the mechanism of vesicle translocation from cell body to terminal. What, however, is still largely *terra incognita*, in any system, is the mechanism of vesicle exocytosis and retrieval. Attempts that have been made to reconstitute this process, central to transmitter release, have not been very successful. As so often in neurobiology, the answers may come from the study of analogous secretion processes in quite different cell types.

Role of neuropeptides Another problem still far from being understood is the role of the neuropeptides. At some central terminals they seem to be transmitters in their own right; at others, they are associated with 'classical' transmitters and are then thought to have a modulating action. Vasoactive intestinal polypeptide (VIP) is often found in cholinergic terminals, and our own work (D.V. Agoston et al, to be published) and that of others indicate that it is preferentially released at high rates of stimulus while acetylcholine is preferentially released at lower rates. Does this give us a clue as to its possible mode of action? Could VIP regulate the supply of acetylcholine, e.g. by stimulating the synthesis of choline acetyltransferase, during periods of intense cholinergic activity (Collier et al 1987)?

We have recently found (Agoston and Conlon 1986) that a type of VIP is present in my favourite model system, the electromotor neurone. This will give the model a new lease of life and provides an excellent system on which to study the role of VIP.

CONCLUDING REMARKS

I hope this overview will have helped us to appreciate the solid advances made during the past 25 years in our understanding of the cell biology of the synapse and also to define the 'grey areas' that still remain. In pursuing our quest for a fuller understanding of synaptic function, the newer techniques of molecular genetics are contributing greatly to determining the structure of receptors, channels and transporters, just as, in an earlier phase, the technique of electron microscopy contributed - and continues to contribute - essential information. Techniques alone, however, cannot specify goals and in the pursuit of an understanding of synaptic function at the fine-structural and molecular level, an integrated approach is necessary. This workshop will contribute greatly to this goal.

REFERENCES

Agoston DV, Conlon JM (1986) Presence of vasoactive intestinal polypeptide-like immunoreactivity in the cholinergic electromotor system of *Torpedo marmorata*. J Neurochem **47**:445-453

Agoston DV, Dowe GHC, Fiedler W, Giompres PE, Roed IS, Walker JH, Whittaker VP, Yamaguchi T (1986) A kinetic study of stimulus-induced vesicle recycling in electromotor nerve terminals using labile and stable vesicle markers. J Neurochem **47**:1584-1592

Barker LA, Dowdall MJ, Whittaker VP (1972) Choline metabolism in the cerebral cortex of guinea pigs: stable bound acetylcholine. Biochem J **130**:1063-1080

Chakrin LW, Marchbanks RM, Mitchell JF, Whittaker VP (1972) The origin of the acetylcholine released from the surface of the cortex. J Neurochem **19**:2727-2736

Cleugh J, Gaddum JH, Mitchell AA, Smith MW, Whittaker VP (1964) Substance P in brain extracts. J Physiol **170**:69-85

Collier B, Araujo DM, Lapchak PA (1987) Presynaptic effects of peptides at cholinergic synapses. In: Dowdall MJ, Hawthorne JN (eds) Cellular and Molecular Basis of Cholinergic Function. Horwood, Chichester, pp 454-459

De Robertis E, Arnaiz GR de L, de Iraldi AP (1962) Isolation of synaptic vesicles from nerve endings of the rat brain. Nature (Lond) **194**:794-795

Dowdall MJ, Zimmermann H (1977) The isolation of pure cholinergic nerve terminl sacs (T-sacs) from the electric organ of juvenile *Torpedo*. Neuroscience **2**:405-421

Ducis I, Whittaker VP (1985) High affinity sodium-gradient-dependent transport of choline into vesiculated presynaptic plasma membrane fragments from the electric organ of *Torpedo marmorata* and reconstitution of the solubilized transporter into liposomes. Biochim biophys Acta **815**:109-127

Fiedler W, Borroni E, Ferretti P (1986) An immunohistochemical study of synaptogenesis in the electric organ of *Torpedo marmorata* by use of antisera to vesicular and presynaptic plasma membrane components. Cell Tiss Res **246**:439-446

Giompres PE, Whittaker VP (1986) The density and free water of cholinergic synaptic vesicles as a function of osmotic pressure. Biochim biophys Acta **882**:398-409

Giompres PE, Zimmermann H, Whittaker VP (1981a) Purification of small dense vesicles from stimulated *Torpedo* electric tissue by glass bead column chromatography. Neuroscience **6**:765-774

Giompres PE, Zimmermann H, Whittaker VP (1981b) Changes in the biochemical and biophysical parameters of cholinergic synaptic vesicles on transmitter release and during a subsequent period of rest. Neuroscience **6**:775-785

Gray EG, Whittaker VP (1960) The isolation of synaptic vesicles from the central nervous system. J Physiol **153**:35-37P

Gray EG, Whittaker VP (1962) The isolation of nerve endings from brain: an electron-microscopic study of cell fragments derived by homogenization and centrifugation. J Anat (Lond) **96**:79-88

Gray EG, Whittaker VP (1981) This week's citation classic. Current Contents **24**:16

Horne RW, Whittaker VP (1962) The use of the negative staining method for the electron-microscopic study of subcellular particles from animal tissues. Z Zellforsch **58**:1-16

Israël M, Whittaker VP (1965) The isolation of mossy fibre endings from the granular layer of the cerebellar cortex. Experientia **21**:325-326

Israël M, Gautron J, Lesbats B (1970) Fractionnement de l'organe électrique de la torpille: localisation subcellulaire de l'acetylcholine. J Neurochem **17**:1441-1450

Israël M, Manaranche R, Mastour-Franchon P, Morel N (1976) Isolation of pure cholinergic nerve endings from the electric organ of *Torpedo marmorata*. Biochem J **160**:113-115

Jones RT, Walker JH, Richardson PJ, Fox GQ, Whittaker VP (1981) Immunohistochemical localization of cholinergic nerve terminals. Cell Tiss Res **218**:355-373

Kiene ML, Stadler H (1987) Synaptic vesicles in electromotoneurones: I. Axonal transport, site of transmitter uptake and processing of a core proteoglycan during maturation. EMBO J **6**:2209-2215

Kosh JW, Whittaker VP (1985) Is propionylcholine present in or synthesized by electric organ? J Neurochem **45**:1148-1153

Lapetina EG, Soto EF, De Robertis E (1968) Lipids and proteolipids in isolated subcellular membranes of rat brain cortex. J Neurochem **15**:437-445

Luqmani YA, Sudlow G, Whittaker VP (1980) Homocholine and acetylhomocholine: false transmitters in the cholinergic electromotor system of *Torpedo*. Neuroscience **5**:153-160

Meunier FM (1984) Relationship between presynaptic membrane potential and acetylcholine release in synaptosomes from *Torpedo* electric organ. J Physiol **354**:121-137

Morris SJ (1973) Removal of residual amounts of acetylcholinesterase and membrane contamination from synaptic vesicles isolated from the electric organ of *Torpedo*. J Neurochem **21**:713-715

Nagy A, Baker RR, Morris SJ, Whittaker VP (1976) The preparation and characterization of synaptic vesicles of high purity. Brain Res **109**:285-309

Ohsawa K, Dowe GHC, Morris SJ, Whittaker VP (1979) The lipid and protein content of cholinergic synaptic vesicles from the electric organ of *Torpedo marmorata* purified to constant composition: implications for vesicle structure. Brain Res **161**:447-457

Richardson PJ, Whittaker VP (1981) The Na^+ and K^+ content of isolated *Torpedo* synaptosomes and its effect on choline uptake. J Neurochem **36**:1536-1542

Richardson PJ, Walker JH, Jones RT, Whittaker VP (1982) Identification of a cholinergic-specific antigen Chol-1 as a ganglioside. J Neurochem **38**:1605-1614

Richardson PJ, Siddle K, Luzio PJ (1984) Immunoaffinity purification of intact, metabolically active, cholinergic nerve terminals from mammalian brain. Biochem J **219**:647-654

Rylett RJ, Whittaker VP (1987) Identification of the high-affinity choline transporter of *Torpedo* electromotor nerve terminals using a ^3H-choline mustard ligand. J Neurochem **48**:S66A

Schwarzenfeld I von (1979) Origin of transmitters released by electrical stimulation from a small, metabolically very active vesicular pool of cholinergic synapses in guinea-pig cerebral cortex. Neuroscience **4**:477-493

Sheridan MN, Whittaker VP (1964) Isolated synaptic vesicles: morphology and acetylcholine content. J Physiol **175**:25-26P

Sheridan MN, Whittaker VP and Israël M (1966) The subcellular fractionation of the electric organ of *Torpedo*. Z Zellforsch **74**:291-307

Stadler H, Tsukita S (1984) Synaptic vesicles contain an ATP-dependent proton pump and show 'knob-like' protrusions on their surface. EMBO J **3**:3333-3337

Suszkiw JB, Zimmermann H, Whittaker VP (1978) Vesicular storage and release of acetylcholine in *Torpedo* electroplaque synapses. J Neurochem **30**:1269-1280

Tashiro T, Stadler H (1978) Chemical composition of cholinergic synaptic vesicles from *Torpedo marmorata* based on improved purification. Eur J Biochem **90**:479-487

Weiler M, Roed IS, Whittaker VP (1982) The kinetics of acetylcholine turnover in a resting cholinergic nerve terminal and the magnitude of the cytoplasmic compartment. J Neurochem **38**:1187-1191

Whittaker VP (1959) The isolation and characterization of acetylcholine-containing particles from brain. Biochem J **72**:694-706

Whittaker VP (1960) The binding of neurohormones by subcellular particles of brain tissue. Proc 4th int Neurochem Symp Varenna June 1960. In: Kety SS, Elkes J (eds) Regional Neurochemistry: the Regional Chemistry, Physiology and Pharmacology of the Nervous System. Pergamon, Oxford 1962, pp 259-263

Whittaker VP (1963) The separation of subcellular structures from brain tissue. Biochem Soc Symp **23**:109-126

Whittaker VP (1964) Investigations on the storage sites of biogenic amines in the central nervous system. Prog Brain Res **8**:90-117 (see also General Discussion, pp 145-147)

Whittaker VP (1976) Tissue fractionation methods in brain research. Prog Brain Res **45**:45-65

Whittaker VP (1984a) The synaptosome. In: Lajtha A (ed) Handbook of Neurochemistry, 2nd edn, vol 7. Plenum, New York, pp 1-40

Whittaker VP (1984b) The synaptic vesicle. ibid, pp 41-69

Whittaker VP (1984c) The structure and function of cholinergic synaptic vesicles. Biochem Soc Trans **12**:561-576

Whittaker VP, Luqmani YA (1980) False transmitters in the cholinergic system: implications for the vesicle theory of transmitter storage and release. Gen Pharmacol **11**:7-14

Whittaker VP, Michaelson IA, Kirkland RJA (1964) The separation of synaptic vesicles from nerve-ending particles ('synaptosomes'). Biochem J **90**:293-303

Whittaker VP, Essman WB, Dowe GHC (1972) The isolation of pure cholinergic synaptic vesicles from the electric organs of elasmobranch fish of the family Torpedinidae. Biochem J **128**:833-846

Zimmermann H, Denston CR (1972) Separation of synaptic vesicles of different functional states from the cholinergic synapses of the *Torpedo* electric organ. Neuroscience **2**:715-730

SPECIALIZATIONS OF SUBSYNAPTIC CYTOPLASMS. COMPARISON OF AXOSPINOUS SYNAPSES AND NEUROMUSCULAR JUNCTIONS

René Couteaux
Institut des Neurosciences du C.N.R.S., Département de Cytologie
Université Pierre-et-Marie-Curie,75005 Paris, France

and Josef Špaček
Charles University Hospital, Department of Pathology
CS-500 Hradec Králové,Czechoslovakia

INTRODUCTION

When chemical synapses of the axosomatic,axodendritic and neuromuscular types are compared at ultrastructural level,the pattern of their organization seems fairly uniform,as regards the axon terminals and the complexes formed by pre- and postsynaptic membranes, and by the electron-dense structures attached to the cytoplasmic surfaces of these membranes. However,the same does not always apply to the deep part of the subsynaptic cytoplasm,which may display particularities of various kinds,relating,for example,to the smooth endoplasmic reticulum or to the cytoskeleton. The subsynaptic cytoplasm may also contain,under the postsynaptic density,an electron-dense material whose distribution varies,depending on the type of synapse.

The cytoplasmic specializations observed below synapses of different types appear so heterogeneous that at first sight it seems difficult to draw any parallel between them.

The structural particularities discovered in the portions of cytoplasm situated below the synapses of mammaliam dendritic spines and below the frog neuromuscular junctions are particularly striking examples of these specializations. These subsynaptic specializations have already been the subject of research,firstly in the dendritic spines, when Gray (1959) described the spine apparatus, and, more recently,at neuromuscular junctions (Couteaux,1981). We intend in this article to compare the two types of specialization.

Despite the considerable differences between the organization of neurons and that of muscle fibres, certain ultrastructural similarities may be found between the subsynaptic cytoplasms which respectively underlie the interneuronal and neuromuscular junctions. Their morphological comparison is

NATO ASI Series, Vol. H21
Cellular and Molecular Basis of Synaptic Transmission
Edited by H. Zimmermann

facilitated by the fact that the electron-dense bodies observed in the subsynaptic cytoplasm of mammalian dendritic spines and in the sarcoplasm of the interfolds of frog neuromuscular junctions can be detected by the same technique (Špaček,1986). Although this comparison was made by means of unspecific staining, we hope that it will help to orientate efficiently immunocytochemical and physiological research, which alone can clarify the puzzling roles of the subsynaptic specializations.

MATERIALS AND METHODS

The dendritic spines were studied on the cerebral visual cortex of young adult albino mice and the neuromuscular junctions, on different types of muscles from the frog (Rana esculenta).

The observations were carried out on three series of electron micrographs. Two series were obtained using the different fixation techniques previously described in studies of mammalian dendritic spines (Špaček,1985a) and of frog muscle neuromuscular junctions (Couteaux,1981).In the third series the treatment was the same for both materials, as follows : animals were anaesthetized with an intraperitoneal injection of chloral hydrate or by inhalation of ethylether, and perfused through the heart with a solution of 1% glutaraldehyde and 1% paraformaldehyde in phosphate buffer, pH 7.3 (Palay and Chan-Palay,1974). Tissue blocks were postfixed with saturated aqueous solution of uranyl acetate on ice, dehydrated in ethanol, passed through acetone and embedded in an Epon-Durcupan mixture. Ultrathin sections were stained with uranyl acetate and lead citrate. The technique used for three-dimensional reconstruction of dendritic spines is described in Špaček and Lieberman (1974).

SPECIALIZATIONS OF THE SUBSYNAPTIC CYTOPLASM IN MOUSE DENDRITIC SPINES

Dendritic spines from the cerebral visual cortex of the mouse were studied mainly with respect to the structure of the spine apparatus and its relationship with other components of the subsynaptic cytoplasm.

The spine apparatus is composed of two or more flattened sacs or cisternae,separated by plates of dense material (Fig.1). It was first described in dendritic spines of the cerebral cortex by Gray (1959) and later was also found in spines of other parts of the brain (for reviews,see Whittaker and Gray,1962; Peters et al.,1976). It was occasionally observed in the dendritic

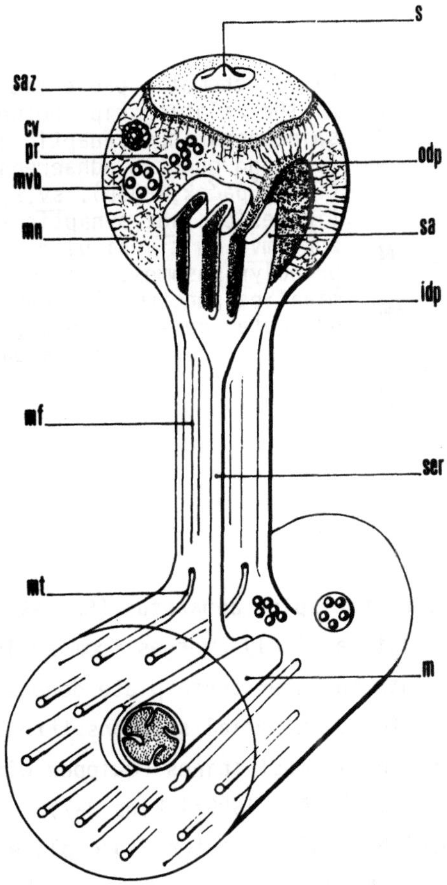

Fig.1.Schematic presentation of the ultrastructure of a dendritic spine with spine apparatus (from Špaček,1985a). s,spinule;saz,synaptic active zone;cv,coated vesicle;pr,polyribosomes;mvb,multivesicular body;mt,microtubule;odp,outer dense plate;mn,microfilamentous network;sa,spine apparatus;idp,inner dense plate;ser, smooth endoplasmic reticulum; m,mitochondrion

trunks (Gray and Guillery,1963; Jones and Powell,1969; Westrum,1970) and often in the initial segment of axons (Palay et al.,1968;Peters et al.,1968).

In the dendritic spines of the mouse visual cortex,the relationship between cisterns of the spine apparatus and the smooth endoplasmic reticulum was shown in three-dimensional reconstructions (Špaček,1985a). A tubule of smooth endoplasmic reticulum connecting these spine apparatus cisterns with the smooth endoplasmic reticulum of the dendritic trunk was regularly found in serial sections. The cistern of the endoplasmic reticulum of this trunk giving rise to the connecting tubule was often juxtaposed to mitochondria. The dense material between the spine apparatus cisterns formed plates about 10-20 nm wide. This interposed dense material, well-known from the descriptions of previous authors,has been designated "inner dense plate" to distinguish it from "outer dense plate",a dense material in continuity with one or

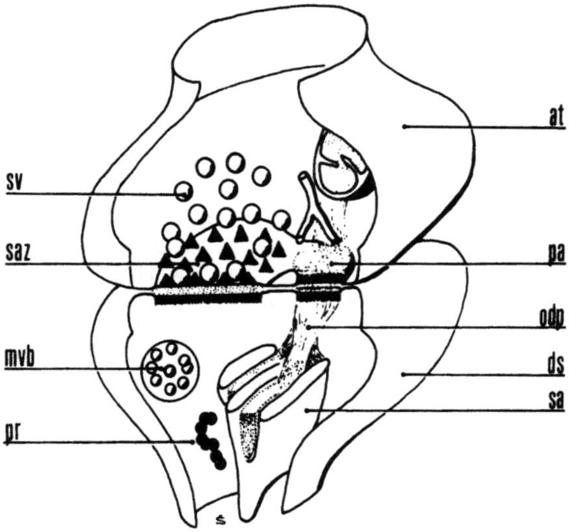

Fig.2.Schematic presentation of the relationships between spine apparatus,synaptic junction and punctum adhaerens (from Špaček,1985b). sv,synaptic vesicle;saz,synaptic active zone;mvb,multivesicular body; pr,polyribosomes;at,axon terminal;pa,punctum adhaerens; odp,outer dense plate;ds,dendritic spine;sa,spine apparatus

more inner dense plates but located outside the spine apparatus (Spacek,1985a). It is not clear whether the outer dense plate actively emerges from an inner dense plate or whether it in fact represents an inner dense plate denuded by cistern retraction. Outer dense plate was found in serial sections through 40 dendritic spines, and seemed to be present in most spines equipped with a spine apparatus and perhaps in all of them (Špaček,1985a)(Figs.3-7).

Thin filaments radiated from the periphery of the outer dense plate and joined a three-dimensional lattice filling the dendritic spine head. When the filaments were located near synaptic active zone,they radiated directly into the postsynaptic density. Sometimes such a radiation was found from the outer dense plate into a attachment plaque between a dendritic spine and axon terminal or glial process.

Figs.3-7. Electron micrographs of dendritic spines. Mouse visual cortex.
Figs.3,4,5. Dendritic spines with subsynaptic dense plate. Scale=0.15 µm (Fig.3) and 0.2 µm (Figs.4,5).
Fig.6. Dendritic spine with spine apparatus,synaptic junction and punctum adhaerens. An outer dense plate (arrow) connects the spine apparatus with the punctum adhaerens. Scale=0.15 µm.
Fig.7. Dendritic spine with spine apparatus. An outer dense plate (arrow) connects the spine apparatus with the postsynaptic density.Scale=0.15 µm.

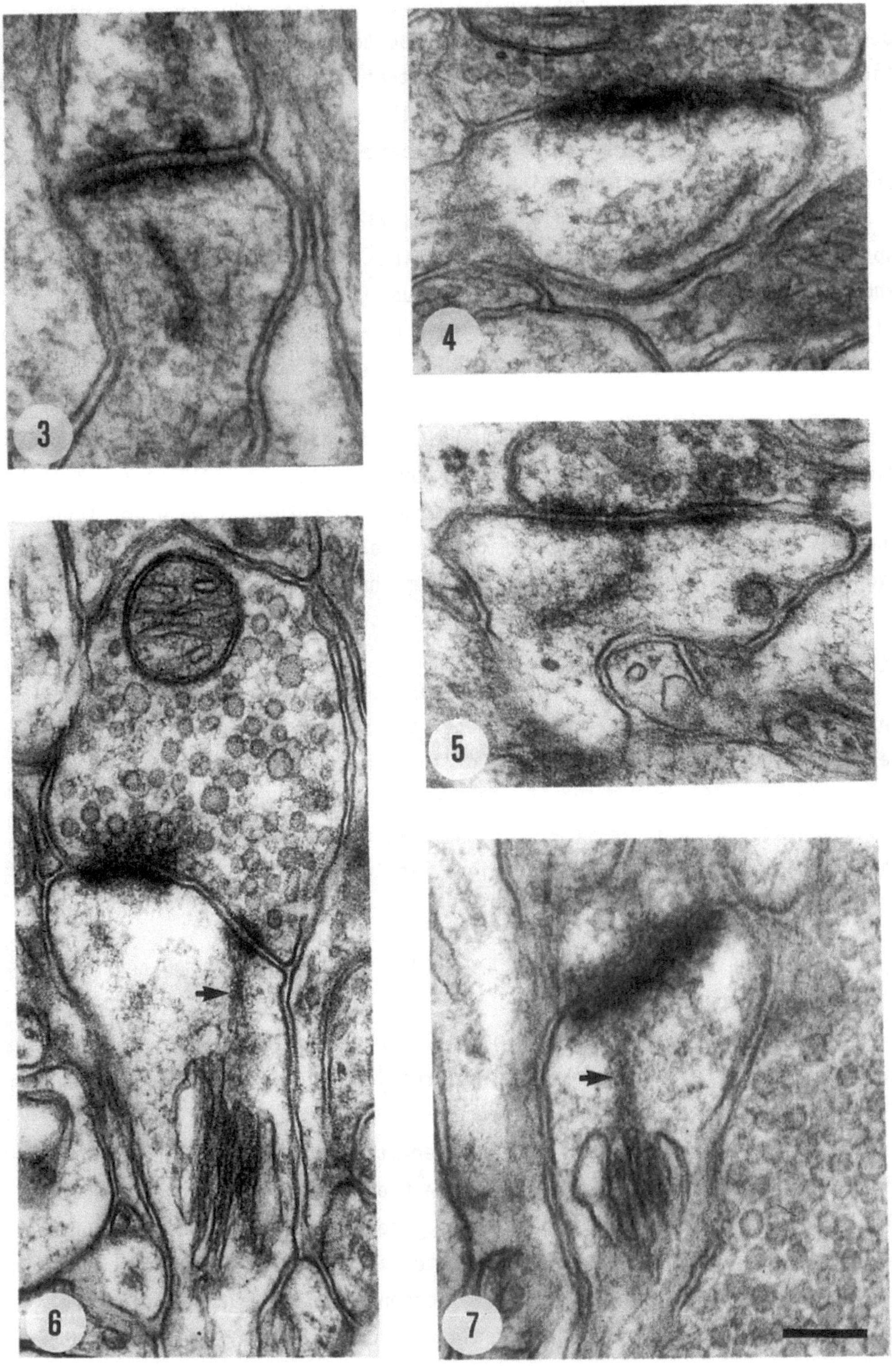
3
4
5
6
7

A close relationship between the spine apparatus and the postsynaptic density has been described (Tarrant and Routtenberg,1979) and this description was very soon followed by the demonstration of direct continuity between the material forming the dense plates of the spine apparatus and the postsynaptic density or an adjacent punctum adhaerens (Westrum et al.,1980; Dyson and Jones,1984; Špaček,1985a,b)(Figs.2,6,7).

A filamentous network is a characteristic component of all dendritic spines. It is organized as a three-dimensional lattice-work in the heads, whereas the filaments in the stalk are usually parallel. Very fine microfilaments 2-4 nm in diameter are the main constituents of this network,which also includes fewer microfilaments with a diameter about 8 nm and intermediate filaments about 10-12 in diameter.

Continuity was observed between this filamentous network,the postsynaptic density,inner dense plate and outer dense plate. Some filaments of the network were anchored in the plasma membrane. Microtubules were not found in the spine heads. Clusters of polyribosomes were always found in serial sections through 20 dendritic spines of the mushroom-shaped type. They were not so frequent in 25 other spines analyzed, which were of the thin and stubby types (Špaček,1985a).

The presence of a spine apparatus was not constant in all dendritic spines of the mouse visual cortex. Dendritic spines,small in size and surface area,with a synaptic active zone of the simple type (round or oval) did not contain a spine apparatus (Špaček and Hartmann,1983). This apparatus was associated with at least 91% of the complex junctions, i.e. horseshoe-shaped,annulate or perforated junctions (Špaček,1985b).

SPECIALIZATIONS OF THE SUBSYNAPTIC CYTOPLASM IN FROG NEUROMUSCULAR JUNCTIONS

The frog skeletal muscle are mainly composed of muscle fibres belonging to three categories : fast,slow and intermediate (for reviews,see Lännergren, 1975; Salpeter,1987).

The fast muscle fibres , singly-innervated with long branched axon terminals (end-bush of Kühne)(for review,see Peper et al.,1974) are capable of eliciting action potentials and twitches. The slow or tonic muscle fibres, multiply-innervated (en-grappe type)(Couteaux,1952,1955; Page,1965; Verma and Reese,1984) are unable to generate action potentials (for review,see Morgan and Proske,1984). The intermediate muscle fibres are multiply-inner-

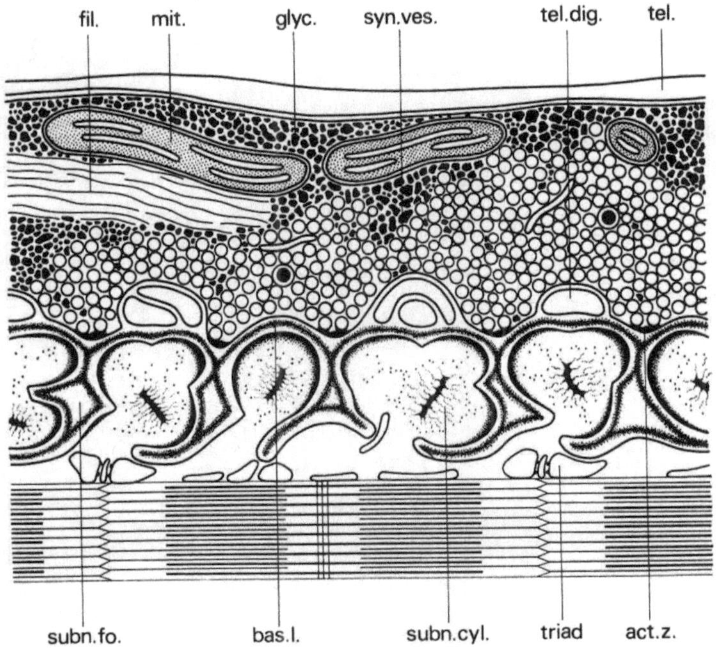

Fig.8. Diagram of a frog neuromuscular junction as seen in a longitudinal section of the muscle fibre (from Couteaux,1980). fil,neurofilaments;mit, mitochondrion;glyc,glycogen;syn ves,synaptic vesicle;tel dig,teloglial digitation;tel,teloglia; subn fol,subneural folds; bas l,basal lamina; subn cyl, subneural cylinder;triad,triad; act z ,active zone

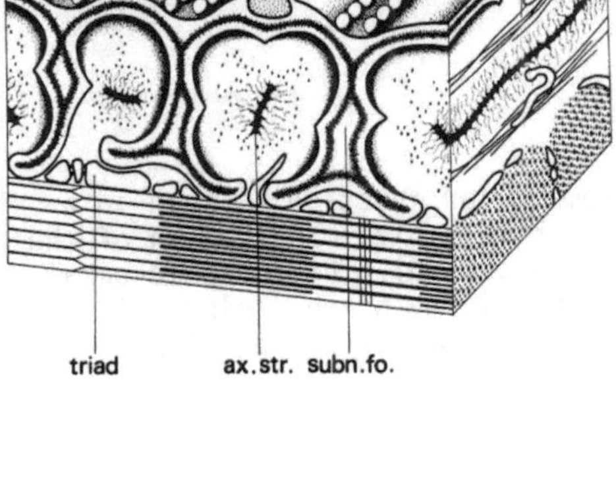

Fig.9. Diagrammatic reconstitution of a portion of a synaptic gutter at a frog neuromuscular junction,showing the disposition of subneural cylinders within the interfolds,and their relationship with subneural filaments and sarcoplasmic reticulum tubules (from Couteaux,1980). syn cl,synaptic cleft; tel fing,teloglial finger; act z,active zone; subn fil,subneural filaments; triad,triad; ax str,axial strip; subn fo, subneural fold

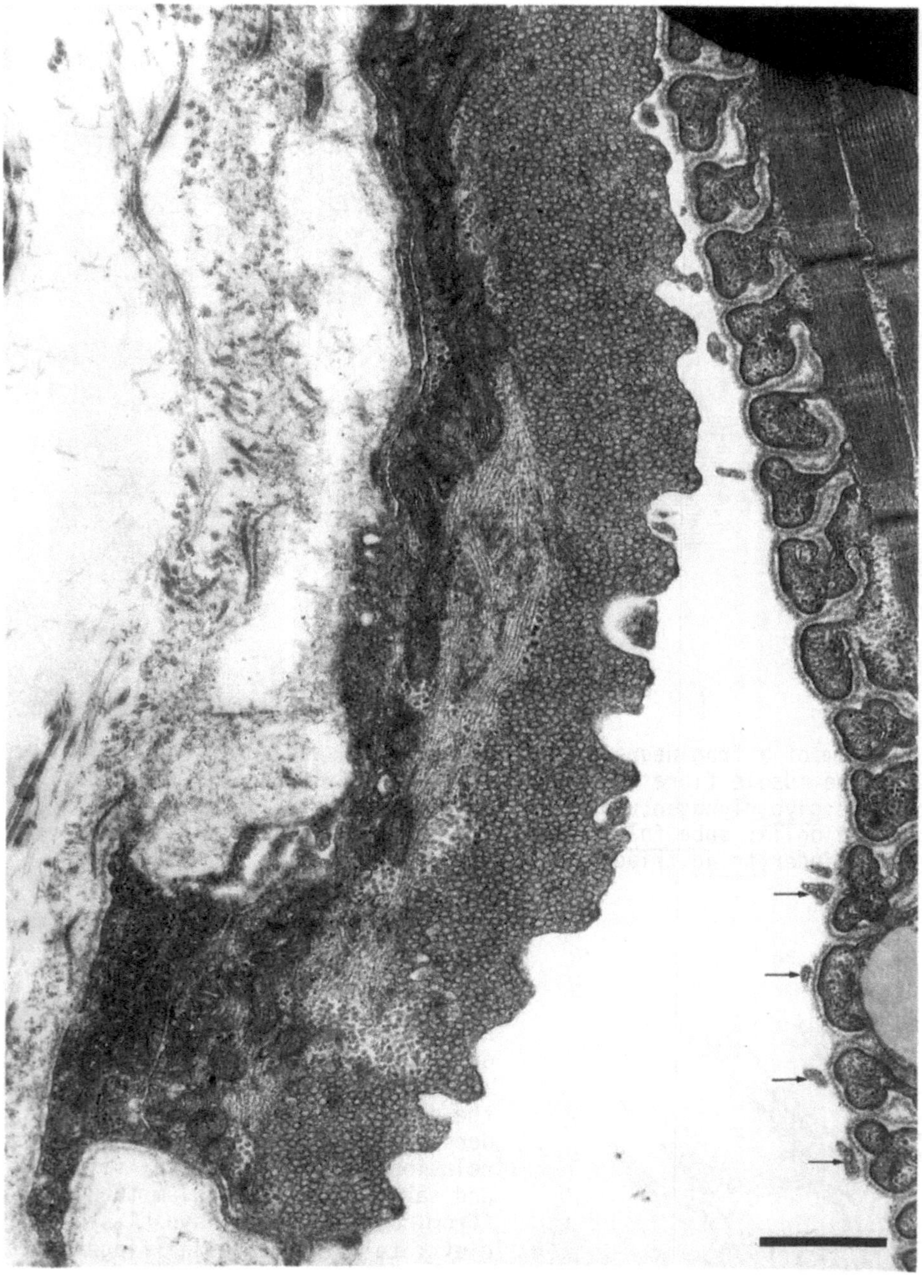

Fig.10. Electron micrograph of a frog neuromuscular junction as seen in a longitudinal section of the muscle fibre after mechanical separation of the pre- and postsynaptic portions (see the diagram,Fig.8). The synaptic basal lamina remains attached to the folded postsynaptic membrane. Teloglial fingers (arrows) often remain attached to the basal lamina. Scale=0.8 µm

vated,like the slow muscle fibres,but they also resemble the fast muscle fibres because they are capable of eliciting action potentials and twitches (Smith and Ovalle,1973; Kordylewski,1979; Lännergren,1979; Miledi and Uchitel,1981;Verma,1984a,b; Uchitel and Miledi,1987).

In the fast muscle fibres of the frog,the postsynaptic membrane beneath the axon terminal branches develops deep invaginations - the subneural or subsynaptic folds- which segment the subsynaptic sarcoplasm. We will deal mainly with the structural specializations observed in the portions of sarcoplasm located between the folds - the interfolds (Figs.18-18).

As in the case of the chemical interneuronal synapses,the most superficial specialization of the subsynaptic cytoplasm is the postsynaptic density,consisting of a dense layer,which is contiguous with the postsynaptic membrane and looks "fuzzy" on its cytoplasmic side. This intracellular coating of the postsynaptic membrane is coextensive with the portions of the membrane which are rich in nicotinic acetylcholine receptors - the receptive zones - ,but does not line the membrane right to the bottom of the folds, where the membrane is very poor in receptors (Fertuck and Salpeter,1974,1976).

The sarcoplasm of the interfolds contains bundles of intermediate-size filaments - the subneural or subsynaptic filaments - ,which run parallel to the folds (Couteaux and Pécot-Dechavassine,1968; Heuser,1980; Couteaux,1981;

Figs 11-18. Electron micrographs of frog neuromuscular junctions. Muscles of the fast type.
Fig.11. Tangential section of a nerve terminal branch showing three active zones. Synaptic vesicles are lined up in a double row along each active zone. Scale=0.47 µm
Fig.12. Longitudinal section of the electron-dense bar of an active zone. Synaptic vesicles are attached to the presynaptic membrane along this bar. Scale=0.74 µm
Figs. 13 and 14. Cross sections of active zones showing synaptic vesicles of the double rows opening into the synaptic cleft (from Couteaux and Pécot-Dechavassine,1970). Scale=0.17 µm
Fig.15. In the interfolds,the subneural cylinders and the filaments surrounding them are seen in cross section. Tenuous trabeculae radiate from the electron-dense axial strip; on the right,we can see an interfold containing two cylinders. Scale=0.45 µm
Fig.16. Two subneural cylinders in a longitudinal section,parallel to the primary synaptic cleft.fo,fold; fi,subneural filaments.Scale=0.47 µm
Fig.17. Bundles of subneural filaments in a longitudinal section,parallel to the primary synaptic cleft. These filaments run parallel to the folds,then curve and turn towards the postsynaptic density which underlies the postsynaptic membrane. Scale=0.40 µm
Fig.18. Oblique section of a subneural cylinder showing tubules of sarcoplasmic reticulum surrounding the cylinder. ax,axial strip; fo,fold.Scale=0.51 µm

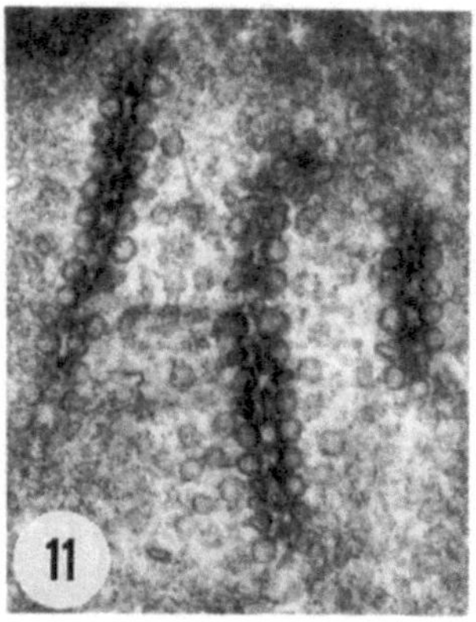
11

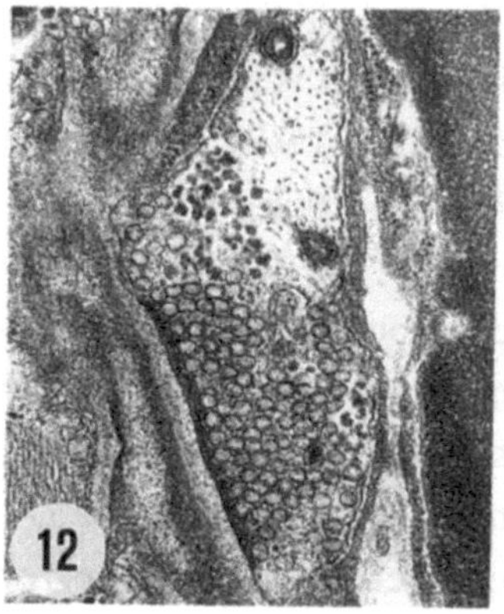
12

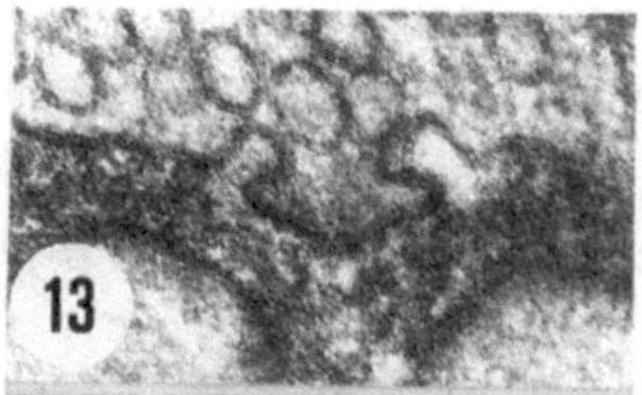
13

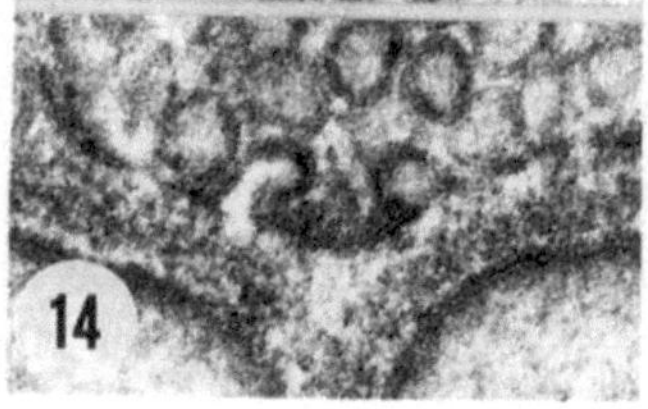
14

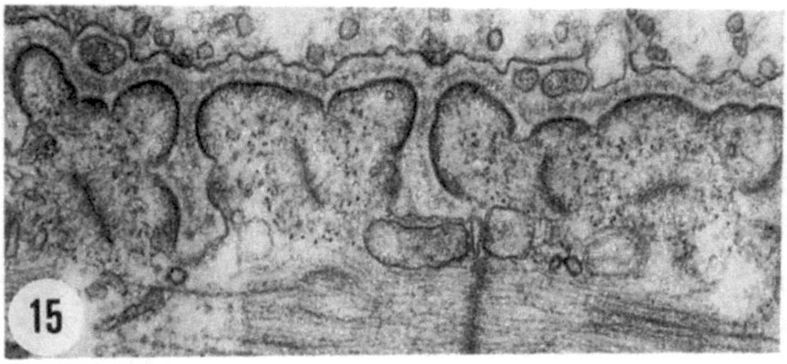
15

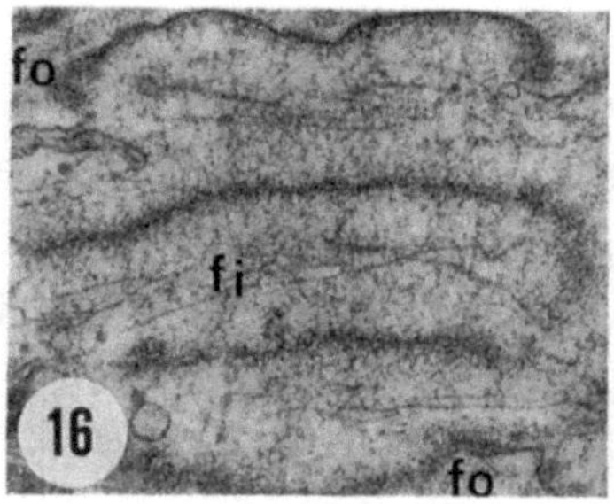
fo
fi
16
fo

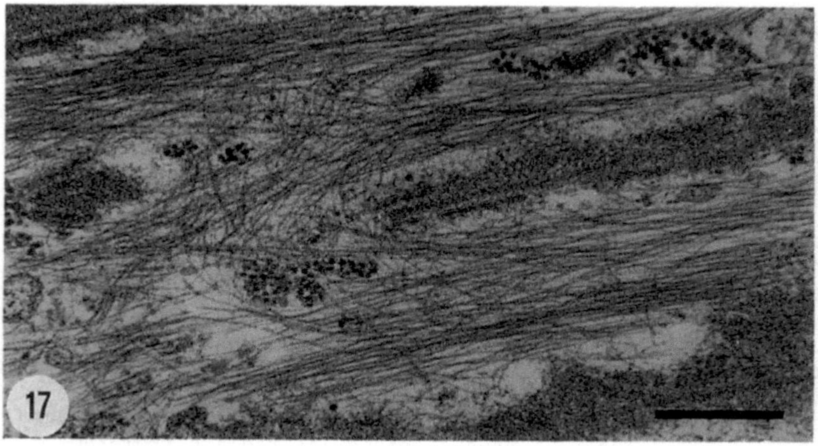
17

ax
18

Hirokawa and Heuser,1982; Burden,1982,1987). These filaments remain inside the bundles over most of their length,and then curve outwards,leave the bundles and run towards the postsynaptic density (Couteaux,1980). It seems that the filaments approach this density tangentially,so that it is not easy to define their relationship with the plasma membrane precisely. Similar intermediate-size filaments have been described in the interfolds of the rat motor endplate (Ellisman et al.,1976).

The interfolds also contain microtubules which are generally superficial. However,these microtubules are apparently far less numerous in frog interfolds than in the interfolds of mammalian endplates.

In the fast muscle fibres of the frog,complex organelles which appear to be specific to the subsynaptic sarcoplasm,were observed in the middle of the subsynaptic filament bundles. These organelles have a cylindrical shape; their diameter generally ranges between 150 and 300 nm and their length,between 500 nm and several micrometres. Like the folds and filament bundles of the interfolds,the cylindrical organelles are perpendicular to the axis of the nerve terminal branches. They are not delimited by a membrane,but their cylindrical form is outlined by the interfold filaments which are applied to their surface and constitute a sort of openwork muff. Most of the interfolds contain only a single subsynaptic cylinder. Occasionally,two,three or more may be seen in the same interfold; in this case,the cylinders remain distinct and are generally separated by filaments. Each cylinder contains an electron-dense axial strip from which tenuous trabeculae radiate. These trabeculae fill the space between the strip and the subsynaptic filaments. When cut along their longitudinal axis,the cylinders have a bottle-brush appearance. The axial strip is connected to the plasma membrane at both ends.

Tubules belonging to the sarcoplasmic reticulum can be seen on the surface of the cylinders. Some of them even penetrate the cylinders,by passing their axial strip. These sarcotubules form part of a canalicular system which occurs in all the interfolds and is continuous with the sarcoplasmic reticulum surrounding the myofibrils.

Figs.19 and 20.Electron micrographs of frog neuromuscular junctions. Muscle fibres of the slow type,multiply-innervated, without M line or folds.In subsynaptic and juxtasynaptic regions of sarcoplasm,we can observe dense bodies (arrows) of very irregular shape and variable orientation. Scale=0.76 µm(Fig. 19);0.88(Fig.20) and 0.44 µm (inset)

Fig.21. Electron micrograph of a frog neuromuscular junction belonging to a multiply-innervated muscle fibre,without M line and with irregular folds. Scale=0.93 µm

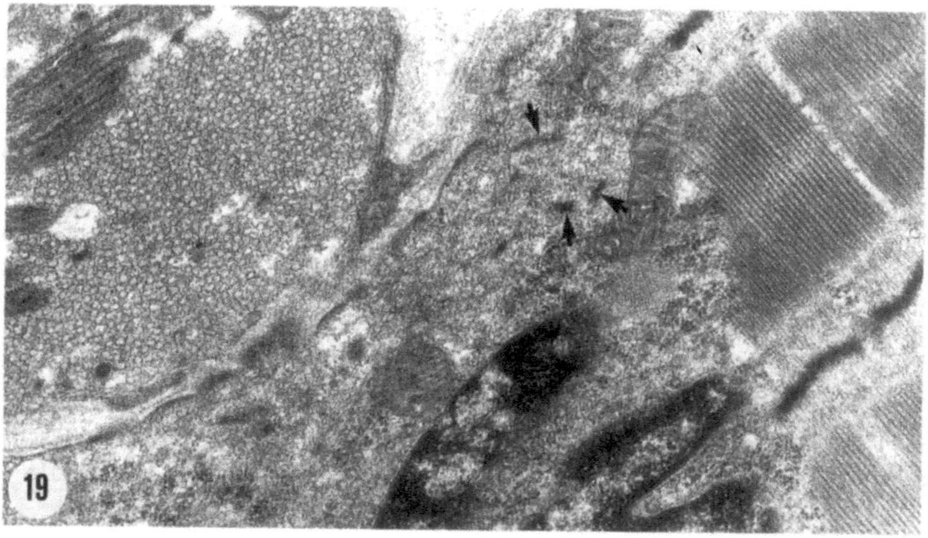

19

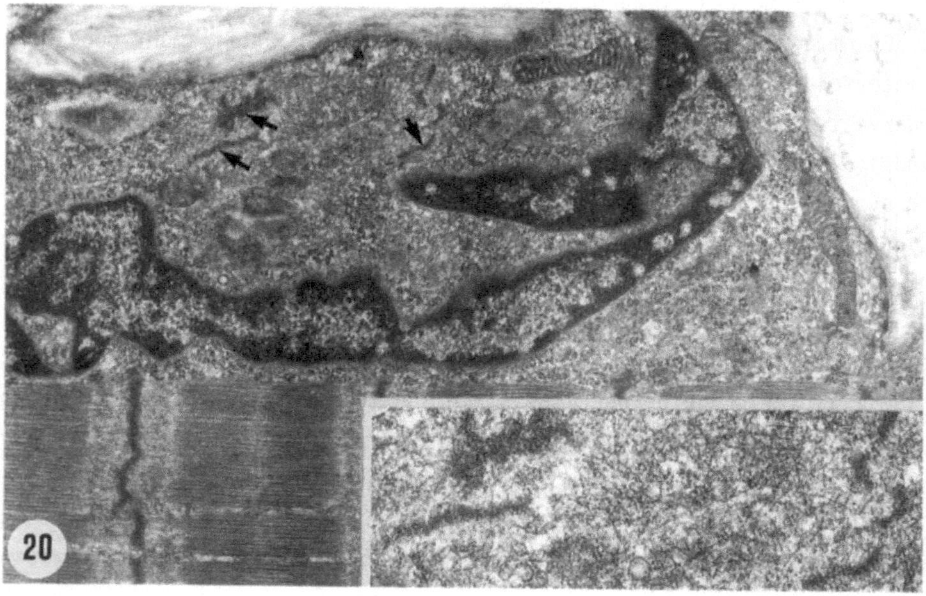

20

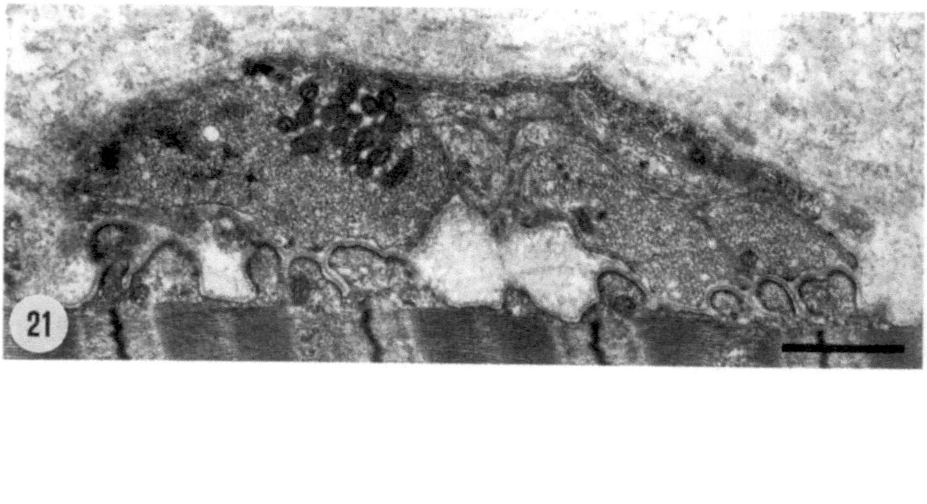

21

The slow muscle fibres can be distinguished from the fast fibres by the absence of an M line,the characteristic zigzag pattern of the Z line,the ill-defined myofibrils and the sarcity of triads. In these slow muscle fibres,the organization of the subsynaptic sarcoplasm differs considerably from that of the sarcoplasm underlying the large end-bushes of the fast fibres.

The neuromuscular junctions of slow fibres are very small,round,often grouped in clusters and generally devoid of subsynaptic folds (Figs.19,20). The subsynaptic cylinders are very short : their length may reach several hundred nanometres,but ordinarily it ranges much less. The axial strips have a shape and orientation which are extremely variable,so that they appear on thin sections as dense bodies of very irregular outline. Their distribution is disordered. They are often located in regions of the subsynaptic sarcoplasm close to what may be termed the junctional muscle nuclei,which are satellites of the nerve terminal branches in frog muscles (Couteaux,1945, 1947) as in mammalian muscles.

Tubules of smooth endoplasmic reticulum are visible close to the cylinders but they are few in number and apparently have no specific relationship with the cylinders. The intermediate-size filaments do not form bundles as in the fast muscle fibre interfolds,but are orientated in all directions. They are entangled and unlike the fast fibre filaments do not form muffs surrounding the cylinders.

In the intermediate muscle fibres,which possess M and Z lines resembling those of the fast muscle fibres,the neuromuscular junctions are of different sizes and their shape varies from the shape of the end-bush to that of the en-grappe junction. They are often of different types on the same muscle fibre. The subsynaptic folds are of various depths or absent. These differences in the folding of the postsynaptic membrane may be observed not only between the junctions,but even between different points of the same junction.

Figs.22-24. Electron micrographs of frog neuromuscular junctions. Muscle fibres of the intermediate type,multiply-innervated, with M line and irregular folds. The subsynaptic cylinders are very short and located in the meshes of a polygonal network formed by sarcotubules.
Fig.22. In this section perpendicular to the postsynaptic membrane,its folding appears extremely irregular. Scale=0.86 µm
Fig.23. The section passes trough the subsynaptic layer of sarcoplasm,which contains the cylinders and the polygonal network of sarcotubules separating them. Scale=1 µm
Fig.24. Synaptic gutter in a section parallel to the surface of the muscle fibre. Near a fold (fo) the subneural filaments form bundles running along both sides of the fold. Scale=0.50 µm

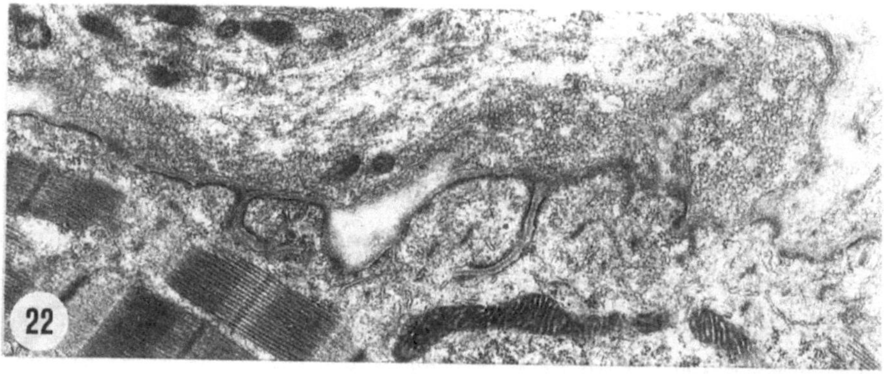
22

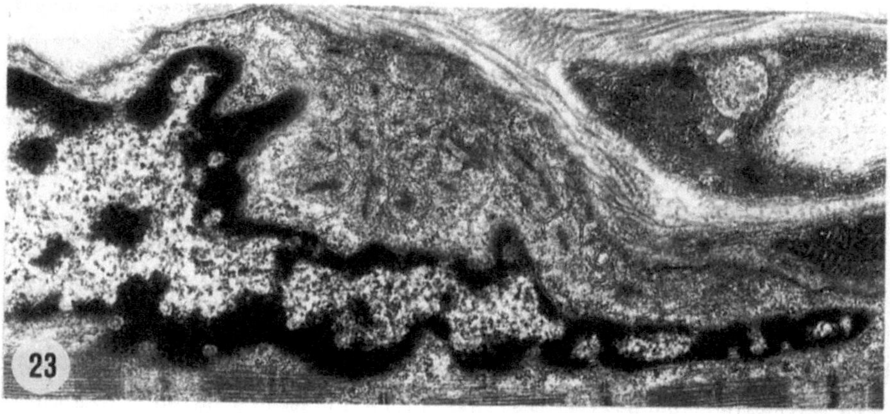
23

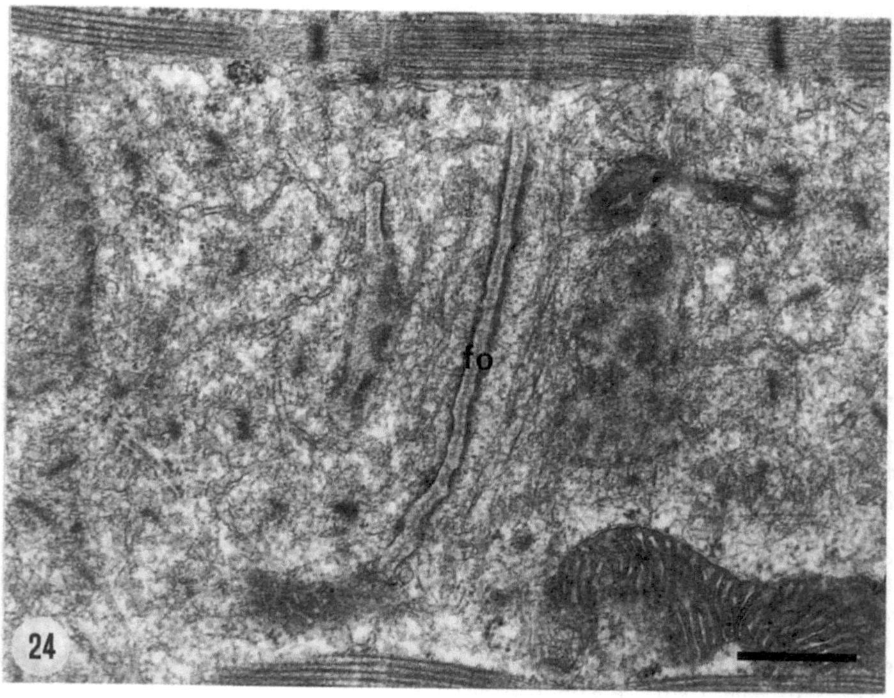
fo
24

The cylinders are of the same type as those of the slow muscle fibres, but their relationship with the smooth endoplasmic reticulum is quite diffe-rent since they are located in the meshes of a polygonal network formed by sarcotubules which separate and delimit the cylinders (Figs. 22-24). In the parts of the neuromuscular junctions where there are no subsynaptic folds, the orientation of the subsynaptic filaments is as disordered as in the slow muscle fibres,but near the folds,these filaments form bundles which run par-alell to the folds as in the fast muscle fibres (Fig. 24).

Besides the above-mentioned types of muscle fibres and neuromuscular junctions,there are other less common types,found in frog muscles. Verma (1984b),for example,described muscle fibres without an M line whose neuro-muscular junctions have subsynaptic folds , and we have also had occasion to observe this type of muscle fibre (Fig. 21).

DISCUSSION

Two points emerge from the structural paralell drawn here between the res-pective subsynaptic cytoplasms of mammalian dendritic spines and frog neuro-muscular junctions : firstly,the presence in both of electron-dense bodies,and secondly,the close relationships of these bodies with the postsynaptic mem-brane and smooth endoplasmic reticulum. Many conjectures have been made regar-ding the functional significance of the subsynaptic electron-dense bodies. We shall consider here only those which have been to some extent supported by mor-phological observations.

Relationships of the subsynaptic dense bodies with the postsynaptic membrane and smooth endoplasmic reticulum

The demonstration in the dendritic spines of the continuity of the sub-synaptic dense plates and the postsynaptic density suggested that the dense material emerging from the spine apparatus constitutes a source of macromo-lecules for the complex formed by the postsynaptic membrane and the postsy-naptic density (Tarrant and Routtenberg,1977,1979; Westrum et al.,1980). These macromolecules might consist , not only of the proteins which,like nicotinic acetylcholine receptors , are involved in the mechanism of synap-tic transmission,but also of other proteins,since the outer dense plate may be continuous both with the postsynaptic density and with an adjacent atta-chment plaque (punctum adhaerens). It follows that the spine apparatus might be a repository or a point of protein release ,or even perhaps a post-synaptic protein synthesizing centre (Špaček,1985a).

At neuromuscular junctions,a similar possibility may be considered regarding the functional significance of the subsynaptic cylinders, in view of their relationships with the other postsynaptic components.

These cylinders are attached by their axial strip to the postsynaptic membrane and the connections thus formed are particularly obvious at the neuromuscular junctions of the fast muscle fibres,where each cylinder is attached to the membrane at both extremities. The cylinders are located in the centre of the interfolds, under the postsynaptic membrane in which acetylcholine receptors are concentrated , and which edges the primary and secondary clefts where most of the junctional cholinesterase is located (McMahan et al., 1978; for review,see Rotundo,1987).

The special relationship of the subsynaptic dense bodies with the smooth endoplasmic reticulum,which in the spine apparatus is illustrated by the apposition of dense plates to cisterns , is also clearly visible at the neuromuscular junctions of the frog. In the case of the fast muscle fibres,the sarcotubules are arranged around the cylinders in networks which in some ways resemble those surrounding the myofibrils. It is conceivable that the pericylindrical tubules,which are continuous with the perimyofibrillar tubules, are,like the latter,capable of sequestering Ca^{2+} and releasing it during muscle contraction.

The connections of the subsynaptic cylinders with both the postsynaptic membrane and the tubules of the sarcoplasmic reticulum argue in favour of their participation in the local regulation of subsynaptic metabolism,notably in terms of receptor protein and enzyme synthesis (Couteaux,1981).

Cytochemical research (Burgoyne et al.,1983) has shown that the cisterns of the spine apparatus,like the smooth endoplasmic reticulum of the presynaptic nerve terminal (McGraw et al.,,1980), is also a Ca^{2+}sequestering organelle. As mentioned above,spine apparatus cisterns are connected by a tubule to a cistern of the endoplasmic reticulum of the dendritic trunk. It seems of interest to emphasize that the latter cistern of the parent dendrite is often apposed to mitochondria,which are themselves able to sequester Ca^{2+} and supply energy.

The association of the smooth endoplasmic reticulum with dense bodies of the subsynaptic cytoplasm does not seem to be constant. In particular,the subsynaptic bodies described by Taxi (1961,1963), Milhaud and Pappas (1966) and Akert et al.(1967) were apparently not accompanied by a specialization of smooth endoplasmic reticulum.

Present orientation of research on the functional significance of subsynaptic structural specializations

a) Axospinous synapses

As pointed out above, the presence of the spine apparatus in the subsynaptic cytoplasm depends on the size of the dendritic spine and the type of synapse. Most large mushroom-shaped spines with a large active zone that is complex or multifocal contain a spine apparatus (Špaček and Hartmann,1983; Špaček,1985b).

The number of cisterns and dense plates in a spine apparatus is extremely variable. The most complex spine apparatus in our material possessed 8 cisterns. It seems that the degree of development of the spine apparatus is directly linked to the surface area of the active zone,what would fit well with the hypothesis of spine apparatus participations in the synthesis of proteins,particularly of the receptor proteins and enzymes involved in synaptic transmission mechanism.

The size of the spine seems also to depend on the surface area of the active zone (Špaček and Hartmann,1983) and in addition the shape of the spine might change in response to variations in the synaptic activity. After the observations of Fifková and Van Harreveld (1977),who described long-lasting morphological changes in the dendritic spines of dentate granular cells following stimulation of the entorhinal area,the results of many other experiments led to the assumption that the dendritic spines are dynamic structures which significantly change their shape correlatively with repeated afferent activity.

The question therefore arises as to whether the different shapes of the dendritic spines in the cerebral cortex (mushroom-shaped,thin or stubby)are just transient appearances expressing an instantaneous state of the spine's synaptic activity. It would however be premature to attempt to answer this question now.

The discovery by immunocytochemical methods that dendritic spines contain a high concentration of actin (Matus et al.,1982; Caceres et al.,1983) confirmed the observations of Le Beux and Willemot (1975a,b) who demonstrated,by heavy meromyosin labeling,the presence in spines of actin-like filaments;however,the exact part played by these filaments in the mechanism that changes the shape of dendritic spines is still unclear (for review,see Coss and Perkel,1985).

b) Neuromuscular junctions

In the frog muscles,examination of the neuromuscular junctions on fast, intermediate and slow fibres shows that the folding of the postsynaptic membrane is maximal in the fast fibres,declines in the intermediate fibres and is reduced to almost nothing in the slow fibres. Since we still do not know what part the subsynaptic folds play in the functioning of the neuromuscular junction,the significance of this gradation is not clear.

Whether the three types of fibres have folds or not,their subsynaptic cytoplasm always displays structural specializations,in particular,the cylinders with their electron-dense axial strips,and the numerous intermediate-size filaments;however,appreciable differences exist between these specializations,depending on the type of fibre.

It may be noted that in the fast and intermediate fibres,which are both capable of eliciting action potentials and twitches,the subsynaptic cylinders are surrounded by tubules of the smooth endoplasmic reticulum,while in the slow fibres,which cannot elicit action potentials,there is no comparable relationship between the cylinders and smooth endoplasmic reticulum.

Besides the differences observed between the subsynaptic cytoplasms, others have been established between the neuromuscular junctions of fast and slow fibres concerning,in particular,their cholinesterase activities. Like the fast fibre junctions,the slow fibre junctions contain cholinesterases (Couteaux,1955;Salpeter et al.,1974;Verma and Reese,1984;Walrond and Reese, 1985), but this content seems much lower in the slow fibre junctions (Couteaux,1958;Lännergren and Smith,1966;Verma and Reese,1984;Morgan and Proske, 1984). With histochemical detection by the acetylthiocholine method,much longer incubation is necessary to reveal the cholinesterase activities of the slow fibre junctions than of the fast fibre junctions.

It is possible that the difference between the cholinesterase site densities in the junctions of fast and slow fibres might be correlated with the difference observed between the densities of the large intramembraneous particles which characterize the presynaptic active zones on these fibres.In freeze-fractured replicas of neuromuscular junctions from frog fast fibres, these particles appear on the P face of the presynaptic membrane as pairs of double rows (Dreyer et al.,1973; Heuser et al.,1974). However,on frog slow fibres,the rows are not always in pairs and their spacing and layout are irregular; the average length of the particle rows per unit length of nerve terminal is half that of the terminal on fast fibres (Verma and Reese,1984;

Verma,1984). The difference in the density of active zone particles in fast and slow muscle fibres is still more obvious at the neuromuscular junctions of the lizard since in this species the active zones always exhibit paired rows of particles on the fast fibres and single rows on the slow fibres.

The large particles of active zones have been tentatively identified as calcium channels (Pumplin et al.,1981;for review,see Atwood and Lnenicka, 1986),so that if the correlation between the densities of the active zone particles and cholinesterase activities is quantitatively established,it might constitute a new and interesting link between structure and function in the study of neuromuscular relationships.

Research on the postsynaptic organization of neuromuscular junctions has raised two other problems whose solution may depend on the structural specialization of the subsynaptic cytoplasm.

The first problem is how to explain the maintenance of the subsynaptic folds even though they undergo considerable deformation during muscle contraction. It is quite conceivable that the intermediate-size filaments and cylinders of the subsynaptic cytoplasm have a supportive function and constitute the framework of the sarcoplasm in each interfold. In that case they would contribute to maintain the folds and their elasticity would also help to re-establish the resting length and width of the synaptic gutters during muscle relaxation. When there are no folds,subsynaptic filaments and cylinders are still present,but the arrangement of the filaments in bundles is dependent on the presence of folds in the vicinity.

The second problem raised by the study of the postsynaptic organization of the neuromuscular junction concerns the remarkable stability of the nicotinic acetylcholine receptors in the postsynaptic membrane. Spectroscopic studies using paramagnetic resonance on Torpedo electrocyte membrane fragments showed that these receptors are firmly immobilized in the postsynaptic membrane (Rousselet and Devaux,1977; Rousselet et al.,1982).

Ultrastructural observations on mammalian motor endplates suggested acetylcholine receptors are directly connected to the filaments located in the interfolds, and such connections of the receptors with the cytoskeleton of the subsynaptic sarcoplasm would explain their stability (Ellisman et al., 1976). Studying the subsynaptic filaments of the Torpedo electrocytes,Heuser and Salpeter (1979) and Heuser (1980) concurred in this view.

Using direct freezing and deep etching,Hirokawa and Heuser (1982) described a submembranous meshwork just beneath the postsynaptic membrane, which might anchor receptor molecules to underlying bundles of cytoskeletal intermediate filaments and thus immobilize these receptors in the plane of the postsynaptic membrane. According to these authors,this meshwork is presumably equivalent to the postsynaptic density seen in thin sections,and similar to the meshwork described with the same technique at interneuronal synapses (Gulley and Reese,1981; Landis and Reese,1983).

More recently,on further ultrastructural investigations carried out by freeze-etching technique,the cytoplasmic surface of the postsynaptic membrane appeared to be devoid of filamentous structure in the *Torpedo* electrocytes (Bridgman *et al.*,1987).

At present,the ultrastructural data available do not allow us to reach definite conclusions on the relationship between the receptor molecules and the cytoskeleton of the subsynaptic cytoplasm. Nevertheless an important progress has been made in the approach to this difficult problem by the identification and localization of proteins associated with the postsynaptic membrane,in particular the protein of Mr 43,000 daltons (Sobel *et al.*,1978; Hamilton *et al.*, 1979).

The localization of the 43 Kd protein in the *Torpedo* electrocytes was a turning point in the study of the relationship between receptors and the subsynaptic cytoplasm. After removal of the 43 Kd protein by alkaline treatment (pH 11) from the receptor rich-membranes,the kinetics of agonist-induced ion permeability of the membranes is unaltered (Neubig *et al.*,1979),but this treatment increases rotational and translational mobility of the receptors (Barrantes *et al.*,1980; Lo *et al.*,1980;Rousselet *et al.*,1982) and enhances their sensitivity to heat denaturation (Saitoh *et al.*,1979) and to the proteolysis (Klymkowsky *et al.*,1980). These experiments led to think that the 43 Kd protein interacts with the receptor and may contribute to its stabilization.

The postsynaptic density,whose staining has been enhanced by a tannic acid-glutaraldehyde fixation,is no longer visible after the alkaline extraction of the 43 Kd protein (Cartaud,1980; Cartaud *et al.*,1981;Sealock,1982). The disappearance of the postsynaptic density after the removal of the 43 Kd protein suggested that this protein constitutes a major component of the postsynaptic density,which is closely apposed to the cytoplasmic surface of the postsynaptic membrane in the receptor-rich regions of the membrane (Sealock,1980;Sealock and Kavookjian,1980).

Other observations,biochemical (St.John *et al.*, 1982; Wennogle and Changeux,1980) and immunocytochemical at ultrastructural level (Nghiêm *et al.*,

1983; Sealock et al.,1984; Bridgman et al.,1987),led to similar conclusions.

Using a double labeling technique,Sealock et al.(1984) were able to establish that the distributions of the 43 Kd protein and acetylcholine receptor are coextensive. In addition,cross-linking studies demonstrate that the 43 Kd protein is intimately associated with the acetylcholine receptor and within 1 nm of the beta-subunit (Burden et al.,1983).

This association of the acetylcholine with a non-receptor protein,demonstrated in the *Torpedo* electrocytes,changes the terms of the problems raised by the study of the relationships between the acetylcholine receptors and the cytoskeleton of the subsynaptic cytoplasm. It seems indeed possible that the 43 Kd protein is interposed between the acetylcholine receptors and cytoskeletal filaments (Cartaud et al.,1983; Sealock et al.,1984). The anti-43 Kd (mouse monoclonal antibody) labeling of the free end of some detached intermediate-size filaments is in agreement with this assumption (Cartaud et al., 1983). These intermediate-size filaments are labeled by anti-desmine antibodies (Kordeli et al., 1986).

As for the muscle fibres,a protein immunologically related to the *Torpedo* 43 Kd protein has been detected at the neuromuscular junctions (Froehner et al., 1981; Froehner,1984; Burden,1985) and it is probable that its location is the same as in the *Torpedo* electrocytes.

Other proteins are also concentrated at neuromuscular junctions. A cytoplasmic isoform of actin (Hall et al., 1981),several proteins associated with actin in other cell types, such as alpha-actinin,filamin and vinculin (Bloch and Hall,1983), and talin (Sealock et al., 1984) are probably located in the postsynaptic portion of the neuromuscular junctions since the antibody staining at these junctions persist after denervation.

Recently Woodruff et al.(1987) have shown that a non-receptor protein of Mr 300,000 daltons ,copurifying with acetylcholine receptor-rich membranes (Burden et al., 1983),is tightly associated with the postsynaptic membrane. The 300 Kd protein,as the 43 Kd protein,appears to be a peripheral protein located on the intracellular surface of the postsynaptic membrane,both in the *Torpedo* electrocyte and at neuromuscular junction.

All these data suggest that the cytoplasmic projections of the receptor molecules form complexes with one or several non-receptor proteins associated with the postsynaptic membrane. These complexes have not yet been well defined and consequently the relationships of the acetylcholine receptors with the cytoskeleton of the subsynaptic sarcoplasm have yet to be determined.

REFERENCES

Akert K,Pfenniger K and Sandri C (1967) The fine structure of synapses in subfornical organ of the cat. Z.Zellforsch. 81:537-556

Atwood HL and Lnenicka GA (1986) Structure and function in synapses: emerging correlations. Trends NeuroSci. 7:229-233

Barrantes FJ,Neugebauer D-Ch,Zingsheim HP (1980) Peptide extraction by alkaline treatment is accompanied by rearrangement of the membrane-bound acetylcholine receptor from Torpedo marmorata.FEBS Lett. 112:73-78

Bloch RJ and Hall ZW (1983) Cytoskeletal components of the vertebrate neuromuscular junction : vinculin,alpha-actinin, and filamin. J. Cell Biol. 97:217-223

Bridgman PC,Carr C,Pedersen SE and Cohen JB (1987) Visualization of the cytoplasmic surface of Torpedo postsynaptic membranes by freeze-etch and immunoelectron microscopy. J. Cell Biol. 105:1829-1846

Burden SJ (1982) Identification of an intracellular postsynaptic antigen at the frog neuromuscular junction. J. Cell Biol. 94:521-530

Burden SJ (1985) The subsynaptic 43-kDa protein is concentrated at developing nerve-muscle synapses in vitro . Proc.Natl.Acad.Sci.USA 82:8270-8273

Burden SJ (1987) The extracellular matrix and subsynaptic sarcoplasm at nerve-muscle synapses. In: Salpeter MM (ed).The vertebrate neuromuscular junction. Alan R Liss New York :163-186

Burden SJ,DePalma RL and Gottesman GS (1983) Crosslinking of proteins in acetylcholine receptor-rich membranes :association between the beta-subunit and the 43 kd subsynaptic protein. Cell 35:687-692

Burgoyne RD, Gray EG and Barron J (1983) Cytochemical localization of calcium in the dendritic spine apparatus of the cerebral cortex and at synaptic sites in the cerebellar cortex. J.Anat.(Lond) 136:634-635

Caceres A,Payne MR,Binder LI and Stewart O (1983) Immunocytochemical localization of actin and microtubule-associated protein MAP2 in dendritic spines. Proc.Natl.Acad.Sci.USA 80:1738-1742

Cartaud J (1980) A critical re-evaluation of the structural organization of the excitable membrane in Torpedo marmorata electric organ. In: Taxi J (ed). Ontogenesis and functional mechanisms of peripheral synapses. Elsevier,North Holland. Amsterdam.New York: 199-210

Cartaud J,Kordeli C, Nghiêm H-O and Changeux J-P (1983) La protéine de 43,000 dalton : pièce intermédiaire assurant l'ancrage du récepteur cholinergique au cytosquelette sous-neural ? C.R.Acad.Sci.Paris 297:285-289

Cartaud J,Sobel A,Rousselet A,Devaux PF and Changeux J-P (1981) Consequences of alkaline treatment for the ultrastructure of the acetylcholine-receptor-rich membranes from Torpedo marmorata electric organ.J.Cell Biol. 90: 418-426

Coss RG and Perkel DH (1985) The function of dendritic spines : a review of theoretical issues. Behav. and neur.biol. 44:151-185

Couteaux R (1945) Rapports du buisson de Kühne et des noyaux musculaires chez la Grenouille. C.R.Soc.Biol. 139:641-644

Couteaux R (1947) Contribution à l'étude de la synapse myo-neurale. Rev.canad. Biol. 6:563-711

Couteaux R (1952) Le système à "petites" fibres nerveuses et à contraction "lente";contribution à son identification histologique sur les muscles de la Grenouille. C.R.Assoc.Anat. 39:264-269

Couteaux R (1955) Localization of cholinesterases at neuromuscular junctions. Internat. Rev. Cytol. 4:335-375

Couteaux R (1958) Morphological and cytochemical observations on the postsynaptic membrane at motor end-plates and ganglionic synapses. Exper.Cell Res.,Suppl.5 :294-322

Couteaux R (1980) L'organisation postsynaptique de la jonction neuromusculaire. In:Fondation Singer-Polignac(ed). La transmission neuromusculaire. Les médiateurs et le "milieu intérieur".(Coll.internat.1978 Paris) Masson 1980 Paris New York Barcelona Milano : 39-70

Couteaux R (1981) Structure of the subsynaptic sarcoplasm in the interfold of the frog neuromuscular junction. J.Neurocyt. 10:947-962

Couteaux R and Pécot-Dechavassine M (1968) Particularités structurales du sarcoplasme sous-neural. C.R.Acad.Sci. 266D:8-10

Couteaux R and Pécot-Dechavassine M (1970) Vésicules synaptiques et poches au niveau des "zones actives" de la jonction neuromusculaire. C.R. Acad. Sci. 271D:2346-2349

Dreyer F,Peper K,Akert K,Sandri C and Moor H (1973) Ultrastructure of the 'active zone' in the frog neuromuscular junction. Brain Res. 62:373-380

Dyson SE and Jones DG (1984) Synaptic remodelling during development and maturation : junction differentiation and splitting as a mechanism for modifying connectivity. Dev.Brain Res. 13:125-137

Ellisman MH,Rash JE,Staehelin LA and Porter KR (1976) Studies of excitable membranes II.A comparison of specializations at neuromuscular junctions and non-junctional sarcolemmas of mammalian fast and slow twitch muscle fibers. J. Cell Biol. 68:752-774

Fertuck HC and Salpeter MM (1974) Localization of acetylcholine receptor by ^{125}I-alpha-bungarotoxin binding at mouse motor endplates. Proc.Natl.Acad. Sci.USA 71:1376-1378

Fertuck HC and Salpeter MM (1976) Quantitation of junctional and extrajunctional acetylcholine receptors by electron microscope autoradiography after ^{125}I-alpha-bungarotoxin binding at mouse neuromuscular junctions.J. Cell Biol. 69:144-158

Fifková E and Van Harreveld A (1977) Long-lasting morphological changes in dendritic spines of dentate granular cells following stimulation of the entorhinal area. J.Neurocytol. 6:211-230

Froehner SC (1984) Peripheral proteins of postsynaptic membranes from Torpedo electric organ identified with monoclonal antibodies. J. Cell Biol. 99:88-96

Froehner SC,Gulbrandsen V,Hyman C,Jeng AY,Neubig RR and Cohen JB (1981) Immunofluorescence localization at the mammalian neuromuscular junction of the Mr43,000 protein of Torpedo postsynaptic membranes. Proc.Natl.Acad. Sci.USA 78:5230-5234

Gray EG (1959) Axo-somatic and axo-dendritic synapses of the cerebral cortex: an electron microscope study. J. Anat.(Lond) 93:420-433

Gulley RL and Reese TS (1981) Cytoskeletal organization at the postsynaptic complex. J. Cell Biol. 91:298-302

Hall ZW,Lubit BW and Schwartz JH (1981) Cytoplasmic actin in postsynaptic structures at the neuromuscular junction. J.Cell Biol. 90:789-792

Hamilton SL,McLaughlin M and Karlin A (1979) Formation of disulfide-linked oligomers of acetylcholine receptor in membranes from Torpedo electric tissue. Biochemistry 18:155-163

Heuser JE (1980) 3-D visualization of membrane and cytoplasmic specializations of the frog neuromuscular junction. In: Taxi J (ed). Ontogenesis and functional mechanisms of peripheral synapses. INSERM symposium n°13. Elsevier,North-Holland. Amsterdam : 139-155

Heuser JE,Reese TS and Landis DMD (1974) Functional changes in frog neuromuscular junctions studied with freeze-fracture. J.Neurocytol. 3:109-131

Heuser JE and Salpeter SR (1979) Organization of acetylcholine receptors in quick-frozen,deep-etched, and rotary-replicated Torpedo postsynaptic membranes. J.Cell Biol. 82:150-173

Hirokawa N and Heuser JE (1982) Internal and external differentiations of the postsynaptic membrane at the neuromuscular junction. J.Neurocytol. 11:487-510

Jones EG and Powell TPS (1969) Morphological variations in the dendritic spines of the cortex. J.Cell Sci. 5:509-529

Klymkowsky MW,Heuser JE and Stroud RM (1980) Protease effects on the structure of acetylcholine receptor membrane from Torpedo californica. J.Cell Biol. 85:823-838

Kordeli E,Cartaud J,Nghiêm H-O,Pradel L-A,Dubreuil C,Paulin D and Changeux J-P (1986) Evidence for a polarity in the distribution of proteins from the cytoskeleton in Torpedo marmorata electrocytes. J.Cell Biol. 102:748-761

Landis DMD and Reese TS (1983) Cytoplasmic organization in cerebellar dendritic spines. J.Cell Biol. 97:1169-1178

Lännergren J (1975) Structure and function of twitch and slow fibres in amphibian skeletal muscle. In: Lennerstrand G and Bach-y-Rita P (eds). Basic mechanisms of ocular motility and their clinical implications.Pergamon 1975 Oxford : 63-84

Lännergren J (1979) An intermediate type of muscle fibre in Xenopus laevis. Nature 279:254-256

Lännergren J and Smith RS (1966) Types of muscle fibers in toad skeletal muscle. Acta Physiol. Scand. 68:263-274

Le Beux YJ and Willemot J (1975a) An ultrastructural study of the microfilaments in rat brain by means of heavy meromyosin labelling. I.The perikaryon,the dendrites and the axon. Cell Tissue Res. 160:1-36

Le Beux YJ and Willemot J (1975b) An ultrastructural study of the microfilaments in rat brain by means of E-PTA staining and heavy meromyosin labelling. II. The synapses. Cell Tissue Res. 160:37-68

Lo MMS,Garlang PB,Lamprecht J and Barnard EA (1980) Rotational mobility of the membrane-bound acetylcholine receptor of Torpedo electric organ measured by phosphorescence depolarization. FEBS Lett. 111:407-412

McGraw F,Somlyo AV,Blaustein MP (1980) Localization of calcium in presynaptic nerve terminals. An ultrastructural and electron microprobe analysis. J.Cell Biol. 85:228-241

McMahan UJ,Sanes JR and Marshall LM (1978) Cholinesterase is associated with the basal lamina at the neuromuscular junction. Nature 271:172-174

Matus A,Ackerman M,Pehling G,Byers HR and Fusinava K (1982) High actin concentrations in brain dendritic spines and postsynaptic densities.Proc. Natl.Acad.Sci.USA 79:7590-7594

Miledi R and Uchitel OD (1981) Properties of postsynaptic channels induced by acetylcholine in different frog muscle fibres. Nature 291:162-165

Milhaud M and Pappas GD (1966) Post-synaptic bodies in the habenula and interpeduncular nuclei of the cat. J.Cell Biol. 30:437-441

Morgan DL and Proske U (1984) Vertebrate slow muscle : its structure,pattern of innervation,and mechanical properties. Physiol.Rev. 64:103-169

Neubig RR,Krodel EK,Boyd ND and Cohen JB (1979) Acetylcholine and local anesthetic binding to Torpedo nicotinic post-synaptic membranes after removal of non-receptor peptides. Proc.Natl.Acad.Sci.USA 76:690-694

Nghiêm H-O,Cartaud J,Dubreuil C,Kordeli C,Buttin G and Changeux J-P (1983) Production and characterization of a monoclonal antibody directed against the 43,000-dalton ν 1 polypeptide from Torpedo marmorata electric organ. Proc.Natl.Acad.Sci.USA 80:6403-6407

Page SG (1965)A comparison of the fine structures of frog slow and twitch muscle fibres. J.Cell Biol. 26:477-497

Palay SL and Chan-Palay V (1974) Cerebellar cortex. Cytology and organization. Springer, Berlin,Heidelberg,New York

Palay SL,Sotelo C, Peters A and Orkand PM (1968) The axon hillock and the initial segment. J.Cell Biol. 38:193-201
Peters A,Palay SL and Webster H de F (1976) The fine structure of the nervous system. The neurons and supporting cells.Saunders,Philadelphia,London,Toronto
Peters A,Proskauer CC and Kaiserman-Abramof IR (1968) The small pyramidal neuron of the rat cerebral cortex. The axon hillock and initial segment. J.Cell Biol. 39:604-619
Peper K,Dreyer F,Sandri C,Akert K and Moor H (1974) Structure and ultrastructure of the frog motor endplate. A freeze-etching study. Cell Tiss.Res. 149:437-455
Pumplin DW,Reese TS and Llinás R (1981) Are the presynaptic membrane particles the calcium channels ? Proc.Natl.Acad.Sci.USA 78:7210-7213
Rotundo RL (1987) Biogenesis and regulation of acetylcholinesterase. In: Salpeter MM (ed).The vertebrate neuromuscular junction.Alan R Liss,New York : 247-284
Rousselet A,Cartaud J,Devaux PF and Changeux J-P(1982) The rotational diffusion of the acetylcholine receptor in Torpedo marmorata membrane fragments studied with a spin-labelled alpha-toxin : importance of the 43,000 protein(s). EMBO J. 1:439-445
Rousselet A and Devaux PF (1977) Saturation transfer electron paramagnetic resonance on membrane-bound proteins. II. Absence of rotational diffusion of the cholinergic receptor protein in Torpedo marmorata membrane fragments. Biochem.Biophys.Res.Comm. 78:448-454
St.John PA,Froehner SC,Goodenough DA and Cohen JB (1982) Nicotinic postsynaptic membranes from Torpedo :sideness,permeability to macromolecules,and topography of major polypeptides. J.Cell Biol. 92: 333-342
Saitoh T,Wennogle LP and Changeux J-P (1979) Factors regulating the susceptibility of the acetylcholine receptor protein to heat inactivation.FEBS Lett. 108:489-494
Salpeter MM (1987) Vertebrate neuromuscular junctions : general morphology, molecular organization,and functional consequences. In: Salpeter MM (ed). The vertebrate neuromuscular junction. Alan R Liss,New york : 1-54
Salpeter MM,McHenry FA and Feng H (1974) Myoneural junctions in the extraocular muscles of the mouse. Anat.Rec. 179:201-224
Sealock R (1980) Identification of regions of high acetylcholine receptor density in tannic acid-fixed postsynaptic membranes from electric tissue. Brain Res. 199:267-281
Sealock R (1982) Cytoplasmic surface structure in postsynaptic membrane from electric tissue visualized by tannic-acid-mediated negative contrasting. J.Cell Biol. 92:514-522
Sealock R and Kavookjian A (1980) Postsynaptic distribution of acetylcholine receptors in electroplax of the torpedine ray,Narcine brasiliensis.Brain Res. 190:81-93
Sealock R,Wray BE and Froehner SC (1984) Ultrastructural localization of the Mr 43,000 protein and the acetylcholine receptor in Torpedo postsynaptic membranes using monoclonal antibodies. J.Cell Biol. 98:2239-2244
Smith RS and Ovalle WK (1973) Varieties of fast and slow extrafusal muscle fibres in amphibian hind limb muscles. J.Anat. 116:1-24
Sobel A,Heidmann T,Hofler J and Changeux J-P (1978) Distinct protein components of Torpedo membranes carry the acetylcholine receptor site and the binding site for local anesthetics and histrionicotoxin. Proc.Natl.Acad. Sci.USA 75:510-514
Špaček J (1985a) Three dimensional analysis of dendritic spines. II.Spine apparatus and other cytoplasmic components. Anat.Embryol. 171:235-243
Špaček J (1985b) Relationship between synaptic junctions,puncta adhaerentia

and the spine apparatus at neocortical axo-spinous synapses. A serial section study. Anat.Embryol. 173:129-135

Špaček J (1986) Similarities between organization of subsynaptic cytoplasm at frog neuromuscular and mammalian axospinous junctions. In:Proceedings of the IInd Czechoslovak-East German bilateral symposia of anatomists, histologists and embryologists.Bratislava. p.88

Špaček J and Hartmann M (1983) Three-dimensional analysis of dendritic spines. I. Quantitative observations related to dendritic spine and synaptic morphology in cerebral and cerebellar cortices. Anat.Embryol. 167:289-310

Špaček J and Lieberman AR (1974) Three-dimensional reconstruction in electron microscopy of central nervous system. Sbornik Věd Praci Hradec Králové. 17:203-222

Taxi J (1961) Etude de l'ultrastructure des zones synaptiques dans les ganglions sympathiques de la Grenouille. C.R.Acad.Sci.(Paris) 252:174-176

Taxi J (1965) Contribution à l'étude des connexions des neurones moteurs du système nerveux autonome. Ann.Sci.naturelles,Zool.,12e s. 7:413-674

Tarrant SB and Routtenberg A (1977) The synaptic spinule in the dendritic spine : electron microscopic study of the hippocampal dentate gyrus.Tiss. Cell 9:461-473

Tarrant SB and Routtenberg A (1979) Postsynaptic membrane and spine apparatus : proximity in dendritic spines. Neurosci.Lett. 11:289-294

Uchitel OD and Miledi R (1987) Characteristics of synaptic currents in frog muscle fibers of different types. J.Neurosci.Res. 17:189-198

Verma V (1984a) The presynaptic active zones in three different types of fibres in frog muscle. Proc.R.Soc.Lond. B221. 369-373

Verma V (1984b) Innervation and membrane specializations at neuromuscular junctions in submaxillaris muscle of the frog. J.Ultrastruct.Res. 87: 136-148

Verma V and Reese TS (1984) Structure and distribution of neuromuscular junctions on slow muscle fibers in the frog. Neuroscience 12:647-662

Walrond JP and Reese TS (1985) Structure of axon terminals and active zones at synapses on lizard twitch and tonic muscle fibers. J.Neuroscience 5: 1118-1131

Wennogle LP and Changeux J-P 1980) Transmembrane orientation of proteins present in acetylcholine receptor-rich membranes from Torpedo studied by selective proteolysis. Eur.J.Biochem. 106:381-393

Westrum LE (1970) Observations on initial segments of axons in the prepyriform cortex of the rat. J.Comp.Neurol. 139:337-356

Westrum LE,Jones DH,Gray EG and Barron J (1980) Microtubules,dendritic spines and spine apparatus. Cell Tissue Res. 208:171-181

Whittaker VP and Gray EG (1962) The synapse : biology and morphology. Brit. Med.Bull. 18:223-228

Woodruff ML,Theriot J and Burden SJ (1987) 300-kD subsynaptic protein copurifies with acetylcholine receptor-rich membranes and is concentrated at neuromuscular synapses. J.Cell Biol. 104:939-946

NEW EVIDENCE SUPPORTING THE VESICLE HYPOTHESIS FOR QUANTAL SECRETION AT THE NEUROMUSCULAR JUNCTION

Bruno Ceccarelli, Flavia Valtorta and William P. Hurlbut *

Department of Medical Pharmacology, Center for the Study of Peripheral Neuropathies and Neuromuscular Diseases, CNR Center of Cytopharmacology, University of Milan, Via Vanvitelli 32, 20129 Milano, Italy.
* The Rockefeller University, New York, N.Y. 10021, USA

Introduction

Acetylcholine (ACh) is stored in two compartments within vertebrate motor terminals: the cytoplasm and the synaptic vesicles (Whittaker et al., 1964; Dunant et al., 1972); it is released from terminals in two ways: continuously in a relatively steady molecular stream (Katz and Miledi, 1977; Vyskocil and Illes, 1979), and intermittantly in pulses, or quanta, (Fatt and Katz, 1952, del Castillo and Katz, 1954) which contain about 10^4 molecules (Kuffler and Yoshikami, 1975). The function of the continuous leak is not understood (Edwards et al., 1985), but the spontaneous or neurally evoked release of quanta generates the discrete, transient miniature endplate potentials (mepps) or endplate potentials (epps) that mediate neuromuscular transmission. Two hypotheses have been proposed for the origin of the quanta: a) the cytoplasmic hypothesis which postulates that quanta are comprised of cytoplasmic ACh which diffuses in pulses through channels in the axolemma which become intermittantly permeable to it (Israel and Manaranche, 1985), and b) the vesicle hypothesis which postulates that quanta are comprised of ACh which is released by exocytosis from the interiors of synaptic vesicles whose membranes have fused with the axolemma (del Castillo and Katz, 1956). In our opinion, the vast preponderance of evidence supports the vesicle hypothesis of quantal secretion, though its proof

NATO ASI Series, Vol. H21
Cellular and Molecular Basis of Synaptic Transmission
Edited by H. Zimmermann

is not yet absolute and most of the mechanistic details of the fusion and recovery processes have yet to be worked out.

The major pieces of evidence acquired before 1979 in favor of the vesicle hypothesis for synapses which secrete quantally are:

1) Vesicles are present in the presynaptic terminals of these synapses (de Robertis and Bennet, 1955, Palay, 1956; Birks et al., 1960) and they contain neurotransmitters (Whittaker et al., 1964; Dunant et al., 1972).

2) Vesicles fuse with the presynaptic membrane during prolonged vigorous quantal secretion (Heuser et al., 1974; Ceccarelli et al., 1979 a and b). The fusion usually occurs at the active zones, but need not be restricted to these zones (Ceccarelli et al., 1979b; Ceccarelli et al., 1988a).

3) During prolonged, vigorous secretion, vesicles take up extracellular tracers such as horseradish peroxidase (HRP) suggesting that they are recovered by endocytosis from the axolemma (Ceccarelli et al., 1972 and 1973; Heuser and Reese, 1973; Ceccarelli and Hurlbut, 1980a). This recovery, or recycling, of vesicles from the axolemma sustains their number in the axoplasm during prolonged periods of secretion. When the synthesis of ACh is blocked while recycling is active, then the terminals will contain many vesicles which lack transmitter (Ceccarelli and Hurlbut, 1975).

4) Vesicles are depleted from the axoplasm if quantal secretion is stimulated while recycling is blocked (Ceccarelli and Hurlbut, 1980b), and their loss is accompanied by expansion of the axolemma. Terminals that have lost all their vesicles do not release quanta of transmitter (Gorio et al., 1978; Ceccarelli and Hurlbut, 1980a).

Since 1979, additional evidence in favor of the vesicle hypothesis has been accumulated: 5) Vesicles fuse with the axolemma at the moment transmitter is released. (Heuser et al., 1979; Torri-Tarelli et al., 1985). 6) When recycling is blocked and no HRP is taken up, then the number of quanta secreted is approximately equal to the number of vesicles lost from the axoplasm, and the maximum number secreted approaches the number of vesicles in resting terminals (Haimann et al.,

1985; Fesce et al., 1986a; Ceccarelli et al., 1988b). When recycling is active and HRP is taken up, then the number of quanta secreted greatly exceeds both the number of vesicles lost and the number in resting terminals (Segal et al., 1985; Fesce et al., 1986a, Valtorta et al., 1988). 7) Synaptophysin (p38), a specific integral protein of the vesicle membrane (Jahn et al., 1985, Wiedenmann and Franke, 1985) becomes permanently incorporated into the axolemma of terminals that are stimulated to exhaustion while recycling is blocked. When recycling is active, however, synaptophysin is not permanently incorporated into the axolemma (Valtorta et al., 1988).

The findings published before 1979 have been reviewed by several workers (Israel et al., 1979; Marchbanks, 1979; Zimmermann, 1979; Ceccarelli and Hurlbut, 1980b; MacIntosh, 1980; Tauc, 1982; Reichardt and Kelly, 1983), some of whom drew conclusions quite different from ours. We discuss below our recent work which has contributed to the last three points listed above.

Temporal Coincidence between Vesicle Fusion and Quantal Secretion

The quick-freezing technique has been used to demonstrate that vesicles fuse with the presynaptic membrane at the moment quanta are released (Torri-Tarelli et al., 1985). In this technique a neuromuscular preparation is dropped onto a cold (≈ 15° K) copper block a known interval of time after a single suprathreshold electric shock has been delivered to the nerve. A layer of tissue about 10 μm deep at the surface of contact with the block freezes within a msec. Structural and chemical changes occurring in this layer are arrested within the freezing time and can be preserved for later examination. Heuser et al. (1979) applied the technique to frog muscles, but they found few images of fused vesicles in terminals that had been stimulated in standard Ringer's solution. This negative result is not surprising, since the 200 or so quanta released from a terminal by a single action potential would be scattered over the 600 micron length of its presynaptic membrane, making it unlikely that many vesicle fusions would be found on freeze-fracture replicas which are usually a few microns long. They did find many fusions on terminals that had been treated with 1 mM 4-aminopyridine (4-AP), a drug which enhances evoked quantal release about 100 fold (Lundh et al., 1977; Heuser et al., 1979; Katz and Miledi, 1979), but the earliest fusions occurred about 5 msec after the stimulus, a delay which exceeds the sum of the conduction (≈ 1 msec) and synaptic delay (≈ 1 msec) times at 20° C. This discrepancy has been cited as evidence that vesicles serve a function other than storing and secreting quanta of transmitter (Israel and Manaranche, 1985). We believe that the excessive delay observed by Heuser et al. (1979) arose from two sources: a) precooling of the preparation during its descent toward the cold block (Heuser et al., 1979; Ceccarelli and Hurlbut, 1980b), and b) the use of freeze-fracture. The latter technique is unreliable in this instance because the fracture plane cannot be directed to pass through the most superficial 10 µm layer of tissue that is properly frozen. It can easily pass through deeper tissue that is improperly frozen, even though the inner domain of the presynaptic membranes appears well preserved. Indeed, Heuser et al. (1979) showed only replicas of P faces of nerve terminals, and such images could not come from terminals on the outer surfaces of the most superficial layer of muscle fibers, i.e. the surface which contacts the cold block and the only region that would have frozen within 1 msec (see Torri-Tarelli et al., 1985 for a critical discussion).

We applied this technique to frog cutaneous pectoris muscles treated with 1 mM 4-AP, used an apparatus which minimized precooling (Escaig, 1981), confined our observations to junctions at the thin medial edge of the muscle in order to standardize the latency between the electric stimulus and the beginning of the epp (≈ 3 msec; Torri-Tarelli et al., 1985), examined thin sections cut from cryosubstituted muscles, and studied junctions only in the most superficial layer of tissue that was properly frozen. The earliest fusions occurred 2.5 msec after the stimulus and about 0.5 msec before the beginning of the epp, i.e., at the time quanta are released (Fig.1). This demonstration of temporal coincidence between vesicle fusion and quantal secretion is strong evidence in favor of the vesicle hypothesis.

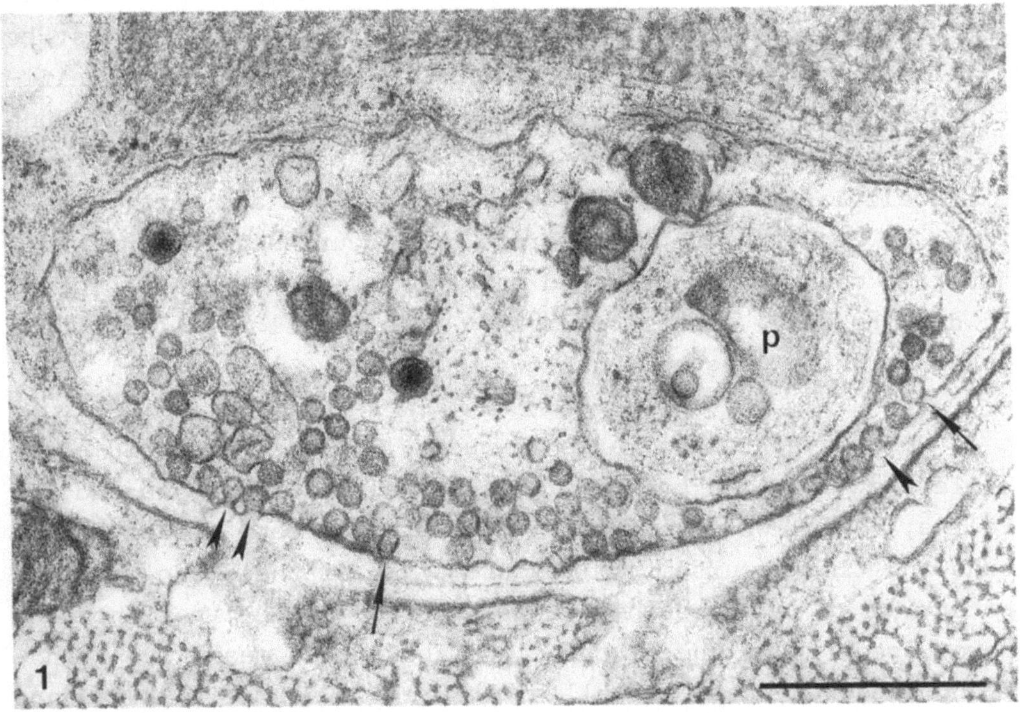

Fig.1. Electron micrograph of a cryosubstituted frog neuromuscular junction quick-frozen 2.5 msec after a single electric stimulus in 1 mM 4-AP. In this cross-section at the level of an active zone different degrees of association between synaptic vesicles and the prejunctional membrane are evident. Arrowheads indicate clear openings whereas arrows indicate images that suggest intermediate states between the fusion and fission steps as proposed by Palade (1975). p, invagination of a Schwann cell process. Scale marker: 0.5 µm (from Torri-Tarelli et al., 1985).

Correlations between Quantal Counts and Vesicle Counts

Ever since mepps were first discovered, quantal secretion has been measured by counting individual mepps, or by calculating the ratio:

(mean epp amplitude)/(mean mepp amplitude),

after correcting the former for non linear summation. Individual mepps can be recognized and counted easily at rates of occurrence up to a few hundred/sec, but not at higher rates because then the individual events overlap extensively. Rates exceeding 1000/sec are encountered routinely when terminals are treated with a variety of agents, and such rates must be attained if the number of vesicles is to be lowered significantly in a reasonable length of time (≈ 1 hr). The much briefer miniature endplate currents (mepcs) recorded by an extracellular micropipette can be resolved at much higher rates of occurrence, but the rate of occurrence for the whole junction cannot be determined reliably by this method because it detects only events occurring at a small portion of the junction whose length is unknown.

High asynchronous rates of quantal release can, in principle, be measured by fluctuation analysis (Verveen and De Felice, 1974), which provides estimates of mepp rate, amplitude and time course from measurements of the mean membrane potential at a secreting junction, and of the variance and power spectrum of the fluctuations about that mean. The standard procedure uses Campbell's theorem (Campbell, 1909) and assumes several ideal conditions which are seldom fulfilled at real junctions. These assumptions are: 1) changes in potential at an endplate are due only to mepps, 2) mepps occur randomly at a stationary rate, 3) mepps sum linearly and 4) at a given endplate mepps are identical in amplitude and time course. Rice (1945) has shown that relations analogous to Campbell's theorem exist for all the central moments (more strictly, the cumulants or semi-invariants) of the fluctuations, so that it is possible, in principle, to determine mepp rate and amplitude from any pair. We have used the skew (third central moment) and variance of high-pass filtered records of the fluctuations and determined the mepp time course, as is usual, from the power spectrum of unfiltered records (Segal et al., 1985; Fesce et al., 1986b). Since this method is based entirely on the characteristics of the fluctuations, it gives several advantages: 1) changes in the mean potential caused by extraneous factors do not affect its results, 2) it avoids the major error caused by nonlinear summation, i.e. the one which arises from the difference between the chord and slope of the potential-conductance curve at an endplate, 3) the regions of the power spectrum containing spectral components arising from nonstationary mepp rates can be suppressed by appropriately filtering the data before they are analyzed, 4) the residual second order effects of nonlinear summation, i.e. those which arise from the slight convexity of the potential-conductance curve, can be easily corrected, and 5) it provides a numerical index which monitors the spread of the distribution of mepp amplitudes and which can be used to correct the estimates of mepp rate and amplitude for that spread.

The method provides reliable measures of mepp rate and amplitude over a range from $10\text{-}10^4$ quanta/sec, and it has been applied to nerve terminals treated with 0.05-0.1 mM ouabain (Haimann, et al., 1985), 0.1 mM La^{3+} (Segal et al., 1985) or low doses (0.03-0.3 glands/ml) of black widow spider venom (BWSV) (Fesce et al., 1986a), all at room temperature, or high concentrations (2 µg/ml) of α-latrotoxin (α-LTx) at 1-3° C (Ceccarelli et al., 1988b). Each agent was tested in a Ca^{2+}-free solution and in a solution with the normal concentration of Ca^{2+} (1.8 mM). We also fixed control muscles and muscles that had been treated for 1-3 hrs under one of the experimental conditions, counted the number of vesicles in longitudinal or cross sections of the terminals and calculated the total number of vesicles that remained in terminals with an average length of 600 µm. The results of the quantal counts and vesicle counts are summarized in Table 1, together with the results of experiments in which HRP was added to the bathing solution to monitor recycling. With 0.1 mM La^{3+}, or low doses of BWSV with Ca^{2+}, HRP uptake is abundant (i.e., recycling is active), the final mepp rates are high and the number of quanta secreted far exceeds the number of vesicles contained in the control terminals. In the other conditions HRP uptake is nil (i.e., recycling impaired), the final mepp rates are low, the number of quanta secreted roughly equals

TABLE 1. **Number of vesicles lost from the axoplasm and number of quanta secreted under various conditions.**

Conditions		Quanta secreted: Peak rate (δ^{-1})	Quanta secreted: End rate (δ^{-1})	Quanta secreted: Total ($X10^{-3}$)	Vesicles/terminal ($X10^{-3}$): Remaining	Vesicles/terminal ($X10^{-3}$): Lost*	HRP uptake	References
La^{3+} 0.1 mM, R.T., @ 1 hr.	+Ca	1,900	300	3,500	200	540	+	Segal, et al, 1985;
	-Ca	3,800	60	2,700	97	640	+	Fesce, et al, 1986a
BWSV, 0.03-0.3 gland/ml, R.T., @ 1 hr.	+Ca	1,900	440	1,700	N.D#	N.D.#	+	Fesce, et al, 1986a;
	-Ca	2,300	17	700	≈ 0	740	NO	Ceccarelli and Hurlbut, 1980a
Ouabain, 0.5-1 mM, R.T., @ 3 hr.	+Ca	1,300	<10	740	170	570	NO	Haimann, et al, 1985
	-Ca	700	<10	880	90	650	NO	
α-LTx, 2 μg/ml, 1-3° C, 2 hr.	+Ca	560	39	710	210	530	NO	Ceccarelli, et al, 1988b
	-Ca	440	46	580	230	510	NO	

Mean values only.

* The average number of vesicles in the control terminals was 740 X 10^3. The number of vesicles lost from the axoplasm is the difference between this value and the average number of vesicles remaining in the experimental terminals.

\# N.D. = not determined
@ R.T. = room temperature (21-25° C)

the number of vesicles lost and it approaches the number of vesicles in control terminals. The close correspondence between vesicles lost and quanta secreted when recycling is blocked suggests that each vesicle secretes a single quantum.

We do not know what factors control recycling. Ca^{2+} seems not to be required under all circumstances, and ouabain, BWSV or α-LTx may act indirectly by changing the ionic composition of the axoplasm, the one by inhibiting the Na^{+}-pump (Birks and Cohen, 1968), the others by increasing the ionic permeability of the nerve terminal (Hurlbut and Ceccarelli, 1979; Wanke et al., 1986).

The fluctuation method also estimates the mean mepp amplitude and monitors the spread of the distribution of amplitudes. That distribution usually broadens during intense secretion and small mepps are recorded often at the ends of the experiments (Kriebel and Gross, 1974; Fesce et al., 1986a). The average amplitudes of the mepps we recorded usually decreased by 50%, and most of the decrease was due to postsynaptic factors such as decreases in membrane potential and resistance or (possibly) changes in sensitivity of ACh-receptors to ACh. Florey and Kriebel (1983) and Kriebel and Florey (1983) observed many small mepps (about one tenth the normal size) at junctions treated with 1 mM La^{3+}, and they suggested that these mepps represented subunits from which the normal mepp is built. However, Lambert and Parsons (1970) have reported that La^{3+} first increased the postsynaptic sensitivity of the frog neuromuscular junction to ACh and then accelerated its "desensitization". Thus it is possible that the small mepps recorded by Florey and Kriebel (1983) were due to (nearly) fullsized quanta acting upon desensitized endplates. As pointed out above, our data suggest that a quantum is secreted from a single vesicle, rather than from several vesicles as the sub-mepp hypothesis implies.

However, we emphasize that the vesicle hypothesis can easily account for real reductions in quantal size (i.e., the number of molecules of ACh in a quantum) at junctions which are rapidly recycling their vesicles, since such vesicles may not be fully loaded with transmitter. Indeed, reductions in the size of the quanta secreted by such junctions contains potentially useful information about the rate at which recycled vesicles are refilled with ACh. However, this kind of measurement is better made at voltage clamped junctions, since the clamp eliminates the effects of spurious changes in membrane potential and resistance on the response of an endplate to a quantum of ACh.

Identification of Molecular Components of the Membranes of Synaptic Vesicles at the Frog Neuromuscular Junction.

HRP is a marker for the extracellular space, and although its presence within vesicles proves that their interiors were once in contact with that space, it provides no information on the nature of their membranes. Thus, it is possible that the membranes of the HRP-labelled vesicles differ from those of vesicles that have not recycled: some of their peripheral proteins could have been lost or others could have been acquired. Moreover, if intermixing of molecular components of the fused membranes occurs, then the membranes of the recycled vesicles could also contain integral components of the axolemma or lack some of the integral components of mature vesicle membranes. Selective markers for vesicle or axolemmal components are needed to identify them and trace their movements through an exo-endocytotic cycle. Two potential markers of components of frog synaptic vesicles are the proteins, synapsin I and synaptophysin (p38), which have been identified in mammalian synaptic vesicles. The former is a phosphoprotein that is bound to the cytoplasmic surface of the vesicle membrane (Ueda et al., 1977; Huttner et al., 1983; Schiebler et al., 1986; De Camilli and Greengard, 1986); the latter is an integral, transmembrane glycoprotein of the synaptic vesicle (Jahn et al., 1985; Wiedemann and Franke, 1985).

We have used antibodies against mammalian synapsin I to demonstrate that this protein is present in frog motor nerve terminals, and to show, at the E.M. level of resolution, that it is associated with the cytoplasmic surface of their synaptic vesicles (Fig.2).

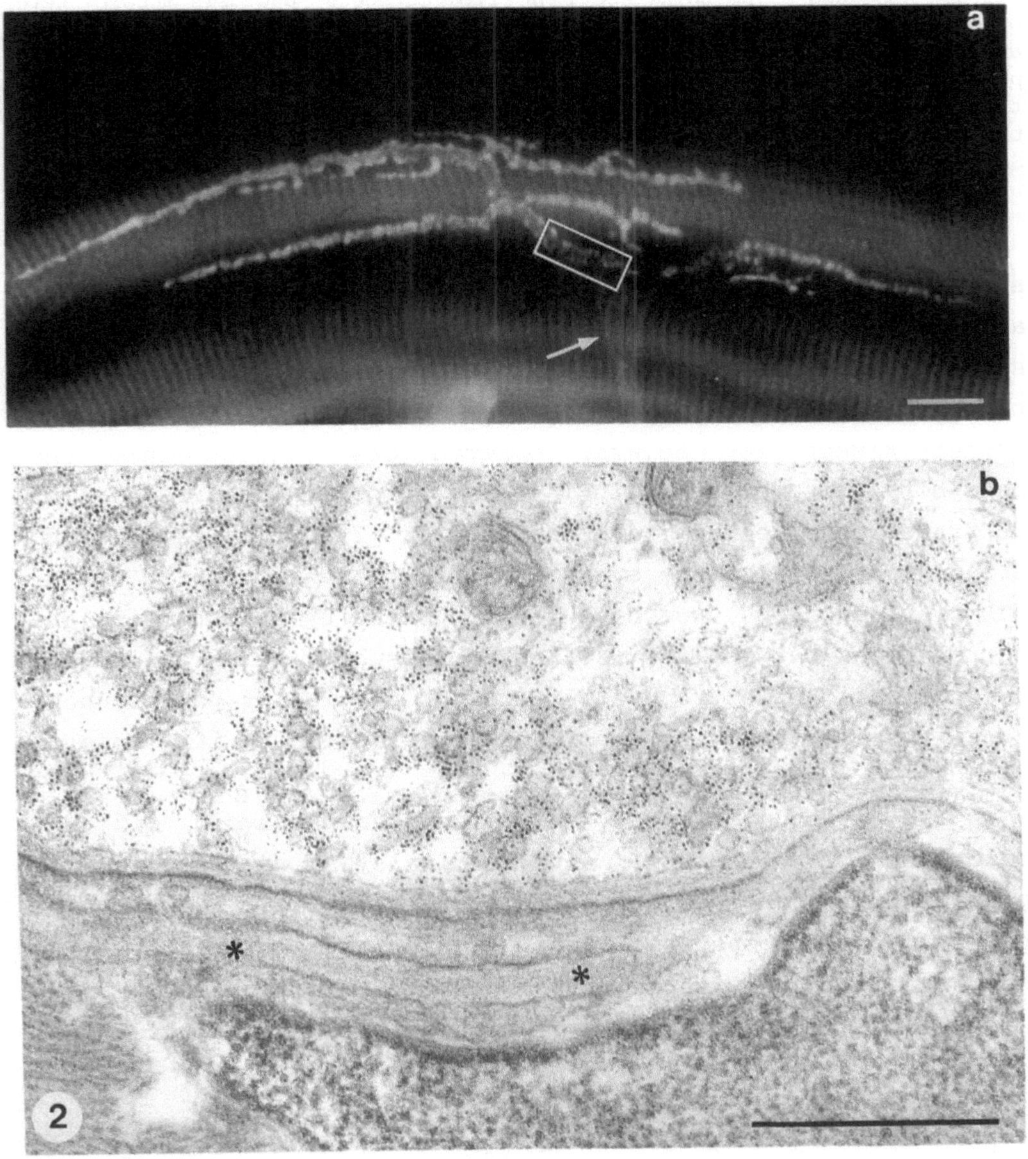

Fig.2.
a) Immunofluorescence micrograph showing the distribution of immunoreactivity for synapsin I at a neuromuscular junction on a single fiber teased from frog cutaneus pectoris muscle. Immunoreactivity is highly concentrated in the nerve terminal region and virtually no fluorescence is associated with the muscle fiber or with the unmyelinated preterminal axon (arrow). The region of intense fluorescence are arranged in the branching pattern characteristic of frog motor nerve endings. The rectangle delineate a terminal branch where segmented staining, transverse to the long axis of the branch, is particularly evident. This segmented pattern probably

reflects the clustering of synaptic vesicles in regions of the nerve terminals between Schwann cell invaginations.
b) Electron micrograph showing the distribution of synapsin I immunoreactivity at a neuromuscular junction as revealed by immunoferritin labeling. Agarose-embedded 25-µm-thick sections from cutaneus pectoris muscle were processed for immunoelectron microscopy using anti-synapsin I antibodies and ferritin-conjugated goat anti-rabbit $F(ab)_2$s. Most of the dark ferritin particles are located in regions where synaptic vesicles are highly concentrated. The synaptic cleft, the postjunctional infolding (asterisks) and the muscle fiber are virtually devoided of ferritin particles.
Scale markers: 20 µm in a and 0.5 µm in b. (from Valtorta et al., 1987).

We have also studied the biochemical properties of synapsin I from frog brain and neuromuscular junction (Valtorta et al., 1987). Frog synapsin I can be phosphorylated at different sites by the catalytic subunit of mammalian cAMP-dependent protein kinase and by mammalian Ca^{2+}/calmodulin-dependent protein kinase II. Several observations suggest that the phosphorylation of synapsin I plays a role in the regulation of transmitter release (Llinas et al., 1985; De Camilli and Greengard, 1986). Since synapsin I is present at frog neuromuscular junctions, we have the opportunity of studying the correlations between its states of phosphorylation and other biochemical or electrophysiological properties of these junctions, thereby obtaining important information on its role in synaptic transmission.

The antibody against mammalian synaptophysin did not react strongly with frog nerve terminals, so we purified the protein from frog brain and raised antibodies to it. These polyclonal, monospecific antibodies reacted strongly with components within resting frog nerve terminals (Valtorta et al., 1988), proving that synaptophysin is an intraterminal protein.

Movements of Vesicle Membrane during Quantal Secretion

Antibodies are large molecules which usually cannot penetrate the intact plasmalemma. When they are used to study the distribution of intracellular proteins in fixed but intact terminals, the plasmalemma must first be permeabilized with detergents to permit antibodies to enter the cytoplasm. This procedure was used to demonstrate the presence of synapsin I and synaptophysin in frog motor nerve terminals. However, if synaptic vesicles become permanently incorporated into the axolemma, then their inner surface should be exposed to the extracellular space and antibodies should gain access to them even when the axolemma has not been permeabilized.

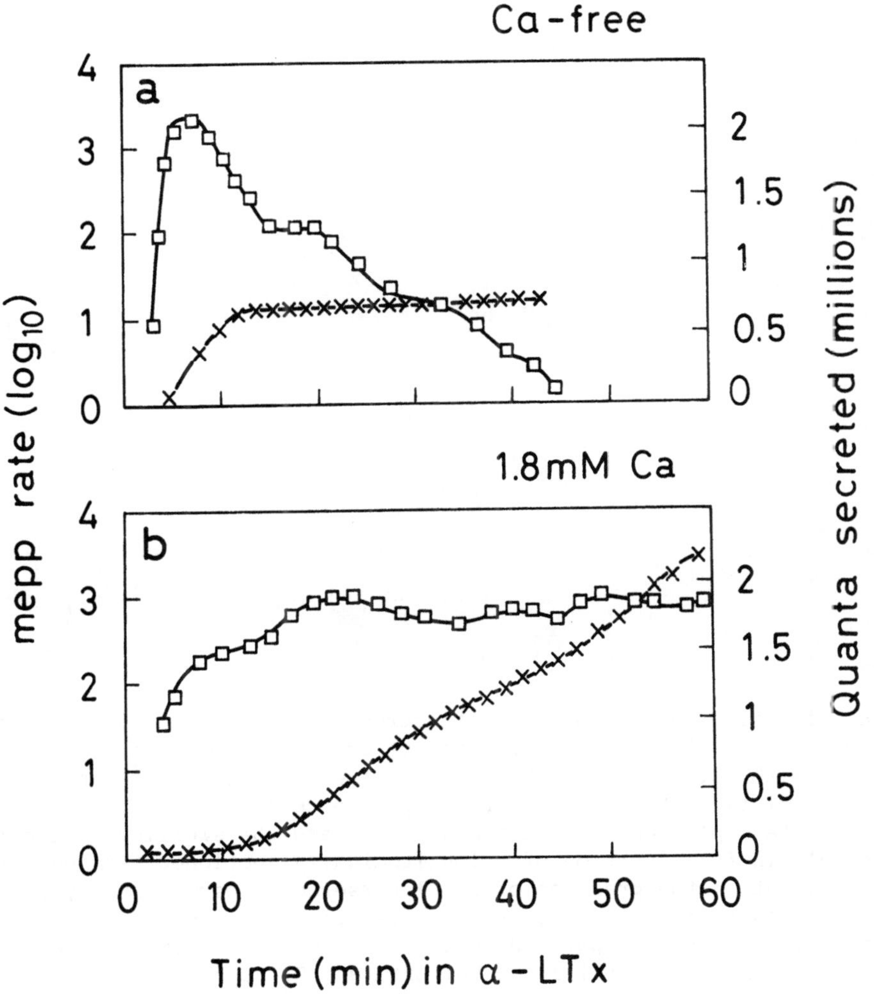

Fig.3. Time course of mepp rate (□-□) and cumulative quantal secretion (X-X) computed by fluctuation analysis of intracellular endplate recordings from a pair of frog cutaneus pectoris muscles exposed to 0.2 µg/ml α-LTx in either Ca^{2+}-free (a) or Ca^{2+}-containing (b) Ringer's solution. Mepp rate (sec^{-1}) on logarithmic scale. (Original from R. Fesce, F. Valtorta and B. Ceccarelli)

We exploited this possibility by applying antibodies against frog synaptophysin to two sets of neuromuscular preparations. One set was treated with a low dose (0.2 µg/ml) of α-LTx in a Ca^{2+}-free solution supplemented with 4 mM Mg^{2+}, and the other was treated with an equal dose of toxin in standard Ringer's solution supplemented with 4 mM Mg^{2+}. Recycling is blocked under the first condition: the terminals stop secreting after they have released their initial store of quanta (Fig. 3a), and they lose their vesicles and swell (Fig. 4a).

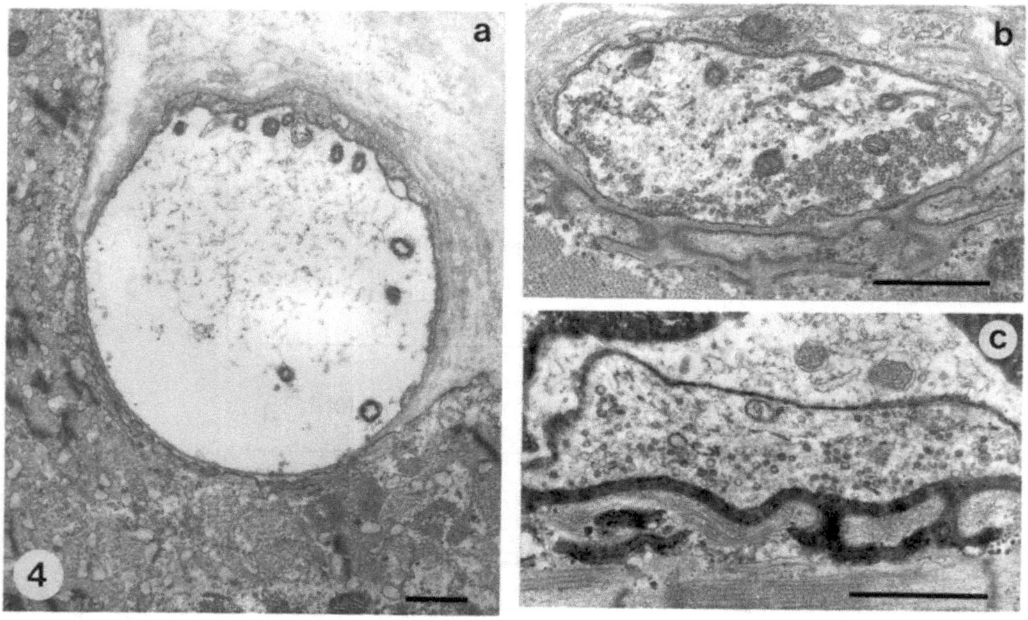

Fig.4. Effects of α-LTx on the ultrastructure of neuromuscular junctions. Panels a and b: electron micrographs of cross-sectioned terminals treated for 1 h with 0.2 µg/ml α-LTx in Ca^{2+}-free (a) or Ca^{2+}-containing (b) Ringer's solution. Notice the normal appearance of the terminal in (b) and the complete depletion of synaptic vesicles and the swelling of the terminal in (a). In (c) a portion of a longitudinally sectioned terminal treated as in (b) but in the presence of 1.6 % horseradish peroxidase is illustrated. Notice that most of the synaptic vesicles present in the field are loaded with horseradish peroxidase reaction product. Scale markers: 1 µm. (Original from B. Ceccarelli, F. Valtorta and R. Fesce)

Recycling is vigorous in the second condition (Fig.4b and c): the terminals retain most of their synaptic vesicles, maintain high rates of secretion for at least an hour and release many more quanta than their initial store (Fig. 3b).

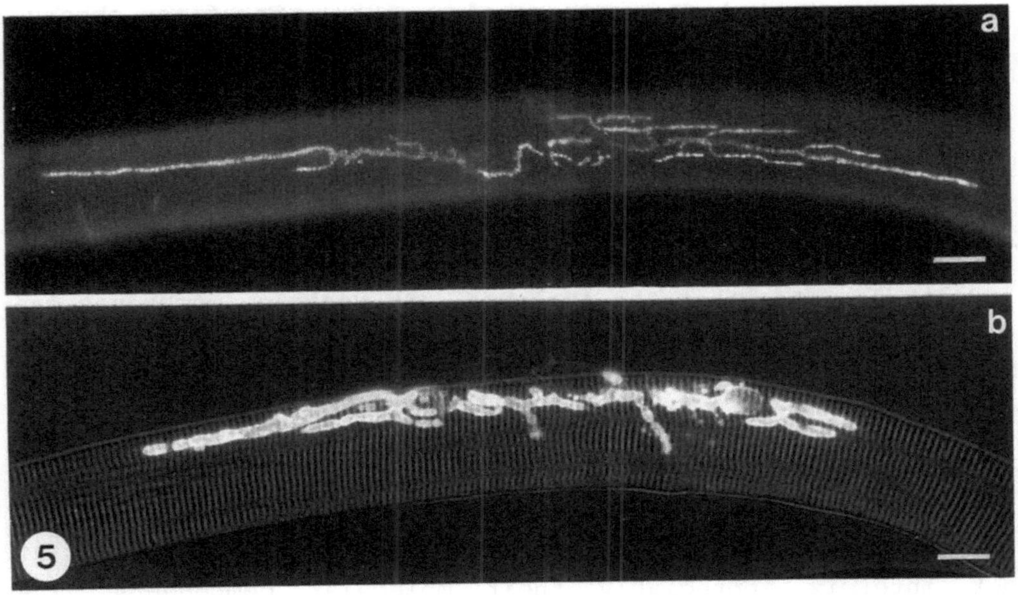

Fig.5. Fluorescence micrographs of neuromuscular junctions stained to reveal synaptophysin immunoreactivity. Single muscle fibers were teased apart from frog cutaneus pectoris muscle.
a) resting preparation treated with 0.1 % Triton-X-100 after fixation and then exposed to anti-synaptophysin antiserum followed by rhodamine-conjugated goat anti-rabbit IgGs. The pattern of immunoreactivity revealed by this presynaptic marker is similar to the one observed for synapsin I (compare with Fig.2a). When detergent was not used under this condition synaptophysin immunoreactivity was undetectable and the nerve terminal region could be identified only by the postsynaptic distribution of α-bungarotoxin labeling (see Valtorta et al., 1988).
b) neuromuscular junction fixed 1 hr after exposure to 0.2 μg/ml α-LTx in Ca^{2+}-free Ringer's solution. No detergent was used. In this condition no permeabilization was necessary to reveal synaptophysin immunoreactivity and the nerve terminal branches show a marked increase in their transverse dimension. This experimental condition is similar to that shown in Fig.4a. Scale markers: 20 μm (Original from F. Valtorta, R. Fesce and B. Ceccarelli)

When the antibodies to frog synaptophysin were applied, without prior permeabilization, to the preparations that had lost their vesicles, they bound to their antigen and their spatial distribution was revealed by immunofluorescence (Fig. 5). When these antibodies were applied, without prior permeabilization, to the preparations that had recycled their vesicles, they did not bind to their antigens and no specific fluorescence was detected (Valtorta et al., 1988). The first observation shows that when vesicles are lost from the axoplasm, then some of the antigenic determinants of synaptophysin are exposed to the extracellular space, and it supports the idea that the vesicle membrane has been incorporated into the axolemma. The second observation shows that synaptophysin is not incorporated into the axolemma when recycling is active, and it suggests that the axolemma and vesicle membranes can fuse with no extensive intermixing of their molecular components.

Our interpretation of this constellation of findings is that a quantum of transmitter is secreted by exocytosis from a synaptic vesicle which has fused with the axolemma, usually, but not necessarily, at an active zone. The fused membranes do not intermix appreciably their molecular components. The vesicles are quickly recovered from the axolemma by an endocytic process which does not necessarily involve clathrin coats (Torri-Tarelli et al., 1987), and they are refilled with ACh and used again to secrete additional quanta. The quanta accumulated by the recycled vesicles need not be the size of quanta stored in the vesicles of resting terminals, and if ACh synthesis is blocked, the recycled vesicles may contain no releasable transmitter. We do not wish to imply, however, that vesicle-dependent quantal release is the only functionally significant means of release, to claim that all synapses must have vesicles (Lasansky, 1980), or to deny that at some synapses (those between cells which do not produce action potentials, in particular) transmission, or modulation of the excitability of the postsynaptic cell, may occur by regulating the nonquantal release of neurotransmitters directly from the cytoplasm of the presynaptic element (Schwartz, 1987).

Recent Evidence from other Cells for Fusion between Secretory Granules and Plasmalemma

Direct evidence for fusion between the secretory granules and the plasmalemma of adrenal chromaffin cells or mast cells from mutant mice has been obtained using the patch-clamp technique to measure minute, stepwise changes in the electric capacitance of the surface membrane of these cells (Neher and Marty, 1982; Zimmerberg et al., 1987; Breckenridge and Almers, 1987 a and b). The size of the capacitative steps is consistent with the area of membrane which envelops the secretory granules, and in the chromaffin cells both increases and decreases in capacitance were observed (Neher and Marty, 1982). The mast cells from the mutant mice contain small numbers of giant secretory granules large enough to be seen with a light microscope, and in these cells it has been possible to correlate increases in capacitance with secretion from individual granules. The change in capacitance precedes by about 100 msec both the visible swelling of the granules which accompanies exocytosis and the measurable loss of fluorescent material from the interior of the granules. The increase in surface capacitance usually develops over tens of msec, and its time course has been interpreted as reflecting the widening of the "fusion pore" which connects the interior of the granule with the extracellular space (Zimmerberg, et al., 1987; Breckenridge and Almers, 1987a). "Flicker" is sometimes observed before the increase in capacitance becomes permanent, and recently it has been observed that, before the increase in capacitance begins, an ion-sized connection forms between the granule interior and the extracellular space (Breckenridge and Almers, 1987). These observations are compatible with the idea that the initial steps in the "fusion" between the much smaller synaptic vesicles and the presynaptic membrane could quickly reverse but still allow the escape of small molecules like ACh and the intake of larger molecules like extracellular tracers.

Acknowledgements

This review was prepared while the authors were recipients of grants from the Muscular Dystrophy Association (B.C.) and NIH NS-18354 (W.P.H.).

References

Birks R, Huxley HE and Katz B (1960) The fine structure of the neuromuscular junction of the frog. J Physiol Lond 150:134-144.

Birks RI and Cohen MW (1968) The action of sodium pump inhibitors on neuromuscular transmission. Proc R Soc Lond B 170:381-399.

Breckenridge LJ and Almers W (1987a) Currents through the fusion pore that forms during exocytosis of a secretory vesicle. Nature 328:814-817.

Breckenridge LJ and Almers W (1987b) Final steps in exocytosis observed in a cell with giant secretory granules. Proc Natl Acad Sci USA 84:1945-1949.

Campbell N (1909) The study of discontinuous phenomena. Proc Cambridge Phil Soc 15:117-136.

Ceccarelli B, Fesce R, Grohovaz F and Haimann C (1988a) The effect of potassium on exocytosis of transmitter at the frog neuromuscular junction. J Physiol Lond, in press.

Ceccarelli B, Grohovaz F and Hurlbut WP (1979a) Freeze-fracture studies of frog neuromuscular junctions during intense release of neurotransmitter. I. Effects of Black Widow Spider Venom and Ca^{2+}-free solutions on the structure of the active zone. J Cell Biol 81:163-177.

Ceccarelli B, Grohovaz F and Hurlbut WP (1979b) Freeze-fracture studies of frog neuromuscular junctions during intense release of neurotransmitter. II. Effects of electrical stimulation and high potassium. J Cell Biol 81:178-192.

Ceccarelli B and Hurlbut WP (1975) The effects of prolonged repetitive stimulation in hemicholinium on the frog neuromuscular junction. J Physiol Lond 247:163-188.

Ceccarelli B and Hurlbut WP (1980a) Ca^{2+}-dependent recycling of synaptic vesicles at the frog neuromuscular junction. J Cell Biol 87:297-303.

Ceccarelli B and Hurlbut WP (1980b) Vesicle hypothesis of the release of quanta of acetylcholine. Physiol Rev 60:396-441.

Ceccarelli B, Hurlbut WP and Iezzi N (1988b) Effect of α-latrotoxin on the frog neuromuscular junction at low temperature. J Physiol Lond, submitted.

Ceccarelli B, Hurlbut WP and Mauro A (1972) Depletion of vesicles from frog neuromuscular junctions by prolonged tetanic stimulation. J Cell Biol 54: 30-38.

Ceccarelli B, Hurlbut WP and Mauro A (1973) Turnover of transmitter and synaptic vesicles at the frog neuromuscular junction. J Cell Biol 57:499-524.

De Camilli P and Greengard P (1986) Synapsin I: a synaptic vesicle-associated neuronal phosphoprotein. Biochem Pharmacol 35:4349-4357.

de Robertis EDP and Bennet HS (1955) Some features of the submicroscopic morphology of synapses in frog and earthworm. J Biophys Biochem Cytol 1:47-58.

del Castillo J and Katz B (1954) Quantal components of the end-plate potential. J Physiol Lond 124:560-573.

del Castillo J and Katz B (1956) Biophysical aspects of neuro-muscular transmission. Prog Biophys Biophys Chem 6:121-170.

Dunant Y, Gautron J, Israel M, Lesbats B and Manaranche R (1972) Les compartiments d'acetylcholine de l'organe electrique de la Torpille et leurs modifications par la stimulation. J Neurochem 19:1987-2002.

Edwards C, Dolezal V, Tucek S, Zemkova H and Vyskocil R (1985) Is an acetylcholine transport system responsible for nonquantal release of acetylcholine at the rodent myoneural junction? Proc Natl Acad Sci USA 82:3514-3518.

Escaig J (1981) New instruments which facilitate rapid freezing at 83 K and 6 K. J Microsc Paris 126:221-229.

Fatt P and Katz B (1952) Spontaneous subthreshold activity at motor nerve endings. J Physiol Lond 117:109-128.

Fesce R, Segal JR, Ceccarelli B and Hurlbut WP (1986a) Effects of Black Widow Spider Venom and Ca^{2+} on quantal secretion at the frog neuromuscular junction. J Gen Physiol 88:59-81.

Fesce R, Segal JR and Hurlbut WP (1986b) Fluctuation analysis of nonideal shot noise. Application to the neuromuscular junction. J Gen Physiol 88:25-57.

Florey E and Kriebel ME (1983) Changes in acetylcholine concentration, miniature end-plate potentials and synaptic vesicles in frog neuromuscular preparations during lanthanum treatment. Comp Biochem Physiol 75c:285-294.

Gorio A, Hurlbut WP and Ceccarelli B (1978) Acetylcholine compartments in mouse diaphragm. Comparison of the effects of Black Widow Spider Venom, electrical stimulation, and high concentrations of potassium. J Cell Biol 78:716-733.

Haimann C, Torri-Tarelli F, Fesce R and Ceccarelli B (1985) Measurement of quantal secretion induced by ouabain and its correlation with depletion of synaptic vesicles. J Cell Biol 101:1953-1965.

Heuser JE and Reese TS (1973) Evidence for recycling of synaptic vesicle membrane during transmitter release at the frog neuromuscular junction. J Cell Biol 57:315-344.

Heuser JE, Reese T, Dennis MJ, Jan Y and Evans L (1979) Synaptic vesicle exocytosis captured by quick-freezing and correlated with quantal transmitter release. J Cell Biol 81:275-300.

Heuser JE, Reese TS and Landis DMD (1974) Functional changes in frog neuromuscular junctions studied with freeze-fracture. J Neurocytol 3:109-131.

Hurlbut WP and Ceccarelli B (1979) Use of Black Widow Spider Venom to study the release of neurotransmitter. In: Ceccarelli B and Clementi F (eds). Advances in Cytopharmacology. Raven Press New York 3:87-115.

Huttner WB, Schiebler W, Greengard P and De Camilli P (1983) Synapsin I (Protein I), a nerve terminal-specific phosphoprotein. III. Its association with synaptic vesicles studied in a highly purified synaptic vesicle preparation. J Cell Biol 96:1374-1388.

Israel M, Dunant Y and Manaranche R (1979) The present status of the vesicular hypothesis. Progr Neurobiol 13:237-275.

Israel M and Manaranche R (1985) The release of acetylcholine: from a cellular towards a molecular mechanism. Biol Cell 55:1-14.

Jahn R, Schiebler W, Ouimet C and Greengard P (1985) A 38,000-dalton membrane protein (p38) present in synaptic vesicles. Proc Natl Acad Sci USA 82: 4137-4141.

Katz B and Miledi R (1977) Transmitter leakage from motor nerve endings. Proc R Soc Lond B 196:59-72.

Katz B and Miledi R (1979) Estimates of quantal content during "chemical potentiation" of transmitter release. Proc R Soc Lond B 205:369-378.

Kriebel ME and Florey E (1983) Effect of lanthanum ions on the amplitude distribution of miniature endplate potentials and on synaptic vesicles in frog neuromuscular junctions. Neurosci 9:535-547.

Kriebel ME and Gross CE (1974) Multimodal distribution of frog miniature endplate potentials in adult, denervated and tadpole leg muscle. J Gen Physiol 64:85-103.

Kuffler SW and Yoshikami D (1975) The number of transmitter molecules in a quantum: an estimate from iontophoretic application of acetylcholine at the neuromuscular synapse. J Physiol Lond 251:465-482.

Lambert DH and Parsons RL (1970) Influence of polyvalent cations on the activation of muscle end plate receptors. J Gen Physiol 56:309-321.

Lasansky A (1980) Lateral contacts and interactions of horizontal cell dendrites in the retina of the larval tiger salamander. J Physiol Lond 301: 59-68.

Llinas R, McGuinness TL, Leonard CS, Sugimori M and Greengard P (1985) Intraterminal injection of synapsin I or calcium/calmodulin dependent protein kinase II alters neurotransmitter release at the squid giant synapse. Proc Natl Acad Sci USA 82:3035-3039.

Lundh H, Leander S and Thesleff S (1977) Antagonism of the paralysis produced by botulinum toxin in the rat. J Neurol Sci 32:29-43.

MacIntosh FC (1980) The role of vesicles in cholinergic systems. In: Brzin M, Sket D and Bachelard H (eds). Synaptic constituents in Health and Disease. Pergamon Press Oxford: 11-52.

Marchbanks RM (1979) Role of storage vesicles in synaptic transmission. In: Hopkins and Duncan (eds). Secretory Mechanisms. Soc Exp Biol Symp XXXII: 251-276.

Neher E and Marty A (1982) Discrete changes of cell membrane capacitance observed under conditions of enhanced secretion in bovine adrenal chromaffin cells. Proc Natl Acad Sci USA 79:6712-6716.

Palade GE (1975) Intracellular aspects of the process of protein synthesis. Science Wash DC 189:347-358.

Palay SL (1956) Synapses in the central nervous system. J Biophys Biochem Cytol Suppl 2:193-202.

Reichardt LF and Kelly RB (1983) A molecular description of nerve terminal function. Ann Rev Biochem 52:871-926.

Rice SO (1944) Mathematical analysis of random noise. Bell Tech Syst J 23:282-332.

Schiebler W, Jahn R, Doucet JP, Rothlein J and Greengard P (1986) Characterization of synapsin I binding to small synaptic vesicles. J Biol Chem 261:8383-8390.

Segal JR, Ceccarelli B, Fesce R and Hurlbut WP (1985) Miniature endplate potential frequency and amplitude determined by an extension of Campbell's theorem. Biophys J 47:183-202.

Schwarz EA (1987) Depolarization without calcium can release γ-aminobutyric acid from a retinal neuron. Science 238:350-355.

Tauc L (1982) Non-vesicular release of neurotransmitter. Physiol Rev 62:857-893.

Torri-Tarelli F, Grohovaz F, Fesce R and Ceccarelli B (1985) Temporal coincidence between synaptic vesicle fusion and quantal secretion of acetylcholine. J Cell Biol 101:1386-1399.

Torri-Tarelli F, Haimann C and Ceccarelli B (1987) Coated vesicles and pits during enhanced quantal release of acetylcholine at the neuromuscular junction. J Neurocytol 16:205-214.

Ueda T and Greengard P (1977) Adenosine 3':5'- monophosphate-regulated phosphoprotein system of neuronal membranes. I. Solubilization, purification, and some properties of an endogenous phosphoprotein. J Biol Chem 252:5155-5163.

Valtorta F, Jahn R, Fesce R, Greengard P and Ceccarelli B (1988) Synaptophysin (p38) at the frog neuromuscular junction: its incorporation into the axolemma and recycling after intense quantal secretion. J Cell Biol, submitted.

Valtorta F, Villa A, Jahn R, De Camilli P, Greengard P and Ceccarelli B (1987) Localization of synapsin I at the frog neuromuscular junction. Neurosci, in press.

Verveen AA and De Felice LJ (1974) Membrane noise. Progr Biophys Mol Biol 28:189-268.

Vyskocil R and Illes P (1979) Non-quantal release of transmitter at mouse neuromuscular junction and its dependence on the activity of Na^+-K^+-ATPase. Pfluegers Arch Europ J Physiol 370:295-297.

Wanke E, Ferroni A, Gattanini P and Meldolesi J (1986) Alpha-latrotoxin of Black Widow Spider Venom opens a small, non-closing cation channel. Bioch Biophys Res Comm 134:320-325.

Whittaker VP, Michaelson IA and Kirkland RJ (1964) The separation of synaptic vesicles from nerve ending particles ("synaptosomes"). Biochem J 90:293-303.

Wiedemann B and Franke W (1985) Identification and localization of synaptophysin, an integral membrane glycoprotein of M_r 38,000 characteristic of presynaptic vesicles. Cell 41:1017-1028.

Zimmerberg J, Curran M, Cohen FS and Brodwick M (1987) Simultaneous electrical and optical measurements show that membrane fusion precedes secretory granule swelling during exocytosis of beige mouse mast cells. Proc Natl Acad Sci USA 84:1585-1589.

Zimmermann H (1979) Vesicle recycling and transmitter release. Neurosci 4:1773-1804.

POSTSTIMULATION INCREASE OF SYNAPTIC VESICLE NUMBER IN THE PREGANGLIONIC NERVE TERMINALS OF THE CAT SYMPATHETIC GANGLION IN VIVO

Tomoko Kadota, *Ken Kadota, Hozumi Tatsuoka, **Muneaki Mizote and ***Tomoichiro Yamaai
Department of Anatomy, School of Medicine, Chiba University, Chiba 280, *Department of Neurochemistry, Psychiatric Research Institute of Tokyo, Tokyo 156, **Department of Information Engineering, Faculty of Informations, Teikyo University of Technology, and ***Department of Anatomy, School of Dentistry, Okayama University, Okayama 700, Japan

INTRODUCTION

Vesicular release of neurotransmitter has been suggested in a variety of nerve endings by morphological reports in which tetanic stimulation is shown to cause the reduction of synaptic vesicles (SVs) in number and simultaneously to depress the postsynaptic response (Ceccarelli et al., 1973; Heuser and Reese, 1973; Pysh and Wiley, 1974; Zimmerman and Whittaker, 1974a, b). Vesicular release of transmitter via exocytosis is supposed to be followed by the subsequent reformation of vesicles by endocytotic retrieval of terminal plasmamembrane by coated vesicles (recycling hypothesis of synaptic-vesicle membrane) (Heuser and Reese, 1973). Questions remains, however, concerning the vesicle hypothesis of the release of quanta of neurotransmitter (Zimmerman, 1979; Ceccarelli and Hurlbut, 1980; Tauc, 1979; Israël and Manarche, 1985). In addition, recent articles have shown the low rate of coated vesicles in the SV reformation, having implied the necessities of examinations for other mechanisms for supplying SVs during transmitter release (Kadota and Kadota, 1982; Meshul and Pappas, 1984; Parducz, 1986; Torri-Tarelli et al., 1987). The purpose of this study is to examine the ultra-structural changes in an neuro-neuronal synapse under normal

NATO ASI Series, Vol. H21
Cellular and Molecular Basis of Synaptic Transmission
Edited by H. Zimmermann

supply of blood. The preganglionic nerve terminal of the cat superior cervical ganglion (SCG) was employed as the experimental material in the present experiment. It was simple to maintain this ganglion under intact blood sypply and to fix rapidly by perfusion via the lingual artery (Kadota and Kadota, 1982).

MATERIALS AND METHODS

General Procedures

The experiments were carried out with the cat SCG *in vivo* as previously described (Kadota and Kadota, 1982). Briefly, the animal was anaesthesized with pentobarbital, i.p., at a dosis of 30 - 50 mg/kg. Both sides of SCG and the carotid artery were prepared, and submerged in liquid paraffin. Either side of the preganglionic trunks was stimulated at 10 Hz with the 1 msec square pulses of 3 - 5 V for various periods up to 30 min. After the predetermined periods of stimulation, the ganglion was rapidly fixed by local perfusion while electrical stimulation was continued. The chemical fixatives were flushed into the plastic tubes connected to the lingual arteries simultaneously occluding the common and external carotid arteries. Most ganglia were fixed with a combined aldehyde-osmium mixture consisting of glutaraldehyde (2%), formaldehyde (1%), osmium tetroxide (0.5%) and potassium ferrocyanide (0.4%) in 0.1 M phosphate buffer at pH 7.4. Then, the ganglia were excised, sliced, postfixed in 1% OsO_4 in the same buffer for 1 h at 4°C, dehydrated, and embedded in Epon 812 for electron microscopy.

Morphometrical Study

The results of each experimental treatment were determined for three to four preparations. Electron micrographs were taken at x 10,000 to x 20,000. Well contrasted terminals were printed at x 50,000 for quantitative measurement. The nerve terminals cut transversely to the dendrite, were used for morphometry.

The poststimulation changes of SV density were topographically studied in the nerve terminal. The presynaptic terminal was subdivided into two areas by a half-circle centering at the middle point of the presynaptic membrane. The area within the half-circle apposing the synaptic contact was termed as the first zone (Zone I) and the lateral area outside the half-circle, the second zone (Zone II). Profiles of the presynaptic terminal and the Zone I were measured with a Muto digigrammer system (Model - G) programmed to compute irregular areas and line lengths. The SV density was obtained by dividing the total number of SVs in the Zone I or II by the area of each zone.

RESULTS

1. Presynaptic fine structures in control (unstimulated) nerve terminals

The preganglionic nerve fibers form axo-dendritic synapses with the principal cells. An axodendritic synapse in a cross section was shown in Fig. 1. Electron-lucent SVs, spherical in shape and about 56 nm in a mean diameter, fill the presynaptic terminal. A small number of cored vesicles, 70 - 100 nm in diameter, and a few mitochondria scattered in between the clusters of SVs. The synaptic complex was straight in contour, and the pre- and postsynaptic membranes apposed each other nearly in parallel.

Vacuolar profiles, 100 - 150 nm in diameter, and coated vesicles (CVs) with a diameter of 80 - 100 nm were rarely found in axon endings. Profiles of smooth endoplasmic reticulum (SER), a few in number and tubular in shape, were occassionally observed to be present in the nerve endings adjacent to the presynaptic terminals.

2. Ultrastructural changes in electrically stimulated nerve terminals

Electrical stimulation resulted in a varaiety of morphological alterations in the presynaptic terminal. One of the significant changes was the increase in the number of vacuoles and CVs. The other was the augmentation in population of the

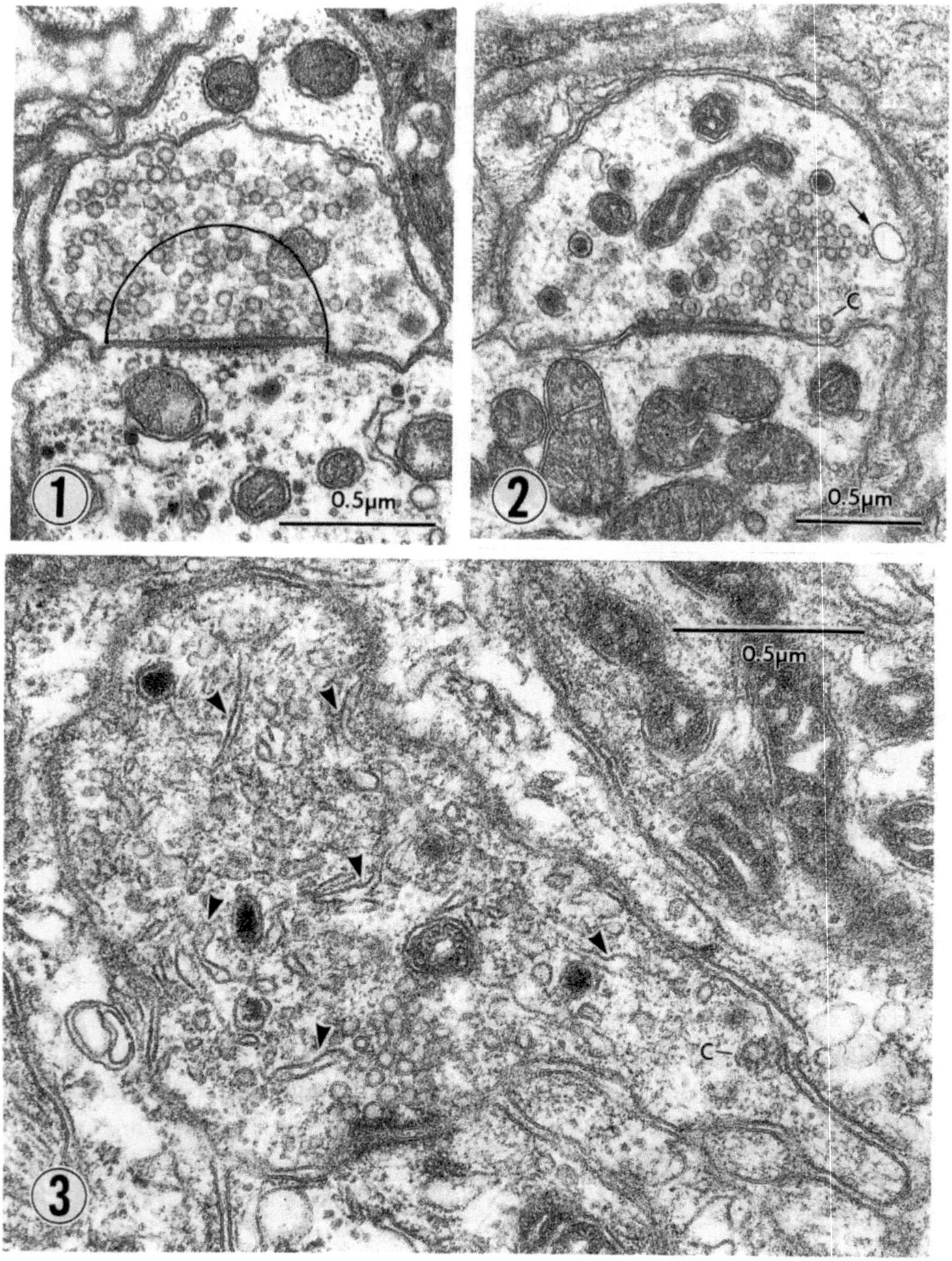
1
0.5μm
2
C
0.5μm
3
0.5μm
C

profiles of SER (Figs. 2 and 3). The vacuoles with a diameter of 100 - 150 nm, produced by macropinocytosis, increased in number 1 - 3 sec after the onset of stimulation. The macropinocytotic vacuoles migrated retrogradedly to the preterminal area as shown in our previous study (Kadota and Kadota, 1982). The CVs with a diameter of 80 - 100 nm, formed by coated endocytosis via the coated pits, augmented in population after about 10 sec of activation, slower in formation rate than the macropinocytotic vacuoles. Horseradish peroxidase (HRP) was incorporated into many macropinocytotic vacuoles and CVs. The marker substance, in contrast, only very rarely taken up into SVs. No vesicular profiles, similar in size and appearances to SVs, were observed to be budded off from the macropinocytotic vacuoles. The inner vesicles of CVs were morphologically indistinguishable from SVs. The findings suggested the possible reformation of SVs from the terminal axolemma via CVs in a very low rate but not through the macropinocytotic vacuoles.

Profiles of SER increased in population in presynaptic terminals. The tubular structures of SER were observed to be distributed in between the clusters of SVs. The poststimulation increase in the density of SER profiles was quantitatively shown in our previous report (Kadota and Kadota, 1985). Vesicular expansions, similar to SVs, were often found to be formed at the ends of the tubules of SER profiles.

The number of SVs appeared to be augmented. The morphometrical study showed the increase of SV density in Zone I. The SV density in the Zone I increased by about 50% of the control value after c.a. 20 sec of stimulatin, maintained this level

Fig. 1-3 Axo-dendritic synapse between the preganglionic axon terminal and the ganglion cell dendrite. Transverse view. Fig. 1 Control (unstimulated). The half-circle having a diameter equal to the length of the presynaptic membrane, means the zone I. The intraterminal area outside this semicircle is the zone II. Fig. 2 Stimulated for 10 sec at 10 Hz. Arrow, macropinocytotic vacuole in zone II. C; coated vesicle in zone I. Fig. 3 Stimulated for 5 min at 10 Hz. Arrow-head, tubules of smooth endoplasmic reticulum (SER).

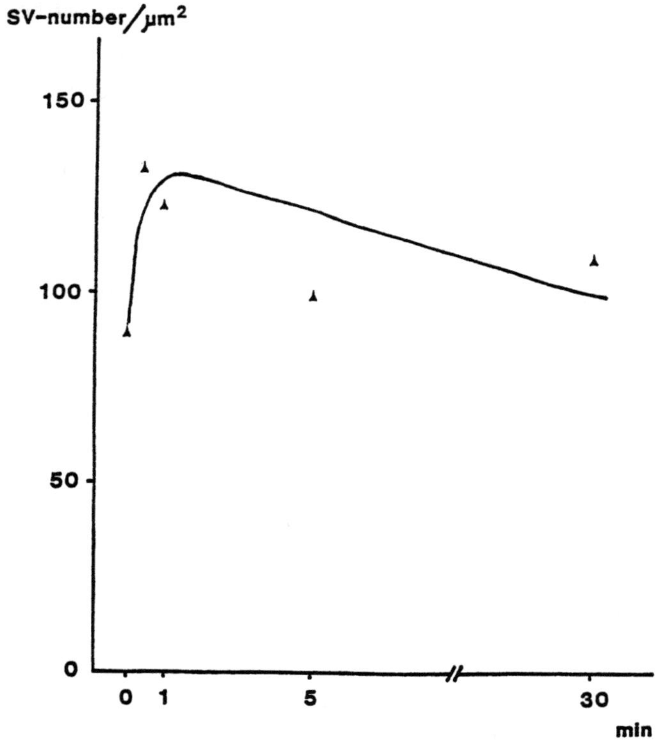

Fig. 4 Synaptic-vesicle densities in number in zone I during the periods of stimulation at 10 Hz. The density of synaptic vesicles was estimated by dividing the synaptic vesicle number by the area of zone I (vesicle number/μm^2).

until approximately 1 min, and then decreased to reach a plateau level around roughly 5 min. The SV density at this plateau level was still about 10% higher than that at the control.

DISCUSSION

The aim of the present study was to examine the fine structures of the cat preganglionic nerve terminals electrically stimulated in vivo. For this, the preganglionic trunk of the cat SCG under intact blood supply was stimulated at 10 Hz for various periods, and the ganglion was rapidly fixed by perfusion with a quick chemical fixative, an osmium-aldehyde mixture.

One of the remarkable changes in the stimulated terminals

was an increase in the SV number per square micrometer, especially the SV density in the Zone I near the presynaptic membrane. This contrasts with many previous works which have shown the poststimulation reduction of SV number in the preganglionic nerve terminals (Pysh and Wiley, 1974; Kadota and Kadota, 1978; Tremblay and Phillipe, 1981; Dicknson-Nelson and Reese, 1983; Wiley et al., 1987), in the axon endings in neuromuscular junctions (Korneliussen, 1972; Ceccarelli et al., 1973; Heuser and Reese, 1973; Rose et al., 1975; Brewer and Lynch, 1986) and in presynaptic terminals of the Torpedo elctromotor synapses (Zimmerman and whittalur, 1974a, b; Boyne et al., 1975; Suszkiw, 1980). The discrepancy between the preceding works and our present results remain obscure. Comparisons with the observations of other investigators is a tangled problem for the differences in preparations and in fixation techniques employed. However, the disparity might be explained by our perfusion fixation procedure in which the osmium-aldehyde mixture was injected with the simultaneous occulusion of the common and external carotid arteries while stimulation was still in progress. The postganglionic action potential disappeared within 3 - 4 sec after the onset of the perfusion fixation as shown previously (Kadota and Kadota, 1982). The observations of Quilliam and Tamarind (1973a, b) on the rat SCG in vitro appear similar to our present results. Quilliam and Tamarind (1973a, b) showed the increase of synaptic-vesicle population within a zone 0.25 μm wide adjacent to the presynaptic membrane. The investigators fixed the ganglia 1 min after 3 min 10 Hz tetanic stimulation of the preganglionic nerve.

Recent articles have shown low level of HRP labeling of coated vesicles and SVs in the stimulated nerve terminals of the cat SCG or frog neuromuscular junctions, implying the low rate of SV reformation along the recycling route involving coated vesicles (Parducz, 1986; Meshul and Pappas, 1984; Torri-Tarelli et al., 1987). The macropinocytotic vacuoles as displayed in Fig. 2, were shown in our previous work to be retrogradely transported to the axon in the stimulated cat SCG in vivo. No SVs were formed from these vacuoles with endocytotic origin (Kadota and Kadota, 1982). Then, a question comes arise that some part of SVs may be produced not from the axolemma but from other membranous sources such as the SER during transmitter release in the cat SCG under intact blood supply. The SER is displayed to extend continuously from the neuronal body to the nerve terminal in the chick parasympathetic ganglion (Droz et al., 1975). However, in many neurons it is poor in the preterminal area as well as the axon ending. Poststimulation increase of SER profiles have been shown in the nerve terminals in neuromuscular junctions (Korneliussen, 1972; Reinecke and Walther, 1978). The tubular structures of SER increase in population in the activated presynaptic terminals in the present experimental material (Fig. 3). Our previous experiments quantitatively showed the increase of the SER density during transmitter release in the cat SCG *in vivo*. Vesicles, similar in size and appearances to SVs, were observed to bud off from the tubules of SER in the presynaptic terminal (Kadota and Kadota, 1985). From these preceding observations and the present results it is likely that SVs are rapidly produced from SER, actively flowing from the axon into the terminal, to result in increase in their number during transmitter release in the cat SCG *in vivo*.

REFERENCES

Brewer PA and Lynch K (1986) Stimulation-associated changes in frog neuromuscular junctions. A quantitative ultrastructural comparison of rapid-frozen and chemically fixed nerve terminals. Neuroscience 17: 881-895

Boyne AF, Bohan TP and Williams TH (1975) Changes in cholinergic synaptic vesicle populations and the ultrastructure of the nerve terminal membranes of *Narcine brasiliensis* electric organ stimulted to fatigue *in vivo*. J Cell Biol 67: 814-825

Ceccarelli B and Hurlbut WP (1980) Vesicle hypothesis of the release of quanta of acetylcholine. Physiol Rev 60: 396-441

Ceccarelli B, Hurlbut WP and Mauro A (1973) Turnover of transmitter and synaptic vesicles at the frog neuromuscular junction. J Cell Biol 57: 499-524

Dickinson-Nelson A and Reese TS (1983) Structural changes during transmitter release at synapses in the frog sympathetic ganglion. J Neuroscience 3: 42-52
Droz B, Rambourg A and Koenig HL (1975) The smooth endoplasmic reticulum: structure and role in the renewal of axonal membrane and synaptic vesicles by fast axonal transport. Brain Resesarch 93: 1-13
Heuser J and Reese TS (1973) Evidence for recycling of synaptic vesicle membrane during transmitter release at the frog neuromuscular junction. J Cell Biol 57: 315-344
Israel M and Manarache R (1985) The release of acetylcholine: from a cellular towards a molecular. Biology of the Cell 55: 1-14
Korneliussen H (1972) Ultrastructure of normal and stimulated motor endplates. With comments on the origin and fate of synaptic vesicles. Z Zellforsch Mikrosk Anat 130: 28-57
Kadota T and Kadota K (1982) Membrane retrieval by maqcropinocytosis in presynaptic terminals during transmitter release in cat sympathetic ganglia in situ. J Electron Microscopy 31: 73-80
Kadota T and Kadota K (1985) Tubular network of the smooth endoplasmic reticulum which appears in the axon terminal following stimulation of the cat sympathetic ganglion in situ. Biomedical Research 6: 13-22
Pysh JJ and Wiley RG (1974) Synaptic vesicle depletion and recovery in cat sympathetic ganglia electrically stimulated in vivo. J Cell Biol 60: 365-374
Quilliam JP and Tamarind DL (1973a) Some effects of preganglionic nerve stimulation on synaptic vesicle populations in the rat superior cervical ganglion. J Physiol 235: 317-331
Quilliam JP and Tamarind DL (1973b) Local vesicle populations in rat superior cervical ganglia and the vesicle hypothesis. J Neurocytol 2: 59-75
Reinecke M and Walther C (1978) Aspects of turnover and biogenesis of synaptic vesicles at locust neuromuscular junctions as revealed by zinc iodide-osmium tetroxide (ZIO) reaction with intravesicular SH-groups. J Cell Biol 78: 839-855
Rose SJ, Pappas GD and Kriebel ME (1978) The fine structure of identified frog neuromuscular junctions in relation to synaptic activity. Brain Res 144: 213-239
Suszkiw JB (1980) Kinetics of acetylcholine recovary in Torpedo electromotor synapses depleted of synaptic vesicles. Neuroscience 5: 1341-1349
Tauc L (1979) Are vesicles necessary for release of acetylcholine at cholinergic synapses. Biocem Pharmacol 27: 3493-3498
Torri-Tarelli F, Haimann C and Ceccarelli B (1987) Coated vesicles and pits during enhanced quantal release of acetylcholine at the neuromuscular junction. J Neurocytol 16: 205-214
Tremblay JP and Phillipe E (1981) Morphological changes in presynaptic terminals of the chick ciliary ganglion after stimulation in vivo. Exp Brain Res 43: 439-446

Wiley PG, Spencer C and Pysh JJ (1987) Time course and frequency dependence of synaptic vesicle depletion and recovery in electrically stimulated sympathetic ganglia. J Neurocytol 16: 359-372

Zimmermann H (1979) Vesicle recycling and transmitter release. Neuroscience 4: 1773-1804

Zimmerman H and denston CR (1977) Recycling of synaptic vesicles in cholinergic synapses of Torpedo electric organ during induced transmitter release. Neuroscience 2: 695-714

Zimmerman H and Whittaker VP (1974a) Effect of electrical stimulation on the yield and composition of synaptic vesicles from the cholinergic synapses of the electric organ of Torpedo: a combined biochemical,electrophysiological and morpholgical study. J Neurochemistry 22:435-450

Zimmerman H and Whittaher VP (1974b) Different recovery rates of the electrophysiological biochemical and morphological parameters in the cholinergic synapses of the Torpedo electric organ after stimulation. J Neurochemistry 22: 1109-1114

SYNAPTIC VESICLE CLASSES IN TORPEDO AND SKATE ELECTRIC ORGAN AND MUSCLE

G.Q. Fox, M.E. Kriebel* and D. Kötting, Department of Neurochemistry, Max-Planck-Institut für biophysikalische Chemie, 3400 Göttingen, Federal Republic of Germany and *Department of Physiology, SUNY Health Science Center at Syracuse, 750 E. Adams Street, Syracuse, New York, U.S.A.

Introduction

Presynaptic nerve terminals contain spherical membrane-bound structures termed synaptic vesicles, which contain transmitter substances (Whittaker, 1984, Zimmermann, 1982). A calcium mediated mechanism fuses vesicles with the presynaptic membrane resulting in the release of transmitter into the synaptic cleft. The increase in internal calcium results from an inward calcium current through a voltage dependent calcium channel which is activated by the action potential. The vesicular shell either pulls back (Ceccarelli and Hurlbut, 1973) or is later internalized by a membrane recycling mechanism (Heuser and Reese, 1973) and subsequently is filled with transmitter for another cycle. Physiologically, transmitter is spontaneously released in the form of quanta which produce miniature end-plate potentials (MEPP); and, these quanta are synchronously released with a nerve action potential to generate the evoked end-plate potential (Fatt and Katz, 1952; del Castillo & Katz, 1954; Boyd and Martin, 1956). These observations form the basis of the quantum of transmitter release. There is evidence from freeze fracture studies with potentiated release that the fusion of 1 vesicle generates 1 quantum (Heuser, Reese, Jan, Jan, and Evans, 1978; Katz and Miledi, 1979) However, several problems have arisen with the hypothesis that one vesicle generates one MEPP. One has been the identification of the sub-MEPP class of quanta which appears in normal preparations as 1/10 the size of the larger class and both classes have the same time characteristics (Erxleben & Kriebel, 1988a; Kriebel and Gross, 1974; Kriebel et al., 1976, 1982). In addition, MEPP amplitude histograms show integral peaks which suggest an alternate hypothesis whereby each MEPP contains 7 to 10 sub-units that are the size of sub-MEPPs (Kriebel and Gross, 1974; Kriebel et al. 1976, 1982; Carlson & Kriebel, 1985; Erxleben & Kriebel, 1988b). Anatomically, this means that each MEPP may represent either the simultaneous release of 10 vesicles or that sub-units are contained within one vesicle. Potentiated release with an action

NATO ASI Series, Vol. H21
Cellular and Molecular Basis of Synaptic Transmission
Edited by H. Zimmermann

potential shows that quantal content is in excess of the number of vesicles contacting the presynaptic membrane (Katz and Miledi, 1979) which implies that it may be impossible to correlate vesicles and quantal content of an end-plate potential. The sub-MEPPs have been observed in the skate (Kriebel, Gross & Pappas, 1986 and in Torpedo (Muller & Dunant, 1986; Kriebel, Fox, and Kötting this issue) and there is some indication of a subunit composition of MEPPs. The classes of MEPP quanta have not been correlated with different classes of vesicles (Kriebel & Florey, 1983; Kriebel, Hanna, Muniak, 1986; Kriebel & Pappas, 1987). However in the Torpedo electric organ a smaller class of vesicles is thought to represent the refilled vesicles (Zimmermann and Whittaker, 1974). More recent studies indicate that this population may represent those vesicles actively involved in transmission and recycling (Stadler & Kiene, 1987) with the larger diameter vesicles functioning more in a storage capacity. Stadler and Kiene (1987) have demonstrated three classes of vesicles based on ACh content with a ratio of 0:1:5-10. The two classes containing ACh may correspond to the MEPP and sub-MEPP class, but only one vesicle class was found in the skate (Kriebel, Gross, and Pappas, 1987).

We have begun testing the subunit hypothesis of the quantum of transmitter release using the electric organs of Torpedo marmorata and skate (Raja undulata). For comparative purposes to the neuromuscular junction of the mouse and frog we have used the Torpedo extrinsic eye muscle. We define several vesicle size classes in the Torpedo electric organ and compare these to the single class found in eye muscle junctions, skate electric organ and the neuromuscular junctions of mouse and frog. In contrast to the other cholinergic junctions studied we report that Torpedo electric terminals contain 4-6 classes of vesicles based on diameter.

Materials and Methods

Animals and Preparation: Torpedo marmorata and Raja undulata were obtained from the Institute de Biologie Marine, Arcachon, France. They were kept in artificial sea water at 18°C. Fish were either anesthetized with MS222 and pithed or killed by a blow to the head. Torpedo electric organs with the skin attached were cut from the animal and sliced into 1 cm wide strips and pinned dorsal skin down to a cork board. The exposed electric tissue was then sliced to within 1-2 mm of the skin and cut into 4x4 mm squares. The 'top' electrocytes have innervated surfaces facing upwards for physiological studies. We found these cells to have a -70 mV resting potential and to generate end-plate potentials with field stimulation. Moreover, the surface terminals did not show spontaneous MEPPs allowing us to conclude that the cut nerves quickly sealed leaving an electrophysiologically normal terminal.

Skate electric organs were freed from isolated tails by blunt dissection. In contrast to the Torpedo electrocytes, skate electrocytes showed spontaneous MEPPs. Extrinsic eye muscles of Torpedo were isolated with both origin and insertion intact as well as a cm long length of the efferent nerve. All tissue was rinsed and maintained in Torpedo saline consisting of (in mM): NaCl, 200; KCl,8; $MgCl_2$, 1.8; $CaCl_2$, 3.4; Tris, 10; $NaHCO_3$, 5; glucose, 5.5;Urea, 300; sucrose, 100.

Specimen preparation: Several fixatives were examined because much previous work has been based on the use of multiple aldehyde fixatives (Karnovsky, 1965) which are hypertonic and cause extremely high rates of quantal transmitter release (see Kriebel, et al this volume). We used an isosmotic fixative (810 mosMol) consisting of 2% glutaraldehyde and 0.3M sodium cacodylate (pH 7.2-7.4), that was found not to induce MEPPs. All tissue was initially superfused with fixative at room temperature for a few minutes and then fixed from 1-2 hr to several days at 4°C. Sodium cacodylate was used throughout due to the high osmolarity of Torpedo and skate serum. Following primary fixation, tissue blocks were rinsed in buffer, postfixed in 1% osmium tetroxide, and en bloc stained with 1% uranyl acetate in 0.25 M sodium acetate buffered at pH 5.5. Ethanol, propylene oxide and Epon 812 were used as dehydrating, transferal and embedding agents.

Sections were cut by D.K. to given interference colors. For methacrylate embedding media, gold signifies 100 nm thickness (Sorval interference color reference guide, Hayat: 1979, Vol. I). The section fold method of section thickness determination (Weibel, 1979) yielded a thickness of 118 nm which is the figure used in our calculations. Sections were mounted without film support on 150 mesh grids. Generally, they were not counterstained with uranyl or lead salts.

A Jeol 100B electron microscope equipped with a goniometer stage was used at the 24,000x setting. Exact magnification was routinely determined with a cross grating replica of 2160 lines/mm. Photographic

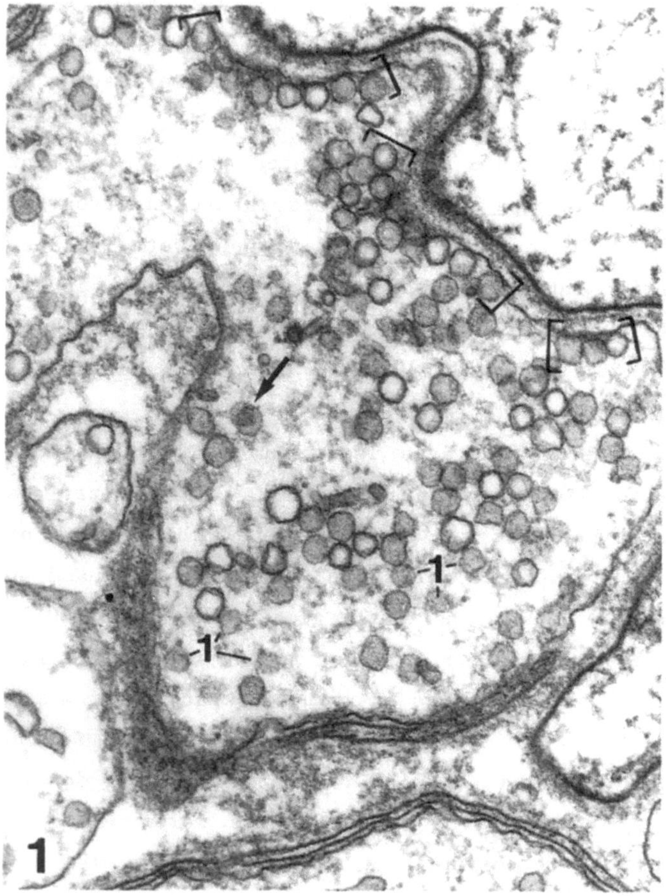

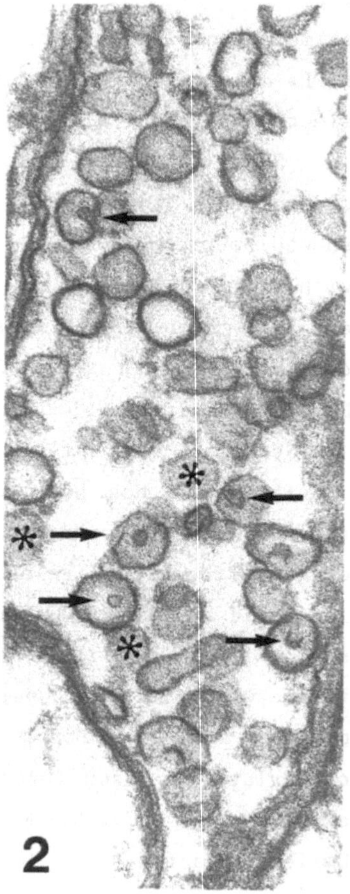

Figure 1. A presynaptic terminal from the electric organ of an adult Torpedo marmorata. The synaptic vesicles appear to be distributed throughout most of the cytoplasm. At the presynaptic membrane they may line-up in rows (brackets). Uniform grey disks show no unit membrane and represent vesicle caps (#1). Vesicles range from 38 to 109 nm in diameter. The interior composition of vesicles varies from dark grey to translucent. The uniform grey interior is due to the plasma membrane as can be demonstrated by increased density in regions of overlap (arrow). The translucence results from either the loss of both caps or to non-contrasted spots on the plasma membrane. Section counterstained with uranyl acetate and lead citrate. Mag = 50,620X

Figure 2. A high power electron micrograph from a 118 nm thick section of a Torpedo terminal from the top electrocyte of the tissue sample. This section has been stained only en bloc with uranyl acetate and illustrates a number of dimpled vesicles (arrows). Caps of vesicles (unit membranes not distinct) are indicated(*). Mag = 96,000X.

cross grating prints were enlarged 3 times (within 1%). Final magnifications were 74,300 to 80,600x. For vesicle diameter analyses ten presynaptic terminals from each experiment were used producing vesicle counts of 500-1000. Vesicles touching the presynaptic membrane were also measured. The presynaptic terminal area was measured with a Haff 315E planimeter. Synaptic vesicle densities were determined by the formula: $N_V = Na/(t+D-2h)$ (Weibel, 1979 where: Nv= computed number/unit volume (um^3), Na=number of vesicle profiles/unit area, t=section thickness, D=mean vesicle diameter, 2h=correction to account for the loss of caps (see Abercrombie, 1946 and Weibel, 1979). Here $h = R \sqrt{(R^2-r^2)}$ where R=maximum radius observed, r= minimum radius observed. We found that 4 readers varied in their vesicle counts by 14%. Reader bias was also evaluated for 4 readers. Each reader calibrated the Zeiss particle analyzer using 1 mm graph paper. The slope (m) and the intercept (b) were determined from the formula $y = m(x)+b$ to convert classes (48) to nm. Synaptic vesicle diameter means measured from the same set of 10 micrographs ranged from 76-84 nm with the 4 readers yielding a 15% reader variation. Tested statistically by either the parametric Student t-test or the non-parametric Mann-Whitney test, these differences were highly significant ($p < .001$). Thus, all vesicles of one experiment or preparation were measured by one person. Since the number of events in each bin is relatively small, random variations produce a ragged profile on the sides of the histograms. We smoothed the histogram profiles with a 3 bin moving average analysis which aids in identifying multiple peaks (Erxleben and Kriebel, 1986b.)

Results

A. Torpedo electric organ nerve terminals:

(1) General structure: The electric organ of *Torpedo marmorata* is innervated by cholinergic motoneurons located in the electric lobes. Innervation occurs exclusively upon 70-80% of the ventral surface of each electrocyte (Fox and Kötting, 1984). Nerve terminals contain synaptic vesicles, mitochondria, neuro-tubules, filaments and glycogen granules (Fig. 1). A basal lamina separates the terminal from the ventral electrocyte surface. Finger-like projections of the terminals covered by basal lamina are found protuding into postsynaptic pits when fixed with a paraformaldehyde fixative. These projections are lost with a 2% glutaraldehyde + 0.3 M sodium cacodylate buffer fixative. A thin Schwann cell process covers the presynaptic terminal.

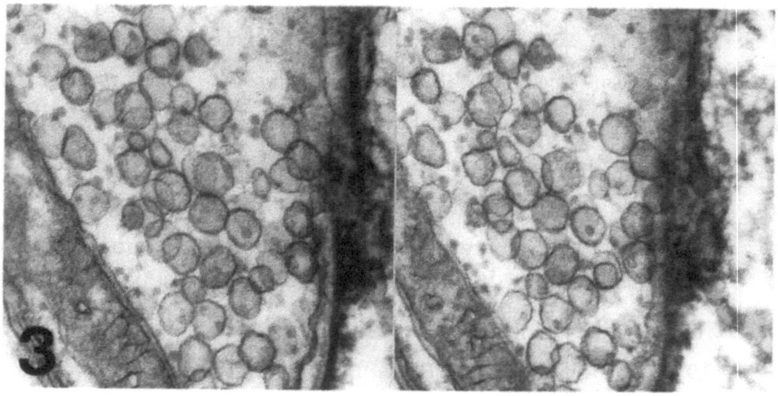

Figure 3. A pair of micrographs used for stereo observation of a Torpedo electric organ terminal. The angle of stage-tilt between the two micrographs is 60°. The section is 118 nm thick and is en bloc stained Note the numerous overlapping vesicles which show different amounts of overlap in the two micrographs. Mag = 45,000X.

(2) Synaptic vesicles: A two percent glutaraldehyde and 0.3 M sodium cacodylate fixative postfixed in 1% osmium tetroxide and en bloc stained in 1% uranyl acetate produces synaptic vesicles which are circular to slightly ovoid with a 4-fold range of diameters (Fig. 1). Vesicle interiors range from clear to opaque (not to be confused with 'dense-core'). Since increasing section thickness decreases the percentage of lucent cores and increases the number of overlapping images there is only one class of synaptic vesicle based on anatomical appearance (Figs. 1-3). Smaller diameter discs devoid of unit membrane are present and these represent caps of vesicles (fig. 2). A small percentage (usually 1%) of vesicles contain a dimple or a finger-like pocket (Fig. 2). Linear and two-dimensional arrays of vesicles are occasionally seen touching the presynaptic membrane (Fig. 1) Exo-or endo-cytotic omega figures along the plasma membrane are rarely seen. Micrographs of terminals were taken for stereo imaging at stage angles of ±30° (Fig. 3). At both section angles vesicle membranes are seen as dark rings and the apparent diameter of the rings is at the maximal vesicle diameter within the section. Overlapping rings represent hemispheres of two adjacent vesicles.

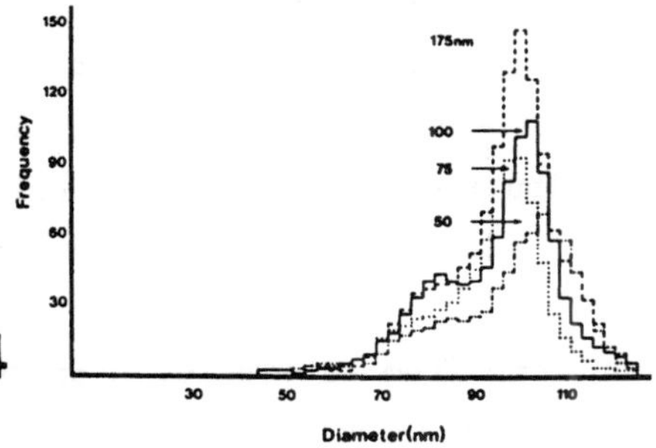

4

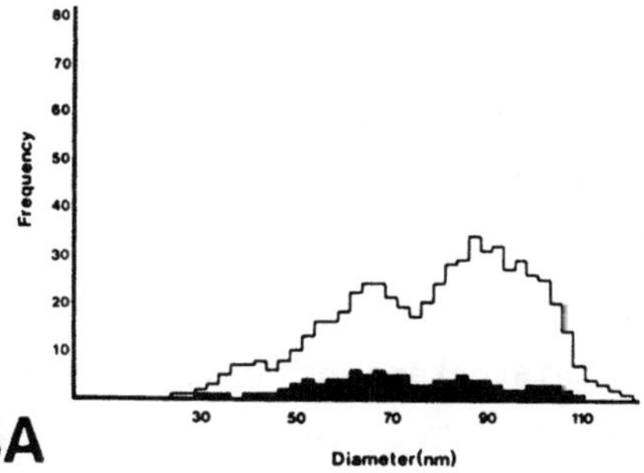

5A

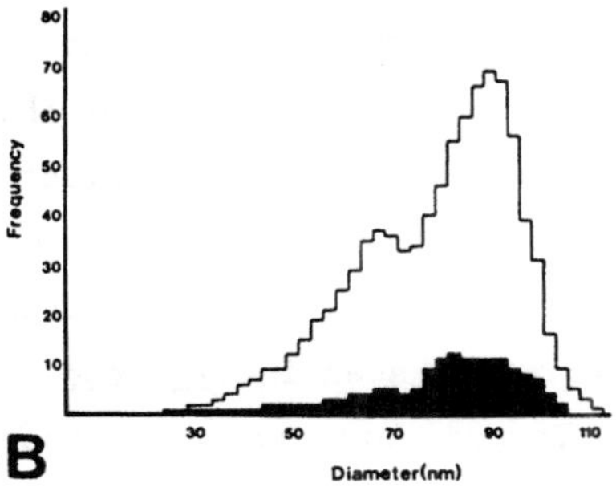

5B

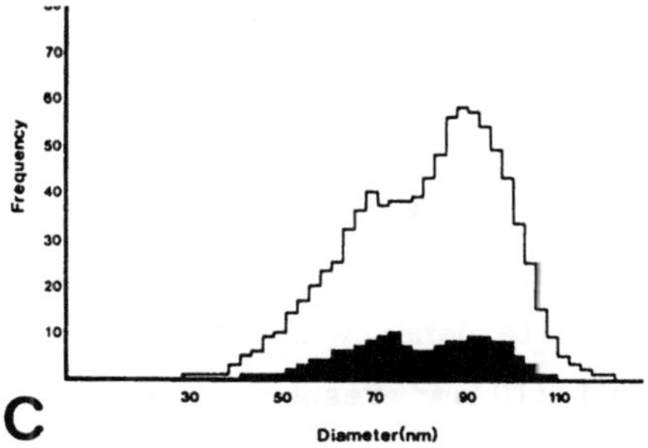

5C

Figure 4. Distribution of Torpedo electric organ synaptic vesicle diameters from sections cut at thicknesses of 175, 100, 75 and 50 nm. The curves overlap reasonably well and only the numbers, which are a function of section thickness, are affected.

Figure 5. Vesicle diameter distributions from 3 different methods of sampling Torpedo electric organ terminals. (A) Vesicles from 10 micrographs of terminals from 3 electrocytes of 3 electrocyte columns, (B) vesicles from 10 micrographs of terminals from 3 non-consecutive sections of the same electrocyte, and (C) vesicles from 10 micrographs from serial sections of the same terminal. The filled distributions represent vesicles touching the presynaptic membrane.

B. Effect of section thickness on vesicle size distribution:

Sections were cut from the same block at four interference colors, dark grey (est. 50 nm), silver (est. 75 nm), gold (est. 100 nm) and purple (est. 175 nm) (Fig. 4). As expected, thicker sections yielded more vesicle counts (712, 761, 947 and 1212 respectively) but calculated densities were statistically the same (300-325/μm^3; Student t-test). Additionally, all histograms show a left (negative) skewed distribution with the major mode between 100 nm (175 nm thickness) and 106 nm (50 nm thickness) and a smaller secondary peak.

C. Vesicle densities and diameters: Total population and touching:

Vesicle diameters were determined from 10 terminals of 3 adjacent electrocyte columns; from 10 different terminals on the same electrocyte (3 adjacent sections) and from 10 serial sections of the same terminal. All distributions are skewed to the left, have major modes at 90 nm, and have prominent left-sided shoulders or secondary peaks at 65 nm (Fig. 5). The mean size is 77-80 nm (Table 1). The distributions of vesicles touching the presynaptic membrane are similar to the total population and comprise about 15-20%.

Histograms of vesicle diameter distributions from embryonic, neonate, and adult Torpedo electric organ, adult Torpedo extrinsic eye muscle, adult skate electric organ, and polystyrene beads were examined. Log-normal, Gaussian and left or negative-skewed distributions were found from these sources (Fig. 6 and Table II). Vesicle diameters of the skate electric organ terminals produced a slightly log-normal distribution (Kriebel et al., 1986; 60 nm for R. erinacea) with a mode and mean of 74 nm (Fig. 6A). Vesicle density is also somewhat higher than normally seen for Torpedo (387 and 310/μm^3 respectively). Distributions of Torpedo extrinsic eye muscle synaptic vesicles are either slightly log-normal or Gaussian (Fig. 6B; these distributions are from 2 muscles of 2 animals, modes ranging from 62 - 75 nm). Polystyrene beads, which were whole-mounted, produced a slight left handed tail on the distribution which is nearly Gaussian (Fig. 6C). The beads (S130-1) yielded a mean of 87 nm which is within reader bias of the 91 nm value determined by the supplier (Plannet, GmbH, Marburg, FRD).

TOTAL

	3-Column	3-Grid	Serial
Size (nm):	78.9±4.1	77.7±3.9	79.8±3.3
Density :	371±99	382±93	571±143
N :	10	10	10
Counts :	605	901	897

TOUCHING

	3-Column	3-Grid	Serial
Size (nm):	74.0±7.0	77.2±6.6	78.6±5.1
Density :	N/A	N/A	N/A
N :	10	10	10
Counts :	101	155	152

Table I. Statistical summary of data from sampling survey.

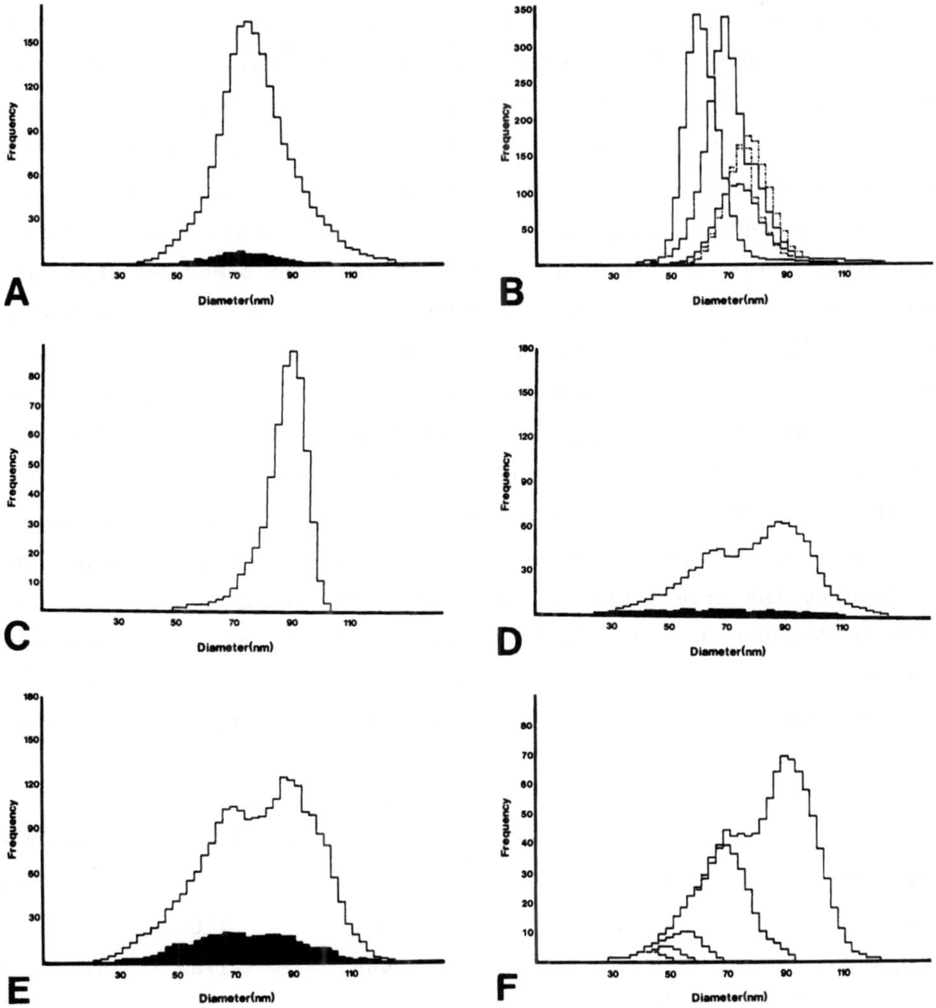

Figure 6: Vesicle diameter and touching vesicle distributions:
A. Skate electric organ vesicles, curve nearly Gaussian but shows a slight log-normal or right skew.
B. Torpedo eye muscle vesicles. 5 distributions from two animals. All curves are slightly log-normal with no secondary peaks or shoulders.
C. Diameters of whole polystyrene beads.
D. Vesicles from electric organs (growth cones) of 74 mm long embryo.
E. Vesicles from electric organ of neonate Torpedo.
F. Vesicles from electric organ of adult Torpedo. The distribution of the class producing the smaller peak and the distributions of the smaller classes giving rise to the left-skew were determined from deconvolution (all classes assumed to be Gaussian).

Vesicle distributions from Torpedo electric organ at different stages of development to the adult typically produced curves skewed to the left (Figs. 6D, 6E, 6F). The 74 mm embryo (Fig. 6D) represents vesicle populations contained within growth cones (Fox and Richardson, 1979). By the neonate stage, the adult innervation pattern has been established (Fox and Kötting, 1984). There are no apparent differences in the distributions or mean size of vesicles from these 3 stages of development (Figs. 6D-F and Table II). Vesicle density, however, increases dramatically following embryogenesis then falls somewhat by adulthood (Table II). Since the overall diameter distribution shows two or more peaks and a marked left-hand skew, there are several classes of vesicles based on diameter. We have deconvoluted the overall distribution by assuming that each class is Gaussian (Fig. 6F). This produces classes with modes at 90 nm and 70 nm as well as numerous smaller sizes to 40 nm. All vesicle distributions from 26 adult, control Torpedo marmorata electric organs have a mean size of 83 nm ± 3.9 and a density of $266/\mu m^3$ ±121 (Table II Fig. 6).

Since the distributions and means of synaptic vesicles are essentially independent of section thicknesses (50–175 nm) most measured diameters must be close to the true spherical diameter. Thin slices of vesicles are not possible because our standard section thickness equals or exceeds the vesicle diameter. Grey or opaque circles devoid of unit membrane structure represent caps of vesicles whose centers lie outside the section. Caps

	Size (nm)	(std)	Density (μm^3)	(std.)	N*	Counts
Polystyrene	88	1	-	-	3	572
Skate	74	3	387	110	10	1902
Torpedo Eye	70	2	468	178	41	7349
74 mm Torpedo Embryo	78	4	97	45	30	1124
Neonate Torpedo	79	4	397	115	30	2324
Adult Torpedo	83	4	266	121	26	18893

Table II. Statistical summary of synaptic vesicle distributions from different sources and polystyrene beads.

*Number of micrographs except Adult Torpedo (26 = animals i.e. 260 micrographs)

must be included for purposes of density determinations but will skew the distribution to the left. Since the overall profiles of the distributions were independent of section thickness we conclude that many caps were not recognized as such and many were obscured by vesicles (Fig.3). Thus, the number of obscured caps would increase as the section thickness increases.

Discussion:

The three vesicle distribution patterns observed in this study represent single classes (log-normal and Gaussian) or multiple classes (multiple peaked and/or left-skew; Underwood, 1970). Log-normal and Gaussian distributions from skate electric organ and Torpedo eye muscle display relatively smooth curves in contrast to the multi-peaked and/or shoulders present on the left-skewed Torpedo electric organ distributions. Distributions that show multiple peaks and/or distinct shoulders are definitely composed of a corresponding number of classes of vesicles because of the discontinuities in the profile and increasing the same size enhanced the multiple peaks. The sample sizes are relatively small in the left hand part of the distributions so one would not expect to consistently see 4-6 multiple peaks. Several classes can also be described with deconvolution of the parent distribution in which case larger diameter means and more subpopulations are produced in log-normal distributions than with Gaussian distributions. Thus we chose to use Gaussian distributions as the most conservative estimate of 4-6 subpopulations in the Torpedo electric organ.

Vesicle distributions from single-peaked distributions produced coefficients of variance under 10% indicating a low variability in vesicle diameters. An average mean of 70 nm was obtained from Torpedo eye muscle and 74 nm from skate electric organ. In contrast, the Torpedo electric organ gave a mean size of 82 nm, but these distributions are composed of several sub-populations. This means that the average vesicle diameter changes as the ratios of the sub-populations change. The left secondary peak at about 70 nm has been named VP2 by Zimmermann and Whittaker (1974), and they concluded it represents a recycling vesicle population during long term stimulation following the initial fusion of the larger-sized class of 90 nm vesicles (VP_1). Vesicles of the Torpedo extrinsic eye muscle neuromuscular junctions and the skate electric organ (Kriebel et al., 1985) are thus relevant because they are of the 60-70 nm class. The frog and mouse neuromuscular vesicles are also 60 nm in diameter with a coefficient of variance of 10% (Kriebel and Florey, 1983; Kriebel and Pappas, 1987). Thus, cholinergic vesicle distributions representing the releasable store of quanta appear to be relatively uniform across species. On the other hand, mean diameters of the various sub-populations from Torpedo vary between 40-90 nm. This represents an enormous difference with respect to internal volumes. The Torpedo electric organ as used by Whittaker and coworkers over the years has shown that there are different classes of synaptic vesicles based on ACh content and density as well. There are also two classes of quanta (Muller & Dunant, 1987; Kriebel et al, 1986; Kriebel et al, this volume) but it remains to be demonstrated in Torpedo which vesicle sub-population generates the MEPP or sub-MEPP class of quanta.

Supported by the Alexander von Humboldt Foundation and N.S.F. 19694. We thank Ms. Erika Kriebel for careful vesicle analysis.

References

Abercrombie, M., 1946. Estimation of nuclear population from microtomic sections. Anat. Rec. 94:239.

Boyd, I.A. and A.R. Martin, 1956. The end-plate potential in mammalian muscle. J. Physiol. 132: 74-91.

Carlson, G.D. and M.E. Kriebel, 1985. Neostigmine increases the size of subunits composing the quantum of transmitter release at the mouse neuromuscular junction. J. Physiol. 367:489-502.

Ceccarelli B., W.P. Hurlbut, and A. Mauro, 1973. Turnover of transmitter and synaptic vesicles at the frog neuromuscular junction. J. Cell Biol 57:499-524.

del Castillo, J. and B. Katz, 1954. Quantal components of the end-plate potential. J. Physiol. 124: 560-73.

Erxleben, C. and M.E. Kriebel, in press. Characteristics of spontaneous miniature and subminiature end-plate currents at the neonate and adult mouse neuromuscular junction. J. Physiol.

Erxleben, C. and M.E. Kriebel, in press. Subunit composition of the spontaneous miniature end-plate currents at the mouse neuromuscular junction. J. Physiol.

Fatt, P. and B. Katz. 1952. Spontaneous subthreshold activity at motor nerve endings. J. Physiol. 117: 109-128.

Fox G.Q. and D. Kötting, 1984. Torpedo electromotor system development: A quantitative analysis of synaptogenesis. J. Comp. Neurol 224:337-343.

Fox, G.Q. and G.P. Richardson, 1979. The developmental morphology of Torpedo Marmorata: Electric organ-electrogenic phase. J. Comp. Neurol. 185: 293-316.

Hayat, M.A., 1970. Principles and techniques of electron microscopy. Vol. I. Van Nostrand Reinhold, New York.

Heuser, J. and T.M. Reese, 1973. Evidence for recycling of synaptic vesicle membrane during transmitter release at the frog neuromuscular junction. J. Cell Biol. 57:315-344.

Heuser, J.E., T.S. Reese, M.J. Dennis, Y. Jan, L. Jan and L. Evans. 1979. Synaptic vesicle exocytosis captured by quick freezing and correlated with quantal transmitter release. J. Cell Biol. 81: 275-300.

Karnovsky, M.J., 1965. A formaldehyde-glutaraldehyde fixative of high osmolarity for use in electron microscopy. J. Cell Biol. 27: 137A-138A.

Katz, B. and R. Miledi, 1979. Estimates of quantal content during 'chemical potentiation' of transmitter release. Proc. R. Soc. Lond. B. 205; 369-378.

Kriebel, M.E. and E. Florey, 1983. Effect of lanthanum ions on MEPP amplitude distributions and vesicles in frog neuromuscular junctions. Neuroscience 9:535-547.

Kriebel, M.E., G.Q. Fox & D. Kötting. 1988. Effect of nerve stimulation K+ saline and hypertonic saline on classes of quanta, quantal content and synaptic vesicle size distribution of Torpedo electric organ. This volume.

Kriebel, M.E. and C.E. Gross, 1974. Multimodal distribution of frog miniature endplate potentials in adult, denervated and tadpole leg muscle. J. Gen. Physiol. 64:85-103.

Kriebel, M.E., C. Gross and G.D. Pappas, 1987. Two classes of spontaneous miniature excitatory junction potentials and one synaptic vesicle class are present in the ray electrocyte. J. Comp. Physiol. 160:331-340.

Kriebel, M.E., R. Hanna and C. Muniak, 1986. Synaptic vesicle diameters and synaptic cleft widths at the mouse diaphragm in neonates and adults. Devel. Brain Research. 27:19-29.

Kriebel, M.E., F. Llados and D.R. Matteson, 1976. Spontaneous subminiature end-plate potentials in mouse diaphram muscle: Evidence for synchronous release. J. Physiol. 262:553-581.

Kriebel, M.E., F. Llados, and D.R. Matteson, 1982. Histograms of the unitary evoked potentials of the mouse diaphragm show multiple peaks. J. Physiol. 322:211-222.

Kriebel, M.E. and G.D. Pappas, in press. Effect of hypertonic saline on quantal size and synaptic vesicles in identified neuromuscular junction of the frog. Neuroscience.

Muller, D. and Y. Dunant, 1987. Spontaneous quantal and subquantal transmitter release at the Torpedo nerve - electroplaque junction. Neuroscience 20:911-921.

Stadler H. and M.-L. Kiene, 1987. Synaptic vesicles in electromotoneurones. II. Heterogeneity of populations is expressed in uptake properties; exocytosis and insertion of a core proteoglycan into the extracellular matrix. EMBO J. 6:2217-2221.

Underwood E.E., 1970. Quantitative stereology. Addison-Wesley, Boston.

Weibel, E.R., 1979. Stereological Methods. Vol. 1. Academic Press, London. pp. 415.

Whittaker, V.P., 1984. The synaptic vesicle. in Handbook of Neurochemistry. Vol. 7. ed. Abel Lajtha. Plenum.

Zimmermann, H., 1982. Insights into the functional role of cholinergic vesicles. In Neurotransmitter vesicles. eds: R.L. Klein, H. Lagercrantz, H. Zimmermann, Academic Press, London.

Zimmermann, H. and V.P. Whittaker, 1974. Effect of electrical stimulation on the yield and composition of synaptic vesicles from the cholinergic synapses of the electric organ of Torpedo: a combined biochemical, electrophysiological and morphological study. J. Neurochem. 22: 435-450.

Effect of Nerve Stimulation, K^+ Saline and Hypertonic Saline on Classes of Quanta, Quantal Content and Synaptic Vesicle Size Distribution of Torpedo Electric Organ

M.E. Kriebel, G.Q. Fox* and D. Kötting*

Department of Physiology, SUNY Health Science Center, 750 E. Adams St. Syracuse, N.Y. and *Department of Neurochemistry, Max-Planck-Institut für biophysikalische Chemie, 3400 Göttingen, Fed. Rep. Germany

INTRODUCTION

The Torpedo electric organ is an extremely useful preparation for the study of cholinergic mechanisms because of its easy accessibility to biochemical, morphological and physiological methods. A kilogram of electric organ can be obtained from one animal of which 10% are synaptic terminals. At least 3 classes of synaptic vesicles have been described based on density (Stadler and Kiene, 1987) and 2 classes based on diameter (Zimmermann and Whittaker, 1974, 1977). Its unique structure of vertical columns of stacked electrocytes enables one to uniformly field stimulate the tissue and obtain reliable quantitative end-plate potential.

Physiologically the large dimensions of the single electrocytes present several disadvantages of quantal size and quantal content studies of end-plate potentials. Miniature end plate potentials (MEPPS), recorded intracellularly, are very small due to the low input resistance of the cell, which is essentially that of saline. For this reason and because of the small ratio of cell thickness to diameter the electrical space constant is short. This means that MEPPs are rapidly attenuated in the radial axis from the site of generation. However, in the thin axis of the cell this electrical circuit is the basis for the voltaic pile effect of the stack of electrocytes which develops 40 volts in the dorsal-ventral axis. The low transverse input resistance relative to the high radial resistance electrically couples the electrocytes in the vertical columns (not by means of gap junctions). Together these considerations make it difficult to determine quantal content (number of quanta in EPP/terminal) and quantal size (MEPP) and thus account for the paucity of these electrophysiological studies on the Torpedo electric organ. On the other hand, these same

NATO ASI Series, Vol. H21
Cellular and Molecular Basis of Synaptic Transmission
Edited by H. Zimmermann

properties enable the investigator to electrically isolate an area for extracellular microelectrode studies (Muller & Dunant, 1987).

Earlier studies have demonstrated the occurrence of MEPPs with focal recording (the recording electrode being pressed against the innervated membrane; Miledi et al 1971; Dunant, 1972, 1973; Walther, 1974). Erdelyi and Krenz (1984) estimated the quantal content of reduced EPPs obtained with a high Mg^{++} and low Ca^{++} saline. With a loose-patch voltage clamp, Dunant and Muller (1986) report quantal contents of 1.3 quanta/μm^2 of presynaptic terminal. Muller and Dunant (1987) also determined that quantal size is about that of frog and mouse junctions (1986) and described two classes of spontaneous quanta. The smaller class is similar to the sub-MEPP class reported in frog (Kriebel and Gross, 1974) and mouse (Kriebel et al., 1976, 1982) neuromuscular junctions and in skate electric organ (Kriebel, et al 1987). In this study we also report the presence of the sub-MEPP class of quanta in the Torpedo electrocyte as well as in the extra-ocular eye muscles. Our major objective, however, has been to determine the area of focal recording so that quantal content per terminal volume could be defined. It is necessary to know either normal quantal content of EPPs or total quantal release (number of MEPPs) per terminal volume in order to correlate morphology to the physiological state of the preparation. We quantitate total quantal release during various treatments to synapic vesicle size, classes and densities. The coefficient of variance determined from the focally recorded MEPPs and the synaptic vesicle diameter distributions following various periods of nerve stimulation and treatments known to induce high rates of MEPP release is used as the basis of comparison. Surface electrocytes were fixed within 20 seconds with superfusion of fixative but 2 minutes were required to fix terminals one electrocyte deep and 10 to 30 minutes were needed to fix terminals 4 or 5 electrocytes deep. Thus, we have studied only surface terminals which had been characterized electrophysiologically in order to compare physiological states with classes of synaptic vesicles as defined by Zimmermann & Whittaker (1974 a, b, 1977).

Methods:

Torpedo which were either awake or anesthetized with MS222 were killed by a blow to the head or with brain section. The electromotor nerves were severed and the entire organ was quickly removed by cutting around the organ through the ventral and dorsal skin. The isolated organ was then sectioned into 1 cm wide strips. These were placed skin down onto a cork-board and by slicing parallel to the skin with a long razor blade thin layers of electric organ were left attached to the skin such that areas of one electrocyte deep could be subsequently identified with an electrode. The skin with the thin overlay of electric tissue was diced into 4x4 mm squares, and these were pinned to a Silgard dish and covered with saline (in mM: NaCl, 200; KCl 8; MgCl 1.8; $CaCl_2$ 3.4; Tris, 10; $NaHCO_3$,5; glucose, 5.5; urea, 300; sucrose, 100). The intracellular recording techniques were standard and signals were recorded onto magnetic tape. The tape was either played back at one-fourth speed and signals recorded with a chart recorder or signals were filmed with an oscilloscope camera. Micropipetts (1-5 Megohms, 3M KCl) were held in a micromanipulator at about 70° to the tissue surface and recordings were made by slowly lowering the manipulator until focal extracellular MEPPs were recorded.

In order to study the effective recording area of the electrode, two electrodes were employed, and these were independently positioned with a manipulator. The electrode tips were about 2 μm in diameter and were visualized with horizontal lighting. Electrode tips were positioned adjacent to each other and then separately lowered perpendicular to the tissue surface until MEPPs were induced. Preparations with the dorsal skin down have exposed terminals and in most experiments we did not immerse the tissue in saline and the electrodes were not advanced into the tissue once electrical contact was made. Thus, the adjacent position of the electrodes was maintained and the initial contacts were sufficient to generate small bursts of MEPPs, some of which were recorded at both electrodes, and these were utilized to determine effective recording area.

Potassium chloride and hyperosmotic treatments were achieved by adding the appropriate volume of a 1M KCl solution or a 2M sucrose solution to the bath saline. The number of MEPPs/sec was determined during treatment with

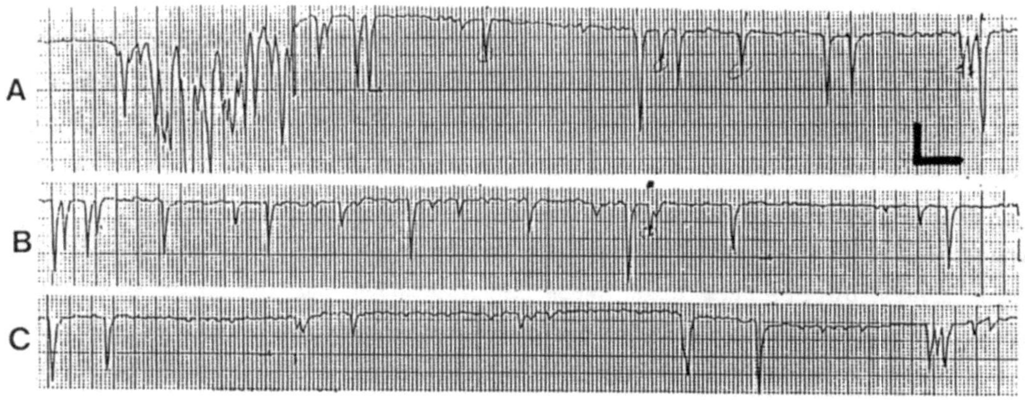

Fig. 1. Focal extracellular MEPPs. A. Initial burst resulting from electrode pressure. B & C. Continuation of (A) showing that the ratio of larger to smaller MEPPs decreased as the frequency fell. The larger and smaller MEPPs are probably from different release sites. Calibration: 2mV and 20 msec.

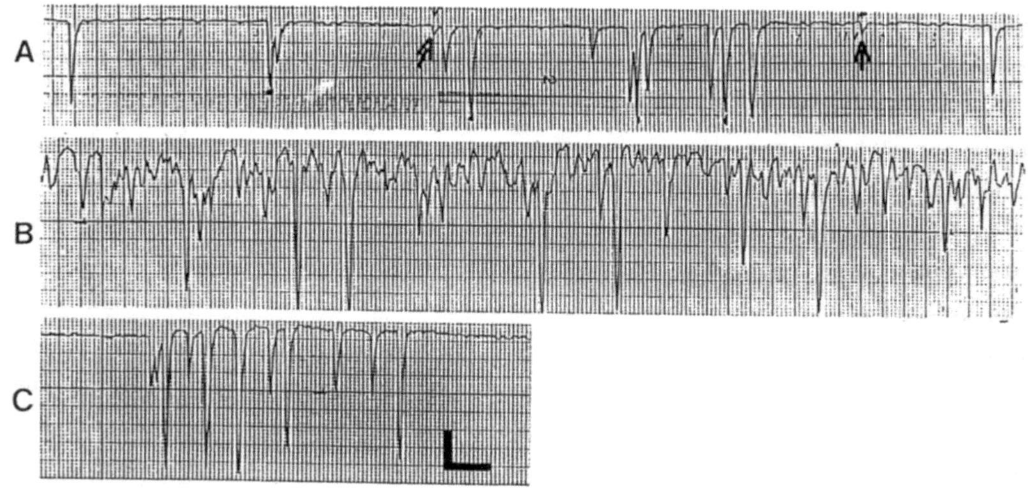

Fig. 2. Focal extracellular MEPPs. A. Low frequency record showing two classes based on amplitude (see Fig. 6D). In this example, the ratio of sub-MEPPs (arrows) to the larger MEPPs remained constant for several minutes demonstrating that both classes were probably generated at the same site. B. This trace also shows two classes of MEPPs. The larger class (giants) shows little variance indicating that they may represent summed MEPPs due to nerve injury. C. Small burst of MEPPs induced with electrode pressure which shows two size classes. Calibration: 1 mV and 20 msec.

relatively fast sweep speeds of the oscilloscope. Thus the total number of MEPPs per electrode site (ca. 12 μm^2) could be calculated. We studied the effect of tonicity of the fixation solutions because we found that hypertonic salines (with sucrose) and a "Karnovsky" fixative (1.7 osmol) induced very high rates of MEPPs. We used fixatives with osmolarities of 0.3, 0.7, 0.8, 0.9, 1.4, 1.7, and 2.1. Nerve stimulation was accomplished with field stimulation (3x threshold V, 0.2msec.). We used frequencies of 10, 1 and 0.1 Hz and fatigued the preparations to 66%, 37% and 5% of the initial EPP values. In some preparations, a period of recovery was permitted before fixation (isosmotic). The EPPs generated in thin blocks were not of sufficient magnitude for self-stimulation. However, blocks as thin as 2 mm generated large enough EPPs (ca, 1 V) to produce 3-5 repetitive, self-induced EPPs. Thus, regions from thin blocks that were one electrocyte deep were used for quantal content studies. It should be emphasized that for effective field stimulation there must be relatively little saline between electrodes.

Results:

I. Intracellularly and focally recorded MEPPs:

A. Dorsal skin down, terminals on top surface: The innervated surfaces of the electrocytes are exposed so that the advancing electrode first comes into contact with nerve terminals. When the saline level is below the tissue surface, focally recorded extracellular MEPPs are induced as soon as electrical contact is made with the tissue (Figs. 1, 2). These MEPPs subside within minutes and the region remains silent until the electrode is advanced, or the table tapped, or the electrode vibrated with the negative capacitance control where upon subsequent bursts of MEPPs can be induced. Sometimes relatively large potentials were pressure evoked and since these showed little variance they could represent summed quanta released with nerve injury (Fig. 2B). These larger potentials resemble those published by Soria (1983) and they sometimes show a regular frequency (Fig. 3). Large potentials may occur either with normal sized, focal MEPPs (Fig. 2B) or by themselves (Figs. 2B and 3). The same results were found with 3M KCl, 8M $CoCl_2$, 2M NaCl and 2M $CaCl_2$ electrodes. Terminals, therefore, do not normally release spontaneous quanta and MEPPs should be referred to as pressure induced. With further electrode advance, a resting

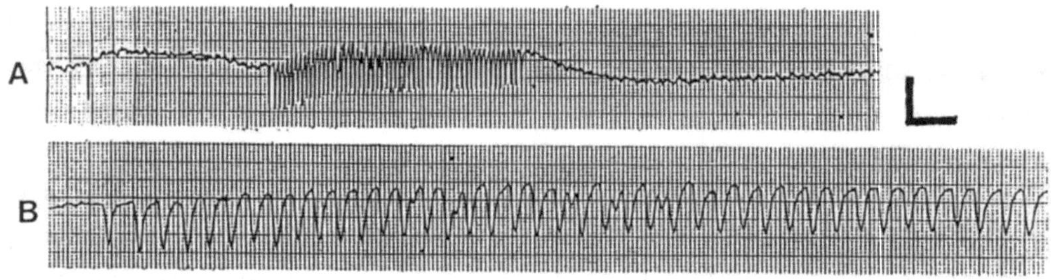

Fig. 3. Focal extracellular "MEPPs". These potentials were pressure evoked with the electrode. They show a frequency of 100 Hz and since they have such a small coefficient of variance amplitude (8%) they may represent release of summed quanta coupled with nerve injury which is driven by 50 Hz noise. These potentials are similar to those reported by Soria (1983). A. Slow chart speed. B. Same signals at a faster chart speed. Calibration: 2 mV and 0.1 sec. (A), 20 msec. (B)

potential of -60 mV was recorded from the surface electrocyte. Penetration was usually accompanied by a burst of intracellularly recorded MEPPs locally induced by the electrode. Intracellularly recorded MEPPs varied from 1 mV (focal, inside) to 50 μV depending on the closeness of the tip to the innervated surface. Intracellularly recorded MEPPs were quite small (due to the low input resistance of the electrocyte, signal to noise 4:1) with a mode of about 40 μV. The transition from extracellular focal MEPPs to intracellular focal MEPPs was not sharp and partial penetration yielded both intracellular and extra- cellular focal potentials.

MEPPs were induced at 80% of the surface which is to be expected considering the dense nerve terminal innervation (Fig. 4A & 4B; Fox & Kötting, 1984). High resting potentials (-50 to -70mV) were recorded at all surfaces so we conclude that cut electrocytes seal. We further verified this by cutting strips of tissue 0.25 x 0.25 by 4 mm long, and we found that all surfaces yielded -50 mV resting potentials (Morel, 1976). Moreover, these small pieces did not generate spontaneous MEPPs but did generate MEPPs induced with electrode pressure and showed EPPs with field stimulation, thus small pieces function normally (Suszkiw, 1980). When the electrode was advanced into the extracellular space the MEPPs generated by the surface electrocyte were positive and looked like intracellularly recorded MEPPs. The "point-resistance" of cytoplasm and extracellulary space between two electrocytes was about the same so axial and radial attenuation is the same. As the electrode was advanced against the second electrocyte, focal extracellularly recorded, negative MEPPs were observed. This position showed large, negative focal MEPPs generated at the second electrocyte and small positive MEPPs generated at the surface electrocyte. All MEPPs have about the same time characteristics. Penetration of the second electrocyte yielded intracellular, positive MEPPs from both the innervated surface and the overlaying electrocyte (Fig. 5).

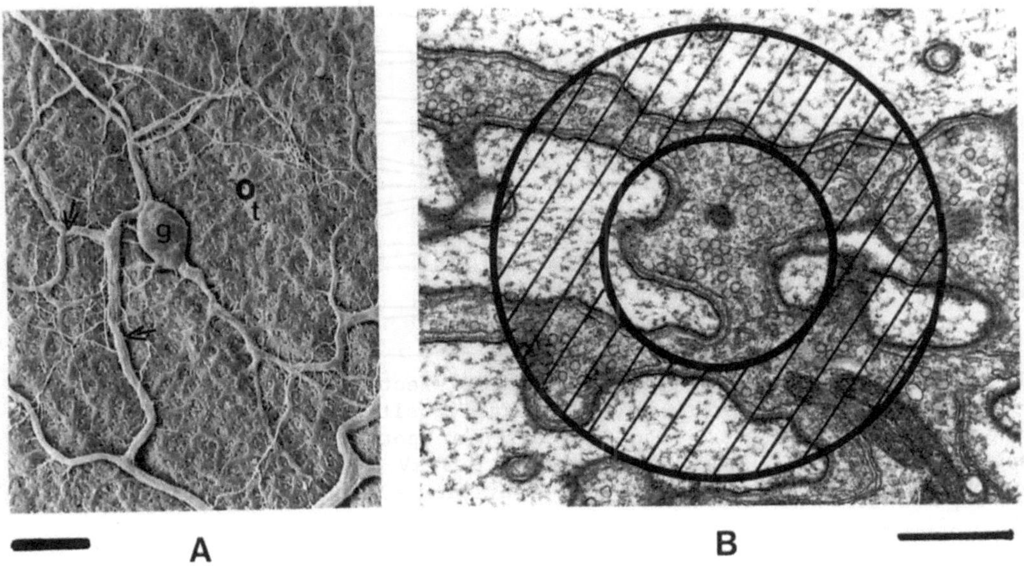

Fig. 4: A. Scanning electron micrograph of innervated surface showing the ramifying nerve filaments and felt-like network of terminals. Nerve filaments indicated with arrows. t= nerve terminals. g= glial cell. The round circle represents the patch of surface recorded from with an electrode. Calibration: 10μm. B. Electron micrograph of section through surface showing nerve terminals in relationship to electrode diameter. Calibration: 1 μm.

B. Ventral skin down, uninnervated surface on top of tissue block: Intracellular resting potentials were either recorded as soon as the electrode made contact with the tissue or upon "ringing" the amplifier with the negative capacitance control. With small, sharp electrodes which promoted penetration, we did not initially find spontaneous intracellular MEPPs when penetration was from the dorsal surface. Focal, large intracellular MEPPs were recorded with further electrode advancement and these occurred in short-lived bursts. Subsequent bursts were induced with further electrode advancement until the electrode tip entered the extracellular space at which time small, negative extracellular potentials (40 μV) were recorded. Further advancement into the second electrocyte also yielded small negative potentials generated by the overlying electrocyte as well as small, intracellular positive MEPPs generated in the second electrocyte (Fig. 5; Bennett et al, 1961).

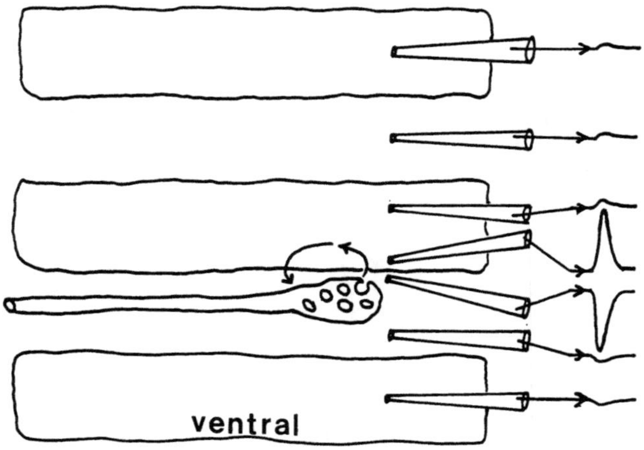

Fig. 5: Schematic cross section of three electrocytes in a column showing relative size and polarity of a MEPP at 7 positions which was generated in the middle electrocyte. The intracellular potential would be only 50 μV whereas the focal potentials would be 1-3 mV. Note that potentials are observed in both the overlying and underlying electrocytes.

II. MEPP intervals (focal, extracellular):

The observation that electrode pressure is required to induce MEPPs means that they are not spontaneous; and, the fact that they usually decrease in frequency and occur in bursts shows interdependencies in the release mechanism. However, we analyzed a few records (which showed stationarity in frequency) for randomness of MEPPs and indeed found that the number of MEPPs in given time intervals fit Poisson statistics (see Erdelyi and Krenz, 1984). For example 431 MEPPs that occured in 314 bins (of 10 msec duration) yielded the following frequencies (% observed) which are compared to predicted Poisson values (Px): percentage of bins with zero MEPPs - % obs. = 0.245, P_x 0.253; % 1 MEPP bins- % obs. 0.344; P_x 0.347; 2 MEPP bins- % obs. 0.248, P_x 0.238; 3 MEPP bins- % obs. 0.127, P_x= 0.108; % 4-6 MEPP bins - % obs. 0.034, P_x=0.047.

III. MEPP amplitude distributions:

Pressure evoked focal potentials were best studied with the dorsal skin down with terminals facing upwards. Therefore, the electrode first came into contact with the terminals and bursts (20-200) of focal MEPPs were induced by gentle tapping of either the table or manipulator (Fig. 1A). The MEPP amplitude distributions showed a wide range of profiles from single peaks to multiple peaks (Figs. 6 & 7). The single peaked amplitude distributions were usually the largest (1-3 mV) and showed the least variance (Fig. 6). The MEPP amplitude profiles which showed multiple peaks changed with time such that peaks dropped from the distribution until usually one peak remained and this one showed a narrow distribution (Fig. 7). Some distributions showed a sub-MEPP class (Figs. 2A & Fig. 6D) and these had the same time characteristics as the larger MEPPs (Kriebel & Gross, 1974, Kriebel et al, 1976; Erxleben & Kriebel, 1987). Some MEPP amplitude distributions did show multiple peaks suggesting a sub-unit composition of the MEPP as we have found in the frog and mouse

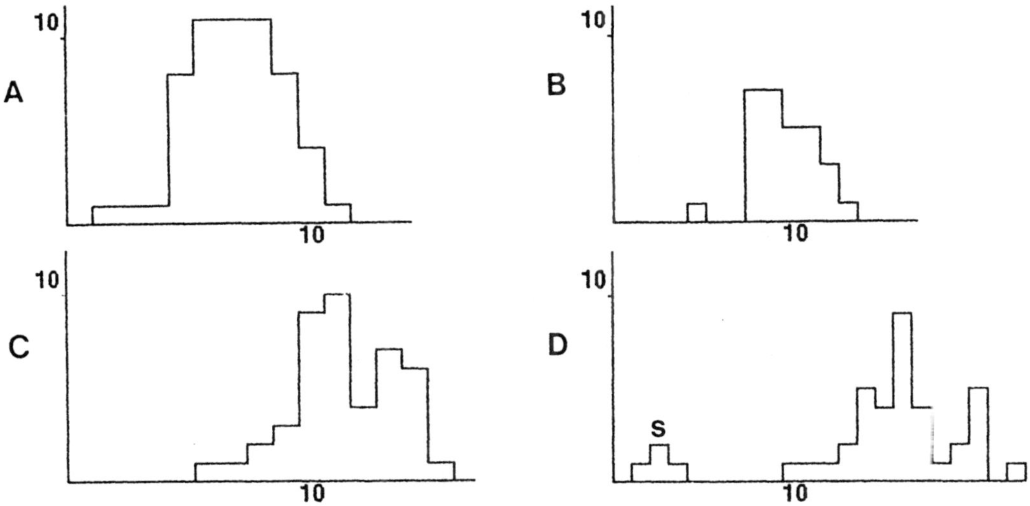

Fig. 6: MEPP amplitude histograms from records showing relatively stable frequencies and mean amplitudes. A. mean 0.7 mV±0.18, coeff. var. 25%. B. mean 0.95 mV ± 0.16, coeff. var. 17%. C. mean 1.1 mV ± 2.2, coeff. var. 20%. D. Two classes of MEPPs. Sub-MEPPs at 0.3mv. Histobar 10 = 1mV.

neuromuscular junction (Kriebel and Gross, 1974; Kriebel et al 1976; Erxleben and Kriebel, 1987). However, the decreasing MEPP frequencies make this analysis difficult because of small sample sizes obtained at one release site. The multiple MEPP exposures published by Muller and Dunant

(1987) certainly suggest sub-units. We did not find mainly sub-MEPPs in the embryo or neonate tissue as is the case for the mouse junction even though the developing junctions have similarities (Fox & Kötting, 1984; Muniak et al 1982; Kriebel, et al 1986). Focal, pressure induced MEPPs became larger as the electrode was advanced against the tissue and the frequency also increased. The burst of pressure induced MEPPs increased in number with electrode pressure until persistent showers were induced (Fig. 2B). These high frequencies (20-100/sec) persisted for 5-15 minutes and showed a decrease in MEPP amplitude. The showers were composed of 2-5x10^3 MEPPs (Fig. 2B) but bursts contained as few as 10 (Fig.2C).

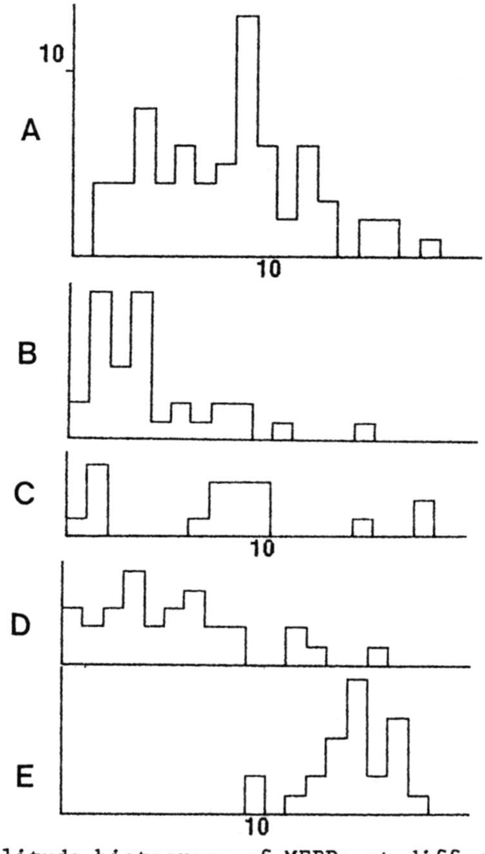

Fig. 7: MEPP amplitude histograms of MEPPs at different times after electrode contact. The MEPP frequencies and amplitudes were changing from A to D but were stable for E. A. MEPP amplitudes recorded for first 5 sec period. B. MEPP amplitudes recorded for next 80 sec period. C. & D. MEPP amplitudes recorded for subsequent 24 and 32 sec periods. E. Final 61 sec period. This section had a low but stable MEPP frequency. The large mean of 1.5 mv ± 0.2 and small coeff. of variance (14%) indicates that only one release area was active. Histobar 10= 1mV.

IV. Area of Recording Site:

Since 80% of the electrocyte surface is covered by terminals (Fox & Kötting, 1984), almost all explored surfaces produced pressure-evoked MEPPs (Figs. 4A & 4B). In order to determine quantal content and induced MEPP release per terminal volume, it was necessary to know the recording area of the electrode and the extent of the "edge-effect", i.e., how far from the edge of the electrode are potentials detected. When two electrode tips were positioned within 1-2 μm and then lowered to the surface, most (95%) MEPPs were recorded at only one electrode. Large focal MEPPs seen at one electrode were not observed even as small MEPPs at the neighboring electrode. Since only a few small MEPPs were recorded at both electrodes most recorded MEPPs are generated only at the electrode tip. The tips were 2-3 μm in diameter and the "edge recording field" was 1 μm (cf. Zucker & Landò, 1986) which gives a recording area of 12 μm^2.

V. Effect of different fixative aldehydes and osmolarities on MEPP frequencies and synaptic vesicles:

(1) Aldehydes: Three fixatives containing different concentrations of glutaraldehyde and formaldehyde were investigated. A hypertonic fixative based on the Karnovsky formulation (Fox et al. this volume) consisted of 5% glutaraldehyde, 4% paraformaldehyde and 0.2M Na cacodylate (1.7 osmolar) induced $2\text{-}5 \times 10^3$ focal MEPPs/μm^2 at 100-200/sec (Fig. 8B). A fixative containing 0.5% glutaraldehyde, 2% paraformaldehyde and 0.15M cacodylate buffer (910 mosMol) induced no MEPPs but changed vesicle densities when compared to an isosmotic glutaraldehyde fixative (810 mosMol: 2% glutaraldehyde and 0.3 M Na cacodylate, Table I). Even though the "Karnovsky" fixative induced a sizeable loss of vesicles (perhaps related to the high MEPP frequency) the profile of the distribution and mean did not change (Table I). The slightly hyperosmotic fixative produced an intermediate distribution profile and vesicle density with no change in mean size (Table I). We conclude from these experiments that the addition of paraformaldehyde to the fixative induces a decline in vesicle numbers and produces transient high MEPP frequencies. Even though more MEPPS were generated per terminal than vesicles present at the start of fixation, there was surprisingly little depletion of vesicles. Moreover, the fact, that the overall profile of the vesicle diameter distributions did not

Fixative		Total	Touching
mosM	Size*	84.0±5.9	82.7±1.2
	Density**	232±74.5	N/A
810	N	20	20
	Counts	1205	207
	Size*	85.2±15.5	83.2+20.7
	Density**	156±58.5	N/A
1700	N	20	20
	Counts	579	111
	Size*	82.3±12.1	82.2±12.5
	Density**	200±55.5	N/A
910	N	20	19
	Counts	814	150

Table I. Effect of aldehydes on synaptic vesicle diameter and densities (* nm; ** μm^3; ± standard deviation, N= number of terminals)
810 = 2% glutaraldehyde + 0.3M sodium cacodylate
1700 = 5% glutaraldehyde + 4% paraformaldehyde + 0.2M sodium cacodylate
910 = 0.5% glutaraldehyde + 2% paraformaldehyde + 0.15M sodium cacodylate

change indicates that no single size class is responsible for generating MEPPs.

(2) Osmolarity: Terminals from electrocyte #1 (topmost) and electrocyte #10 in the sample block of tissue were used to evaluate the effect of fixative osmolarities ranging between 300-2100 mosM (Table II). Only glutaraldehyde was used as a fixative agent. No differences between surface and deep (electrocyte #1 and #10) vesicle size distributions and means occurred with the hypotonic 300 mosM fixative. The hypertonic fixatives caused high MEPP frequencies (Figs. 8B and 8C; over 10^3 MEPPs/μm^3/terminal) and yielded either Gaussian or log-normal distributions with a major mode at 78nm in terminals on the surface electrocyte. At the deeper electrocyte (#10) a left-skewed distribution was found that is typical of published results with 2 modes at 90 and 66 nm. With hypertonic fixation, both surface and deep distributions showed a decreased vesicle density compared to the 300 mosM preparations suggesting that a large indiscriminent loss of vesicles had occurred. There is a statistically

Fixative mosM		#1 Total	#1 Touching	#10 Total	#10 Touching
300	Size	82.9$\pm$3.8	79.6$\pm$7.6	86.7$\pm$2.6	83.6$\pm$6.6
	Variance	14.1	57.1	7.6	44.0
	Density	327$\pm$181.6	N/A	222.8$\pm$76.7	N/A
	N	29	29	29	29
	Counts	3041	340	2402	208
700	Size	80.0$\pm$3.5	74.7$\pm$8.5	84.7$\pm$3.1	81.6$\pm$7.8
	Variance	12.2	72.0	8.5	61.3
	Density	213.2$\pm$79.9	N/A	209$\pm$67.9	N/A
	N	30	30	30	30
	Counts	1892	205	1953	215
810	Size	83.5$\pm$4.5	79.8$\pm$7.9	79.3$\pm$4.8	81.7$\pm$10.6
	Variance	19.3	61.8	22.8	111.4
	Density	212.9$\pm$54.7	N/A	171$\pm$28	N/A
	N	30	30	10	10
	Counts	1788	212	538	49
1300	Size	79.8$\pm$3.3	79.2$\pm$6.6	86.8$\pm$5.1	82.9$\pm$8.0
	Variance	8.6	43.9	26.5	64.4
	Density	331.6$\pm$95.5	N/A	290.0$\pm$73.8	N/A
	N	30	30	30	30
	Counts	2418	270	2609	365
2100	Size	77.5$\pm$3.9	-	84.9$\pm$3.5	82.5$\pm$3.1
	Variance	15.4	-	12.4	9.7
	Density	278$\pm$124	-	266$\pm$64	N/A
	N	10	-	10	2
	Counts	534	-	534	10

Table II. Effect of fixation osmolarity on synaptic vesicle diameters (nm) and densities (μm^3) from surface terminals (#1) and from those on electrocyte #10 deep. ($\pm$ standard deviation)

300 = 1% glutaraldehyde + 0.1M sodium cacodylate
700 = 5% glutaraldehyde + 0.1M sodium cacodylate
810 = 2% glutaraldehyde.+ 0.3M sodium cacodylate
1300 = 2.5% glutaraldehyde + 0.4M sodium cacodylate
2100 = 5% glutaraldehyde + 0.8M sodium cacodylate

significant difference in the mean size of vesicles between the top electrocytes (#1) of the 300 and 2100 mosM treated preparations ($p<.001$ as determined by Student's t-test); however, no differences exist between

	Control	0.1 Hz (210 min.)	1 Hz (10 min.)	10Hz (2 min.)	Hypertonic Saline (10 min.)	K^+Cl^- (3 min.)
Quantal Release $x10^3/\mu m^3$ terminal 3		0	1.1	0.9	1-3	2-4
Vesicle Density $x10^2/\mu m^3$ terminal	2.5	2.5	2.5	2.5	1.7	2.5
Vesicle Mean Dia. nm	84	66	84	84	84	70

TABLE III: Effect of different stimulation frequencies and treatments on quantal release, synaptic vesicle density and mean synaptic vesicle diameter. attributable to tonicity.

vesicles in the deeper terminals (83.6 and 82.5 nm). The density values, although ranging between 158-325 vesicles/μm^3 show no discernible pattern attributable to tonicity. The histogram distributions of osmolarities from 300-1300 mosMol are statistically identical to each other (Table II). Thus, there is no correlation to MEPPs generated and changes in vesicle classes. We conclude from these experiments that synaptic vesicles are osmotically inactive prior to fixation(cf.Breer,et al.1978). It should be pointed out, however, that these fixatives caused only a slow depolarization of the electrocyte membrane potential indicating that glutaraldehyde does not tear unit membranes. Moreover, synaptic vesicles have active membrane properties (Carpenter and Parsons, 1977) and it has been proposed that reloading is accompanied by osmotically induced rehydration (Giompres et al 1981b). We conclude that meaningful correlation studies between quantal release and vesicle diameter distributions (class means and numbers) must utilize only the surface terminals because of long fixation times of deeper terminals.

VI. Electrical Stimulation:

Small blocks of electric tissue were field stimulated at 10, 1, and 0.1 Hz to reduce evoked responses (EPP) to 66%, 37% or 5% of initial values. Tissue was either fixed (isosmotic) immediately or permitted to recover

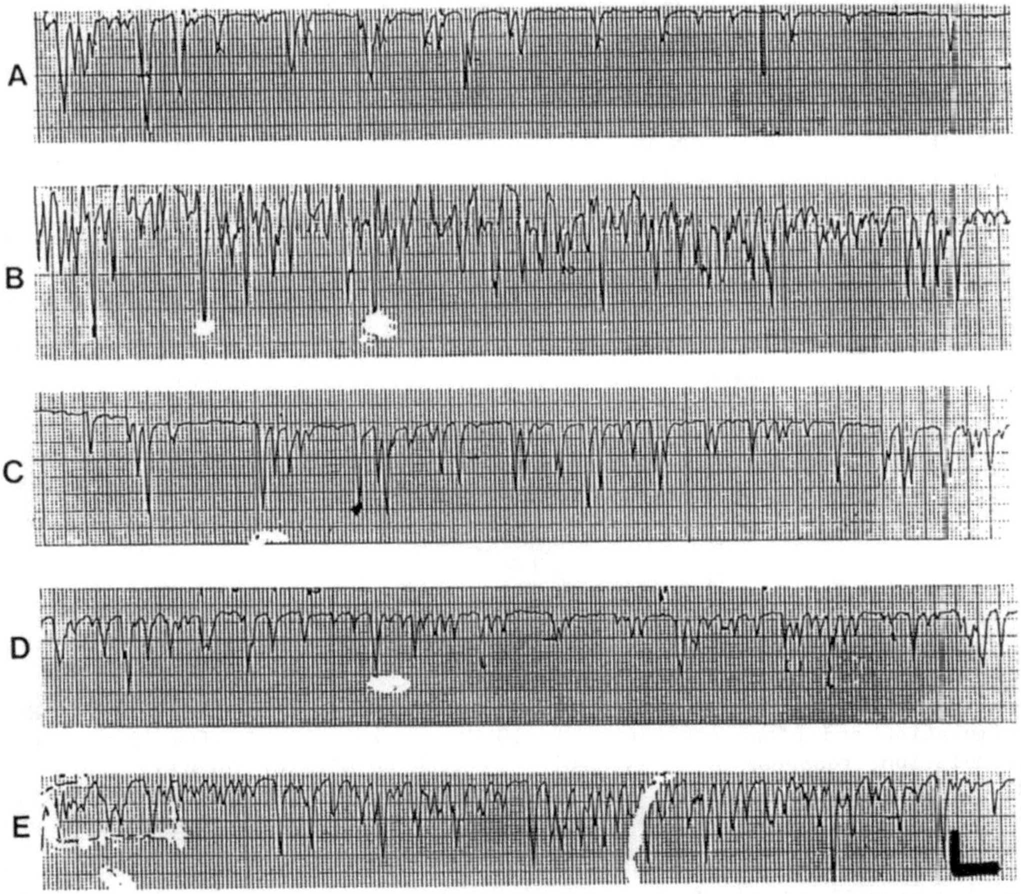

Fig. 8: Sample records of high MEPP frequencies induced by different treatments. A. Pressure evolked burst. B. 1.7 osM fixative. C. 1.4 osM fixative. D. 0.3M KCl isosmotic saline. E. 1.7 osM saline (with added sucrose). Calibration: 1mv, 20 sec.

before fixation. Decreasing the frequency of stimulation and increasing the time caused the distribution of vesicles to shift to smaller diameters (Zimmermann and Whittaker, 1974a, b; 1977). Increasing the time of stimulation to 3.5 hrs and lowering the frequency to 0.1 Hz decreased the number in the 90 nm class whereas the 65 nm class maintained the same numbers (Fig. 9). The numbers of the 90 nm class recovered in 1 hour. A 24 h control was also examined and a noticeable decline in the numbers of the 90 class occured indicating the susceptibility of vesicle distributions

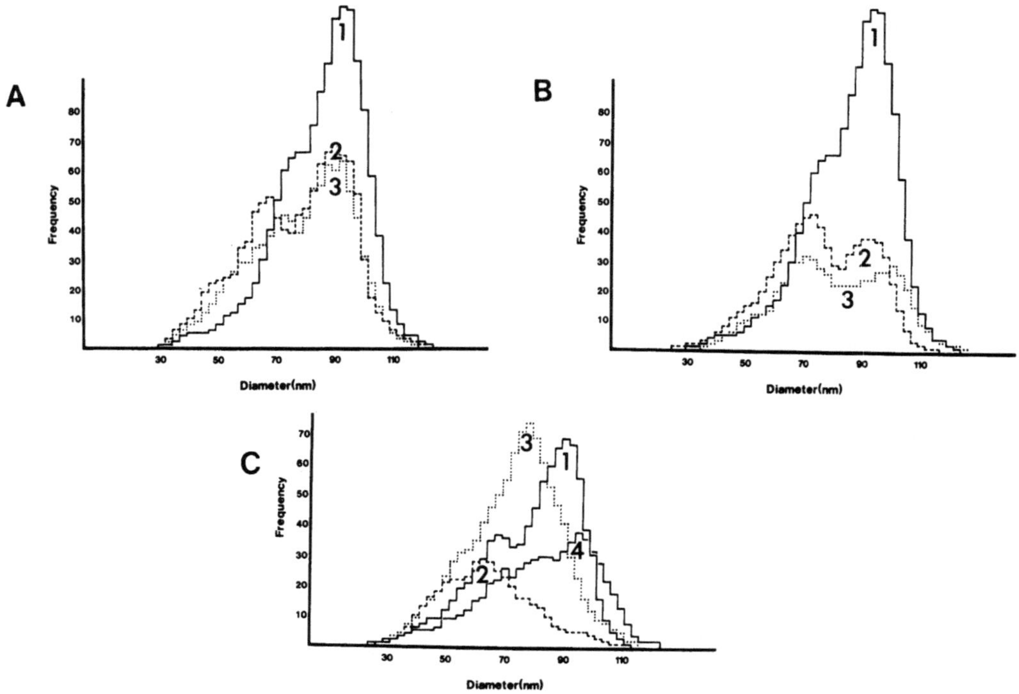

Fig. 9: Distributions of synaptic vesicle diameters with, 10, 1 or 0.1 Hz stimulation and after recovery. A: 10 Hz. 1- control. 2- after 2 min. of stimulation, response 5% of control. 3- with 60 min. of recovery. B: 1 Hz. 1- control. 2- after 18 min of stimulation. 3- after 1 hr of recovery. C: 0.1 Hz. 1- control. 2- after 3 hr of stimulation, response 5% of control. 3- after 1 hr recovery. 4- 24 hr control.

to time as well as stimulation. Mean vesicle size and density of the total control population was 84.5±10.5 nm and $288/\mu m^3$ respectively. Ten Hz stimulation to 5% of control EPP or less reduced the mean to 78.6 nm and slightly increased the density to $333/\mu m^3$. With 1 Hz stimulation to 5% control values, the vesicle mean was further reduced to 72.8 nm and the density was reduced to $235/\mu m^3$. And finally with 0.1 Hz, the mean declined to 65.2 nm and the density to $184/\mu m^3$ (Table III). The three frequencies used (10, 1 & 0.1 Hz) released about the same number of quanta. Thus we conclude from these experiments that electrical stimulation produces a decline in numbers of the 90 nm class which produces a shift in the mean size only with long term stimulation at low frequency. Since the vesicle density remained unperturbed it appears that the total number of quanta released cannot be correlated to changes in vesicle classes.

Quantal Content: The surface of thin blocks of tissue was explored until a region of only 1 electrocyte deep was found. The EPPs ranged from 6-20 mV and the MEPPs were about one mV. Thus quantal content at the electrode was 6-20 which yields 1 to 2 quanta/μm^3 nerve terminal which is the value found by Dunant and Muller (1986). We used this value to calculate total quantal release with the various frequencies and states of fatigue used (Table III).

VII. KCl stimulation.

Pieces of electric organ were stimulated with a 300 mM KCl saline for 3 minutes. Initial MEPP frequencies were 10-30/sec/μm^3 terminal (Fig. 8D) for 5-10 min but then MEPPS show a dramatic decline in frequency and size. This effect is similar to that found in the frog with a high KCl saline (Florey & Kriebel, unpub.) and with lanthanum ions (Kriebel and Florey, 1983) in which cases mainly sub-MEPPs remain. There is also a decrease in synaptic vesicles (Table III) and the larger sub-class of vesicles is noticeably depleted with an accompanying increase in the frequency of smaller-sized vesicles in the 40-65 nm range. Thus, quantal release induced with terminal depolarization has a different influence on vesicle classes than that of nerve stimulation but no difference in density. The decline in MEPP frequency with prolonged KCl treatment (after 5 minutes) can be attributed to inactivation of the Ca^{++} channels. However, Meunier (1984) has evidence against inactivation.

VIII. Osmotic stimulation:

Blocks of electric organ were stimulated for 10 min in hypertonic saline (sucrose added, 1.7 osmolar) which generated 2-4x10^3 MEPPs/μm^3; Fig. 8E, Table III.). The major vesicle mode remained at 85 nm but these distributions differed from the left-skewed curves of controls signifying a loss of smaller-sized vesicle classes. With 30 min of stimulation, the major mode also declined. Tissue allowed to recover for 30 and 60 min following 10 min of stimulation did not return to control distributions.

With longer periods of stimulation, the MEPP means continued to decrease until they were lost into the noise. This effect is different from that with KCl treatment in which case the MEPP frequencies declined after a few minutes even though the MEPP amplitudes tended to remain relatively high. This result is also similar to that found in frog junctions but if the treatment is for a longer duration, mainly sub-MEPPs remain in the frog preparation, (Kriebel and Pappas, 1987).

Discussion:

Synaptic vesicle diameter distributions of Torpedo electric organ are composed of 4-6 subpopulations based of diameter (Fox et al, this volume) and not only two classes as previously reported (Zimmermann & Whittaker, 1974, 1977; Zimmermann, and Denston, 1977). With low frequency (0.1 Hz) stimulation the vesicle diameter distributions changed after 3 1/2 hrs. of stimulation such that the 60 nm class became dominant and since this class became labeled with dextran particles it appeared that the 60 nm class was used for repeated release (Zimmermann & Denston, 1977a). Thus it was proposed that the larger class (90 nm) served as an initial transmitter store. We wondered why the short term, higher rates of stimulation did not show changes in vesicles although it was conceivable that the amount of release varied because previous studies had not determined quantal content and total release per volume of terminal. If, indeed, the changes in vesicle classes result from quantal release, one would expect these changes to be independent of stimulation frequency but dependent on the total number of quanta released. Consequently, we have calculated nerve evoked release per/μm^3 nerve terminal/stimulus so that we could compare total release at different frequencies (10, 1 and 0.1 Hz) to three different physiological states (66%, 37% and 5% or less of initial EPP values). In addition, we either worked with thin blocks of tissue which did not self-stimulate or we calculated total release including self-stimulation. Self-stimulation occurred in tissue blocks over one mm thick so that total release in early stages of stimulation was two to four times that due to nerve stimulation. We also used surface terminals for these studies because of the long times required for a fixative solution to reach deep terminals (over 20 minutes were required to reach those 4 cells deep). Finally, we have induced high rates of MEPP release with KCl or hypertonic saline which yielded total quantal release numbers per μm^3 nerve terminal comparable with those evoked by nerve stimulation (Table III). We correlate physiological state and total quantal release to the morphological picture and biochemical condition so we have used the stimulation parameters developed by Zimmermann and Whittaker, (1974 a, b, 1977). We report that total quantal release at 10 Hz, 1 Hz and 0.1 Hz to a fatigue state of 5% control is roughly 10^3 quanta/ μm^3 nerve terminal. However, changes in vesicle diameter (and ratios of vesicle classes) clearly occur at 0.1 Hz as reported by Zimmermann and Whittaker (1974a, 1977; Suszkiw & Whittaker, 1979); thus, the loss of VPI vesicles and maintenance of the VPII vesicle class is not just a function of release but also has a time dependency. It is interesting to note that Corthay et al (1982) found changes in acetylcholine with only one stimulus during potentiated release. Dunant (1973) has also reported changes in free ACh but no change in bound ACh or in the number of vesicles. Dunant et al (1974) also report that synaptic vesicles may decrease but then recover in numbers and may even exceed their initial numbers with continued stimulation. We found all classes of vesicles present in control terminals and only found the relative increase (same density) in the VPII class at low frequencies so we cannot conclude that the population becomes heterogenous only as a result of nerve stimulation. We find it very interesting that transmission can take place with scarcely any ACh in the vesicular fraction (Zimmermann & Whittaker, 1974b). Our results with hypertonic or high KCl salines are of interest because the KCl treatment decreased the size of the VPI class whereas the hypertonic saline showed no change in mean vesicle diameter. Yet, hypertonic treatment reduced the vesicle density whereas the three nerve stimulation frequencies (10, 1 & 0.1 Hz) and KCl treatment showed no vesicle depletion. In this regard, the hypertonic fixation experiments are of interest because they also generated 1-3x10^3 quanta/ μm^3 terminal at time of fixation although there was only minor vesicular depletion. The different treatments and periods of nerve stimulation were adjusted to yield about the same total quantal release per volume nerve terminal and to be equal to the density of synaptic vesicles at the start of experiments. We found different changes in vesicle sizes and densities (Table III) indicating that the vesicle classes respond to these treatments differently and that changes are not only dependent on transmitter release. We should note that the immediately available store of ACh could be associated with a membrane particle (Israël et al, 1982). Israël and coworkers have also reported that vesicle ACh and the number of vesicles remained stable during physiological stimulation and they note changes in membrane particles (see Israël and Manaranche, 1985; Israël et al, 1987; Dunant, et al 1982).

The Torpedo eye muscle neuromuscular junctions have only the 60 nm class of synaptic vesicle and this class probably represents the class in the electric organ involved in quantal release (Zimmermann and Whittaker, 1974a and b; 1977, Dowdall and Zimmermann, 1974). The smaller vesicles appear to load during nerve stimulation (Giompres et al 1981a) possibly by an osmotically active process due to the macromolecular core material (Ohsawa et al, 1979; Whittaker et al, 1972, 1974). The other vesicle classes are probably used for other functions such as vesicle transport (Stadler & Kiene, 1987; Kiene & Stadler, 1987) or ACh storage or even calcium uptake and release (Michaelson et al, 1980; cf. Perri, et al 1972). It is interesting that the electric organ of the skate has only the 60 nm class (Kriebel et al 1985) so it appears that the electric organ of the Torpedo is very unique and may be the tissue of choice for studies designed to elucidate the different functions of synaptic vesicles. The different classes must certainly represent different functional states (Zimmermann and Denston, 1977b; Stadler & Kiene, 1987). For example, we found that hypertonic salines reduced the VPI class but not a high KCl saline even though both treatments yielded the same quantal release. Dunant and Muller (1986) report quantal size to be similar to that of frog or mammal (cf. Whittaker, et al, 1972).

We also report that a hypertonic fixative is a stimulus which yields high rates of quantal release (for long periods of time in deeper junctions). Therefore changes in vesicle density must be regarded with caution particularly with fixatives containing paraformaldehyde (Table I). This observation probably explains the wide range in results with nerve stimulation from a 50% loss of vesicles (Boyne et al, 1975) to no change (Zimmermann & Whittaker, 1974a). Isosmotic glutaraldehyde fixatives showed little effect on diameters and density (Table II). However, the vesicle densities were greatly increased (but counts decreased) with all hypertonic fixatives which would be expected with terminal shrinkage. A hypotonic fixative, by comparison, increased vesicle counts with little or no effect on density. Thus, some vesicles may be used for terminal volume regulation. We were quite surprised by the influence of the relative depth of the terminal to vesicle numbers and size (Table II). Our studies show that great care must be used to utilize only the surface nerve terminal because of the influence of depth; and, also the time needed for a fixative to diffuse to deeper terminals.

With all fixatives, frequencies and periods of nerve stimulation, a high KCl saline and hypertonic treatments the touching vesicles (surface and deep terminals) were slightly smaller (3-5%). This value is that of error value and thus subject to bias, (Fox, et al, this volume). Vesicles positioned along the unit membrane could be simply measured differently than those which appear to "float" within the terminal. The overall histogram profile of the touching vesicles is nevertheless exactly the same as that profile of the entire complement of vesicles; and, this relationship persisted with changes in ratios of vesicle classes and densities induced by stimulation (cf Boyne et al, 1975). This observation indicates that the different classes of vesicles (based on diameter) are not preferentially selected to be touching vesicles and it forces us to admit that all classes based on diameter have equal probabilities to quantal release.

Supported by the Alexander von Humboldt Foundation and N.S.F. 19694. We thank Ms. Erika Kriebel for careful vesicle analyses.

References

Bennett, M.V.L., Wurzel, M. and Grundfest, H., 1961. The electrophysiology of electric organs of marine electric fishes. I. Properties of electroplaques of *Torpedo nobiliana.* J. gen. Physiol. 44: 757-804.

Boyne, A., T.P. Bohan and T.H. Williams, 1975. Changes in cholinergic synaptic vesicle populations and the ultrastructure of the nerve terminal membranes of *Narcine brasiliensis* electric organ stimulated to fatigue *in vivo*. J. Cell Biol. 67: 814-825.

Breer, H., S.J. Morris and V.P. Whittaker, 1978. A structural model of cholinergic synaptic vesicles from the electric organ of Torpedo marmorata deduced from density measurements at different osmotic pressures. Eur. J. Biochem. 87: 453-458.

Carpenter, R.S. and S.M. Parsons, 1977. Electrogenic behavior of synaptic vesicles from Torpedo californica. J. Biol. Chem. 253: 326-329.

Corthay, J., Y. Dunant and F. Loctin. 1982. Acetylcholine changes underlying transmission of a single nerve impulse in the presence of 4-aminopyridine in Torpedo. J. Physiol. 325: 461-479.

Dowdall, M.J. and Zimmermann, H., 1974. Evidence for heterogenous pool of acetylcholine in isolated cholinergic synaptic vesicles. Brain Research 71: 160-166.

Dunant, Y., 1973. Acetylcholine metabolism and release at the nerve-electroplaque junction. Brain Research 62: 543-549.

Dunant, Y. and D. Muller, 1986. Quantal release of acetylcholine evoked by focal depolarization at the Torpedo nerve-electroplaque junction. J. Physiol. 379:461-478.

Dunant, Y., J. Gautron, M. Israël, B. Lesbats et R. Manaranche,. 1974. Evolution de la decharge de l' organe elecrique de la Torpille et variations simultanees de l'acetylcholine au cours de la stimulation J. Neurochem. 23: 635-643.

Dunant, Y., G.J. Jones and F. Loctin, 1982. Acetylcholine measured at short time intervals during transmission of nerve impulses in the electric organ of Torpedo. J. Physiol. 325: 441-460.

Dunant, Y. and D. Muller, 1986. Quantal release of acetylcholine evoked by focal depolarization at the Torpedo nerve-electroplaque junction. J. Physiol. 379: 461-478.

Erdelyi, L. and W.D. Krenz, 1984. Electrophysiological aspects of synaptic transmission at the electromotor junction of Torpedo marmorata J. Comp. Biochem. Physiol. 79A: 505-511.

Erxleben, C.; M.E. Kriebel, 1987. Characteristics of spontaneous miniature and subminiature end-plate currents at the neonate and adult mouse neuromuscular junction: J. Physiol, in press.

Fox, G.Q. and D. Kötting, 1984. Torpedo electromotor system development: A quantitative analysis of synaptogenesis. J. Comp. Neurol. 224: 337-343.

Giompres, P.E., H. Zimmermann & V.P. Whittaker, 1981a. Purification of small dense vesicles from stimulated Torpedo electric tissue by glass beadlumn chromatography. Neuroscience 6: 765-774.

Giompres, P.E., H. Zimmermann and V.P. Whittaker, 1981b. Changes in the biochemical and biophysical parameters of cholinergic synaptic vesicles on transmitter release and during a subsequent period of rest. Neuroscience: 6: 775-785.

Israël, M., F.M. Meunier; N. Morel and B. Lesbats, 1987. Calcium-induced desensitization of acetylcholine release from synaptosomes or proteoliposomes equipped with mediatophore, a presynaptic membrane protein. J. Neurochem. 49: 975-982.

Israël, M. and R. Manaranche, 1985. The release of acetylcholine: from a cellular towards a molecular mechanism. Biol. Cell 55: 1-14.

Israël, M.,B. Lesbats; R. Manaranche, N. Morel; T. Gulik-Krzywicki and J-C. Dedieu, 1982. Rearrangement of intramembrane particles as a possible mechanism for the release of acetylcholine. J. Physiol. Paris 78: 348-356.

Kiene, M.- L. and H. Stadler, 1987. Synaptic vesicles in electromotoneurones. I. Axonal transport, site of transmitter uptake and processing of a core proteoglycan during maturation. EMBO J. 6: 2209-2215.

Kriebel, M.E. and E. Florey, 1983. Effect of lanthanum ions on MEPP amplitude distributions and vesicles in frog neuromuscular junctions. Neuroscience 9: 535-547.

Kriebel, M.E. and C.E. Gross, 1974. Multimodal distribution of frog miniature end plate potentials in adult, denervated, and tadpole leg muscle. J. Gen. physiol. 64: 85-103.

Kriebel, M.E., C. Gross and G.D. Pappas. 1987. Two classes of spontaneous miniature excitatory junction potentials and one synaptic vesicle class are present in the ray electrocyte. J. Comp. Physiol. A 160: 331-340.

Kriebel, M.E., R. Hanna and C. Muniak, 1986. Synaptic vesicle diameters and synaptic cleft widths at the mouse diaphragm in neonates and adults. Devel. Brain Research 27: 19-29.

Kriebel, M.E., F. Llados and D.R. Matteson, 1976. Spontaneous subminiature end-plate potentials in mouse diaphragm muscle: evidence for synchonous release. J. Physiol. 262: 553-581.

Kriebel, M.E. and G.D. Pappas, 1984. Hyperosmotic solution can change MEPP amplitude distribution of the frog neuromuscular junction to s-MEPPs with little change in synaptic vesicle numbers. Neuroscience, in press.

Meunier, F. - M., 1984. Relaitonship between presynaptic membrane potential and acetylcholine release in synaptosomes from Torpedo electric organ. J. Physiol. 354: 121-137.

Michaelson, D.M., I. Opher and I. Angel, 1980 ATP- Stimulated Ca^{2+} transport into cholinergic Torpedo synaptic vesicles. J. Neurochem. 35: 116-124.

Miledi, R., P. Molinoff & L.T. Potter, 1971. Isolation of the cholinergic receptor protein of Torpedo electric tissue. Nature 229: 554-557.

Morel, N., 1976. Effect of choline on the rates of synthesis and of release of acetylcholine in the electric organ of Torpedo. J. of Neurochem. 27: 779-784.

Muniak, C.G., M.E. Kriebel and C.G. Carlson, 1982. Changes in MEPP and EPP amplitude distributions in the mouse diaphragm during synapse formation and degeneration. Devel. Brain Res. 5: 123-138.

Ohsawa, K., G.H.C. Dowe, S.J. Morris and V.P. Whittaker, 1979. The lipid and protein content of cholinergic synaptic vesicles from the electric organ of Torpedo marmorata. Purified to constant composition: implications for vesicle structure. Brain Res. 161: 447-457.

Perri, V., O. Sacchi, E. Raviola and G. Raviola. 1972. Evaluation of the number and distribution of synaptic vesicles at cholinergic nerve endings after sustained stimulation. Brain Res. 39: 526-529.

Soria, B., 1983. Properties of miniature post-synaptic currents at the Torpedo marmorota nerve-electroplate junction. Quart. J. Exp. Physiol. 68: 189-202.

Stadler, H. and M.-L. Kiene, 1987. Synaptic vesicles in electromotoneurones. II. Heterogeneity of populations is expressed in uptake properties, exocytosis and insertion of a core proteoglycan into the extracellular matrix. EMBO J. 6: 2217-2221.

Suszkiw, J.B., 1976. Acetylcholine translocation in synaptic vesicle ghosts in vitro. J. Neurochem. 27: 853-857.

Suszkiw, J.B. and V.P. Whittaker, 1979. Role of vesicle recycling in vesicular storage are release of acetylcholine in Torpedo electroplaque synapses. Prog. Brain Res. 49: 153-162.

Walther, C., 1974. Effects of potassium, lanthanum, and black widow spider venom on miniature synaptic potentials in the Torpedo electroplax. J. Comp. Physiol. 90: 71-73.

Whittaker, V.P., M.J. Dowdall; G.H.C. Dowe; R.M. Facino and J. Scotto, 1974. Proteins of cholinergic synaptic vesicles from the electric organ of Torpedo: characterization of a low molecular weight acidic protein. Brain Res. 75: 115-131.

Whittaker, V.P., W.B. Essman and G.H.C. Dowe, 1972. The isolation of pure cholinergic synaptic vesicles from the electric organs of elasmobranch fish of the family Torpedinidae. Biochem. J. 128: 833-846.

Zimmermann, H. and C.R. Denston, 1977a. Recycling of synaptic vesicles in the cholinergic synapses of the Torpedo electric organ during induced transmitter release. Neuroscience 2: 695-714.

Zimmermann, H. and C.R. Denston, 1977b. Separation of synaptic vesicles of different functional states from the cholinergic synapses of the Torpedo electric organ. Neuroscience 2: 715-730.

Zimmermann, H. and V.P. Whittaker, 1974a. Effect of electrical stimulation on the yield and composition of synaptic vesicles from the cholinergic synapses of the electric organ of Torpedo: A combined biochemical, electrophysiological and morphological study. J. Neurochem. 22: 435-450.

Zimmermann, H. and V.P. Whittaker, 1974b. Different recovery rates of the electrophysiological, biochemical and morphological parameters in the cholinergic synapses of the Torpedo electric organ after stimulation. J. Neurochem. 22: 1109-1114.

Zimmermann, H. and V.P. Whittaker, 1977. Morphological and biochemical heterogeneity of cholinergic synaptic vesicles. Nature 267: 633-635.

Zucker, R. and L. Landò, 1986. Mechanism of transmitter release: voltage hypothesis and calcium hypothesis. Science 231: 574-579.

FINE STRUCTURAL CORRELATES OF CALCIUM DYNAMICS IN THE PRESYNAPTIC TERMINAL

George D. Pappas, Virginia Kriho and Robert P. Becker
Department of Anatomy and Cell Biology, University of Illinois at Chicago, P.O. Box 6998, Chicago, Illinois 60680, USA

INTRODUCTION

Correlative morphological changes brought about by synaptic function have proven to be illusive. We have been interested in the fine structural correlates of synaptic fatigue in the frog sartorius neuromuscular preparation brought about by tetanic stimulation of the sartorius nerve with a suction electrode (Rose, et al., 1978). It has been suggested that the decrement of presynaptic transmitter release may be due to the decrement in the synaptic vesicle population, as the vesicular release of ACh at the neuromuscular junction exceeds the replacement of the vesicles when the presynaptic nerve stimulation is greatly augmented (Heuser and Reese, 1973). Examination with the electron microscope of the same presynaptic nerve terminal from which postsynaptic endplate potentials were recorded does not support the vesicle depletion theory. Extensive studies on identified synapses in the frog sartorius neuromuscular preparation indicated that there was no correlation with the decrease of the end-plate potential following stimulation and the number of vesicles present in the presynaptic ending (Rose, et al., 1978). It has also been reported that the steady decrease in the amount of transmitter being released during tetanic stimulation does not deplete the amount of ACh present in the presynaptic ending (Kriebel, et al., 1976). One can reasonably conclude that if the synaptic vesicle population or the presynaptic stores of ACh are not

NATO ASI Series, Vol. H21
Cellular and Molecular Basis of Synaptic Transmission
Edited by H. Zimmermann

critically affected, then some other factor such as the obligatory role that calcium ions play in depolarization secretion-coupling may be critically affected and may bring about synaptic fatigue (cf. Augustine et al., 1987).

It is well established that depolarization of the presynaptic nerve terminal opens voltage-dependent calcium channels, allowing an influx of calcium ions into the presynaptic ending. In the absence of extracellular Ca^{++} in a neuromuscular preparation, ACh release cannot be evoked (Katz and Miledi, 1968). We propose that the influx of Ca^{++} into the presynaptic terminal, in a concentration-dependent fashion, brings about three distinct actions. The first action is that of transmitter release, in response to a certain critical concentration intracellularly just within the presynaptic membrane. This threshold concentration has not been measured in the neuromuscular junction, but has been approximated for the stellate synapse of the squid (Llinas and Nicholson, 1975). The internal calcium concentration increases greatly during tetanic stimulation since the Ca channels stay open. These higher concentrations lead to the second action of calcium which is the eventual decrement in transmitter secretion. Kusano (1970) reported that an increase in calcium ions in the presynaptic ending decreases the postsynaptic potential. If the intracellular calcium concentration is increased even more, the third action of calcium is the activation of proteases which bring about irreversible proteolysis of the presynaptic terminal (Zimmerman and Schlaepfer, 1982).

Using the Oshman and Wall (1972) procedure of adding excess calcium to the fixative and the Borgers method whereby calcium is initially precipitated with oxalate followed by pyroantimonate (Borgers, et al., 1977, 1981), electron dense deposits representing calcium binding sites have been demonstrated in synaptic vesicles in control preparations and in mitochondria (Politoff et al., 1974; Pappas and Rose, 1976).

In identified synapses that were stimulated (10Hz for 20

min.) so that the end-plate potential was decreased to less than one-tenth its original amplitude, the binding of calcium can no longer be demonstrated. Following one hour of rest, the fatigued synapse recovers and the calcium binding by the synaptic vesicles can again be demonstrated (Rose, et al., 1974; Pappas, et al., 1975).

An initial influx of calcium ions entering through the presynaptic voltage-dependent channels effects transmitter release. The intracellular response to this calcium influx is the activation of Ca^{++} buffering mechanisms which involve calcium binding proteins and sequestration in membrane-bound organelles, the most prevalent being the synaptic vesicles. Indeed, it has been shown that inhibition of the calcium buffering mechanism depresses evoked transmitter release (Adams et al., 1985). Conversely, it has been demonstrated at the squid giant synapse that if EGTA is injected presynaptically, thereby artificially increasing calcium buffering intracellularly, then the postsynaptic potential increases (Adams, et al., 1985). While these buffering mechanisms may be efficient for short term regulation, ultimately the excess cytosolic calcium must be extruded if the presynaptic process is to return to its low resting levels of calcium.

ATP dependent calcium ion extrusion mechanisms have been described in the squid giant axon (Baker and DiPolo, 1984), in synaptic plasma membranes from mammalian brain (Papazian, et al., 1984) and in the hybrid neuronal cell line 108CC5 (Kurzinger et al., 1980). This ATPase activity, in some cases, has been shown to be calmodulin dependent (cf. Klee et al., 1980; DeLorenzo 1982).

While the biochemical evidence for a Ca^{++} pumping ATPase in neuronal plasma membrane is ample, correlative cytochemical demonstration of its presence is lacking. Moreover, at the frog neuromuscular junction, the most commonly used preparation for the study of synaptic function, there has been neither biochemical nor morphological demonstration of the presence of a plasma membrane ATPase. Biochemical evidence for ATPase activity at the neuromuscular junction is particularly difficult to obtain because of the small size of the nerve terminal in relation to the amount of muscle present. We have now obtained EM cytochemical localization of Ca^{++}-Mg^{++} ATPase activity at the frog neuromuscular junction utilizing the histochemical procedure originally developed by Wachstein and Meisel (1957) and recently modified by Salama et al., (1987).

HISTOCHEMICAL PROCEDURES

In this procedure, sartorius muscle of the frog is dissected out, stretched to normal length and pinned out so that terminal areas are visible. The muscle is then fixed in 1.4% glutaraldehyde in 0.1M cacodylate buffer and 5% sucrose, pH 7.2 for 15 min. at 4°C. After 15 min., terminal areas are cut out, placed in vials and fixed for an additional 45 min. in the cold. The tissue is then rinsed in 0.1M cacodylate buffer and refrigerated overnight. For localization of ATPase activity in the modified Wachstein-Meisel procedure, cerium instead of lead is used for the precipitated end-product (Robinson and Karnovsky, 1983; Salama et al., 1987).

The muscle pieces, which were refrigerated overnight, are transferred to Tris-maleate buffer, pH 7.2, for two changes of 5 min. each. To circumvent problems associated with penetration of the cytochemical medium, tissues were pre-incubated at room temperature for 1 hour in the precipitating medium, without substrate. This medium consists of 80mM Tris-maleate buffer, pH 7.2, 5mM $MgSO_4$, 5mM $CaCl_2$ and 2mM $CeCl_3$ as the capture ion. To demonstrate liberated phosphates, tissues were incubated in the same medium, plus 1mM ATP disodium salt as substrate, for 1 hr. at 37°C. The medium was filtered before use and two fresh changes were made within this 1 hour of incubation. Vials were then put on ice to stop the reaction. The tissues were rinsed 2x5 min. in cold Tris-maleate buffer and 2x10 min. in 0.1M cacodylate buffer. Following the rinses, tissues were post-fixed in 1% OsO_4 for 1 hr. After rinsing in 0.1M cacodylate buffer, tissues were dehydrated and embedded in EPON. Thick sections (1-2 μm) were cut and stained with toluidine blue for orientation in the light microscope. Thin sectioning for electron microscopy followed, once terminal areas were located.

We ran three types of control tissue for non-specificity: 1) Incubation in the complete medium without substrate (ATP) 2) Incubation in complete medium, substituting Na_2HPO_4 for the substrate, and 3) Incubation in complete medium after heating specimens to denature the enzyme.

RESULTS AND CONCLUSIONS

When pieces of muscle are incubated in the complete medium with ATPase as substrate, a dense reaction product forms in the area of the Schwann cell-neuronal plasma membrane interface along with some non-specific deposits in the extracellular space (Fig. 1). The intensity of this non-specific labeling can be reduced, but not completely eliminated by running the reaction for only a few minutes at 4^oC.

When tissue was incubated without Ca^{++} or Mg^{++}, the dense reaction product surrounding the nerve terminal was usually absent (Fig 2). In order to determine whether this ATPase reaction product is calmodulin-dependent, tissue was incubated in medium which contained the calmodulin inhibitor, R24571 (10^{-5}M in DMSO) (Van Belle, 1981). Only in about 10% of the terminals did the calmodulin inhibitor block the dense reaction product from forming. This may indicate that other calcium binding proteins are more prevalent or that the ATPase activity at the frog sartorius presynaptic ending is not calmodulin dependent.

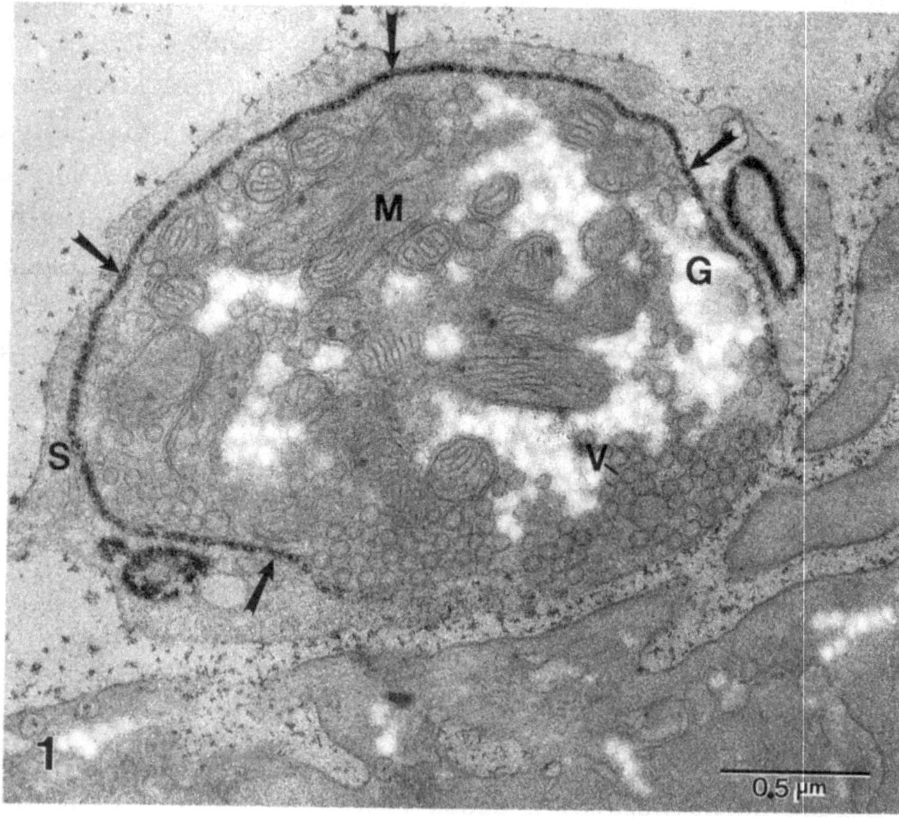
M
G
S
V
1
0.5 µm

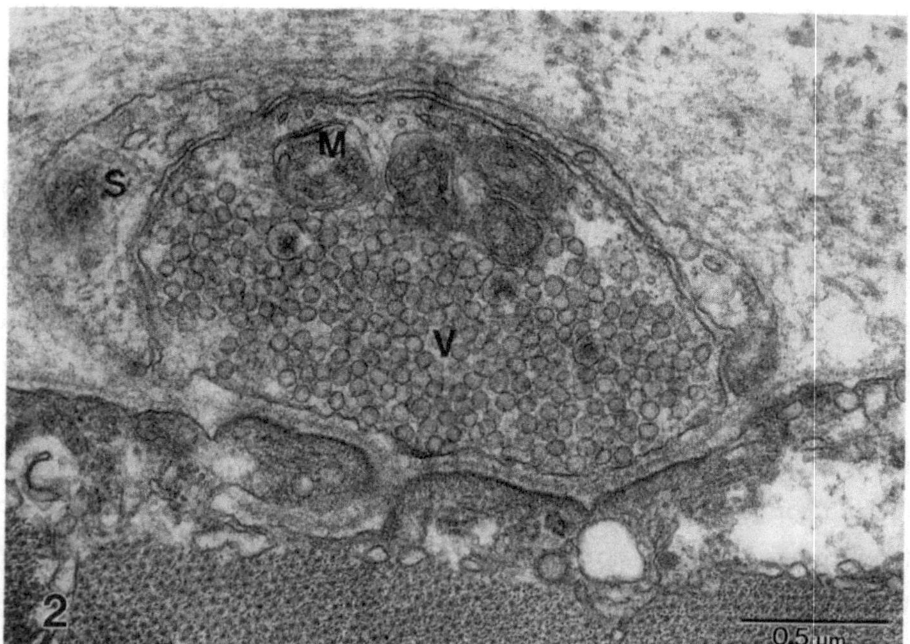
M
S
V
2
0.5 µm

Figure 1. Electron micrograph of a section of frog sartorius neuromuscular synapse. The preparation was uncubated in complete medium with ATP as substrate. The dense reaction product (arrows) at the interface of the Schwann cell and neuronal plasma membrane demonstrates the localization of $Ca^{++}Mg^{++}$ATPase.
G = glycogen M = mitochondria
S = Schwann cell V = synaptic vesicles

Figure 2. Electron micrograph of a similar preparation as Fig. 1, but this section was taken from a preparation that was incubated in the absence of Ca^{++} and/or Mg^{++}. The dense reaction product for $Ca^{++}Mg^{++}$ ATPase is absent.
M = mitochondria S = Schwann cell V = synaptic vesicles

Neither the presence of levamisole (1mM) in the medium as an inhibitor of alkaline phosphatase or ouabain (1mM) as an inhibitor of Na^{+}-K^{+}ATPase affected the amount of labeling present in any terminals.

Since it is difficult to assign the precipitated phosphate reaction product either to the plasma membrane of the nerve terminal or the Schwann cell, we eliminated the nerve terminal in some preparations and in others, the Schwann cell.

It has been shown that when a muscle preparation is allowed to survive after denervation, there is movement of the Schwann cell into the space previously occupied by the nerve terminal, and Schwann cell miniature end-plate potentials can be recorded from the muscle (Kriebel, et al., 1980). When a preparation which has been denervated for 5 days is incubated in the same manner as the control for ATPase activity, the absence of reaction product is obvious (Fig.3).

Direct exposure of axons to the anti-neoplastic agent, doxorubicin results in Schwann cell necrosis (Parnad and Griffin, 1983). When the neuromuscular preparation is incubated for 1.5 hours at room temperature in doxorubicin (2mg/ml), there is a discoloration of the tissue and a disruption of muscle fibers at the light microscope level. At the EM level, the necrosed Schwann cell appears thin and vesiculated. When the tissue is incubated for the demonstration of ATPase activity, a strong reaction product forms at the nerve terminal plasma membrane (Fig. 4).

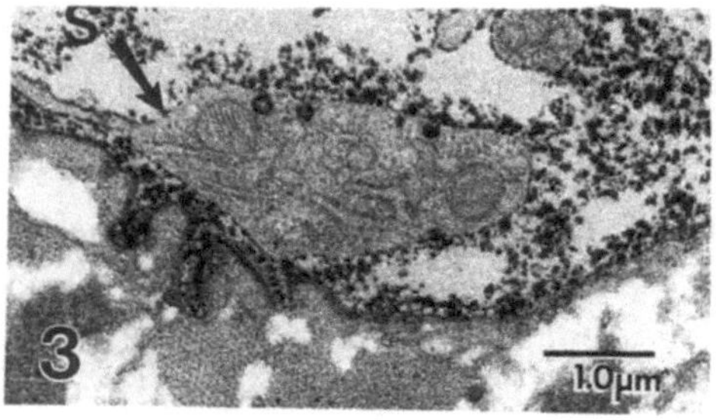

Figure 3. Electron micrograph of a section from a preparation that was denervated for 5 days prior to processing for the Ca++Mg++ATPase reaction. Note that a portion of a Schwann cell(S) process has now occupied the presynaptic area. In the absence of the nerve terminal, there is no specific reaction product related to the Schwann cell plasma membrane.

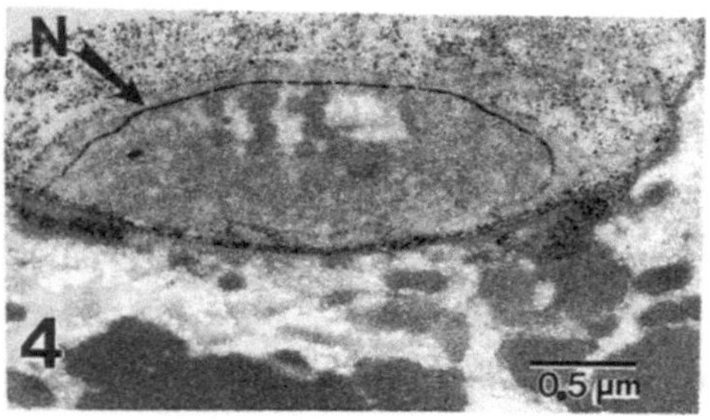

Figure 4. This section is taken from a preparation in which the Schwann cells were preferentially killed by exposure to doxorubicin (see text). A dense reaction product indicates the presence of Ca++Mg++ATPase associated with (N) facing the necrosed Schwann cell.

In order to localize Ca++Mg++ATPase within the nerve terminal of the neuromuscular junction, we treated the tissue initially with 1% triton-X for 2 minutes and then carried out, in the usual manner, with the ATPase reaction procedure. As expected, no reaction product could be localized on the partially digested plasma membrane. However, reaction product was found in the mitochondria between the inner and outer membranes (see Fig. 5) and surrounding the cytoplasmic side of the synaptic vesicles (Fig. 5).

Recent biochemical studies on synaptic vesicles correlate well with our morphological findings. Kiene and Stadler (1987) isolated synaptic vesicles from the cholinergic endings of

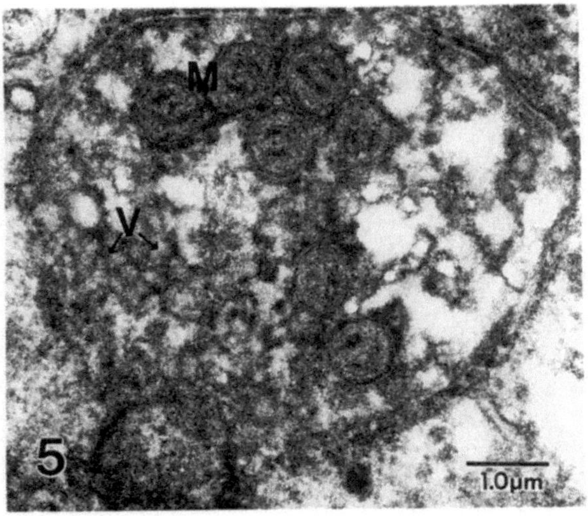

Figure 5. This section is taken from a preparation treated with triton to facilitate penetration of the reacting medium for the localization of $Ca^{++}Mg^{++}$ATPase within the nerve terminal. The membranes are poorly preserved. Reaction product can be seen in the space between the mitochondrial membranes (M). Reaction product on the surface of the synpatic vesicles (V) suggests $Ca^{++}Mg^{++}$ATPase activity at these sites.

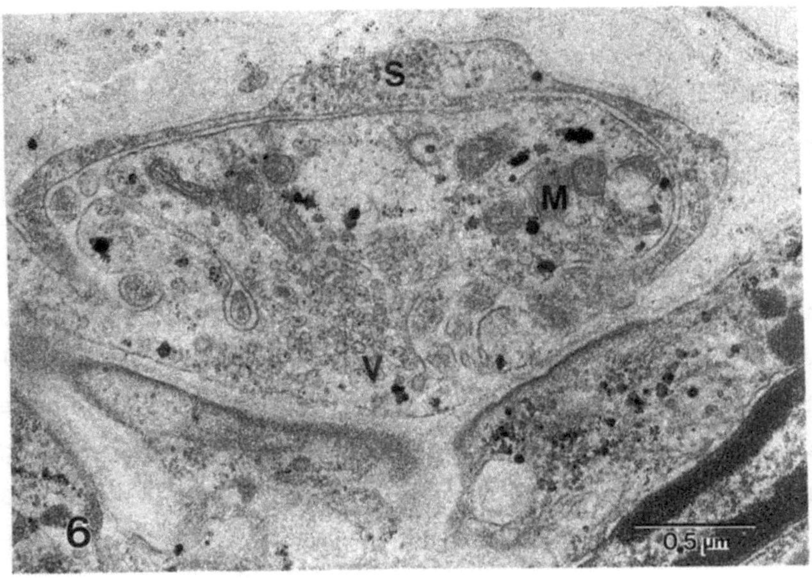

Figure 6. Electron micrograph of a section from a preparation processed according to the Borgers procedure for visualization of the subcellular distribution of calcium. In this unstimulated preparation, single calcium pyroantimonate particles are found in the synaptic vesicles (V) and in mitochondria (M) suggesting that they accumulate calcium ions. S = Schwann cell

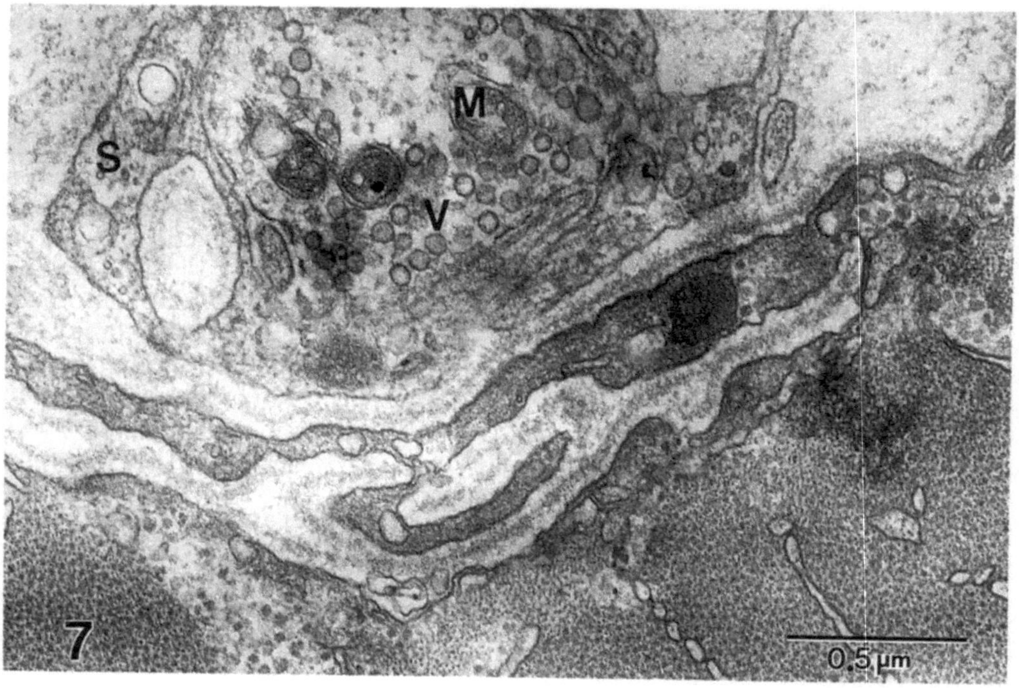

Figure 7. A section from a preparation processed by the Borgers method. This preparation, however, was stimulated (10Hz-30 mins.). The synaptic vesicles (V) do not contain the dense calcium precipitates suggesting that they do not accumulate calcium ions.
M = mitochondria, S = Schwann cell

electromotorneurons of Torpedo marmorata. They showed that *in vitro* surface activity of Mg-ATPase is practically identical for the entire vesicle pool. They identified a core heparansulphate proteoglycan within the vesicles and depending on the amount present classified the isolated vesicles into 3 populations. The predominant population of vesicles present following intense stimulation contains much less core proteoglycan. Of interest here is that this class of vesicles, while it can accumulate ACh and ATP, does not accumulate calcium ions. These in vitro uptake studies of Stadler and Kiene (1987) fit well with our earlier electron microscopic findings (Pappas and Rose, 1976) in that synaptic vesicles in stimulated endings having presumably undergone exo- and endocytosis do not bind calcium and therefore lack electron dense "dots" (See Figs. 6 and 7).

These findings suggest that energy-dependent calcium ion buffering by synaptic vesicles and other mambrane bound organelles is very rapid and efficient for short term regulation. The extrusion of calcium from terminal involves two mechanisms. There is a Na^{++}-Ca^{++} exchange system (DiPolo and Beaugé, 1983). The energy required for this system is derived from the inward movement of sodium ions down their electrochemical gradient. The other, of course, is the ATP-dependent process whereby energy derived from ATP is used to drive calcium ions across the plasma membrane which is the main topic we have considered in this presentation.

In addition to our EM histochemical studies on ATPase and calcium localization, we are also investigating the relative levels of total calcium in nerve terminals by electron microscope X-ray microanalysis. Our preliminary findings based on several identified endings clearly show that the calcium concentration at the presynaptic ending has increased greatly following tetanic stimulation. However, we find that it is very difficult to locate at the EM level the specific terminal from which we have recorded postsynaptically in the frog sartorius neuromuscular preparations. In order to be able to increase our data (N), we have initiated X-ray microprobe microanalysis studies of the calcium content in the frog photoreceptor cells of the retina. Adjacent small clusters of retinal rod cells are presynaptic to the same horizontal cell. These rods are most probably in an identical physiological state. In this preparation the possibility of taking large numbers of sample readings of total calcium concentration under very similar presynaptic conditions in relation to the transmitter release can be accomplished.

Since our X-ray microprobe studies are at their initial phase, our preliminary findings of the amount of Ca present will be qualitative and not quantitative. In these studies (in collaboration with Claudia Tellez and Dr. Harris Ripps), frogs were dark adapted overnight (Ripps, et al., 1979). In the morning retinal eye cups were prepared under normal frog Ringers solution in deep red light. The calcium concentration in photoreceptor presynaptic rod pedicles was measured by X-ray microanalysis in resinless (de-embedded) sections of retinas fixed in the presence of the calcium precipitant, oxalic acid. In this method retinas were rapidly dehydrated and embedded in polyethyleneglycol. Sections were cut dry and de-embedded in acetone prior to critical point drying. Analysis was performed at -100°C. This preparation offers the same level of sensitivity for calcium measurement as is found in the freeze-dried section approach (Becker, et al., 1982). Three conditions were examined: 1) dark adapted (control), 2) dark adapted and 100 mM K^+ Ringer for 20 minutes, 3) dark adapted and intense light for 15 minutes. In parallel experiments, frog retina pedicels were exposed as above and prepared for demonstrating calcium precipitation sites by transmission electron microscopy, according to the method of Borgers for visualization of electron dense deposits indicating the presence of calcium binding sites at the EM level of resolution (Borgers, et al., 1981).

After stimulation of transmitter release in high K^+, an increased calcium content was measured by X-ray microanalysis in pedicel presynaptic areas as compared to the unstimulated, dark adapted state. In addition an increase in calcium dependent deposits in synaptic vesicles was seen in pedicles prepared for transmission electron microscopy (TEM). Exposure of photoreceptors to intense light, which hyperpolarizes the rods, depresses synaptic activity, presumably because the presynaptic Ca channels are closed, resulting in a low intracellular calcium concentration at photoreceptor synaptic pedicles as shown by X-ray microanalysis. Under these conditions it was not surprising that no Ca uptake (no dense deposits) was visualized in synaptic vesicles. Thus the relationship of increased intracellular calcium concentration is correlated with increased transmitter release in the presynaptic endings of the retinal rod photoreceptors and these findings are consistant with results at the frog neuromuscular junction.

The amount of calcium present directly affects not only synaptic vesicles, but also the cytomatrix and formed cytoskeletal elements (i.e., microfilaments, intermediate filaments and microtubules) of the presynaptic nerve ending. A comprehension of the significance of calcium concentration dynamics lies in understanding the interrelationships of these cytoplasmic elements in bringing about transmitter release.

ACKNOWLEDGEMENTS

Supported in part by NSF Grant BNS-84067724. The EM facility of the Research Resources Center of the University of Illinois at Chicago provided the equipment necessary to conduct this study.

REFERENCES

Adams DJ, Takeda K and Umbach JA (1985) Inhibitors of calcium buffering depresses evoked transmitter release at the squid giant synapse. J Physiol 369:145-159

Augustine GJ, Charlton MP and Smith SJ (1987) Calcium action in synaptic transmitter release. Ann Rev Neurosci 10:633-693

Baker PF and Di Polo R (1984) Axonal calcium and magnesium homeostasis. Kleinzeller A and Baker PF (eds) Academic Press 1984 New York: 195-242 In: Current Topics in Membranes and Transport

Becker RP, Canada J and Pappas GD (1982) Calcium localization at the neuromuscular junction: X-ray microanalysis of resinless sections. J Cell Biol 95:101a

Borgers M, de Brabander M, van Reempts J, Awouters F and Jacob WA (1977) Intranuclear microtubules in lung mast cells of guinea pigs in anaphyulactic shock. Lab Invest 37:1-8

Borgers M, Thoné F and van Nueten JM (1981) The subcellular distribution of calcium and the effects of calcium-antagonists as evaluated with a combined oxalate-pyroantimonate technique. Acta histochem Suppl Band XXIV S 327-332

De Lorenzo RJ (1982) Calmodulin in neurotransmitter release and synaptic function. Fed Proc 41:2265:2272

Di Polo R and Beaugé L (1983) The calcium pump and sodium-calcium exchange in squid axons. Ann Rev Physiol 45:313-324

Heuser JE and Reese TS (1973) Evidence for recycling of synaptic vesicle membrane during transmitter release at the frog neuromuscular junction. J Cell Biol 57:315-344

Katz B and Miledi R (1968) The role of calcium in neuromuscular facilitation. J Physiol 195:481-492

Kiene M-L and Stadler H (1987) Synaptic vesicles in electromotoneurones. I. Axonal transport, site of transmitter uptake and processing of a core proteoglycan during maturation. The EMBO Jour 6:2209-2215

Klee CB, Crouch TH and Richman PG (1980) Calmodulin. Ann Rev Biochem 49:489-515

Kriebel ME, Hanna RB and Pappas GD (1980) Spontaneous potentials and fine structure of identified frog denervated neuromuscular junctions. Neurosci 5:97-108

Kriebel ME, Matteson DR and Pappas GD (1978) Acetylcholine content in physiologically fatigued frog nerve-muscle preparations and in denervated muscle. Gen Pharmacol 9:229-234

Kurzinger K, Stadtkus C, Humprecht B (1980) Uptake and energy-dependent extrusion of calcium in neural cells in culture. Eur J Biochem 103:597-611

Kusano K (1970) Influence of ionic environment on the relationship between pre- and postsynaptic potentials. J Neurobiol 1:435-457

Llinás R and Nicholson C (1975) Calcium role in depolarization-secretion coupling: An aequorin study in squid giant synapse. PNAS 72:187-190

Oschman JL and Wall BJ (1972) Calcium binding to intestinal membranes J Cell Biol 55:58-73

Papazian DM, Rahamimoff H and Goldin SM (1984) Partial purification and functional identification of a calmodulin-activated, adenosine 5'-triphosphate-dependent calcium pump from synaptic plasma membranes. J Neurosci 4:1933-1943

Pappas GD and Rose S (1976) Localization of calcium deposits in the frog neuromuscular junction at rest and following stimulation. Brain Res 103:362-365

Pappas GD, Rose S and Kriebel ME (1975) Dynamic aspects of the calcium binding sites in synaptic vesicles in the frog neuromuscular junction. In: Yamada E (ed). Proc 10th Int Congr Anat Science Council of Japan: p 15

Parhad IM and Griffin JW (1983) Segmental demyelination after subperineurial injection of doxorubicin. J Neuropath Exp Neurol 43:317

Politoff AL, Rose S and Pappas GD (1974) The calcium binding sites of synaptic vesicles of the frog sartorius neuromuscular junction. J Cell Biol 61:818-823

Ripps H, Shakib M, Chappell RL and MacDonald ED (1979) Ultrastructural localization and X-ray analysis of calcium-induced electron-dense deposits in the skate retina. Neurosci 4:1689-1703

Robinson JM and Karnovsky MJ (1983) Ultrastructural localization of several phosphatases with cerium. J Histo Cyto 31:1197-1208

Rose S, Pappas GD, Kriebel M and Tousimis AJ (1974) Evidence for the synaptic vesicle calcium binding site at the neuromuscular junction of the frog sartorius. Biol Bull 147:495-496

Rose SJ, Pappas GD and Kriebel ME (1978) The fine structure of identified frog neuromuscular junctions in relation to synaptic activity. Brain Res 144:213-239

Salama AH, Zaki AE-ME and Eisenmann DR (1987) Cytochemical localization of Ca^{2+}-Mg^{2+} adenosine triphosphatase in rat incisor ameloblasts during enamel secretion and maturation. J Histo Cyto 35:471-482

Stadler H and Kiene M-L (1987) Synaptic vesicles in electromotoneurones. II. Heterogeneity of populations is expressed in uptake properties; exocytosis and insertion of a core proteoglycan into the extracellular matrix. The EMBO Jour 6:2217-2221

Van Belle H (1981) R24571: A potent inhibitor of calmodulin activated enzymes. Cell Calcium 2:483-494

Wachstein M and Meisel E (1957) Histochemistry of hepatic phosphatase at a physiological pH. Am J Clin Pathol 27:13-23

Zimmerman U-JP and Schlaepfer WW (1982) Characterization of a brain calcium activated protease that degrades neurofilament proteins. Biochem 21:3977-3983

DIFFERENTIAL RELEASE OF TRANSMITTERS AND NEUROPEPTIDES CO-STORED IN CENTRAL AND PERIPHERAL NEURONS.

Åsa K. Thureson-Klein, Richard L. Klein, Pei-Chun Zhu and Jae-Yang Kong
University of Mississippi Medical Center, Department of Pharmacology
Jackson, MS 39216, USA

INTRODUCTION

During recent years, the prototype single-transmitter neuron concept has been replaced by a multi-messenger model from which several substances may be released for subsequent interaction with receptors on target cells. This change has occurred because numerous informational substances (Schmitt 1984) including nucleotides, neuropeptides and neurohormones that can be synthesized and co-stored with the classical transmitter have been identified in virtually all types of neurons.

While it is still debated to what extent the costored substances function directly in neurotransmission, there is no doubt that neuropeptides and nucleotides can modulate the action of the classical transmitters via pre- or postsynaptic mechanisms (Bartfai 1985; Burnstock 1986; Stjärne *et al* 1986; Campbell 1987; Hökfelt and Terenius 1987). There is also evidence that certain combinations of neuropeptide and transmitter are more common than others. For example, noradrenergic neurons in the periphery as well as in the central nervous system are likely to store and release an opioid peptide or neuropeptide Y (NPY) (Klein *et al* 1982; Hökfelt *et al* 1983; Allen *et al* 1984; Ekblad *et al* 1984; Douglas *et al* 1986; Stjärne *et al* 1986; Edvinsson *et al* 1987). Cholinergic neurons, on the other hand, commonly contain VIP (Johansson *et al* 1981a; Lundberg *et al* 1981; Bartfai 1985) while serotonin (5-HT) is costored with substance P (Johansson *et al* 1981b; Pelletier *et al* 1981; Le Gray *et al* 1984). Whereas a classical transmitter and its co-stored neuropeptide may be released in parallel under certain circumstances, there is evidence that a differential release is more common. With a low frequency of stimulation transmitter release dominates, but with intense stimulation there is an unproportional increase in neuropeptide release. This has been suggested to result from autoinhibition (Bartfai 1985).

The coexistence of several messengers with similar or sometimes opposite actions makes it important for the neuron to respond differentially as needed to

NATO ASI Series, Vol. H21
Cellular and Molecular Basis of Synaptic Transmission
Edited by H. Zimmermann

maintain physiological homeostasis. Therefore, along with presynaptic modulation and changes in the rates of synthesis, the neuron has developed additional mechanisms to regulate the amount of individual messengers being released. The ultrastructural manifestations of two of these mechanisms are the main subject for the present report, namely; (1) the manufacturing of different types of storage vesicles and (2) the formation of specialized zones along the neurolemma which favors recruitment of transmitter storing vesicles over those containing neuropeptide.

DIFFERENT STORAGE FOR TRANSMITTER AND NEUROPEPTIDE

Small vesicles in peripheral nerve terminals:

All neurons in the peripheral and central nervous systems contain one or more distinct populations of vesicles that can be categorized ultrastructurally according to size, shape and electron density. Evidence collected from biochemical analyses of subcellular fractions, immunocytological examination and pharmacological experiments indicate that the small (45-55nm) type of vesicle, universally present in mammalian adrenergic and cholinergic neurons, contains transmitter and ATP (Fried *et al* 1981; Zimmermann 1982; Whittaker 1986) but is devoid of neuropeptides (Fried *et al* 1985, 1986; Lundberg *et al* 1981; De Potter 1987). Ultrastructurally, the absence of neuropeptide is revealed by a lack of electron dense contents in small cholinergic vesicles. Similarly, noradrenergic small vesicles depleted of catecholamine, e.g., by treating an animal with low doses of reserpine, become electron lucent (Fig. 1a). The corresponding small vesicles in untreated animals contain an electron dense precipitate representing noradrenaline, after the same type of chromaffin reaction during fixation (Fig. 1b).

Large vesicles in peripheral nerve terminals:

In addition to small vesicles, the cholinergic and noradrenergic terminals (varicosities) contain large vesicles measuring 75 nm (or more) in diameter. Along with small vesicles, these have been observed to undergo exocytosis in actively secreting noradrenergic neurons. Large vesicles in presumptive cholinergic perivascular neurons also appear to undergo exocytosis in response to intense stimulation such as cerebral ischemia (Thureson-Klein *et al* 1987).

In contrast to the small type vesicles, large (75-85 nm) vesicles are equally electron dense in noradrenergic terminals of control and reserpine treated animals, even though they must have lost their contents of noradrenaline as measured by

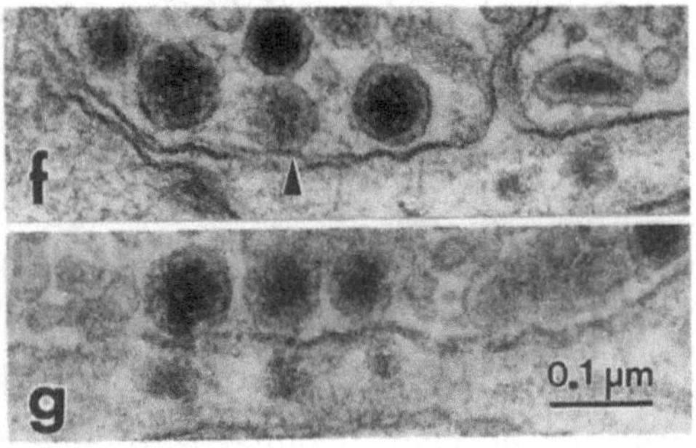

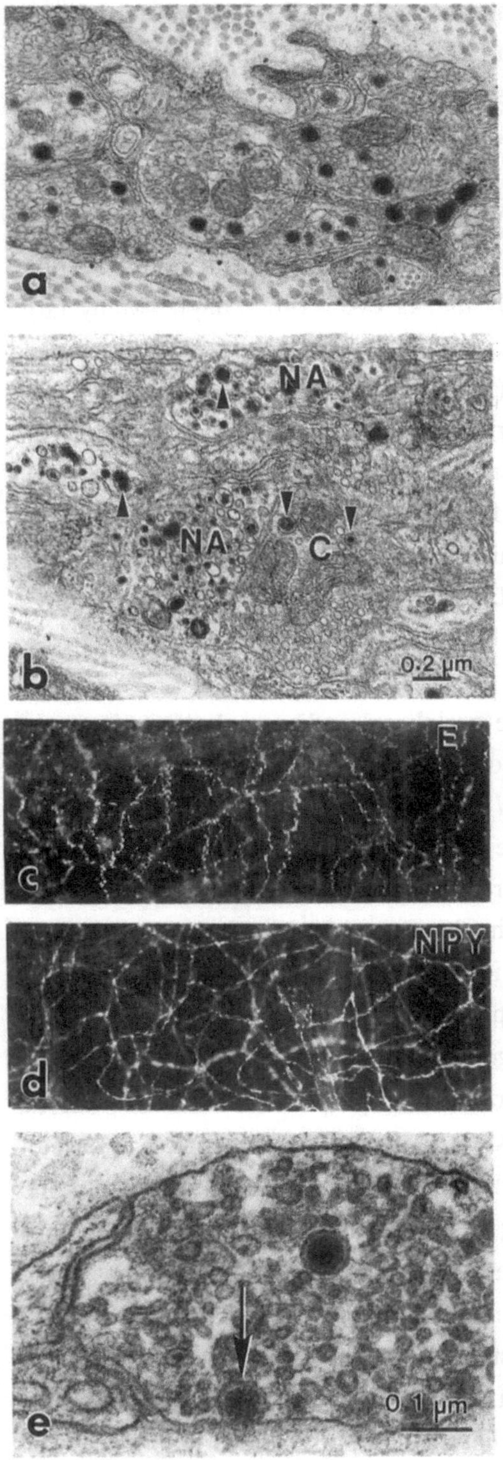

Fig. 1. Nerve fibers in the middle cereberal artery and vas deferens of reserpinized and control pigs (*Sus scrofa*)

a. Middle cerebral artery after 7 days of reserpine (0.05mg/kg every other day). All small vesicles are empty due to loss of noradrenaline but the large vesicles retain their electron density.

b. Cross-section of varicose fibers from the middle cerebral artery of a control pig. Noradrenergic fibers (NA) have small vesicles with dense cores. Cholinergic (C) small vesicles are clear. Large vesicles (arrows) in both types of varicosities have a dense matrix.

c. Enkephalin-like immuno-reactivity in nerve fibers of the middle cerebral artery remains after 7 days of reserpine.

d. NPY-like immunoreactivity also persists after the present doses of reserpine.

e. A large vesicle (arrow) may be in an early phase of exocytosis in the vas deferens after 3 days of reserpine.

f. A large vesicle (arrow) appears to be fusing with the neurolemma after 14 days of reserpine.

g. Large vesicles remain dense after 28 days of reserpine and may continue to fuse with the neurolemma of catecholamine depleted terminals.

biochemical analyses (HPLC) and the absence of histofluorescent nerve fibers (glyoxylic acid). The persisting matrix of catecholamine depleted large vesicles reflects the presence of several substances including ATP, dopamine β-hydroxylase (Smith 1971) and neuropeptides (Klein and Thureson-Klein 1984; Douglas *et al* 1986), usually enkephalin or NPY. Both of these peptides remain in the middle cerebral artery (Figs. 1c and 1d) and vas deferens after reserpine treatment of the pig (*Sus scrofa*) for at least 7 days using a dose that is sufficient to deplete adrenaline from the adrenal medulla by 70-90% (0.05 mg/kg every other day). Many large vesicles, particularly in the vas deferens, remain peripherally distributed and therefore in position to undergo exocytosis (Figs. 1e-g) (Thureson-Klein *et al* 1986).

The large vesicles in cholinergic nerve terminals (Fig. 1b) have an electron-dense matrix similar to that of the noradrenergic large vesicles, most likely because of their neuropeptide content. This type of fiber is sparse in pig vas deferens, except around the lumen, but frequent in the middle cerebral artery. This distribution corresponds to VIP-like immunoreactive fibers. Therefore, the large cholinergic vesicles probably store VIP which would be in analogy to those in other tissues (Johansson and Lundberg 1981a)

Analyses of particles isolated from the peripheral and central nervous systems indicate that large vesicles are the main storage organelles for neuropeptides (von Euler 1963; Lundberg *et al* 1981; Floor *et al* 1982; Klein *et al* 1982, 1984; Fried *et al* 1985, 1986). The physiological importance of these peptides probably varies with tissues and animal species because there are great differences in the numbers and consequently in the storage capacity of the large vesicle population (Klein and Thureson-Klein 1984; Douglas *et al* 1986) i.e., rodents have few, while ox and pig have high numbers of large vesicles in most peripheral tissues.

However, the low number (5%) of large vesicles in noradrenergic terminals of guinea pig vas deferens may still be important for proper function and can be observed to undergo exocytosis (Thureson-Klein 1983, 1984). In theory, a release of small quantities of NPY from these vesicles may serve to inhibit noradrenaline release thereby maintaining the guinea pig vas deferens quiescent. On the other hand, a massive release of NPY from the many large vesicles present in domestic pig vas deferens could potentiate and sustain muscular contractions for up to the 30 minute period required in this species (Sjöstrand, 1965). With regard to the blood vessels, it is important that the neuron can tailor its release of transmitter and neuropeptide to achieve a physiological rather than a pathological response. For example, NPY potentiation of smooth muscle contraction induced by noradrenaline which may be of benefit in some situations, could lead to a harmful vasospasm in

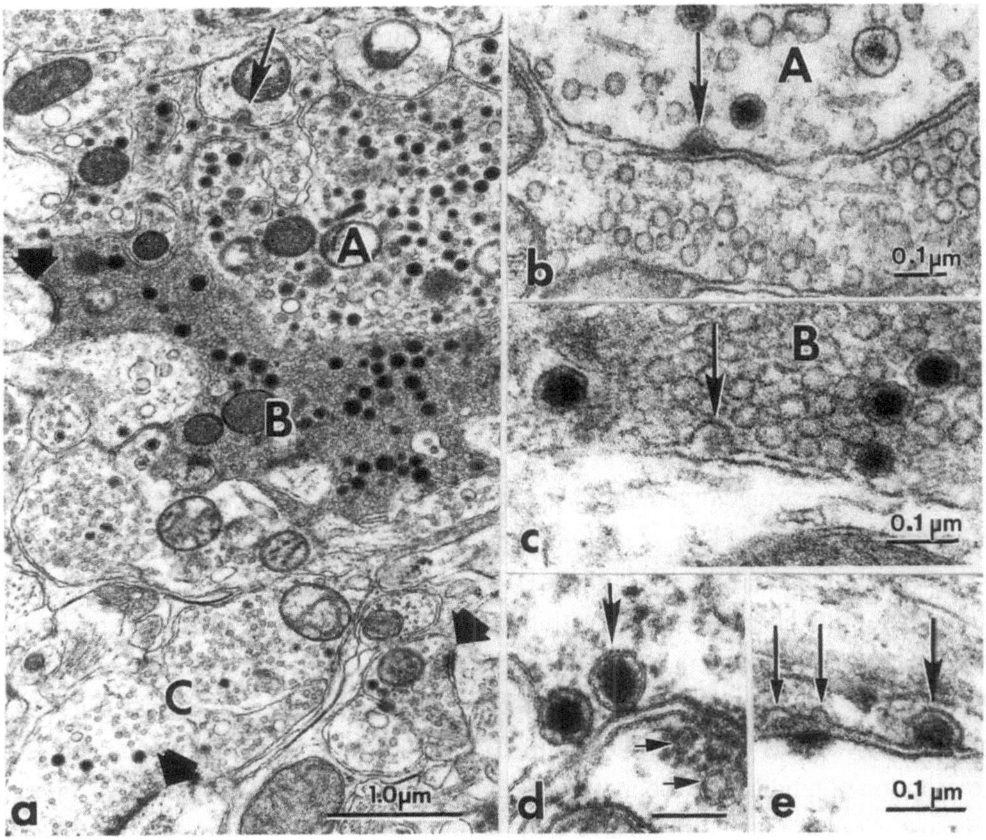

Fig.2 Different categories of terminals and boutons en passant within the neuropil of the ipsilateral rat trigeminal subnucleus caudalis after a lesion in the vibrissae area.

a. Terminals and boutons en passant with large dense cored vesicles in lamina I. One type (A) with small and large vesicles scattered throughout a clear cytoplasm is in close contact with a scallop-shaped terminal (B) that is packed with small and large vesicles giving it a dense appearance. A typical synaptic complex is present in the plane of section (arrow). Most terminals have only a few large vesicles among the many small vesicles (C) and only a few synaptic active zones are present (thick arrows). An omega shaped profile with dense content (arrow) may represent exocytosis.

b. Portion of a terminal (bouton en passant) similar to (A) shows a vesicle releasing its core (arrow) along the structurally non-specialized neurolemma.

c. A terminal similar to (B) with a vesicle undergoing exocytosis (arrow).

d. A large vesicle (large arrow) may be fusing with the neurolemma "postsynaptic" to an active zone where small vesicles (small arrows) have accumulated.

e. Large vesicle (large arrow) exocytosis outside an active zone with two small vesicles (small arrows) undergoing exocytosis.

others (Edvinsson 1985; Allen *et al* 1984, 1986).

Since large vesicles do not loose their neuropeptide content in parallel with noradrenaline depletion during reserpine administration at doses (0.05 mg/kg) sufficient to result in catecholamine depletion in the adrenal chromaffin cells, and since these vesicles appear capable of exocytosis, it is tempting to speculate that release of peptides such as enkephalin and/or NPY continues to play a role in maintaining homeostasis. With excessive doses of reserpine that deplete NPY (Lundberg *et al* 1986), some of the non-specific side-effects including weight loss may involve peptide depletion.

Large dense cored vesicles in the CNS:

Large dense cored vesicles in neurons of the central nervous system also appear to be the main storage organelle for neuropeptides. This belief is based on analyses of isolated particles (Floor *et al* 1982) and immunocytochemical localization at the ultrastructural level (Pelletier *et al* 1981). However, there are very large variations in the populations of large vesicles within neurons in different areas of the brain and spinal cord. One region that is very rich in neurons with large vesicles encompasses the marginal layer and substantia gelatinosa (laminae I and II) of the rat trigeminal subnucleus caudalis (Zhu *et al* 1984, 1986). This portion of the brainstem serves in nociception (Dubner and Bennett 1983) and contains central terminals of bipolar primary afferent neurons that carry sensory impulses from the vibrissae area (Zhu *et al* 1981). In addition second order neurons that relay the signals to higher centra of the brain as well as different types interneurons are present.

Thus, the cyto-architecture of the trigeminal subnucleus caudalis with its many neuronal inter-connections is complex. This is similar to the superficial layers of the dorsal horns in the spinal cord where cutaneous afferent fibers have been shown to form intricate patterns of synaptic contacts (Maxwell and Réthelyi 1987).

In a recent study concerned with the putative role of large vesicles in exocytotic release of neuropeptides, a lesion was placed under the skin in the vibrissae area on one side of the nose of adult rats (Zhu *et al* 1986). At different time intervals after the surgery (1 hr to 4 weeks), the ipsilateral trigeminal subnucleus caudalis was examined by electron microscopy and compared to the corresponding area on the contralateral side. The loss of substance P from the ipsilateral side was followed by immunocytochemical methods. Many categories of terminals with large dense cored vesicles could be distinguished in the neuropil (Figs 2 and 3) which also contained numerous cell bodies of different types, e.g., stalked cells and islet cells. Scallop-shaped terminals packed with small and large vesicles as well as terminals with a

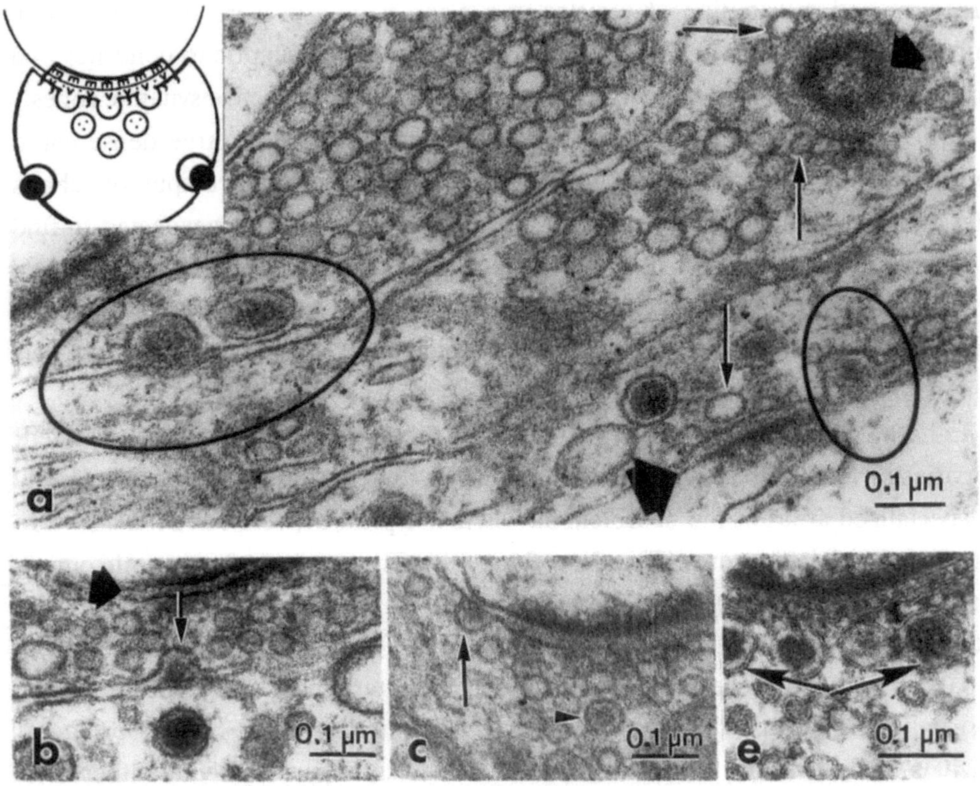

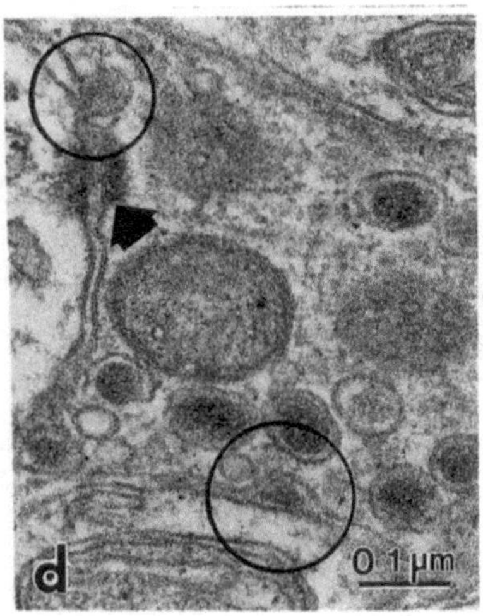

Fig. 3. Diagram and micrographs showing exocytosis from large vesicles in different categories of nerve terminals in the trigeminal subnucleus caudalis.

a. Exocytotic profiles (encircled) at structurally non-specialized portions of the neurolemma either at a distance (left) or close (right) to a typical synaptic complex. Small vesicles (long arrows) are close to the active zones.

b-c. Similar exocytotic profiles (long arrows) at a distance (b) or close (c) to active synaptic zones.

d. Exocytosis from two large vesicles (encircled) in a category of terminal with few, if any, typical active zones. One exocytotic profile is close to dense material along the opposing membranes of an adherent junction (arrow).

e. Large vesicles at an active synaptic zone. Dense material has become evident in the narrow synaptic gap after treatment of the tissue with tannic acid during fixation.

more dispersed distribution of vesicles in a clear cytoplasm were present in both laminae I and II (Fig. 2a). These terminals often were presynaptic to dendrites and axons and usually clusters of small vesicles were present at the active synaptic zones.

It had previously been reported that a selective increase of large dense cored vesicle exocytosis occurred at active synaptic zones in the hippocampus of rabbits undergoing seizures (Nitsch and Rinne 1981). Therefore, it was unexpected to find the reverse situation in the trigeminal subnucleus caudalis of lesioned rats. With a few exceptions (Fig. 3e), large dense cored vesicles were sparse at the pre-synaptic membranes where accumulations of small vesicles were common and exocytosis frequent (Figs. 2e, 4a and c).

In both control and lesioned rats, the large dense cored vesicles in the different categories of terminals appeared to preferentially undergo exocytosis at structurally non-specialized areas of the neurolemma whether typical active synaptic zones were present in the vicinity or not (Figs. 2b-e, 3a-d, 4d). The number of exocytotic omega figures (identified by the dense core material being released into the extracellular space) was significantly increased on the ipsilateral side 1 to 24 hrs after surgery ($p < 0.01$). Because the exocytosis took place from terminals that appeared to be structurally normal and the number of profiles mainly increased on the ipsilateral side, it is likely that the exocytosis reflected changes in sensory input rather than degeneration (Waite and Cragg 1982) or fixation effects (Smith and Reese 1980).

Although an increase in exocytosis within the first day of surgery can be explained by the effects of the lesion on primary sensory terminals and interacting inhibitory and/or amplifying neurons, a second increase in exocytosis 14 days post surgery is difficult to interpret. At this time much of the substance P-like immunoreactivity had been lost, but there was no significant increase in the number of degenerated terminals and/or dendrites on the ipsilateral side.

SMALL VESICLE EXOCYTOSIS AT ACTIVE ZONES VERSUS LARGE VESICLE EXOCYTOSIS AT NON-SYNAPTIC SITES.

Preferential release from small vesicles at active synaptic zones:

The hypothesis that transmitter quanta are released from synaptic vesicles at specialized active zones (Del Castillo and Katz 1957; Couteaux and Pecot-Dechavassine 1970) has gained validity during recent decades even though it is generally accepted that more than one messenger may be released (Zimmermann 1982; Whittaker 1986).

Typical active zones appear to be present where rapid chemical signaling is important, e.g., at the neuromuscular junction, in ganglia and in the brain. It is well known that small vesicles often cluster at the presynaptic membrane where the cytoskeleton forms a lattice (Hirokawa 1983) which may interconnect and entrap the vesicles in an advantageous position for release (Peng 1983). The presence of specific proteins in the membrane of small, but not large, vesicles (Navone et al 1984) may facilitate fusion with the presynaptic membrane. By release into the narrow synaptic cleft, the transmitters are directed to interact rapidly with post-synaptic receptors.

In the trigeminal subnucleus caudalis, small vesicle fusion and release is common (Fig 2e, 4a, c). While the content in the synaptic cleft is relatively electron dense after traditional aldehyde-osmium tetroxide fixation it becomes denser if tannic acid is included (Buma *et al* 1984) and sometimes the presence of a specific substructure is suggested (Fig 3e) which can be speculated to aid in the rapid transfer of molecules.

Infrequent large vesicle exocytosis at active synaptic zones:

Many investigators have noted that large vesicles do not cluster at active synaptic zones (Couteaux and Pecot-Dechavassine 1970; Elfvin 1971; Eldred and Carraway 1987; Peng *et al* 1987). Nevertheless, occasional exocytosis from large vesicles occurs at an active zone (Dickinson-Nelson and Reese 1983; Zhu *et al* 1986) and there may even be circumstances that favor large vesicle release at specialized junctions (Nitsch and Rinne 1981). However, in the trigeminal subnucleus caudalis, only four of nearly 400 exocytotic profiles involving large vesicles were found within active synaptic zones (Fig. 4b), also see Zhu *et al* 1986, although in several instances exocytosis took place close to these membrane specializations (Figs 2e, 3a, c, 4c). It is difficult to accept that the extreme segregation of large dense cored vesicles from the active synaptic zones is accidental rather than functional.

Large vesicle exocytosis at nonsynaptic sites:

At least some neurons in the central nervous system may lack typical chemical synaptic contacts (Baudet and Descarries 1978) and, in that respect, may be similar to the varicose adrenergic and cholinergic fibers innervating vascular smooth muscles. In the trigeminal subnucleus caudalis, the type of terminal with large ovoid to round vesicles (Figs 3d and 4d) appeared to lack or have very few active synaptic zones. This type of terminal sometimes contained more than one exocytotic profile in the same plane of section (Fig 3d), also see Zhu *et al* 1986, which suggests that its large vesicles are important for peptide release.

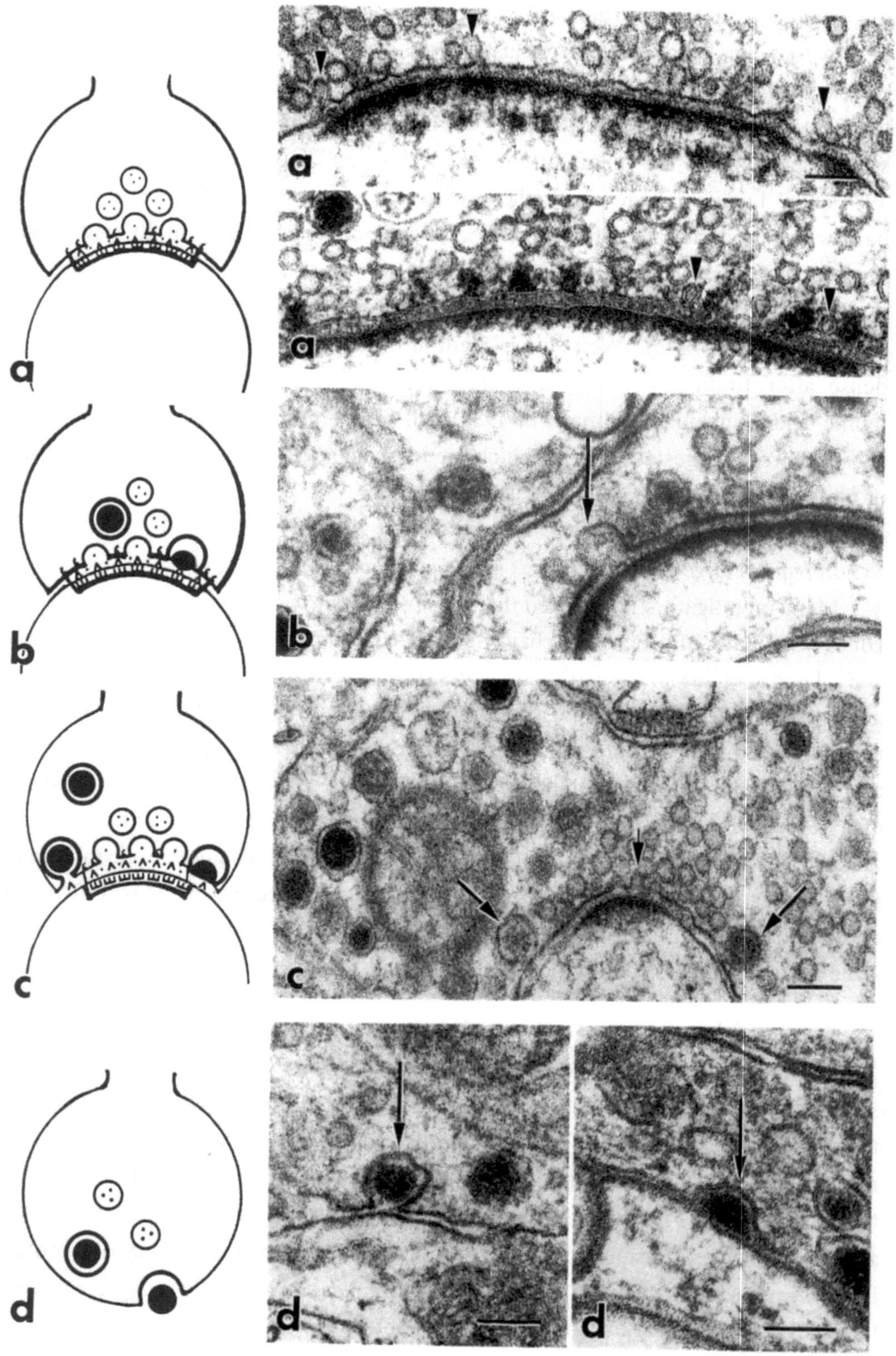
a
a
a
b
b
c
c
d
d
d

Fig. 4. Diagrammatic representations of terminals and the corresponding electron micrographs demonstrating different patterns of exocytosis observed in nerve terminals of the rat trigeminal subnucleus caudalis.

a. The most commonly observed type of exocytosis occurred in terminals with few if any large dense cored vesicles (see diagram and upper panel). After routine aldehyde-osmium fixation followed by uranyl acetate the synaptic complex is delineated by dense postsynaptic material. Small vesicles cluster along the presynaptic membrane. Frequently, fused or open vesicles (arrow heads), can be detected. The content within the synaptic gap is different from the rest of the intracellular space. If tannic acid is included during fixation (lower panel), pre- and postsynaptic densities become more distinct and the material in the synaptic gap appears more structured.

b. The least frequent type of exocytosis involved large vesicles at the presynaptic membrane of typical active zones. In one category of terminal where the size of "small" clear vesicles ranged between 40-70 nm., it was difficult to decide whether large vesicle exocytosis took place at the active zone. Only if some material was present within the releasing vesicle (arrow) was it considered to represent the large dense cored type.

c. Exocytosis from large vesicles at a relatively short distance from the active synaptic zones in a category of terminal with a mixed population of vesicles. While small vesicles (small arrow) underwent exocytosis at the active zone, large vesicles (long arrows) appeared to be excluded from this area.

d. Exocytosis from large vesicles along structurally nonspecialized areas of terminal membranes.

Because the large dense cored vesicles typically undergo exocytosis at structurally non-specialized sites even when typical active synaptic zones are present, they may not be as important as the small synaptic vesicles for rapid communication across a narrow synaptic gap. On the other hand, the large vesicles are likely to be important for presynaptic modulation and autoinhibition by virtue of releasing their contents of peptide(s) and other messenger substances into the intracellular space outside the synaptic complex.

It can be speculated that exocytosis from large vesicles is important for fine regulation of synaptic transmission. It may be one of the mechanisms for non-synaptic transmission observed by neurophysiologists using various types and frequencies of stimulation.

SUMMARY

Many neurons in the peripheral and central nervous systems are multimessenger units that have developed specific storage organelles and characteristic membrane specializations which provide for a differential release of co-existing substances.

Small transmitter-containing vesicles are favored to undergo exocytosis at active synaptic zones developed specifically for rapid chemical signaling, e.g., at the neuromuscular junction and in the central nervous system. Small and large vesicles in neurons concerned with slower tonic responses, e.g., in smooth muscle of blood vessels and vasa deferentia, release their contents by exocytosis at non-specialized sites.

Large neuropeptide and transmitter-containing vesicles appear to undergo exocytosis preferentially at structurally non-specialized sites even when typical active synaptic zones are present. Released substances can diffuse over a wider area than a narrow synaptic gap would allow, which may be essential for pre- and postsynaptic modulation. Under normal circumstances the release of neuropeptides and other substances from large vesicles may be important for "fine tuning" (blood vessel) or for keeping a tissue quiescent (vas deferens). In the case of transmitter depletion, e.g., of catecholamines by reserpine, co-existing neuropeptides may continue to function.

After intense stimulation under stress, the marked increase in release of neuropeptides can amplify or limit the actions of classical transmitters, depending on the type of tissue and messengers involved, and play a critical role in the homeostatic response.

ACKNOWLEDGEMENTS

The support from the American Heart Association, Mississippi Affiliate for portions of this study is gratefully acknowledged. The authors wish to thank Ms. J. Yang and Mr. T. Selva for skillful technical assistance.

REFERENCES

Allen JM, Bircham PMM, Bloom SP and Edwards AV (1984) Release of neuropeptide Y in response to splachnic nerve stimulation in the conscious calf. J Physiol 357: 401-408

Allen JM, Gjörstrup P, Björkman J-A, Ek L, Abrahamsson T and Bloom SR (1986) Studies on cardiac distribution and function of neuropeptide Y. Acta Physiol Scand 126: 405-411

Bartfai T (1985) Presynaptic aspects of the coexistence of classical neurotransmitters and peptides. Trends in Pharmacol Sci 6: 631-634

Baudet A and Descarries L (1978) The monoamine innervation of rat cerebral cortex. Synaptic and nonsynaptic axon terminals. Neuroscience 3: 851-850.

Burnstock C (1986) The changing face of autonomic neurotransmission. Acta Physiol Scand 126: 67-91

Buma P, Roubos EW and Buijs RM (1984) Ultrastructural demonstration of exocytosis of neuronal endocrine and exocrine secretions with an in vitro tannic acid (TARI) method. Histochemistry 80: 247-256

Campbell G (1987) Cotransmission. Ann Rev Pharmacol Toxicol 27: 51-70

Couteaux R and Pecot-Dechavassine M (1970) Vesicules synaptiques et poches au niveau des "zones actives" de la jonction neuromusculaire. C R Seances Acad Sci D 271: 2346-2347

Del Castillo J and Katz B (1957) La base "quantale" de la transmission neuromusculaire, Colloq. Int. CNRS 67:245-258

Del Fiacco M and Cuello AC (1980) Substance P and enkephalin-containing neurons in the rat trigeminal system. Neuroscience 5: 803-815

De Potter WP, Coen EP and De Potter RW (1987) Evidence for the coexistence and co-release of (met)enkephalin and noradrenaline from sympathetic nerves of the bovine vas deferens. Neuroscience 20: 855-866

Dickinson-Nelson A and Reese TS (1983) Structural changes during transmitter release at synapses in the frog sympathetic ganglion. J Neurosci 3: 42-52

Douglas BH, Duff RB, Thureson-Klein AK and Klein RL (1986) Enkephalin contents reflect noradrenergic large dense cored vesicle populations in vasa deferentia. Regulatory Peptides 14: 193-210

Dubner R and Bennett GJ (1983) Spinal and trigeminal mechanisms of nociception. Ann Rev. Neurosci 6:381-418

Edvinsson L (1985) Characterization of the contractile effect of neuropeptide Y in feline cerebral arteries. Acta Physiol Scand 125: 33-41

Edvinsson L, Håkanson R, Wahlestedt C. and Uddman (1987) Effects of neuropeptide Y on the cardiovascular system. Trends in Pharmacol Sci 8: 231-235

Ekblad E, Edvinsson L, Wahlestedt C, Uddman R, Håkanson R and Sundler F (1984) Neuropeptide Y co-exists and co-operates with noradrenaline in perivascular nerve fibers. Regulatory Peptides 8: 225-235

Eldred WD and Carraway RE (1987) Neurocircuitry of two types of neurotensin containing amacrine cells in the turtle retina. Neuroscience 21:603-618

Elfvin L (1971) Ultrastructural studies on the Synaptology of the inferior mesenteric ganglion of the cat. I. J. Ultrastruct Res 37:411-425

Floor E, Grad O and Leeman SE (1982) Synaptic vesicles containing substance P purified by chromatography on controlled pore glass. Neuroscience 7: 1647-1655

Fried G, Thureson-Klein AK, and Lagercrantz H (1981) Noradrenaline content correlated to matrix density in small noradrenergic vesicles from the rat seminal ducts. Neuroscience 6: 787-800

Fried G, Lundberg JM and Theodorsson-Norheim E (1985) Subcellular storage and axonal transport of neuropeptide Y (NPY) in relation to catecholamines in the cat. Acta Physiol Scand 125: 145-154

Fried G, Terenius L, Brodin E, Efendic S, Dockray G, Fahrenkrug J, Goldstein M and Hökfelt T (1986) Neuropeptide Y, enkephalin and noradrenaline coexist in sympathetic neurons innervating the bovine spleen; Biochemical and immunohistochemical evidence. Cell Tissue Res 243: 495-508

Hirokawa N (1983) Membrane specialization and cytoskeletal structures in the synapse and axon revealed by the quick-freeze deep-etch method. In: Chang D, Tasaki I, Adelman WJ and Leuchtag HR (eds.). Structure and function in excitable cells. Plenum Press 1983 New York: 113-141

Hökfelt T and Terenius L (1987) More on receptor mismatch. Trends in Neurosci 10: 22-23

Hökfelt T, Lundberg JM, Lagercrantz H, Tatemoto K, Mutt V, Lindberg J, Terenius L, Everett B, Fuxe K, Agnati L and Goldstein M (1983) Occurrence of neuropeptide Y (NPY)-like immunoreactivities in catecholamine neurons in the human medulla oblongata. Neurosci Lett 36: 217-222

Johansson O and Lundberg JM (1981a) Ultrastructural localization of VIP-like immunoreactivity in large dense cored vesicles of "cholinergic type" terminals in cat exocrine glands. Neuroscience 6: 847-862

Johansson O, Hökfelt T, Pernow B, Jeffcoate SL, White N, Steinbusch HWM, Verhofstad AAJ, Emson PC and Spindel E (1981b) Immunohistochemical support for three putative transmitters in one neuron: coexistence of 5-hydroxytryptamine, substance P and thyrotropin releasing hormone-like immunoreactivity in medullary neurons projecting to the spinal cord. Neuroscience 6: 1857-1881

Klein RL and Thureson-Klein ÅK (1984) Noradrenergic vesicles. In: Lajtha A (ed) Handbook of Neurochemistry Vol. 7, Plenum Press 1984 New York: 71-107

Klein RL, Lemaire S, Thureson-Klein ÅK and Day R (1984) Leu-enkephalin, dynorphin and bombesin contents of a highly purified large dense cored noradrenergic vesicle fraction from bovine splenic nerve. In: Fraiolo R, Isidori A and Mazetti M (eds.) Opioid peptides in the periphery. Elsevier 1984 Amsterdam: 205-212

Klein RL, Wilson SP, Dzielak DJ, Yang W-H, and Viveros OH (1982) Opioid peptides and noradrenaline co-exist in large dense cored vesicles from sympathetic nerves. Neuroscience 7: 2255-2261

Le Gray C, Saffrey M and Burnstock G (1984) Coexistence of immunoreactive substance P and serotonin in neurons of the gut. Brain Res 302:379-382

Lundberg JM, Fried G, Fahrenkrug J, Holmstedt B, Hökfelt T, Lagercrantz H, Lundgren G and Änggård A (1981) Subcellular fractionation of cat submandibular gland: Comparative studies on the distribution of acetylcholine and vasoactive intestinal polypeptide (VIP) Neuroscience 6: 1001-1010

Lundberg JM, Saria A, Hökfelt T, Franco-Cereceda A and Terenius L (1985) Tissue-specific depletion of NPY-like immunoreactivity by reserpine. Acta Physiol Scand 123:363-365

Lundberg JM, Rudehill A, Sollevi A, Theodorsson-Norheim E and Hamberger B (1986) Frequency- and reserpine-dependent chemical coding of sympathetic transmission: differential release of noradrenaline and neuropeptide Y from pig spleen. Neuorsci Lett 63: 96-100

Maxwell DJ and Réthelyi M (1987) Ultrastructure and synaptic connections of cutaneous afferent fibers in the spinal cord. Trends in Neuro Sci 10: 117-123

Navone F, Greengard P and De Camilli P (1984) Synapsin I in nerve terminals: selective association with small synaptic vesicles. Science 226: 1209-1211

Nitsch C and Rinne U (1981) Large dense-core vesicle exocytosis and membrane recycling in the mossy fibre synapses of the rabbit hippocampus during epileptiform seizures. J of Neurocyt. 10:201-219

Pelletier G, Steinbusch HWM and Verhofstad AAJ (1981) Immunoreactive substance P and serotonin present in the same dense-cored vesicles. Nature 293: 71-72

Peng HB (1983) Cytoskeletal organization of the presynaptic nerve terminal and the acetylcholine receptor cluster in cell cultures. J Cell Biol 97: 489-498

Peng HB, Markey DR, Muhlach WL and Pollack ED (1987) Development of presynaptic specializations induced by basic polypeptide-coated latex beads in spinal cord cultures. Synapse 1: 10-19

Schmitt FO (1984) Molecular regulators of brain function: a new view. Neuroscience 13: 991-1001

Sjöstrand NO (1965) The adrenergic innervation of the vas deferens and accessory male genital glands. Acta Physiol Scand 65 (Suppl 257) 1-82

Smith AD (1971) Summing up: some implications of the neuron as a secreting cell. Phil Trans R Soc Lond 261: 423-437

Smith JE and Reese TS (1980) Use of aldehyde fixation to determine the rate of synaptic transmitter release. J Exp Biol 89: 19-29

Stjärne L, Lundberg JM and Åstrand P (1986) Neuropeptide Y - A cotransmitter with noradrenaline and adenosine 5'-triphosphate in the sympathetic nerves of the mouse vas deferens? A biochemical, physiological and electropharmacological study. Neuroscience 18: 151-166

Thureson-Klein ÅK (1983) Exocytosis from large and small dense cored vesicles in noradrenergic nerve terminals. Neuroscience 10: 245-252

Thureson-Klein ÅK (1984) The roles of small and large noradrenergic vesicles in exocytosis. In: Usdin E (ed.) Catecholamines basic and peripheral mechanisms. Alan R. Liss 1984 New York: 79-87

Thureson-Klein ÅK, Klein RL, and Zhu PC (1986) Exocytosis from large dense cored vesicles as a mechanism for neuropeptide release in the peripheral and central nervous system. Scanning Electr Micr I: 179-187

Thureson-Klein ÅK, Zhu PC and Klein RL (1987) Nonsynaptic exocytosis from large dense cored vesicles as a mechanism for non-directional neuropeptide release. In: Nobin A, Owman C and Arneklo-Nobin A (eds.) Neuronal messengers in vascular function. Elsevier Sci Publ 1987 Amsterdam: 211-217

von Euler US (1963) Substance P in subcellular particles in peripheral nerves. Ann NY. Acad Sci 104: 449-461

Waite PME and Cragg BG (1982) The peripheral and central changes from cutting or crushing the afferent nerve supply to the whisker. Proc R Soc Lond 214: 191-211

Whittaker VP (1986) The storage and release of acetylcholine. Trends in Pharmacol Sci 7: 312-315

Zhu PC, Wu HX and Xu H (1981) Somatotropic projections of the cutaneous facial regions of the substantia gelatinosa of the rat. An acid phosphatase study. Acta anat sin 12: 155-161

Zhu PC, Xu H, Tang YP and Wu HX (1984) The ultrastructure of the substantia gelatinosa glomeruli in the spinal trigeminal nucleus of the rat. Acta anat sin 15: 168-173

Zhu PC, Thureson-Klein Å and Klein RL (1986) Exocytosis from large dense cored vesicles outside the active synaptic zones of terminals within the trigeminal subnuclens caudalis: A possible mechanism for neuropeptide release. Neuroscience 19: 43-54

Zimmermann H (1982) Insights into the functional role of cholinergic vesicles. In: Klein RL, Lagercrantz H and Zimmermann H (eds.) Neurotransmitter vesicles 1982 Academic Press, London: 305-360

THE DETECTION, ISOLATION AND PROPERTIES OF SUB-POPULATIONS OF MAMMALIAN BRAIN SYNAPTOSOMES OF DEFINED NEUROTRANSMITTER TYPE

Henry F. Bradford, Maureen Docherty, Tong H.Joh* and Yang-Yen Wu** Department of Biochemistry, Imperial College, London SW7 2AZ, U.K. *Laboratory of Neurobiology, Cornell University Medical College, New York, NY 10021; **Department of Physiology, Pennsylvania State University College of Medicine, Hershey, P.A.17033 (USA)

Abbreviations ChAT, choline acetyltransferase; GAD, glutamate decarboxylase; TH, tyrosine hydroxylase; DBH, dopamine-β-hydroxylase; TPH, tryptophan hydroxylase

INTRODUCTION

Mammalian brain synaptosomes are derived from many populations of neurons each employing a different neurotransmitter. The isolation of highly purified subpopulations of mammalian synaptosomes of specific neurotransmitter type would greatly enhance the value of synaptosome preparations, particularly for studies of presynaptic control of neurotransmitter release, secretion-synthesis coupling, and the co-existence and co-release of neurotransmitters and/or neuropeptides. The detection of specific outer surface markers for sub-categories of synaptosomes would form a basis for a method of separation. Our own recent studies of the ability of antisera recognising biosynthetic enzymes to cause complement-mediated immunolysis of the appropriate synaptosome subpopulation have revealed the presence of such specific surface marker antigens (Docherty et al 1985, 1985a, 1986). These appear to be closely related to a particular key biosynthetic enzyme for the neurotransmitter system involved, and are likely to be the particular enzyme itself in a membrane bound-form (Badamchian et al 1986; Barochovsky et al 1986; Benishin and Carroll 1983; Docherty and Bradford 1987). Pure cholinergic synaptosomes have been isolated from electric organ of Torpedo (Morel et al 1987) and greatly enriched cholinergic synaptosomes have been prepared from mammalian brain (Richardson et al 1984), but no successful method has previously been described which can separate other neurotransmitter-related subtypes of synaptosome. However, using magnetic microspheres (Magnogel)

NATO ASI Series, Vol. H21
Cellular and Molecular Basis of Synaptic Transmission
Edited by H. Zimmermann

coupled to Protein A we have recently found it possible to prepare highly purified and metabolically viable GABAergic, cholinergic and serotonergic synaptosomes from mammalian cerebral cortex, using immunoglobulins recognising glutamate decarboxylase (GAD), choline acetyltransferase (ChAT), and tryptophan hydroxylase (TPH) respectively. These synaptosomes display Ca-dependent neurotransmitter release and are proving excellent preparations for studying the dynamics of neurotransmitter synthesis, release, coexistence and co-release in specific neuronal systems.

SPECIFIC IMMUNOLYSIS OF SYNAPTOSOME SUBPOPULATIONS

Cholinergic Synaptosomes In 1981 during a search for specific outer surface markers for synaptosome sub-populations of particular neurotransmitter category, we discovered that ChAT, or a very similar protein, was present on the outer membrane surface of cholinergic synaptosomes prepared from mammalian brain. This was detected by demonstrating complement-mediated immunolysis of these synaptosomes during their incubation in physiological medium. The criteria for specificity of immunolysis was release of soluble ChAT, and loss of capacity to take up choline by high affinity transport processes (Docherty et al 1982). It

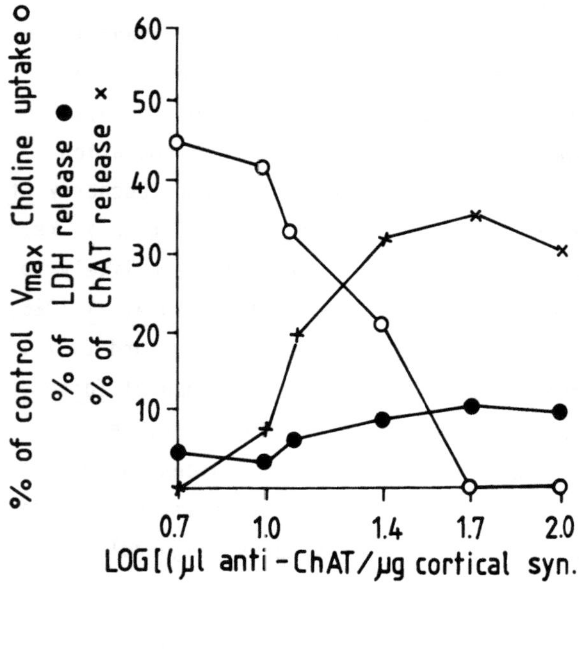

Fig.1.
Specific immunolysis of the cholinergic subfraction of cortical synaptosomes with antiserum to ChAT. A lysis period of 30 min at 37°C produced the maximal effect. Values are mean for 3 - 7 preparations. Experimental details are given in Docherty et al 1985.

was found that the immunolysis was dose-dependent, and on increasing the titre of anti-ChAT antibodies, the extent of synaptosomolysis became maximal as judged by release of LDH and ChAT and inhibition of choline transport. Whereas only approximately 10% of the LDH could be released, 30-40% of ChAT was released and Na^+-dependent choline uptake could be reduced to zero in cortical synaptosomes, and to 20-30% in striatal preparations (Fig 1). There was no release of soluble markers for other synaptosome subtypes, such as GAD or TH, nor loss of transport of dopamine, serotonin or other neurotransmitters tested. Moreover, preimmune, and nonimmune sera were without lytic action, as were complement added alone or antibody added alone. The lysis could be shown to be due to the immunoglobulins present in the sera. Thus, it seemed clear that ChAT antibodies were causing specific complement-mediated immunolysis of cholinergic synaptosomes, indicating the presence of this enzyme, or a closely similar protein, on the outer surface of these synaptosomes.

GABAergic Synaptosomes To test whether biosynthetic enzymes for neurotransmitters, or closely related proteins, might be present on the outer surface of categories of nerve terminal other than the cholinergic type, we investigated whether anti-GAD serum would specifically immunolyse GABAergic synaptosomes (Docherty et al 1983). In the event, both cortical and striatal GABAergic terminals were lysed (Fig 2). In fact some 40% of total cortical, and nearly 30% of total striatal, LDH was released from the whole synaptosome preparations, indicating the probable sizes of the GABAergic subpopulations. Soluble GAD and GABA (Fig 2) were released in parallel in substantial proportions without loss of other enzymes (e.g. ChAT) or neurotransmitters (e.g. monoamines) or loss of choline transport (Docherty et al 1985a).

Immunolysis of Monoaminergic Synaptosomes Specific complement-mediated immunolysis of dopaminergic synaptosomes present in striatal preparations was achieved using tyrosine hydroxylse antibodies. This produced release of TH, LDH and dopamine (Fig 3) with accompanying loss of capacity to transport (^{3}H)-dopamine, without loss of other biosynthetic enzymes or neurotransmitters, or loss of choline or other transport capacity (Docherty et al 1985a). Similarly, we have demonstrated specific immunolysis of serotonergic synaptosomes in striatal preparations by employing antisera to TPH (Docherty et al 1985). This produced substantial and specific release of TPH, and 5HT with loss of 18% of total LDH and 60% of Na-dependent 5HT uptake. Corresponding parameters for other neurotransmitters were unchanged (Fig 4).

Antibodies to DBH caused specific immunolysis of noradrenergic terminals in cortical synaptosome preparations (Docherty et al 1986) with specific release of noradrenaline, DBH and loss of Na-dependent (^{3}H) noradrenaline uptake (Fig 5).

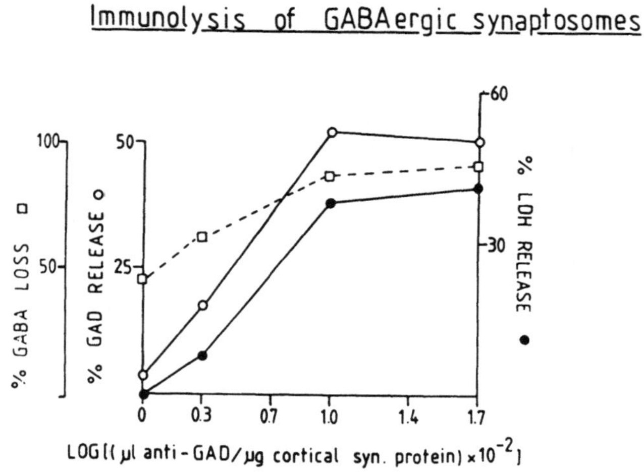

Fig.2.

Immunolysis of the GABAergic subfraction of cortical synaptosomes with antiserum to GAD. Values are mean for 3 separate preparations. Experimental details are given in Docherty et al 1985.

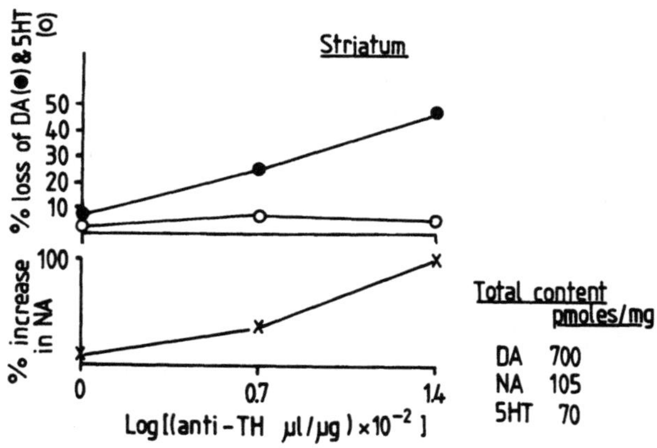

Fig.3.

Immunolysis of the dopaminergic subfraction striatal synaptosomes with antiserum to tyrosine hydroxylase (TH). Release of dopamine (DA) with increasing titre of antiserum. The lower graph shows the parallel increase in noradrenaline (NA) content of striatal synaptosomes which is due to uptake and hydroxylation of dopamine released by immunolysis.

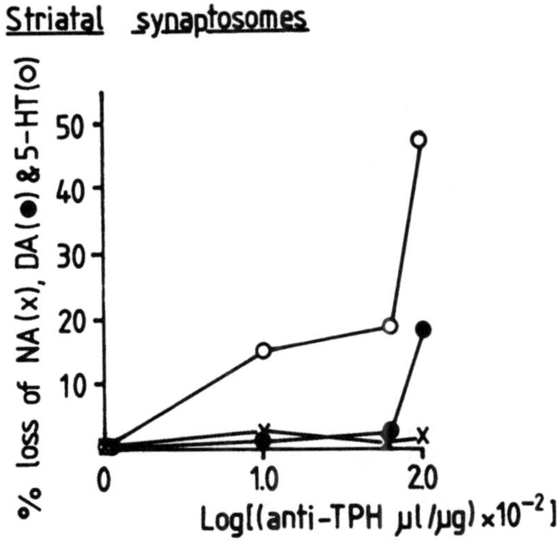

Fig.4.

Immunolysis of the serotonergic subfraction of striatal synaptosomes with antiserum to tryptophan hydroxylase (TPH). Release of 5-HT is specific except at high titres of TPH serum where dopamine is partly released.

Population sizes of synaptosome sub-populations inferred from immunolysis The maximal LDH release caused by these immunolysis procedures provides a useful estimate of the likely contribution of each synaptosome sub-population to the whole cortical or striatal synaptosome preparation, thereby reflecting the relative population densities of these terminals in the two brain regions. Thus GABAergic terminals seen to be by far the most abundant (Table 1). The effects of the separate antisera were found to be additive.

Interestingly, total LDH release due to lysis of the five categories of terminal so far investigated amounts to 66% in cortical preparations and 59% in striatal preparations (Table 1). The remaining 30 plus % must accommodate all other terminal subpopulations remaining unlysed, including glutamatergic, aspartatergic, peptidergic, histaminergic, purinergic as well as other contenders for a neurotransmitter role in these brain regions. No doubt some of these systems (including possibly glutamatergic, see below) will prove to coexist with one or more of the five neurotransmitters of Table 1 as appears to be the case for many neuropeptides.

The fact that 50% or less of many of the biosynthetic enzymes are released by the immunolysis procedures (e.g. 27 - 34% ChAT; 22 - 51% GAD; 28% of TH; Table 1) is most likely due to: (a) binding of these enzymes to the particulate fraction after release (b) the existence of a substantial fraction (e.g. 20%) in the membranes as constituents (c) incomplete lysis due to resealing processes after the action of complement.

Table 1 **Contributions of Synaptosome Subpopulations inferred from Immunolysis**

CEREBRAL CORTEX

	% Release by Immunolysis								% Reduction in Na^+-dependent high-affinity uptake due to Immunolysis		
	LDH	ChAT	GAD	TH	GABA	DA	5HT	NA	Choline	GABA	NA
Anti-ChAT	10%	34%	0%	n.d.	0%	0%	0%	0%	100%	0%	0%
Anti-GAD	41%	0%	51%	n.d.	91%	0%	0%	0%	0%	0%	0%
Anti-DBH	15%	0%	0%	20%	n.d.	0%	0%	40%	0%	n.d.	55%
Anti-TH	0%	0%	n.d.	n.d.	n.d.	0%	0%	0%	0%	n.d.	0%
Anti-GDH	0%	0%	n.d.	n.d.	0%	0%	0%	0%	0%	n.d.	0%
Anti-GLN'ASE	0%	0%	n.d.	n.d.	0%	0%	0%	0%	0%	n.d.	0%
TOTAL	66%										

CORPUS STRIATUM

	% Release by Immunolysis									% Reduction in Na^+-dependent high-affinity uptake due to Immunolysis				
	LDH	ChAT	GAD	TH	TPH	GABA	DA	5HT	NA	Choline	GABA	NA	DA	5HT
Anti-ChAT	10%	27%	0%	0%	0%	0%	0%	0%	0%	70%	0%	0%	0%	n.d.
Anti-GAD	28%	0%	22%	n.d.	n.d.	83%	0%	0%	0%	0%	0%	n.d.	0%	n.d.
Anti-TH	13%	0%	0%	28%	n.d.	0%	45%	0%	*	0%	n.d.	n.d.	50%	n.d.
Anti-TPH	18%	0%	0%	0%	70%	0%	18%	47%	2.5%	44%	n.d.	n.d.	n.d.	67%
TOTAL	69%													

* no release/100% increase in levels; n.d. not determined.

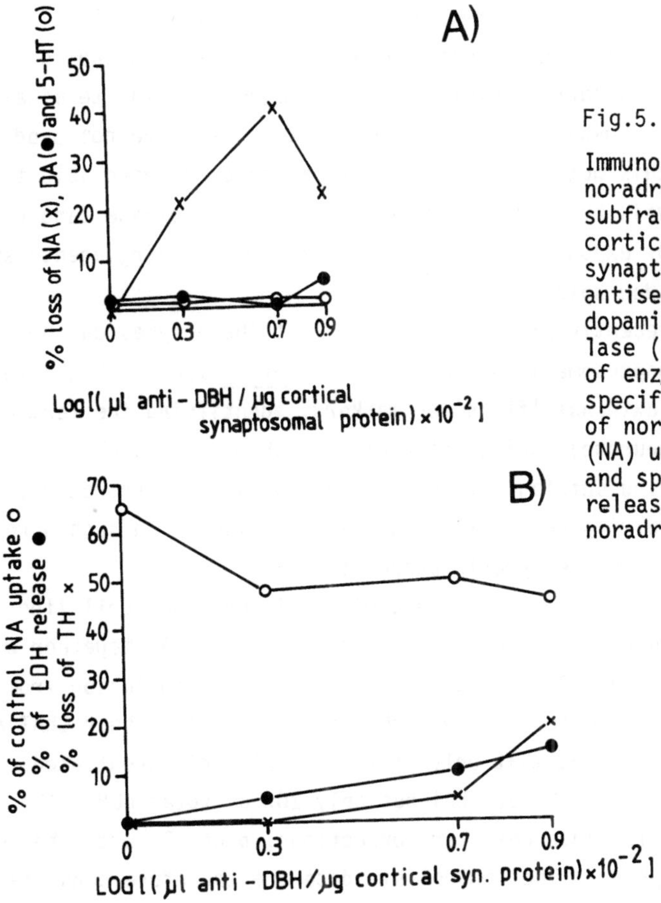

Fig.5.

Immunolysis of the noradrenergic subfraction of cortical synaptosomes with antiserum to dopamine-β-hydroxylase (DBH). Release of enzymes and specific inhibition of noradrenaline (NA) uptake (A) and specific release of noradrenaline (B).

WHY ARE NEUROTRANSMITTER BIOSYNTHETIC ENZYMES PRESENT ON THE SURFACE OF SPECIFIC SYNAPTOSOME SUBPOPULATIONS.

Our own early notion was that perhaps the enzymes were bound artefactually to the outer synaptosome surface following rupture of the various neurones during tissue homogenisation to prepare synaptosomes. Our present view is that they are likely to be integral membrane constituents.

It is difficult to account for the observed specificity of synaptosome immunolysis if the enzyme were simply binding ionically to membrane surfaces in general following tissue homogenisation. There is no clear reason why, say ChAT should bind specifically to cholinergic synaptosomes. In addition, extensive washing of synaptosomes with salt solutions prior to immunolysis did not prevent or reduce the extent of immunolysis (Docherty et al 1985a).

The attractive idea, that exocytosis-endocytosis cycles of synaptic vesicles in nerve terminals carry vesicle-attached enzyme into the terminal plasma membrane remains a possibility. However, to date, we have not produced evidence in support of this notion. For instance, we have not been able to enhance the degree of lysis by stimulating neurotransmitter release (and presumably exocytosis rates) by exposure of synaptosomes to depolarizing agents such as veratrine or black widow spider venom.

We have, however, recently produced evidence that the enzymes can be integral proteins in synaptosome membranes (Docherty et al 1987). Their role in this capacity may be as external 'flags' or markers, identifying the terminals (and perhaps the whole neurone) during interaction with other neural cells such as those which occur during ontological development, growth, regeneration, and other plastic events involving neuronal surface interactions. In such a role they are unlikely to use their enzymatic properties.

It has proved possible to immunolyse a cholinergic neuronal cell line (NS20Y) in tissue culture using antibodies directed against ChAT together with complement (Barochovsky et al 1986). Whereas the simplest conclusion from these findings is that cell bodies of these cholinergic neurones bear ChAT as a surface antigen, our data showed that only plates carrying differentiated neurones would lyse, releasing 70% of ChAT but only 20% of total LDH. These differentiated cells develop intercellular connections, probably with the development of organised synapses. The immunolysis data are therefore consistent with the interpretation that lysis is occurring mainly at nerve-endings, where soluble ChAT would be concentrated.

THE ISOLATION OF SYNAPTOSOME SUBPOPULATIONS

Independent evidence for the existence of the biosynthetic enzymes or their congeners as external surface markers on specific synaptosome subpopulations has come from our finding that it is possible to isolate specific synaptosome subtypes by immunoaffinity methods.

GABAergic and cholinergic synaptosomes Magnetic microspheres (Magnogel) consisting of agarose (4%) and polyacrylamide (4%) forming an envelope around ferric oxide particles (7%), and coupled to Protein A, were mixed with whole synaptosome fractions which had previously been incubated with antibodies to ChAT or GAD to allow binding of the IgG to the cholinergic or GABAergic subpopulations respectively. The Magnogel was then sedimented using a horse-shoe magnet and washed by resuspension in the magnetic field (Docherty et al 1987a).

When cortical synaptosomes were incubated with anti-ChAT serum the synaptosome subfraction subsequently separated by attachment to the affinity gel showed considerable enrichment in ChAT content (27-fold; Table 2 and acetylcholine) content (20 fold) in relation to GAD content. At the same

content per fraction	whole preparation	non-specific adsorption	cholinergic synaptosomes	GABAergic synaptosome
Chat pmol/min	4096 ± 397 (7)	32 ± 8 (14)	1218 ± 209 (14)	46 ± 10 (7)
GAD pmol/min	1297 ± 132 (7)	17 ± 6 (7)	14 ± 4 (5)	304 ± 40 (7)
GABA pmol	16321 ± 1721 (7)	n.d. (8)	326 ± 40 (7)	4162 ± 506 (8)
glutamate pmol	52885 ± 6872 (7)	261 ± 53 (8)	2609 ± 416 (7)	8631 ± 921 (8)
noradrenaline pmol	806 ± 72 (7)	n.d. (5)	n.d. (5)	n.d. (5)
LDH nmol/min	2463 ± 300 (7)	4 ± 2 (14)	84 ± 11 (14)	273 ± 40 (8)
choline transport pmol ^{3}H-choline/2 min	42 ± 3 (4)	n.d. (4)	12 ± 3 (4)	n.d. (2)
Purification factors				
ChAT/LDH	1.7	-	14 (↑x8)	0.2 (↓x8.5)
ChAT/GAD	3.2	-	87 (↑x27)	0.1 (↓x32)
GAD/LDH	0.5	-	0.2 (↓x2.5)	1.1 (↑x2.2)
GAD/ChAT	0.3	-	0.01(↓x30)	6.6 (↑x22)
GABA/LDH	6.6	-	3.9 (↓x1.7)	15 (↑x2.3)
GABA/ChAT	4.0	-	0.2 (↓x20)	90 (↑x22)
glut/LDH	21	-	31 (↑x1.5)	32 (↑x1.5)
glut/ChAT	13	-	2.1 (↓x6.2)	188 (↑x14)

Table 2.

Affinity purification of synaptosomal subpopulations from rat cerebral cortex. Cholinergic and GABAergic synaptosomes were prepared as described in the text using 125μl Magnogel conjugated to Protein A with an incubation time of 45 min. Values given are mean ± S.D. Numbers of synaptosome batches examined are given in parentheses and involved at least 5 separate preparations. ↑ = increase; ↓ = decrease; glut = glutamate, n.d. = not detectable.

time GABA content was much reduced in relation to ChAT content, but no monoamines could be detected (Table 2). The proportion of total ChAT (present in the mother fraction) which was removed by multiple passages of Magnogel was about 58%. When the purified cholinergic synaptosomes attached to Magnogel were exposed to complement, as dscribed previously (Docherty et al 1985a) 42% of the ChAT present was released together with 86% of the LDH. No ChAT release was detected when heat - inactivated complement was used. When Triton X - 100 (1%) was added to the affinity-gel 75% of ChAT was released. This shows that the ChAT present on the affinity gel was largely contained in intact, immunolysable, synaptosomes, or their membranes, and did not represent free ChAT linked to the gel via antibodies to ChAT (Docherty et al, 1987a).

When cortical synaptosomes were incubated with anti-GAD serum the synaptosomes which became attached to the affinity gel were found to be substantially enriched in GAD (22 fold) and GABA (22 fold) in relation to ChAT content. Co-enriched with GABA was glutamate (1.4 fold). Both glutamate and GABA showed a much increased content in relation to other amino acids measured. This purification of GABAergic markers was paralleled by a diminished content of cholinergic markers, (ACh and ChAT) and monoamines (Table 2). Some 40% of total GAD was recovered in the GABAergic synaptosome subfraction after multiple passages of Magnogel, nearly all of which was contained within intact synaptosomes.

These results demonstrate that approximately 1mg (synaptosomal protein as estimated by LDH content) of substantially purified cholinergic (from 30mg protein whole synaptosome fraction) or GABAergic (from 10mg whole fraction) synaptosomes can be prepared by simple affinity-gel procedures.

Using anti-TPH bound to striatal synaptosomes it has also proved possible to isolate apparently pure serotonergic synaptosomes. These have a greatly enriched content of serotonin, other neurotransmitters (e.g. GABA, dopamine) being undetectable (Table 3).

After correction for non-specific adsorption, cholinergic synaptosomes were estimated to consitute 5 - 10% of the whole synaptosome fraction from cerebral cortex based on LDH content. Similar calculations for GABAergic and serotonergic (from corpus striatum) synaptosomes indicate contributions of 40% and 20% respectively. These figures correlate well with our previous estimates of these subfraction sizes based on the extent of complement-mediated immunolysis of whole cortical synaptosome preparations (Table 1).

Table 3.

Content per fraction	Striatal synaptosomes: Whole prep	Non-specific binding	Serotonergic Synaptosomes
5HT (pmol)	347 ± 50 (6)	n.d. (6)	132 ± 40 (6)
DA (pmol)	3100 ± 297 (6)	43 ± 14 (6)	59 ± 12 (6)
NA (pmol)	516 ± 72 (6)	n.d. (6)	n.d. (6)
LDH (nmol/min)	2632 ± 413 (9)	10 ± 6 (9)	150 ± 31 (9)
ChAT (pmol)	7693 ± 821 (3)	69 ± 12 (3)	76 ± 13 (3)
Glutamate (pmol)	73562 ± 893 (6)	301 ± 46 (6)	797 ± 89 (6)
GABA (pmol)	14097 ± 1932 (6)	297 ± 53 (6)	316 ± 70 (6)

The results are from 3 separate synaptosome preparations, 3 batch separations per preparation.

Table 3.

PROPERTIES OF ISOLATED SUBPOPULATIONS OF BRAIN SYNAPTOSOMES

Respiratory properties Oxygen electrode measurements made on the purified synaptosome subpopulations attached to Magnogel demonstrated that oxygen-uptake rates were linear over at least one hour in the presence of glucose (Fig 6). Thus the subfractions were metabolically viable, since their respiration was stimulated by glucose, indicating that the plasma membranes of the synaptosomes were intact, and glycolysis was fully coupled to mitochondrial oxidation. In addition, oxygen uptake by the synaptosomes was completely inhibited by azide (0.2% w/v).

Compositional Properties The enzyme and neurotransmitter content of the cholinergic, GABAergic and serotonergic subpopulations are given in Tables 2 and 3. The acetylcholine GABA and other amino acid content of the former two preparations is given in Table 4. When compared with the mother fraction, the cholinergic subfraction was found to contain both elevated aspartate and glutamate levels (approximately 4-fold higher) compared with the 'metabolic amino acids' e.g. threonine or serine (Table 4). The GABAergic subfraction also showed raised aspartate (2 - 3 fold higher) and glutamate levels (6 - 9 fold higher on this basis). When these figures were expressed in terms of the LDH content of each subfraction (i.e. an index of the number of synaptosomes

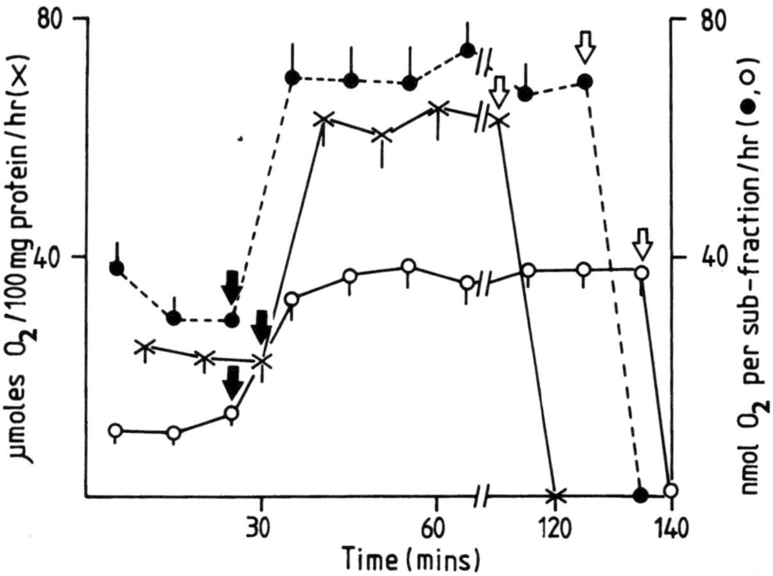

Fig.6.

Respiratory performance of purified synaptosome subpopulations. The synaptosomes were incubated in Krebs-phosphate medium whilst still attached to Magnogel, which, added alone showed no respiratory activity ± glucose. The Rank oxygen electrode was used to monitor O_2 uptake. Glucose (10mM) was added at dark arrows, azide (0.2%) at open arrows. Key: dark circles, GABAergic subpopulations; open circles, cholinergic subpopulations; crosses, whole cortical mother fraction.

present), and compared to this ratio in the mother fraction, it could be seen that the cholinergic synaptosomes were enriched in acetylcholine (ACh), choline, aspartate and glutamate, whilst GABAergic synaptosomes were enriched in both glutamate and GABA but not in aspartate. ChAT was enriched 10 - fold and GAD 3-fold on this basis in cholinergic and GABAergic subpopulations respectively, close to the theoretical maximum for each (see Table 1).

Content per fraction (pmol/separation)	Total Mother Fraction	Cholinergic Subpopulation	% of Total	GABAergic Subpopulation	% of Total
Taurine	28552 ± 3061	731 ± 60	2.5%	696 ± 80	2%
Aspartate	26456 ± 2931	1986 ± 221	8 %	1631 ± 179	6%
Threonine	5459 ± 632	120 ± 15	2 %	142 ± 23	3%
Serine	8321 ± 769	191 ± 21	2 %	187 ± 30	2%
Glutamate	41994 ± 4063	3359 ± 394	8 %	7559 ± 893	18%
Glutamine	7139 ± 921	0	-	0	-
Glycine	8819 ± 763	96 ± 12	1 %	75 ± 12	1%
Alanine	7063 ± 839	106 ± 20	1.5%	103 ± 16	1%
Valine	2870 ± 321	72 ± 15	2.5%	93 ± 12	3%
Isoleucine	1210 ± 213	Low	-	Low	-
Leucine	1096 ± 313	Low	-	Low	-
Tyrosine	809 ± 110	Low	-	Low	-
Phenylalanine	605 ± 109	Low	-	Low	-
GABA	14531 ± 1658	320 ± 43	2 %	3150 ± 429	22%
Histidine	7831 ± 902	121 ± 17	1.5%	262 ± 35	3%
ACh	1380 ± 209	206 ± 31	15 %	41 ± 8	3%
Choline	4200 ± 563	621 ± 109	15 %	504 ± 83	12%

Neuropeptides Preliminary data on the neuropeptide content of cholinergic and GABAergic synaptosomes has revealed substantial differences in their composition. Thus, the cholinergic subpopulation was found to be some eighteen-fold enriched above the mother fraction in VIP (16 and 20-fold in 2 preps.), whilst the GABAergic type were 2.1 (2.0 and 2.2-fold in 2 preps). The latter subpopulation was also enriched 3-fold in Substance P (3.1 and 2.9-fold in 2 preps.), whilst this neuropeptide was not detectable in the cholinergic category. Other neuropeptides were also present in both synaptosome subpopulations, though little enriched above their levels in the mother fraction (Docherty et al 1987c).

Neurotransmitter-releasing properties Neurotransmitters could be released from the incubated subpopulations by stimulation with either potassium (50mM) or Veratrine (75μM) (Fig 7). Thus, in response to these depolarizing treatments,

GABAergic synaptosomes released glutamate and aspartate in addition to GABA, and cholinergic synaptosomes released these two amino acids as well as ACh, though neither subpopulation showed enhanced release of taurine (Fig 7). Release of glutamate and aspartate (though not GABA in these experiments) evoked by high potassium appeared to be calcium-dependent. That evoked by Veratrine was calcium-independent as reported by other workers (Blaustein et al 1972; Norris et al 1983). However, tetrodotoxin (ttx) treatment in most cases abolished Veratrine-stimulated release, whilst, as expected, it was ineffective

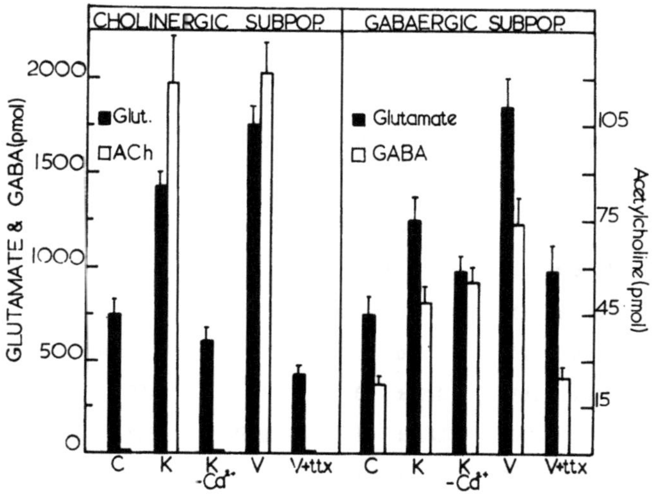

Fig.7.

Transmitter-releasing properties of isolated synaptosome subpopulations. The subpopulations (still attached to Magnogel) were incubated at 37°C in Krebs-phosphate medium. Values are mean ± SEM for 3 separate preparations. Key: C, Control; K, plus KCl, 50mM; K-Ca, KCl 50mM in Ca^{2+} free medium; V = Veratrine 100μM; V+ttx, Veratrine 100μM + Tetrodotoxin 1μM.

against that evoked by high potassium (Fig 7). These findings indicate that the stimulus-induced efflux of these compounds possessed properties typical of those found for conventional synaptosome preparations and brain slices, and usually taken as evidence of its physiological significance (De Belleroche and Bradford 1972; Norris et al 1983).

As the precursor of GABA, glutamate enrichment in GABAergic synaptosomes could be expected, but on an LDH basis it is more enriched in the cholinergic type. However, its stimulus-evoked release from these two categories of synaptosomes suggests it may serve as a neurotransmitter or neuromodulator in these nerve terminals.

It is conceivable that the observed glutamate and aspartate release originates from separate (e.g. glutamatergic) subpopulations which are co- purified. However, the fact that 86% of LDH and 42% of the ChAT in cholinergic synaptosomes could be released by immunolysis employing anti-ChAT and active complement, indicates that most of the synaptosomes carried membrane-bound ChAT and were therefore likely to be cholinergic in nature (cf. 10-15% LDH released from mother fraction (Fig 1). There does not seem to be a higher degree of non-specific binding of aspartate or glutamate-containing terminals since less than 0.1% of these amino acids was found attached to the Magnogel-Protein A microspheres.

These results suggest, therefore, that glutamate can be released from cholinergic and GABAergic terminals. Is it possible that glutamate is released as a neurotransmitter or neuromodulator from a wide category of neurones, part at least being directly from the synaptoplasm (De Belleroche and Bradford 1977; Nicholls and Sihra 1986). Exclusively glutamatergic neurones are perhaps far fewer than has been implied by the widespread occurrence of glutamate receptors, and the high sensitivity of most neurones to iontophoresed glutamate. Electrophysiological findings would not conflict with this proposal (Meyer and Westbrook 1987). In line with such co-release of amino acids and 'classical' neurotransmitters is the recent report that pure cholinergic synaptosomes from Torpedo release glutamate along with ACh (Vyas and Bradford 1987).

Even the apparant anomaly that an excitatory (glutamate) and an inhibitory (GABA) neurotransmitter are released from the same terminal may be encompassed by mechanisms allowing differential release of two neuroactive substances at different stimulus frequencies (Lundberg et al 1982), or the inhibition of the release of one neurotransmitter by the other (Burnstock 1983), or by activation of an appropriate presynaptic receptor linked directly to the transmitter release process.

In addition, our own observations on complement-mediated immunolysis of whole cortical synaptosome preparations have shown that at least 66% of synaptosomes could be lysed as judged by summation of maximal LDH release resulting from immunolysis mediated with anti-ChAT, anti-GAD and anti-dopamine β-hydroxylase (Table 1; Docherty et al 1985a, 1986). This would leave a maximum of 34% of synaptosomes which would have to accommodate the glutamatergic, and other subtypes known to be present in the cerebral cortex. Such calculations would put the contribution of exclusively glutamatergic terminals at a relatively lower figure than would be predicted by the widespread occurrence of glutamate receptors and putative glutamatergic pathways (Monaghan and Cotman 1986). This, too, would therefore argue in favour of a major role for glutamate as a co-transmitter or neuromodulator.

REFERENCES

Badamchian M, Morrow Jr.KJ and Carroll PT (1986) Immunological,isoelectric, hydrophobic and molecular weight differences between soluble and ionically membrane-bound fractions of choline-O-acetyltransferase prepared from mouse and rat brain. Neurochem Int 9: 409-421

Barochovsky O, Docherty M and Bradford HF (1986) Choline acetyltransferase may be a cell-surface marker of the cholinergic NS-20Y cell line. Biochem Soc Trans 14: 759-760

Benishin CG and Carroll PT (1983) Multiple forms of choline-O-acetyltransferase in mouse and rat brain: solubilization and characterization. J Neurochem 41: 1030-1039

de Belleroche JS and Bradford HF (1972) Metabolism of Beds of Mammalian Cortical Synaptosomes: Response to Depolarizing Influences. J Neurochem 19: 585-602

de Belleroche JS and Bradford HF (1977) On the site of origin of transmitter amino acids released by depolarization of nerve terminals In Vitro. J Neurochem 29: 335-343

Blaustein MP, Johnson EM and Needleman P (1972) Calcium dependent norepinephrine release from presynaptic nerve endings in vitro Proc Natl Acad Sci USA 69: 2237-2240

Burnstock G (1983) Recent concepts of chemical communication between excitable cells in Dale's principle and communication between neurones (ed Osborne NN) 7-35 Pergamon Press Oxford

Docherty M, Bradford HF and Anderton B (1982) Lysis of cholinergic synaptosomes by an intiserum to choline acetyltransferase. Febs Letts Vol 144: 47-50

Docherty M, Bradford HF, Anderton B and Jang-Yen Wu (1983) Specific lysis of GABAergic synaptosomes by an antiserum to glutamate decarboxylase. Febs Letts Vol 152: No.1 57-61

Docherty M, Bradford HF, Cash CD and Maitre M (1985) Specific immunolysis of serotonergic nerve terminals using an antiserum against tryptophan hydroxylase. Febs Lett 182: 489-492

Docherty M, Bradford HF, Wu J-Y, Joh TH and Reis DJ (1985a) Evidence for specific immunolysis of nerve terminals using antisera against choline acetyltransferase, glutamate decarboxylase and tyrosine hydroxylase. Brain Res 339: 105-113

Docherty M, Bradford HF and Joh TH (1986) Specific lysis of noradrenergic synaptosomes by an antiserum to dopamine-β-hydroxylase. Febs Lett 202: 37-40

Docherty M and Bradford HF (1987) Chloride ions influence the equilibrium of choline acetyltransferase between the membrane-bound integral form and the soluble form. Biochem Soc Trans 15: 637-640

Docherty M, Bradford HF and J-Y Wu (1987a) The preparation of highly purified GABAergic and cholinergic synaptosomes from mammalian brain. Neurosci Letts 81: 232-238

Docherty M, Bradford HF and J-Y Wu (1987b) Stimulus-evoked co-release of glutamate and aspartate from highly purified cholinergic and GABAergic synaptosomes prepared from mammalian brain. Nature (Lond) 330: 64-66

Docherty M, Bradford HF and Bloom SR (1987c) Neuropeptides in subpopulations of mammalian brain synaptosomes. In Preparation

Lundberg JM, Anggard A, Fahrenkrug J, Lundgren G and Holmstedt B (1982) Acta Physiol Scand 115: 525-528

Meyer ML and Westbrook GL (1987) The physiology of excitatory amino acids in the Vertebrate CNS. Prog Neurobiol 28: 197-276

Monoghan DT and Cotman CW (1986) Anatomical organisation of NMDA, kainate and quisqualate receptors in Excitatory Amino Acids (eds Roberts PJ,

Storm-Mathisen J and Bradford HF) 279-299 MacMillan Hampshire and London
Morel N, Israel M and Manarache R (1978) Determination of ACh concentration in Torpedo synaptosomes. J Neurochem 30: 1553-1557
Nicholls DG and Sihra TS (1986) Synaptosomes possess an exocytotic pool of glutamate. Nature 321: 772-773
Norris PJ, Dhaliwal DK, Druce DP and Bradford HF (1983) The Suppression of Stimulus-Evoked Release of Amino Acid Neurotransmitters from Synaptosomes by Verapamil. J Neurochem 40: No.2 514-521
Richardson PJ, Siddle K and Luzio JP (1984) Immunoaffinity purification of intact, metabolically active, cholinergic nerve terminals from mammalian brain. Biochem J 219: 647-654
Vyas S and Bradford HF (1987) Co-Release of Acetylcholine, Glutamate and Taurine from Synaptosomes of Torpedo Electric Organ. Neurosci Lett 82: 58-64.

EXCITATORY AMINO ACIDS; PHYSIOLOGY, ANATOMY AND BIOCHEMISTRY

Frode Fonnum, Else Marie Fykse and Ragnhild Paulsen

Norwegian Defence Research Establishment
Division for Environmental Toxicology
PO Box 25
N-2007 Kjeller, Norway

PHYSIOLOGY

Several amino acids have been suggested to function as excitatory transmitters in the brain. The strongest candidates are glutamate, aspartate and homocysteate. All 3 candidates are released from brain slices in a Ca^{++} dependent manner. Surprisingly enough there is a good positive correlation between the release of glutamate, aspartate and homocysteate from different regions of the brain (Do et al, 1986), as if they were all released from almost the same structures in the different regions.

Glutamate and aspartate are linked metabolically through the enzyme aspartate aminotransferase, which is widely distributed in the brain. The equilibrium between the two amino acids is highly dependent upon changes in the tricarboxylic acid cycle intermediate oxaloacetate. A small decrease in oxaloacetate immediately shifts the equilibrium towards aspartate (Fonnum, 1985). Both aspartate and glutamate are taken up into synaptosomes and glial cells by the same sodium dependent high affinity uptake system (Fonnum, 1984).

The neuronal physiological response to aspartate, glutamate and homocysteate is similar in potency (Curtis and Watkins, 196o). The response is, however, mediated by at least 3 distinct receptors which have been partially characterized by selective agonists (Table 1). For the EA-1 or NMDA receptor

NATO ASI Series, Vol. H21
Cellular and Molecular Basis of Synaptic Transmission
Edited by H. Zimmermann

TABLE 1. PROPERTIES OF THE 3 EXCITATORY AMINO ACID RECEPTORS IN BRAIN

	EA 1	EA 2	EA 3
AGONISTS	NMDA IBOTENATE	QUISQUALATE AMPA	KAINATE DOMOATE
ANTAGO- NISTS	D-AP5 D-AP7	GDEE KYNURENIC ACID	KYNURENIC ACID
SYNAPTIC ACTIVITY	VOLTAGE DEPENDENT CATION CHANNEL	VOLTAGE INDEP. Na, K CHANNELS	VOLTAGE INDEP. Na, K CHANNELS
MEMBRANE- AFFINITIES			
GLUTAMATE	+ + +	+ + +	+ + +
ASPARTATE	+ +	+	+
HOMO- CYSTEATE	+ +	+ +	+ +
Regions with high levels	HIPPOCAMPUS CA1 NEOSTRIATUM CORTEX LATERAL SEPTUM	CORTEX LAYER 1-3 HIPPOCAMPUS CA3 NEOSTRIATUM OLFACTORY NUCLEI	COMPLEMENTARY TO NMDA CORTEX LAYER 1, 5, 6 NEOSTRIATUM HIPPOCAMPUS CA1 - CA3

D-AP5 = 2-amino-5-phosphono pentanoate
D-AP7 = 2-amino-7-phosphono heptanoate
GDEE = Glutamate diethylester

Data selected from Monaghan and Cotman, 1982; 1985; Monaghan et al, 1984; Rainbow et al, 1984; Cotman and Monaghan, 1987; Foster and Fagg, 1984.

there exist selective antagonists, whereas for the quisqualate and kainate receptors the antagonists are less selective. The 3 amino acid transmitter candidates all have an affinity for the 3 excitatory amino acid receptors, although the binding affinity of aspartate for the quisqualate and kainate receptors is much lower than for the NMDA receptor.

Recently the peptide N-acetyl aspartyl glutamate (NAAG) has been suggested to be an excitatory transmitter in brain based on its electrophysiological response and binding to membranes (Ffrench-Mullen et al, 1985). NAAG is displaced by quisqualate from brain membranes (Zaczek et al, 1986). It is distributed in the brain almost inversely to the high affinity (HA) uptake of glutamate with the highest level in spinal cord, pons and medulla (Koller et al, 1984).

EXCITATORY AMINO ACID PATHWAYS

Since aspartate and glutamate are ubiquitously distributed in brain tissue, it has been a difficult task to identify excitatory amino acids pathways. The loss in three properties of the glutamergic terminal caused by specific lesions have been used to detect such pathways: The loss of HA uptake of glutamate either measured biochemically or autoradiographically, the loss of K^+ stimulated Ca^{++} dependent release of endogenous amino acids and the loss of endogenous amino acids.

Retrograde transport of D-aspartate after microinjection in the terminal region is a method which is very sensitive in tracing excitatory amino acid pathways (Cuénod and Streit, 1983). Most of the pathways shown in Fig 1 have been confirmed by this sensitive technique. When D-aspartate was applied to nucleus accumbens there was labelling in perikarya in anterior olfactory nucleus, medial prefrontal and sulcal cortex, parataenial, paraventricular and several other thalamic nuclei, all amygdaloid nuclei and hippocampal regions (Beart et al, 1986). By biochemical method one has previously shown that a large proportion of the glutamergic terminals (> 70 %) was derived from the hippocampal region and prefrontal cortex (Walaas and Fonnum, 1979). One may wonder whether the retrograde tracing method is showing some false positives due to difficulties in controlling local concentrations necessary for high affinity uptake conditions. Otherwise, it may simply mean that glutamergic fibres are more generally distributed than otherwise believed.

Glutamate is the strongest transmitter candidate among the 3 amino acids for most of the pathways (Fonnum, 1984; Cotman and Monaghan, 1987). Aspartate is a strong candidate for the climbing fibres to the cerebellum, (Wiklund et al, 1983) the the commisural fibres to hippocampus (Nadler et al, 1978) and the auditory nerve (Wenthold, 1980). Homocysteate is not yet identified as a major candidate for any of the fibres.

1A

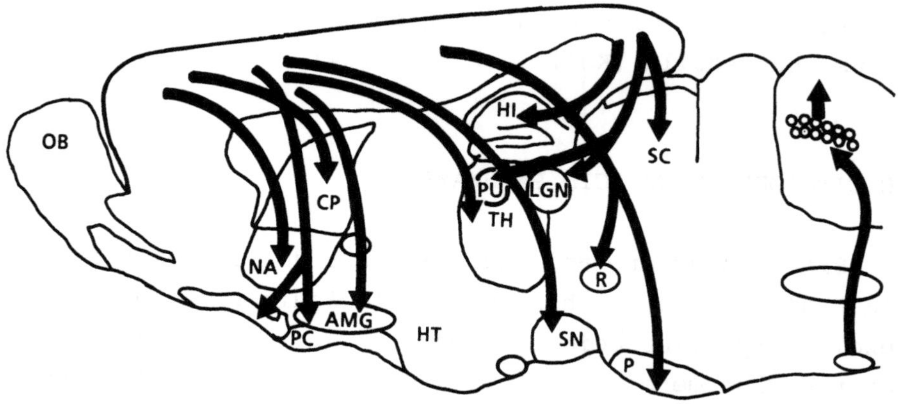

1B

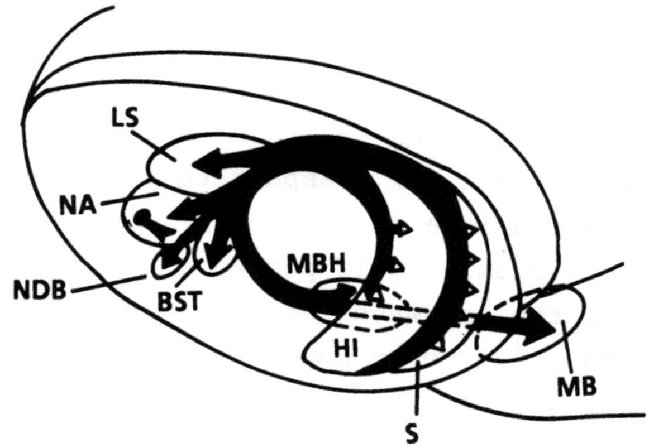

Fig. 1 Summary of the main excitatory amino acid pathways in the brain. In 1A the pathways originate from perikarya in the cerebral cortex. In addition the climbing fibres and the parallel fibres in the cerebellum are shown. In 1B the pathways originate from perikarya in the hippocampus and subiculum. AMG, amygdala; BST, bed nucleus of stria terminals; CP, caudate putamen; HI, hippocampus; LGN, lateral geniculate body; LS, lateral septum; MB, mammilary bodies; NA, nucleus accumbens; NDB, nucleus of the diagonal band; OB, olfactory bulb; P, pontine nuclei; PC, pyriform cortex; PU, pulvinar; R, nucleus ruber; S, subiculum; SC, superior colliculus. Other excitatory amino acid pathways, not shown, are the perforant path, the mossy fibres, the commisural fibres and the Schaffer collaterals of the hippocampus. The data are compiled from Fonnum et al, 1981; Fonnum, 1984; Cotman and Monaghan, 1987.

Recently it has been suggested that NAAG could be the transmitter of the optic nerve and the lateral olfactory tract (Ffrench-Mullen et al, 1985; Tieman et al, 1987). Neither glutamate nor aspartate have really satisfied the criteria to a transmitter for these two pathways.

Due to the difficulties in differentiating between glutamate and aspartate as chemical transmitters, we have studied the uptake of amino acids into synaptic vesicles. Previous attempts to isolate synaptic vesicles enriched in amino acid have all failed (De Belleroche and Bradford, 1973; Lahndesmäke et al, 1977; Rassin, 1972). This has been attributed to the possible leakage of amino acids during the fractionation procedure. The uptake of glutamate, GABA and aspartate into vesicle fractions differed clearly from the uptake into synaptosomes. The vesicular uptake was dependent upon an exogen supply of Mg^{++} and ATP, and the uptake was inhibited by the electrogenic proton carriers CCCP or FCCP in the µM range (Table 2). The vesicular uptake of amino acids displayed low affinity (Km = mM) and was independent upon Na^{+}. (Fykse and Fonnum, 1988; Naito and Ueda, 1985). The uptake of GABA was 7-30 per cent whereas the uptake of aspartate was only 1-5 per cent of that of glutamate. The low uptake of aspartate could either indicate that the aspartate uptake is much more labile than the glutamate uptake or that aspartate does not play a primary role as a neurotransmitter. Surprisingly we found a large regional variation in the uptake of GABA, but not of glutamate, between cortex, hypothalamus, cerebellum and medulla (1:4:2:1 for GABA vs 1:1:1:0.8 for glutamate).

BIOCHEMISTRY

Glutamate (tissue concentration ~ 10 mM) is present in at least 4 different pools in the brain: As transmitter in glutamergic terminals, as precursor for GABA in GABAergic terminals, as a metabolic component in other neuronal structures and in glial cells. We have concentrated our effort on characterizing the different pools of amino acids by pharmacological interaction (Table 3). The studies have been concentrated on neostriatum which receives a heavy glutamergic input from neocortex (Fonnum et al, 1981).

Cortical ablation, leading to a loss of the glutamergic terminals, was accompanied by a loss of 20-30 % of the endogenous glutamate level in neostriatum. Electron microscopic studies of neostriatum after such a lesion show that the

TABLE 2 THE PROPERTIES OF VESICULAR UPTAKE OF TRANSMITTER AMINO ACIDS

THE VESICULAR UPTAKE IS DEPENDENT ON

* Mg^{++} ATPase
* ELECTROCHEMICAL POTENTIAL
* TEMPERATURE
* FREE SH GROUP

THE VESICULAR UPTAKE IS INDEPENDENT ON

* Na^+
* OLIGOMYCIN
* OUABAIN

The conclusion is compiled from Naito and Ueda, 1985; Fykse and Fonnum, 1988.

volume of degenerated terminals only occupies a few (4) per cent of the total volume in this region (Fonnum, Villani and Contestabile, unpublished). This would indicate that the concentration of glutamate in glutamergic terminals could be (10 mM x 20/4) = 50 mM.

In contrast, the concentration of glutamate in GABAergic terminal is probably low. Selective destruction of GABAergic terminals in substantia nigra was accompanied by only a small decrease in glutamate (Minchin and Fonnum, 1979; Korf and Venema, 1983). In agreement, we have found that the level of glutamate in the GABA regions such as globus pallidus and substantia nigra is only half that in regions such as cortex, hippocampus and neostriatum containing high concentration of glutamergic terminals. The main precursors for synthesis of transmitter glutamate are the uptake of glucose into nerve terminals, the transport of 2-oxoglutarate derived from the anaplerotic process in glial cells to the nerve terminal and the transport of glutamine from glial cells to the nerve terminal (Fonnum, 1985).

We have studied the importance of glucose uptake by treating the animals with actrapid insulin. The hypoglycemia led first to a rapid decrease in glutamine, and subsequently decrease in glutamate. Aspartate increased, but not enough to compensate for the loss in glutamate (Table 3, Engelsen and Fonnum, 1983; Engelsen et al, 1986). After cortical abla-

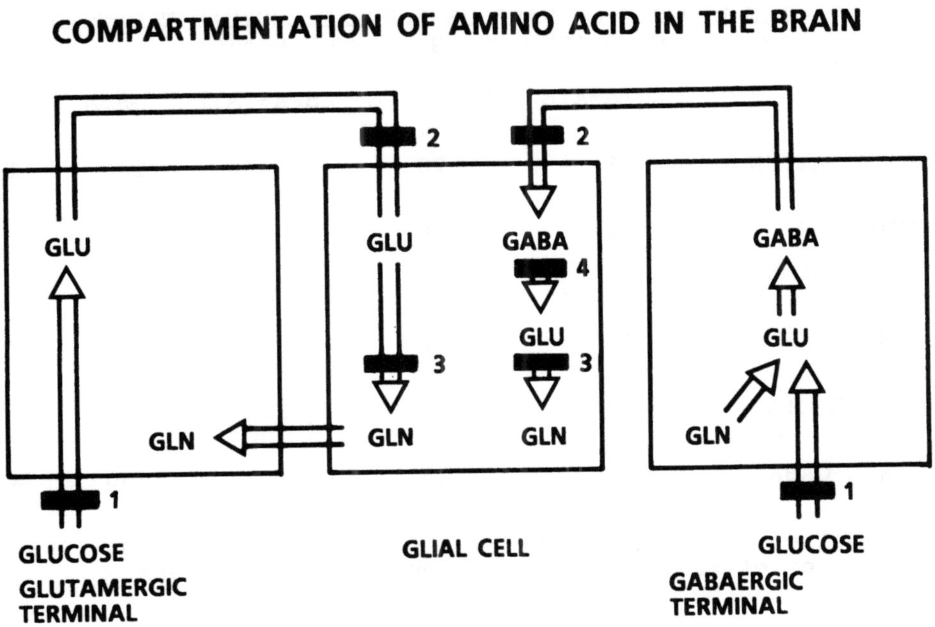

Fig 2. The compartmentation of amino acids in the brain. The numbers show how one can interfer on the compartmentation with different drugs (1) Insulin causes hypoglycemia (Engelsen and Fonnum, 1983) (2) Fluorocitrate inhibits citric acid cycle in glial cells, reduces ATP and probably inhibits transmitter uptake (Paulsen et al, 1987) (3) Methionine sulphoximine inhibits glutamine synthetase (Engelsen and Fonnum, 1985) (4) γ-vinyl GABA inhibits GABA aminotransferase (Paulsen and Fonnum, 1988).

tion, to remove the transmitter glutamate pool in neostriatum, the changes in amino acids were less. The results demonstrated a more pronounced decrease of glutamate in the intact (metabolic and transmitter pools) than in the decorticated (metabolic pool) neostriatum. The reasons are probably a higher release together with a restricted replenishment of

TABLE 3 THE EFFECTS OF VARIOUS PHARMACOLOGICAL TREATMENTS TO DEFINE AMINO ACID POOLS IN NEOSTRIATUM

	GLUTAMATE	GLUTAMINE	ASPARTATE	GABA	REF
DECORTICATION	75	120	95	100	1
METHIONINE SULPHOXIMINE	60	5	-	80	2
FLUOROCITRATE	70	17	70	85	3
HYPOGLYCEMIA	54	15	380	110	1
γ-VINYL GABA	80	58	60	630	4

(1) Engelsen and Fonnum, 1983 (2) Engelsen and Fonnum, 1985 (3) Paulsen et al, 1987 (4) Paulsen and Fonnum, 1988

glutamate due to the lack of glucose on the intact side (Engelsen and Fonnum, 1983; Engelsen et al, 1986).

We have also found a good correlation between the high affinity uptake of 2-oxoglutarate and glutamate in different brain regions (correlation coefficient 0·8). This gives support to the concept that transmitter glutamate can, at least in part, be supplied by metabolism of glucose to 2-oxoglutarate through an anaplerotic process. Since pyruvate carboxylase is specifically localized in the glial cells, 2-oxoglutarate may therefore be transported from the glial cell to the glutamergic nerve terminal (Shank et al, 1985).

After release transmitter glutamate and GABA are taken up into glial cells where they are converted into glutamine which may be circulated back to the nerve terminals to replenish the transmitter (Fig 2). Interaction with this cycle in vivo either by injection of methionine sulphoximine, an inhibitor of glutamine synthetase, or injection of fluorocitrate, an inhibitor of glial cell metabolism, gave a large decrease in the steady state level of glutamine, less decrease in glutamate level and very small effects in GABA level (Table 3). The results could mean that the terminals can in part compensate for the loss of glutamine during synthesis of transmitters. It is also appearant that GABA synthesis was less dependent than glutamate on glutamine (Paulsen and Fonnum, 1988). It has recently been found that the turnover of GABA is more readily affected by losses in glutamine than the steady state level (Paulsen and Fonnum, unpublished).

Injection of γ-vinyl GABA inhibits GABA-aminotransferase and leads to a significant fall in brain glutamine as if GABA should be at least as important as transmitter glutamate in supplying glutamine in glial cells (Table 3, Paulsen and Fonnum, 1988). Thus the pharmacological studies indicate that both glucose, 2-oxoglutarate and glutamine are important precursors for transmitter glutamate synthesis. It may well be that transmitter glutamate is not regulated by a single precursor, but that the level is a result of the general metabolic status in the terminal. Transmitter glutamate is important for maintaining the rate of glutamate turnover in at least neostriatum. This is either due to a rapid turnover of transmitter glutamate or that the glutamergic input is the driving force for metabolism in neostriatum.

As stated in the introduction both glutamate and aspartate are released from brain slices in a Ca^{++} dependent manner by depolarization with KCl (Do et al, 1986). In two regions, the cerebellum after deprivation of the climbing fibres and hippocampus after deprivation of the commisural fibres, there was a larger fall in endogenous aspartate than in glutamate (Wiklund et al, 1983; Nadler et al, 1978).

In neostriatum we found a 5 fold higher Ca^{++} dependent release of glutamate than aspartate. The release was inhibited by μM trifluoperazin which is a calmodulin antagonist. We also found a twofold increase in release by the phorbol ester TPA, the protein kinase C agonist.

The rapid exchange which takes place between aspartate and glutamate during hypoglycemia raises the question whether the two candidates can substitute for each other as neurotransmitters. We therefore depolarized normoglycemic and hypoglycemic slices with KCl. The release of glutamate decreased and the release of aspartate increased 600 per cent in the hypoglycemic slice relative to the normoglycemic slice. Therefore, under these very special conditions, the release of aspartate seemed to substitute for the release of glutamate.

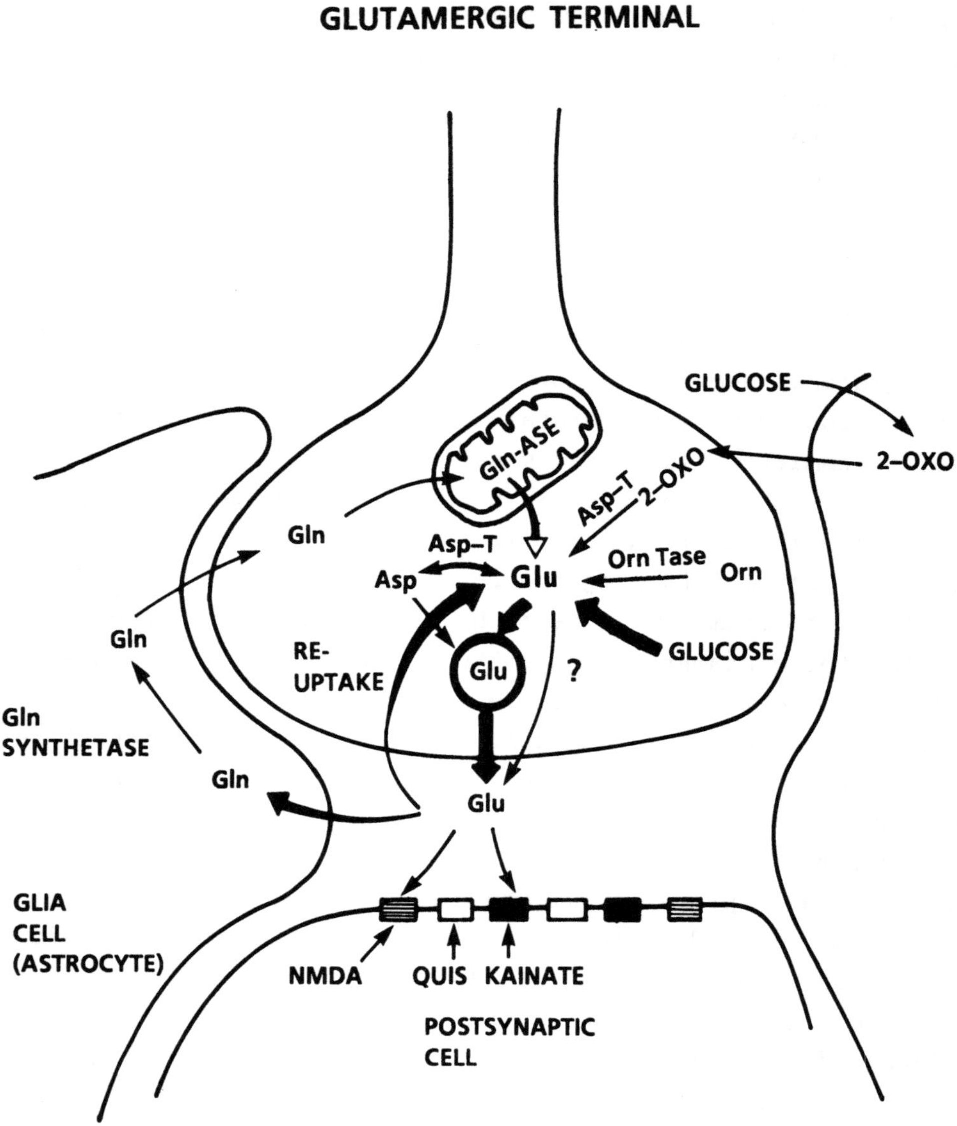

Fig. 3. A summary of the reactions taking place in a glutamergic nerve terminal. The precursors for glutamate synthesis are glucose, 2-oxoglutarate (2-oxo) and glutamine. Glutamate and to a smaller extent aspartate are taken up into synaptic vesicles. The 3 groups of excitatory amino acids receptors are named NMDA, quisqualate and kainate.

CONCLUSIONS

1. Excitatory amino acid receptors can be classified into 3 groups with different agonists and antagonists.
2. Vesicular uptake is important for differentiating between excitatory amino acids.
3. Aspartate and glutamate may substitute for each other as transmitters at least under hypoglycemia.
4. Glutamate turnover is faster in intact than in decorticated neostriatum.
5. Glucose, 2-oxoglutarate and glutamine are important as precursors for glutamate and GABA.
6. Calmodulin and protein kinase C may be involved in the release of exitatory amino acids.
7. Fluorocitrate inactivates the metabolism of glial cells.

REFERENCES

Beart PM, Summers RJ, Christie MJ (1986) Mapping of excitatory amino acid projections to central gray and nucleus accumbens by D-[^{3}H] aspartate retrograde transport. In: Hicks TP, Lodge D, McLennan H (eds) Excitatory amino acid transmission. Alan R Liss Inc, New York, pp 361-369

Cotman CW, Monaghan DT (1987) Chemistry and anatomy of excitatory amino acid systems. In: Meltzer HY (ed) Psychopharmacology: The third generation of progress. Raven Press, New York, pp 192-210

Cuénod M, Streit P (1983) Neuronal tracing using retrograde migration of labeled transmitter-related compounds. In: Bjørklund A, Høkfelt T (eds) Handbook of Chemical Neuroanatomy Vol 1. Elsevier, Amsterdam, pp 365-393

Curtis DR, Watkins JC (1960) The excitation and depression of spinal neurones by structurally related amino acids. J Neurochem 6: 117-141

De Belleroche JS, Bradford HF (1973) Amino acids in synaptic vesicles from mammalian cerebral cortex: a reappraisal. J Neurochem 21: 441-451

Do KQ, Mattenberger M, Streil P, Cuénod M (1986) In vitro release of endogenous excitatory sulfur-containing amino acids from various rat brain regions. J Neurochem 46: 779-786

Engelsen B, Fonnum F (1983) Effects of hypoglycemia on the transmitter pool and the metabolic pool of glutamate in rat brain. Neurosci Lett 42: 317-322

Engelsen B, Fonnum F (1985) The effect of methioninesulphoximine, an inhibitor of glutamine synthetase on levels of amino acids in the intact and decorticated rat neostriatum. Brain Res 338: 165-168

Engelsen B, Westerberg E, Fonnum F, Wieloch T (1986) Effects of insulin-induced hypoglycemia on the concentrations of glutamate and related amino acids and energy metabolites in the intact and decorticated rat neostriatum. J Neurochem 47: 1634-1641

Ffrench-Mullen JMH, Koller K, Zaczek R, Coyle JT, Hori N, Carpenter DO (1985) N-acetyl-aspartyl glutamate: possible role as the neurotransmitter of the lateral olfactory tract. Proc Natl Acad Sci USA 82: 3894-3897

Fonnum F (1984) Glutamate: a neurotransmitter in mammalian brain. J Neurochem 42: 1-11

Fonnum F (1985) Determination of transmitter amino acid turnover. In: Bolton AA, Baker GB, Wood JD (eds) Neuromethods Vol 3 Amino Acids. Humana Press, Clifton, New Jersey, pp 201-237

Fonnum F, Søreide A, Kvale I, Walker J, Walaas I (1981) Glutamate in corticofugal fibres. Advances in Biochem Psychopharmacol 27: 29-41

Foster AC, Fagg GE (1984) Acidic amino acid binding sites in mammalian neuronal membranes: their characteristics and relationship to synaptic receptors. Brain Res Rev 7: 103-200

Fykse EM, Fonnum F (1988) Uptake of GABA by a synaptic vesicle fraction isolated from rat brain. J Neurochem 50: (in press)

Koller KJ, Zaczek R, Coyle JT (1984) N-acetylaspartyl-glutamate: regional levels in rat brain and the effects of brain lesions as determined by a new HPLC method. J Neurochem 43: 1136-1142

Korf J, Venema K (1983) Amino acids in the substantia nigra of rats with striatal lesions produced by kainic acid. J Neurochem 40: 1171-1173

Lahndesmäke P, Karppinen A, Saarni H, Winter R (1977) Amino acids in the synaptic vesicle fraction from calf brain: control and metabolism. Brain Res 138: 295-308

Minchin MCW, Fonnum F (1979) The metabolism of GABA and other amino acids in rat substantia nigra slices following lesions of the striato-nigral pathway. J Neurochem 32: 203-210

Monaghan AT, Cotman CW (1982) The distribution of 3H-kainic acid binding sites in rat CNS as determined by autoradiography. Brain Res 252: 91-100

Monaghan AT, Cotman CW (1985) Distribution of NMDA sensitive L-^{3}H-glutamate binding sites in rat brain. J Neurochem 5: 2909-2919

Monaghan AT, Yao D, Cotman CW (1984) Distribution of ^{3}H-AMPA binding sites in rat brain as determined by quantitative autoradiography. Brain Res 324: 160-164

Nadler JV, White WF, Vaca KW, Perry BW, Cotman CW (1978) Biochemical correlates of transmission mediated by glutamate and aspartate. J Neurochem 31: 147-155

Naito S, Ueda T (1985) Characterization of glutamate uptake into synaptic vesicles. J Neurochem 44: 99-109

Paulsen RE, Contestabile A, Villani L, Fonnum F (1987) An in vivo model for studying function of brain tissue temporarily devoid of glial cell metabolism: the use of flurocitrate. J Neurochem 48: 1377-1385

Paulsen RE, Fonnum F (1988) The regulation of transmitter GABA synthesis and metabolism illustrated by the effect of gammavinyl GABA and hypoglycemia. J Neurochem 50: (in press).
Rainbow TC, Wieczorek CM, Halpain S (1984) Quantitative autoradiography of binding sites for ^{3}H-AMPA, a structural analogue of glutamic acid. Brain Res 309: 173-177
Rassin DH (1972) Amino acids as putative transmitters: failure to bind synaptic vesicles of guinea-pig cerebral cortex. J Neurochem 19: 139-148
Shank RP, Bennett GS, Freytag SO, Campbell GLeM (1985) Pyruvate carboxylase: an astrocyte-specific enzyme implicated in the replenishment of amino acid neurotransmitter pools. Brain Res 329: 364-367
Tieman SB, Kolsdrug JM, Neale JH (1987) N-acetyl-aspartylglutamate immunoreactivity in neurons of the cat visual system. Brain Res 420: 188-193
Walaas I, Fonnum F (1979) The effect of surgical and chemical lesions on neurotransmitter candidates in the nucleus accumbens of the rat. Neuroscience 4: 209-218
Wenthold RJ (1980) Glutaminase and aspartate aminotransferase decrease in cochlear nucleus after lesion of the auditory nerve. Brain Res 190: 293-297
Wiklund L, Toggenburger G, Cuenod M (1983) Aspartate: possible neurotransmitter in cerebellar climbing fibers. Science 216: 78-80
Zaczek R, Koller KJ, Carpenter DO, Fisher R, Ffrench-Mullen JM, Coyle JT (1986) Interactions of acidic peptides: Excitatory amino acid receptors. In: Roberts PJ, Storm-Mathisen J, Bradford HF (eds) Excitatory amino acids. Macmillan, Houndsmills and London, pp 397-409

SYNAPTOSOMAL BIOENERGETICS AND THE MECHANISM OF AMINO ACID TRANSMITTER RELEASE

David Nicholls and Jose Sanchez-Prieto

Department of Biochemistry
University of Dundee
Dundee DD1 4HN, Scotland

INTRODUCTION

The synaptosome can survive in isolation for several hours, maintaining plasma and mitochondrial membrane potentials of some 60mV and 150mV respectively (Scott and Nicholls, 1980), high ATP/ADP ratios (Kauppinen and Nicholls, 1986a) and submicromolar cytoplasmic free Ca^{2+} concentrations (Hansford and Castro, 1985). The guinea-pig cerebral cortical preparation shows a ten-fold increase in glycolysis when oxidative phosphorylation is prevented (Kauppinen and Nicholls, 1986a). This substantial Pasteur effect shows firstly that essentially all the glycolytic capacity is under the control of the mitochondria, i.e. very little glycolysis occurs in membranous structures lacking mitochondria, and secondly that the mitochondria within the synaptosomes are functional. Indeed, the seven-fold respiratory stimulation on addition of uncoupler (Kauppinen and Nicholls, 1986a) proves that this synaptosomal preparation is far more "coupled" and physiologically intact than is sometimes feared.

AMINO ACID TRANSMITTERS

The most important questions which the synaptosomal preparation is capable of elucidating concern the mechanism for the uptake, storage and release of neurotransmitters. In the case of the amino acid transmitters GABA and L-glutamate, the synaptosomal uptake pathways have been characterized in detail, but the nature of their intra-terminal storage and the mechanism of release (whether from the cytoplasm or by vesicle exocytosis) is a matter of great controversy. In this paper we shall review recent work from our laboratory which may throw some light on these problems.

Plasma membrane carriers and Ca^{2+}-independent release

Synaptosomal plasma membranes contain highly active Na^{+}-cotransport pathways for GABA and L-glutamate. The carriers are distinct, and further ions may be involved in the transport stoichiometry, such as chloride in the case of the GABA carrier and K^{+} for the L-glutamate carrier (Kanner, 1983). However these latter ions are close to electrochemical equilibrium across the plasma membrane, and so nearly all the thermodynamic work of amino acid accumulation is provided by the Na^{+}-electrochemical potential

NATO ASI Series, Vol. H21
Cellular and Molecular Basis of Synaptic Transmission
Edited by H. Zimmermann

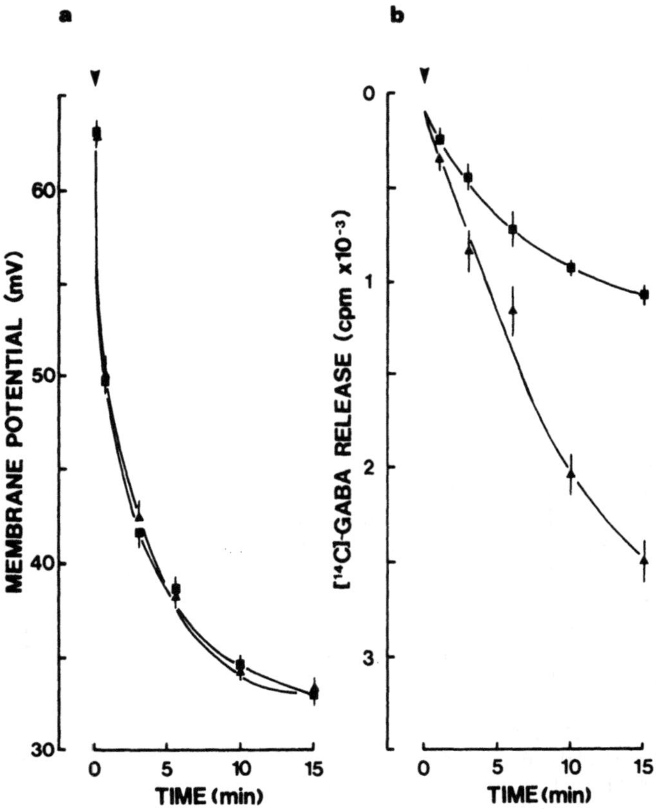

Fig. 1. Plasma membrane depolarisation (a) and long-term Ca^{2+}-independent efflux of ^{14}C-GABA (b) by K^+ or veratridine plus ouabain. Synaptosomes were incubated in the presence of ^{14}C-GABA and ^{86}Rb. 2.6mM-EGTA was added at 59 min to both incubations, followed at 60 min (arrows) by either 30mM KCl (■) or 100µM veratridine plus 100µM ouabain (▲). Plasma membrane potential and ^{14}C-GABA efflux were monitored for the subsequent 15 min. Results are means ± SEM of 4 experiments. For further details see Sihra and Nicholls (1987).

gradient. Since the polarized synaptosome has a plasma membrane potential of some 60mV and a 5-fold Na^+-gradient (out/in), the Na^+-electrochemical gradient amounts to more than 100mV (Akerman and Nicholls, 1981a), and $2Na^+$ per amino acid can thus readily maintain an equilibrium amino acid gradient of several thousand (Pastuszko et al., 1982).

The plasma membrane amino acid carriers are important objects of investigation in their own right, as well as providing a ready means for labelling cytoplasmic pools of GABA and L-glutamate. However, their high activity means that when synaptosomes are depolarized to induce transmitter release, a significant proportion of the released amino acid originates from the cytoplasm by direct Ca^{2+}-independent reversal of the carriers. The carriers reverse because the membrane potential is a major contributor to the Na^+-electrochemical potential maintaining the amino acid gradient, and its loss therefore disturbs the equilibrium across the

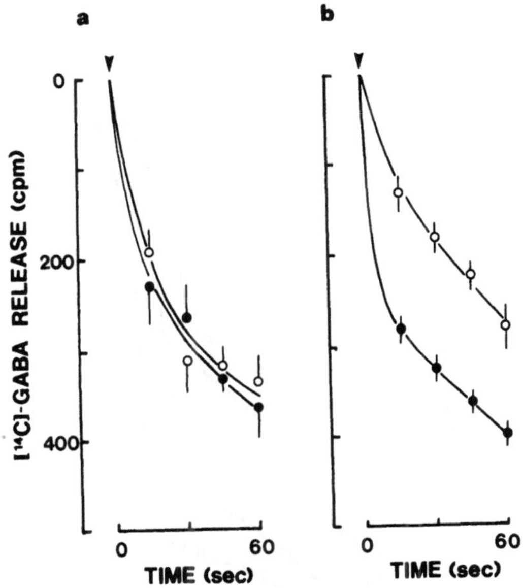

Fig. 2. K^+-induced release of ^{14}C-GABA; increase in Ca^{2+} dependency with time of pre-equilibration. Synaptosomes were preincubated with 0.45μM-^{14}C-GABA (0.1μCi/ml) for either 5 min (a) or 55 min (b) prior to depolarisation (at 60 min) with 30mM KCl (arrow). The increase in extra-synaptosomal ^{14}C-GABA was monitored for the subsequent minute. (●) Ca^{2+} present; (○) Ca^{2+} absent. Results are means ± SEM of 5-10 experiments. The increment in the presence of Ca^{2+} was not significant in (a), but highly significant ($p<0.0005$) in (b). For further details see Sihra and Nicholls (1987).

membrane. This Ca^{2+}-independent release is artifactual in the sense that it would be insignificant during the millisecond duration of a physiological depolarization, but becomes important during the prolonged depolarization to which synaptosomes are subjected by elevated K^+ or by veratridine activation of Na^+-channels. These two means of collapsing the plasma membrane potential decrease the Na^+-electrochemical gradient to differing extents. KCl abolishes the plasma membrane potential without affecting, at least at first, the Na^+-concentration gradient, while veratridine, by effectively permeabilizing the membrane to Na^+, collapses both membrane potential and concentration terms. It would therefore be predicted that veratridine would be more effective in releasing cytoplasmic amino acid than KCl, and this is born out in practice. Fig. 1 shows that concentrations of veratridine which cause less depolarization of the plasma membrane than KCl, nevertheless cause a much more extensive efflux of labelled GABA from the synaptosome (Sihra et al., 1984; Sihra and Nicholls, 1987). Note however that in both cases efflux is slow, with a time-scale measured in minutes rather than seconds.

Ca^{2+} releases GABA from a non-cytoplasmic pool

Numerous isotopic studies (e.g. Ryan and Roskoski, 1975; Cotman and Haycock, 1976; DeBelleroche and Bradford 1977; Haycock et al., 1978; Levi et al., 1978; Abe and Matsuda, 1983) have shown that the specific activity

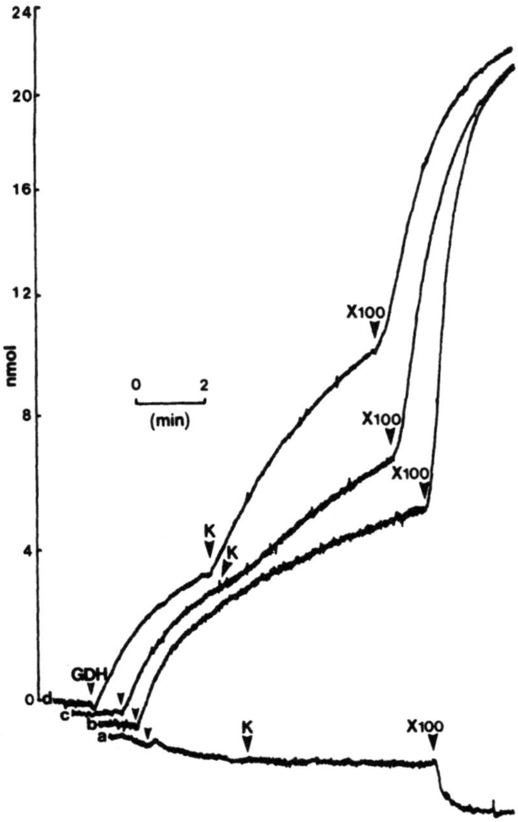

Fig. 3 Representative assays of glutamate release: influence of K^+ and Ca^{2+}. $NADP^+$ was omitted from trace a, 1.3mM $CaCl_2$ was omitted from trace c. Released glutamate was assayed after the addition of glutamate dehydrogenase (GDH). (a) GDH, 30mM KCl (K) and 0.5% (w/v) Triton X100 additions in the absence of added $NADP^+$; (b) release from polarized synaptosomes; (c) effect of 30mM-KCl on release in the presence of EGTA; (d) effect of 30mM KCl on release in the presence of Ca^{2+}. For details see Nicholls et al. (1987).

of GABA released from synaptosomes on depolarization differs from that in the overall preparation, implying compartmentation. With the exception of the Cotman group (e.g. Haycock et al., 1978) the concensus has been that GABA is released directly from the cytosol. However, as discussed by these last authors, many of these studies fail to consider that Ca^{2+}-dependent and Ca^{2+}-independent release might originate from different compartments, and the conflicting conclusions which have been reached reflect this.

Fig. 2 compares the Ca^{2+}-dependency for the release of endogenous GABA (by HPLC) and [^{14}C]-GABA which has been allowed to equilibrate with the synaptosomes for either 5min or 55min prior to deplarization (Sihra and Nicholls, 1987). It is clear that exogenous GABA does not have ready access to the Ca^{2+}-dependent pool, although it can equilibrate with the cytoplasmic pool and be released by agents which lower the Na^+-electrochemical gradient (Fig. 1).

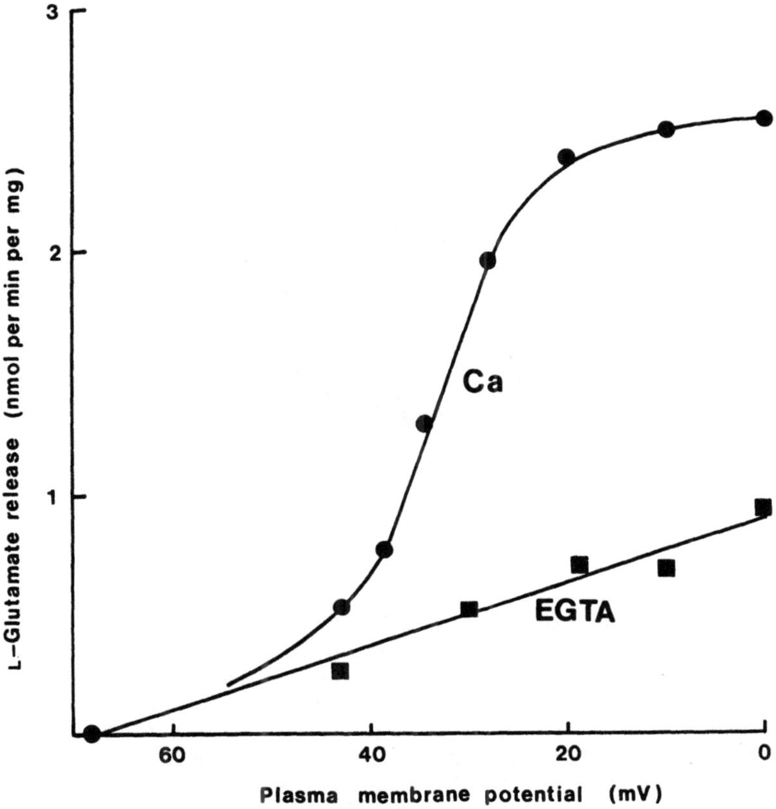

Fig. 4. Release of glutamate and plasma membrane depolarization. Synaptosomes were incubated in the presence or absence of Ca^{2+} and concentrations of KCl from 6mM to 48mM were added to cause a graded depolarization of the membrane. (●) 1.3mM $CaCl_2$ present; (■) 0.5mM EGTA. For details see Nicholls and Sihra (1986).

The results are most easily interpreted in terms of an exocytotic origin for the Ca^{2+}-dependent pool of releasable GABA. The time-dependent increase in Ca^{2+}-dependency for the release of exogenously accumulated GABA argues against a cytosolic origin for Ca^{2+}-dependent GABA, analogous to that proposed for cholinergic termnals (e.g. Israel et al., 1979). If the GABA-rich synaptic vesicles which have been isolated (Angel et al., 1983) were merely for storage of transmitter rather than for release, the Ca^{2+}-dependency for labelled GABA release would decline as label was translocated out of the cytoplasmic "release" pool into the vesicular "storage" pool. This does not occur.

Glutamate release may be monitored continuously

While endogenous glutamate release may be monitored by HPLC, as in the case of GABA, such experiments are time-consuming and do not encourage the detailed investigation of kinetics and mechanism. Our development of a continuous fluorometric assay for released glutamate (Nicholls and Sihra, 1986; Nicholls et al., 1987; Sanchez-Prieto et al., 1987a), in which the inclusion of glutamate dehydrogenase and NADP+ in the incubation medium

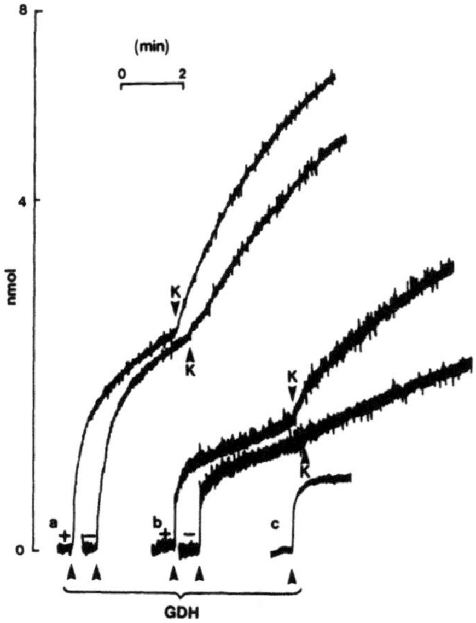

Fig. 5 Preincubation improves the Ca^{2+}-dependency of glutamate release. Synaptosomes were incubated as in Fig. 4, except that for the pair of traces 'a', $NADP^+$ and enzymes were added within 5 min of resuspension, whereas for traces 'b', synaptosomes were preincubated for 120min at 30°C prior to the additions. For each pair, the upper trace is in the presence of $CaCl_2$, added 3 min prior to the enzyme, and the lower trace is in the presence of 0.5mM EGTA; (c) shows the enzyme blank in the absence of synaptosomes. The traces are representative of at least 10 independent experiments. For details see Nicholls et al. (1987).

allows the released glutamate to be trapped and assayed by the fluorescence of the NADPH produced, has greatly increased the ease with which results may be obtained, and in particular has allowed the distinctive kinetics and requirements for Ca^{2+}-independent and Ca^{2+}-dependent release of glutamate to be readily distinguished.

The typical appearance of a set of glutamate release traces is shown in Fig. 3. In the absence of added $NADP^+$ (trace a) no fluorescent changeis seen on addition of either enzyme or KCl. In the presence of $NADP^+$, enzyme addition causes a rapid fluorescence increase due to extra-synaptosomal glutamate, and then a steady release is seen as glutamate exchanging across the plasma membrane carrier is trapped by the assay system. The addition of detergent allows the release of the total internal stores of glutamate. When synaptosomes are depolarized in the absence of Ca^{2+} (trace c) only a slow increase in release is seen, but in the presence of Ca^{2+} depolarization causes a rapid glutamate release, the Ca^{2+}-dependent component of which is largely complete within 2 min.

<u>Ca^{2+}-dependent glutamate release correlates with the opening of voltage-dependent Ca^{2+} channels</u>

Heinonen et al. (1985) have shown that voltage-dependent Ca^{2+} channels in the guinea-pig synaptosomal plasma membrane open when the

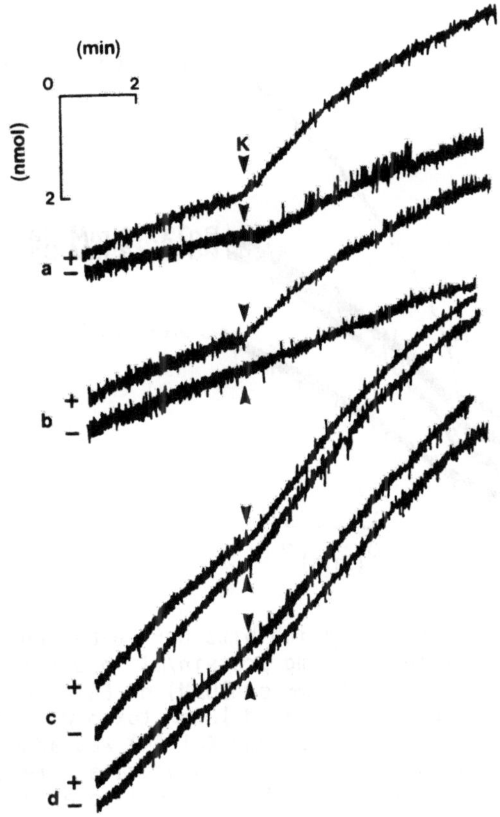

Fig. 6 Energy-dependency of Ca^{2+}-dependent glutamate release. Each pair of traces are respectively in the presence of 1mM Na-EGTA (-) or 1.3mM $CaCl_2$ (+). Metabolic inhibitors were added 8min prior to depolarisation with 30mM KCl (K). (a) controls; (b) 0.4µg/ml oligomycin; (c) 2mM cyanide and (d) 2mM cyanide + 1mM iodoacetate. ATP/ADP ratios at the time of depolarization were respectively 4.3, 2.0, 1.4 and 0.5. For details see Sanchez-Prieto et al., (1987b).

plasma membrane potential is lowered below 40mV. Fig. 4 shows that there is a rapid switch-on of Ca^{2+}-dependent release between 40 and 30mV, when KCl is used to clamp the plasma membrane potential at defined values (Nicholls and Sihra, 1986). In contrast, Ca^{2+}-independent release rates rise in a slow linear fashion as the extent of depolarization is increased. This Ca^{2+}-dependent component is sensitive to high concentrations of the Ca^{2+} channel inhibitor verapamil (Sanchez-Prieto et al., 1987a).

Synaptosomal preincubation enhances Ca^{2+}-dependent release, but inhibits Ca^{2+}-independent release of glutamate

Fig. 5 shows that there is a significant increase in the extent of the Ca^{2+}-dependent release (i.e. the difference $\pm$ Ca^{2+}) when synaptosomes are preincubated prior to depolarization. In contrast, the rate of Ca^{2+}-independent release is greatly inhibited. We have now shown (Pocock and Nicholls, unpublished observations) that this is due to a specific

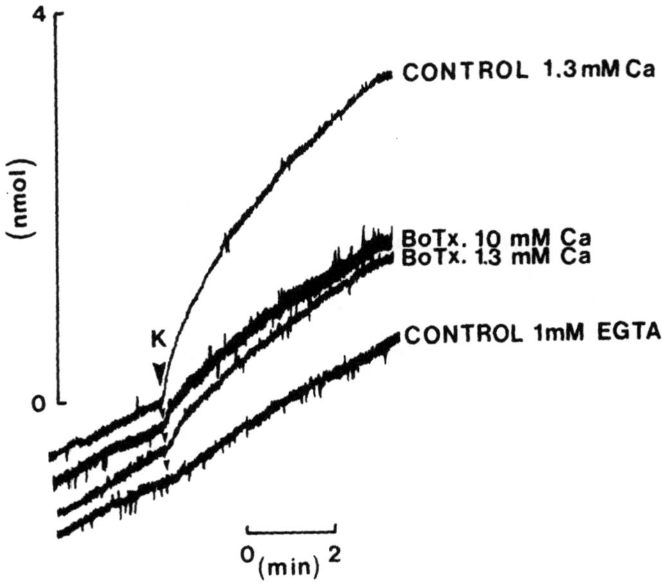

Fig. 7 Botulinum neurotoxin A inhibits Ca^{2+}-dependent glutamate exocytosis. Synaptosomes (0.67mg protein/ml incubation) were preincubated for 2h. $CaCl_2$ 1.3mM or 10mM) or EGTA were added prior to the assay. Where indicated 120nM toxin was present from the start of the preincubation. 30mM-KCl (K) was added at 120min. The vertical scale represents glutamate released (nmol per mg protein). For details see Sanchez-Prieto et al. (1987b).

inhibition of the plasma membrane glutamate carrier, which does not occur, for example, in the case of the GABA carrier.

ATP depletion inhibits Ca^{2+}-dependent release, but enhances Ca^{2+}-independent release

Many forms of secretion require ATP and Fig. 6 shows that Ca^{2+}-dependent glutamate release is no exception. Extensive ATP depletion of synaptosomes by the simultaneous inhibition of oxidative phosphorylation and glycolysis enhances the rate of Ca^{2+}-independent release, but totally blocks any effect of Ca^{2+}.

Botulinum neurotoxin blocks Ca^{2+}-dependent release, but is without effect on Ca^{2+}-independent release

The potent presynaptic botulinum neurotoxin type A is known to block the release of acetylcholine from peripheral terminals (Simpson, 1986). Fig. 7 shows that its action is more generalized, since the toxin almost completely inhibits the Ca^{2+}-dependent release of glutamate from synaptosomes. A period of preincubation at 37°C is required for optimal blockade, due to the need for internalization of the toxin (Simpson, 1986). In view of the ATP requirement for Ca^{2+}-dependent release, controls were performed which showed no decrease in ATP (Sanchez-Prieto et al., 1987b); furthermore there was no inhibition of Ca^{2+} entry upon depolarization. It therefore seems that the toxin acts at a late stage, perhaps the direct

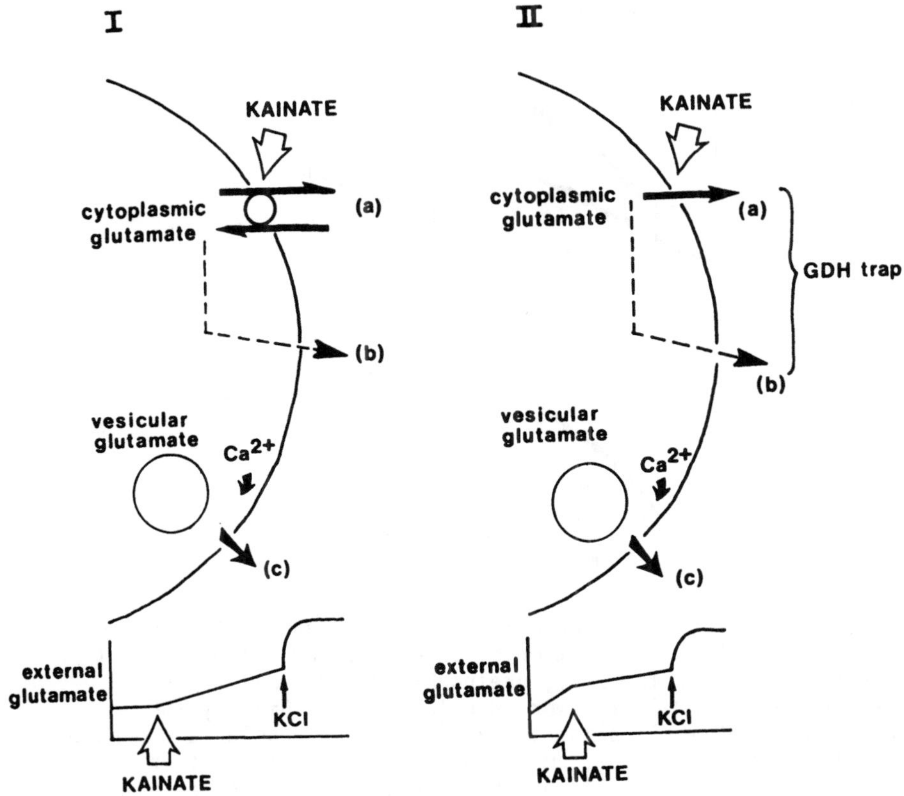

Fig. 8 Summary of the effects of kainate on glutamate/aspartate transport. (I) Steady-state incubation with dynamic exchange of glutamate across the plasma membrane. Kainate inhibits the Na^+-cotransport glutamate or aspartate carrier (a) and therefore inhibits inward exchange of D-aspartate. The observed Ca^{2+}-independent net efflux indicates the presence of a slow leak pathway (b) which becomes more dominant as the Na^+-cotransport pathway becomes inhibited. Ca^{2+}-dependent exocytosis (c) is not affected. (II) Glutamate trapping with glutamate dehydrogenase (GDH). Kainate inhibits the unidirectional efflux of glutamate from the cytoplasm in the presence of the trap, leaving only the slow leak pathway to deplete the cytosol. The overall glutamate efflux is thus inhibited. Again Ca^{2+}-dependent exocytosis is unaffected. For details see Pocock et al. (1988).

interaction of Ca^{2+} with the Ca^{2+}-dependent release apparatus (Sanchez-Prieto et al., 1987b). In contrast, the rate of Ca^{2+}-independent release is completely insensitive to the toxin.

Kainic acids acts presynaptically to inhibit the glutamate carrier, but is without effect on Ca^{2+}-dependent release of glutamate

The reported presynaptic effects of kainic acid are complex and confusing. In the 0.1mM - 1mM range it inhibits the uptake of L-glutamate,

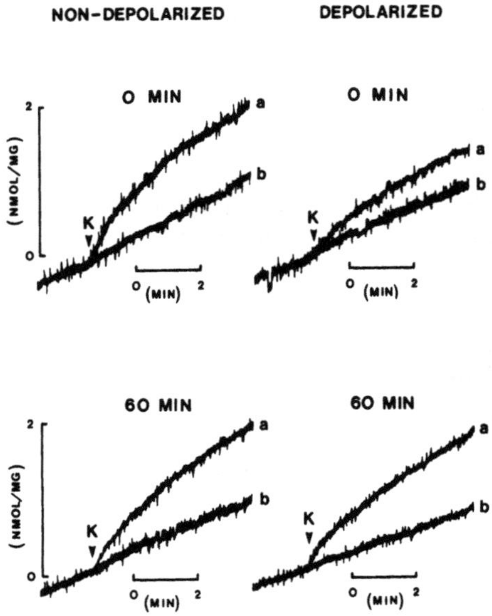

Fig. 9. Time-dependent replenishment of the exocytotic pool of glutamate following depletion by prior depolarization. Synaptosomes (0.67mg protein/ml), preincubated for 60min in the presence of 1.3mM-$CaCl_2$, were depolarized for 5min by the addition of 30mM-KCl prior to washing. A non-depolarized control was treated in parallel, omitting the KCl addition. After resuspension, 1.3mM-$CaCl_2$ or 1mM-EGTA was added. The extent of KCl-induced Ca^{2+}-dependent glutamate release was determined in the fluorimeter either immediately or after 1h.

D-aspartate and L-aspartate (Johnston et al 1979, Krespan et al 1982, Pastuszko et al 1984, Poli et al 1985). In addition, preincubation of brain slices or synaptosomes with kainate causes a net release of endogenous glutamate and aspartate (Krespan et al., 1982; Ferkany & Coyle, 1983; Pastuszko et al., 1984; Poli et al., 1985). However, although Pastuszko et al. (1985) reported that kainate potentiated the release of accumulated [^{3}H]-D-aspartate, Ferkany & Coyle (1983) found no effect of the toxin while Poli et al. (1985) reported that kainate actually inhibited the efflux of [^{3}H]-D-aspartate from superfused synaptosomes.

We have been able to resolve these contradictions by exploiting the fluorometric assay (Pocock et al., 1988). The sole presynaptic action of millimolar kainate on guinea-pig synaptosomes appears to be a blockade of the plasma membrane glutamate carrier (Fig. 8). Ca^{2+}-dependent release of glutamate is not affected, but synaptosomes preincubated in the presence of a glutamate trap (glutamate dehydrogenase or a classical superfusion apparatus) show a decreased rate of Ca^{2+}-independent release of glutamate due to inhibition of the carrier. In contrast, when synaptosomes are preincubated with kainate in the absence of a trap, a large amount of extrasynaptosomal glutamate is found, as a result of inhibition of the carrier preventing the reuptake of glutamate which slowly leaks out of the terminal, perhaps by a carrier-independent mechanism.

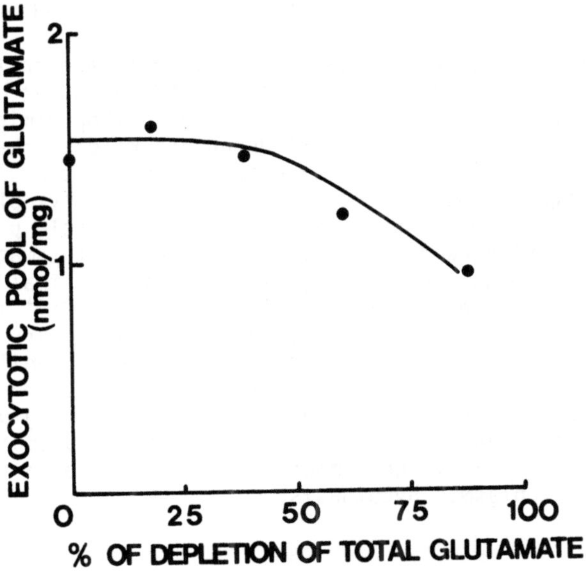

Fig. 10. Depletion of L-glutamate by exchange with D-aspartate does not proportionately decrease the "exocytotic" pool. Synaptosomes were incubated as described in Fig. 3. After depleting by exchange with 100μM-D-aspartate for defined times from 0-50min, the synaptosomes were depolarized by the addition of 30mM-KCl. The exocytotic pool is defined as the difference between the Ca^{2+}-dependent and Ca^{2+}-independent release in the subsequent 5min.

GLUTAMATE COMPARTMENTATION: THE ORIGIN OF THE Ca^{2+}-DEPENDENT POOL

The experiments summarized above indicate that the Ca^{2+}-dependent release of glutamate from guinea-pig synaptosomes shows considerable similarity with the release of other established transmitters and hormones. In particular, an ATP-dependency is common to most, if not all secretory events, while the ability of botulinum neurotoxin to block specifically the Ca^{2+}-dependent release of glutamate indicates a strong homology with acetylcholine release and also with adrenaline exocytosis from chromaffin cells (Knight, 1986; Penner et al., 1986). None of the above experiments, however, prove that glutamate is released by an exocytotic secretory mechanism, rather than by a hypothetical Ca^{2+}-gated glutamate channel which would have to open transiently in the plasma membrane. However, the following two experiments strongly suggest that, as in the case of GABA, the Ca^{2+}-dependent and Ca^{2+}-independent release of glutamate originates from distinct compartments.

Fig. 9 shows that the Ca^{2+}-dependent pool may be extensively depleted by depolarizing synaptosomes for 5min with high KCl. When the synaptosomes are spun down and resuspended in normal, low K^+ medium, the total glutamate content of the synaptosomes is only decreased slightly, but the Ca^{2+}-dependent pool is greatly depleted and takes an hour to return to control levels. Unless one postulates that the hypothetical plasma membrane Ca^{2+}-dependent release channel becomes refractory for such a prolonged period, making it unable to utilize the cytoplasmic glutamate, the results indicate that time is required for the repletion of a functionally distinct transmitter pool after the drastic depletion caused by 5min continuous depolarization.

D-aspartate is transported across the plasma membrane on the same carrier as glutamate, but is not released in a Ca^{2+}-dependent manner on depolarization (unpublished observations). For the "cytoplasmic release" hypothesis this would imply that the Ca^{2+}-dependent release channel cannot transport D-aspartate, while for the "exocytotic" model it would imply that there was no significant uptake of D-aspartate from cytoplasm to vesicles. Interestingly, the highy purified synaptic vesicle preparation of Naito and Ueda (1985) showed the ability to take up glutamate, but not D-aspartate. The experiment shown in Fig. 10 demonstrates that a large external pool of D-aspartate can extensively deplete the Ca^{2+}-independent pool of glutamate from the cytoplasm of synaptosomes with very little effect on the extent of subsequent Ca^{2+}-dependent release. While this would be predicted for two independent pools of glutamate, the cytoplasmic release hypothesis would have to allow a constant extent of glutamate release to occur despite an extensive depletion in the glutamate content of the cytoplasm. While this is not impossible, it provides a further difficulty for the cytoplasmic hypothesis.

CONCLUSIONS

Many of the inconsistencies in the literature concerning synaptosomal amino acid release have been due to difficulties in distinguishing the characteristics of Ca^{2+}-independent release from the cytoplasm by thermodynamic reversal of the plasma membrane carrier, and Ca^{2+}-dependent release, for which we believe there is increasing evidence for an exocytotic mode which may be common to the secretion of a range of transmitters. The complete separation of these two processes, which is now possible for glutamate may provide a basis for further studies of the molecular mechanisms regulating the release of this major excitatory neurotransmitter.

ACKNOWLEDGEMENTS

Work in our laboratory has been supported by the Medical Research Council and the Wellcome Trust. The collaboration of Drs. Talvinder Sihra, Risto Kauppinen and Jennifer Pocock is gratefully acknowledged.

REFERENCES

Abe, M. and Matsuda, M. (1983) On the existence of two GABA pools associated with newly synthesised GABA and with newly taken up GABA in nerve terminals. Neurochem Res. 8, 563-573.

Angel, I., Fleissner A. and Seifert, R. (1983) Synaptic vesicles from hog brain - their isolation and the coupling between synthesis and uptake of v-aminobutyrate by glutamate decarboxylase. Neurochem. Int. 5, 697-712.

Cotman, C. W. and Haycock, J. W. (1976) On the functional coupling of neurotransmitter uptake and release in brain. Br. J. Pharmac. 58, 569-572.

DeBelleroche, J.S. & Bradford, H.F. (1977) On the site of origin of transmitter amino acids released by depolarization of nerve terminals in vitro. J. Neurochem. 29, 335-343.

Ferkany J W & Coyle J T (1983) Evoked release of aspartate and glutamate; disparities between prelabeling and direct measurement Brain Res 278 279-282

Hansford, R.G. and F. Castro. 1985. Role of pyruvate dehydrogenase interconversion in brain mitochondria and synaptosomes. Biochem. J. 227:129-136.

Haycock J. W., Levy, W. B., Denner, L. A. and Cotman, C. W. (1978a) Effects of elevated K^+ on the release of neurotransmitters from cortical synaptosomes: efflux or secretion? J. Neurochem. 30, 1113-1125.

Heinonen, E., Akerman, K.E.O., Kaila, K. and Scott, I.G. (1985) Dependence of cytoplasmic calcium transients on the membrane potential in isolated nerve endings of the guinea-pig. Biochim. Biophys. Acta 815, 203-208.

Israel, M., Dunant, Y. and Manaranche, R. (1979) The present status of the vesicular hypothesis. Prog. Neurobiol. 13, 237-275.

Johnston G A R, Kennedy, S M E & Twitchin B (1979) Action of the neuro toxin kainic acid on high affinity uptake of L-glutamic acid in rat brain slices J Neurochem 32 121-127

Kanner, B. I. (1983) Bioenergetics of neurotransmitter transport. Biochim. Biophys. Acta 726, 293-316.

Kauppinen, R.A. & Nicholls, D.G. (1986a) Synaptosomal bioenergetics: The role of glycolysis, pyruvate oxidation and responses to hypoglycaemia. Eur. J. Biochem. 158, 159-165

Knight,D. (1986) Botulinum toxin types A,B and D inhibit catecholamine secretion from bovine adrenal medullary cells. FEBS Lett. 207, 222-226.

Krespan B, Berl S & Nicklas W J (1982) Alteration in neuronal-glial metab olism of glutamate by the neurotoxin kainic acid J Neurochem 38 509-518

Levi, G., Banay-Schwartz, M. and Raiteri, M. (1978) Uptake, exchange and release of GABA from isolated nerve terminals. In: Amino Acids as Neurotransmitters (Fonnum, F., ed.), pp.327-350, Plenum Press, N.Y.

Martin, D. L. (1976) Carrier-mediated transport and removal of GABA from synaptic regions. In: GABA in Nervous System Function (Roberts, E., Chase, T. N. & Tower, D. B. eds.) pp. 347-386, Raven Press, New York.

Naito, S. and Ueda, T. (1985) Characterization of glutamate uptake into synaptic vesicles. J. Neurochem. 44, 99-109.

Nicholls, D.G. & Sihra, T.S. (1986) Synaptosomes possess an exocytotic pool of glutamate. Nature, 321, 772-773.

Nicholls D G, Sihra T S, Sanchez-Prieto, J (1987) Calcium dependent and independent release of glutamate from synaptosomes monitored by continuous fluorometry J Neurochem 49 50-57

Pastuszko, A., Wilson, D. F. and Erecinska, M. (1982) Energetics of GABA transport in rat brain synaptosomes. J. Biol. Chem. 257, 7514-7519.

Pastuszko A, Wilson D F & Erecinska M (1984) Effects of kainic acid in rat brain synaptosomes: the involvement of calcium J Neurochem 43 747-754

Penner, R., Neher, E. & Dreyer, F. (1986) Intra-cellularly injected tetanus toxin inhibits excoytosis in bovine adrenal chromaffin cells. Nature (Lond.) 324, 76-78.

Pocock, J.M., Murphie, H.M. and Nicholls, D.G. (1988) Kainic acid inhibits the synaptosomal plasma membrane glutamate carrier and allows glutamate leakage from the cytoplasm, but does not affect glutamate exocytosis. J. Neurochem. (in press).

Poli, A., Contestabile, A., Migiani, P., Rossi, L., Virgili, M., Bissoli, R. and Barnebei, O. (1985) Kainic acid differentially affects the synaptosomal release of endogenous and exogenous amino acidic neurotransmitters. J. Neurochem. 45, 1677-1686.

Ryan, L. D. and Roskoski, R. (1975) Selective release of newly synthesised and newly captured GABA from synaptosomes by potassium depolarisation. Nature (London) 258, 254-255.

Sanchez-Prieto J, Sihra T S, Aston A, Evans D, Dolly, J O, Nicholls D G (1987b) Botulinum toxin A and the inhibition of glutamate exocytosis from guinea-pig synaptosomes Eur J Biochem 165, 675-681

Sanchez-Prieto J, Sihra T S, Nicholls D G (1987a) Characterization of the exocytotic release of glutamate from guinea pig cerebral cortical synantosomes J Neurochem 49 58-64
Scott, I.D. and Nicholls, D.G. (1980) Energy transduction in intact synaptosomes: influence of plasma-membrane depolarization on the respiration and membrane potential of internal mitochondria determined in situ. Biochem. J. 186, 21-33.
Sihra, T.S., Scott, I.G. and Nicholls, D.G. (1984) Ionophore A23187, verapamil, protonophores and veratridine influence the release of GABA from synaptosomes by modulation of the plasma membrane potential rather than the cytosolic calcium. J. Neurochem. 43, 1624-1630.
Sihra T S & Nicholls D G (1987) GABA can be released exocytotically from guinea-pig cerebrocortical synaptosomes J Neurochem 49 261-267
Simpson, L.L. (1986) Botulinum toxin. Annu. Rev. Pharmacol. Toxicol. 26, 427-453.

HETEROLOGOUS CARRIERS LOCATED ON RAT CNS NERVE ENDINGS AND THEIR POSSIBLE ROLE IN THE MODULATION OF TRANSMITTER RELEASE

Maurizio Raiteri, Giambattista Bonanno, Anna Pittaluga and Paola Versace
Istituto di Farmacologia e Farmacognosia, Università di Genova, Viale Cembrano 4, I-16148 Genova, Italy.

INTRODUCTION

The existence of presynaptic receptors at the releasing terminal level that mediate the modulation of transmitter release in the central nervous system is now well accepted. The nerve terminals of the central nervous system are also endowed with membrane transporters. It has been suggested that the main function of reuptake processes should be to remove the transmitter released previously into the synaptic cleft and to terminate its action at the postsynaptic receptors.

Very few studies have considered the effects of γ-aminobutyric acid (GABA) on the release of acetylcholine (ACh) (Stoof et al 1979; Scatton and Bartholini 1981; Bianchi et al 1982), dopamine (DA) (Giorguieff et al 1978; Starr 1978,1979; Stoof et al 1979; Ennis and Cox 1981; Reimann et al 1982), noradrenaline (NA) (Starr 1979; Arbilla and Langer 1979; Bowery et al 1980; Fung and Fillenz 1983), and serotonin (5-HT) (Reubi et al 1978; Starr 1979; Balfour 1980; Bowery et al 1980; Schlicker et al 1984). Moreover, the results obtained in these studies have often been controversial as to the direction of the modulation and the type of receptor involved.

Considering the above discrepancies we have reinvestigated the effects of GABA on the release of the various neurotransmitters using the superfused synaptosome technique. The results obtained raise the possibility that GABA might modulate ACh, DA and NA release by a novel mechanism and support the

NATO ASI Series, Vol. H21
Cellular and Molecular Basis of Synaptic Transmission
Edited by H. Zimmermann

hypothesis of a new role for the carrier systems beside those so far acknowledged.

METHODS

Synaptosomes prepared from different rat brain areas according to Gray and Whittaker (1962) and prelabelled with ^{3}H-choline (^{3}H-Ch), ^{3}H-noradrenaline (^{3}H-NA), ^{3}H-dopamine (^{3}H-DA), ^{3}H-5-hydroxytryptamine (^{3}H-5-HT) or ^{3}H-GABA were exposed to the various substances in superfusion (Raiteri et al 1974) at the concentrations indicated. Experimental details can be found in the Legends to the figures.

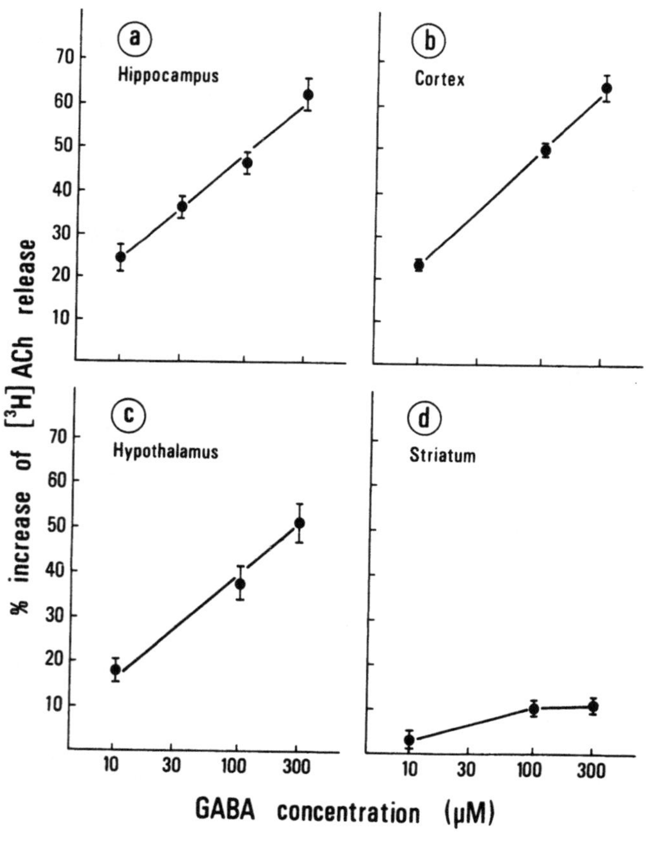

Fig. 1. Potentiation of the basal release of ^{3}H-ACh by GABA in synaptosomes prepared from various areas of the rat brain. Synaptosomes were labelled with ^{3}H-Ch, layered at the bottom of several parallel superfusion chambers and superfused as previously described (Bonanno and Raiteri 1987a,b). Fraction were collected and counted for radioactivity. In some experiments, performed under different experimental conditions, separation of ^{3}H-ACh from ^{3}H-Ch was performed according to Marchi et al (1981). Each point is the mean ± S.E.M. of 4-8 experiments in triplicate.

RESULTS

Effects of GABA on ^{3}H-ACh basal release

When hippocampal synaptosomes prelabelled with ^{3}H-Ch were exposed to GABA in superfusion a dose-dependent long-lasting increase of ^{3}H-acetylcholine (^{3}H-ACh) spontaneous efflux could be observed (Bonanno and Raiteri 1987a; fig. 1, panel a). The effect of GABA was not mimicked by the $GABA_A$ agonist muscimol and by the $GABA_B$ agonist (-)baclofen. Bicuculline and picrotoxin did not antagonize the effect of GABA in rat hippocampus (Fig. 2). These results seem to exclude a possible involvement of the two $GABA_A$ and $GABA_B$ receptors in the effects observed. Unexpectedly the GABA-induced ^{3}H-ACh release was strongly counteracted by new and potent GABA uptake inhibitors, SK&F 89976A, SK&F 100330A, SK&F 100561 (Yunger et al 1984) (Fig. 3).

A similar pattern was observed when the effect of GABA was studied using synaptosomes prepared from different rat brain discrete areas: GABA increased in a dose-dependent way the spontaneous release of ^{3}H-ACh from

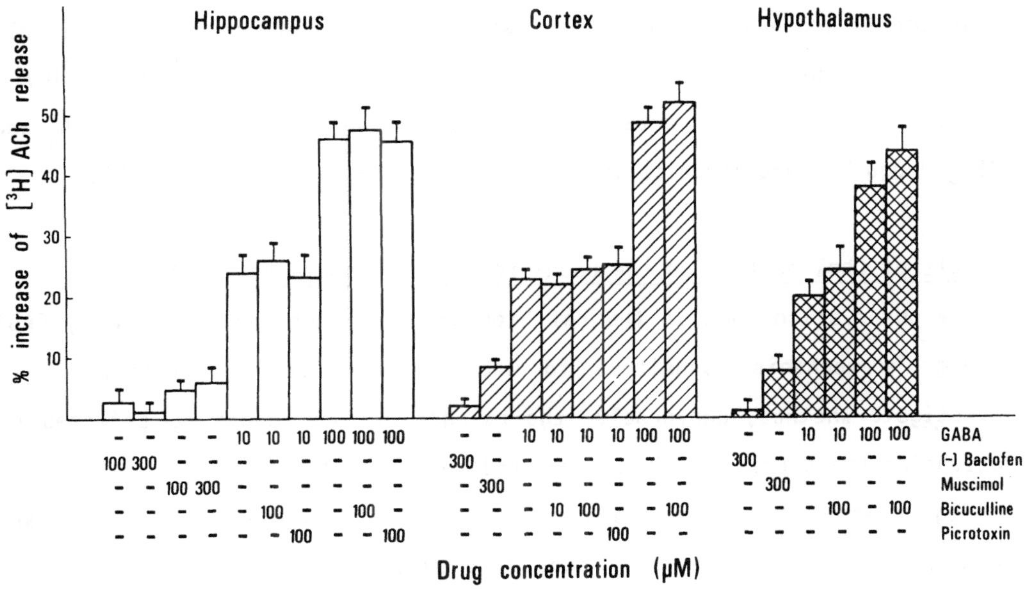

Fig. 2. Study of the GABA-induced ^{3}H-ACh release with drug selective for the GABA receptors. Means ± S.E.M. of 3-6 experiments in triplicate.

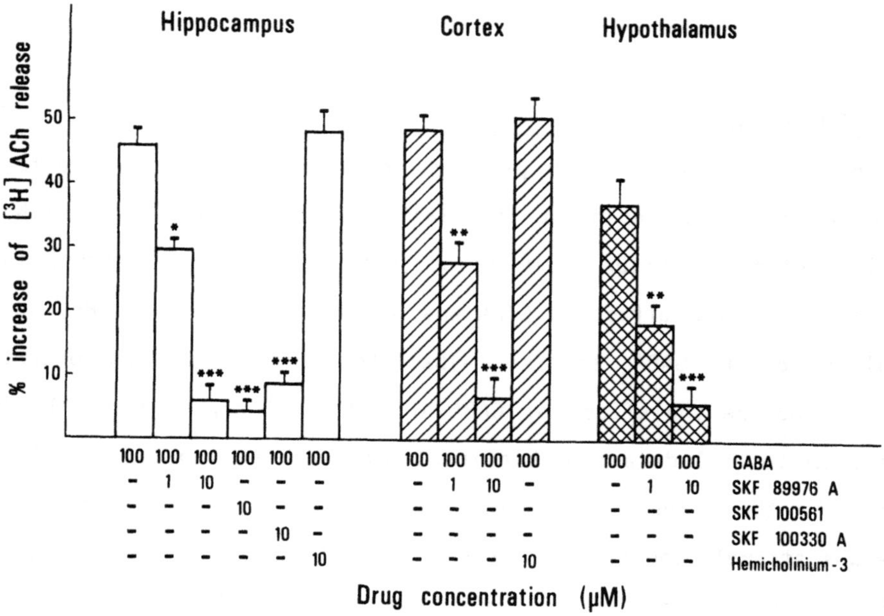

Fig. 3. Effects of inhibitors of GABA or Ch uptake on the GABA-induced ^{3}H-ACh release. Means ± S.E.M. of 3-6 experiments in triplicate. * $P < 0.05$; ** $P < 0.005$; *** $P < 0.001$ using the two-tailed Student's t-test.

brain cortex and hypothalamus synaptosomes while its effect was negligible in the corpus striatum (Fig. 1, panels b,c,d,). Also the GABA-induced ^{3}H-ACh release observed in the cortex and hypothalamus was insensitive to $GABA_A$ and $GABA_B$ drugs but was practically abolished by SK&F 89976A (Figs. 2,3).

Effects of Choline (Ch) on the basal release of endogenous GABA

The observation that in the rat hippocampus ^{3}H-ACh was released by GABA through a mechanism sensitive to GABA uptake blockers prompted us to investigate whether, conversely, the release of GABA could be affected by activating the carrier of Ch on the cholinergic terminal.

Fig. 4 shows that Ch increased dose-dependently the spontaneous release of endogenous GABA from rat hippocampal synaptosomes in superfusion. The effect of Ch was not antagonized by the muscarinic or the nicotinic receptor

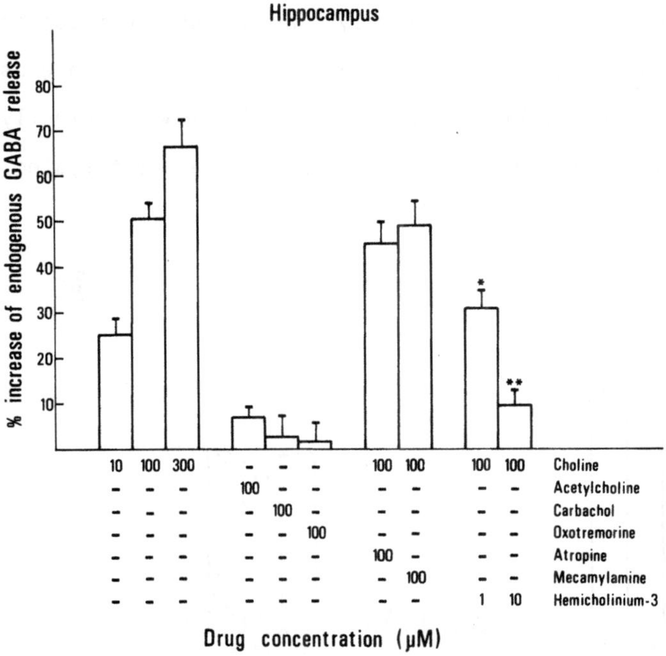

Fig. 4. Dose-dependent Ch-induced increase of endogenous GABA spontaneous release in rat hippocampus: effects of drugs selective for the ACh receptors and hemicholinium-3. Experimental details as previously reported (Pittaluga and Raiteri 1987). Means ± S.E.M. of 3-9 experiments in duplicate or triplicate. * $P < 0.05$; ** $P < 0.001$ using the two-tailed Student's t-test.

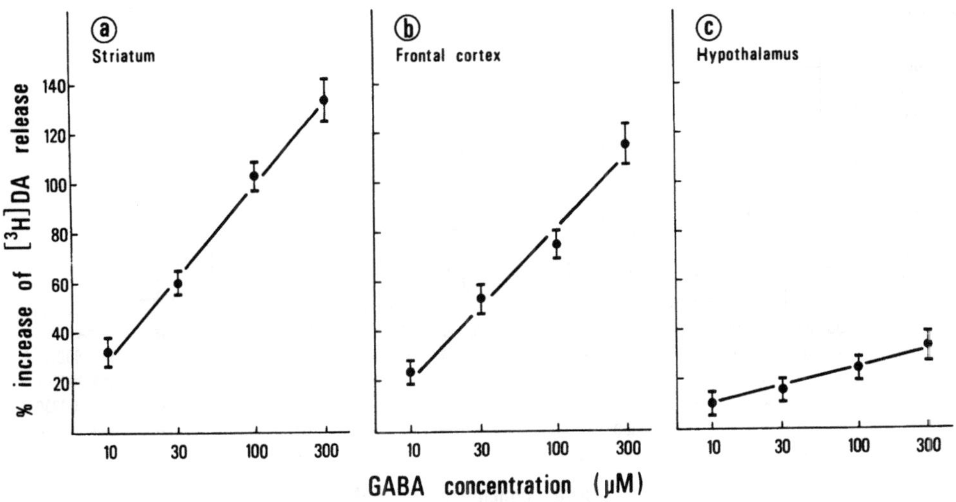

Fig. 5. GABA-induced potentiation of ^{3}H-DA release. For experimental details see the Legend to Fig. 1. In each fraction collected ^{3}H-DA was separated from its ^{3}H-deaminated metabolites according to Smith et al (1975). Means ± S.E.M. of 4-10 experiments run in duplicate.

antagonists atropine and mecamilamine, respectively. The muscarinic receptor agonists ACh, carbachol or oxotremorine showed no effect on the spontaneous efflux of endogenous GABA. Interestingly, the effect of Ch was potently blocked by hemicholinium-3,a selective inhibitor of the Ch transport system.

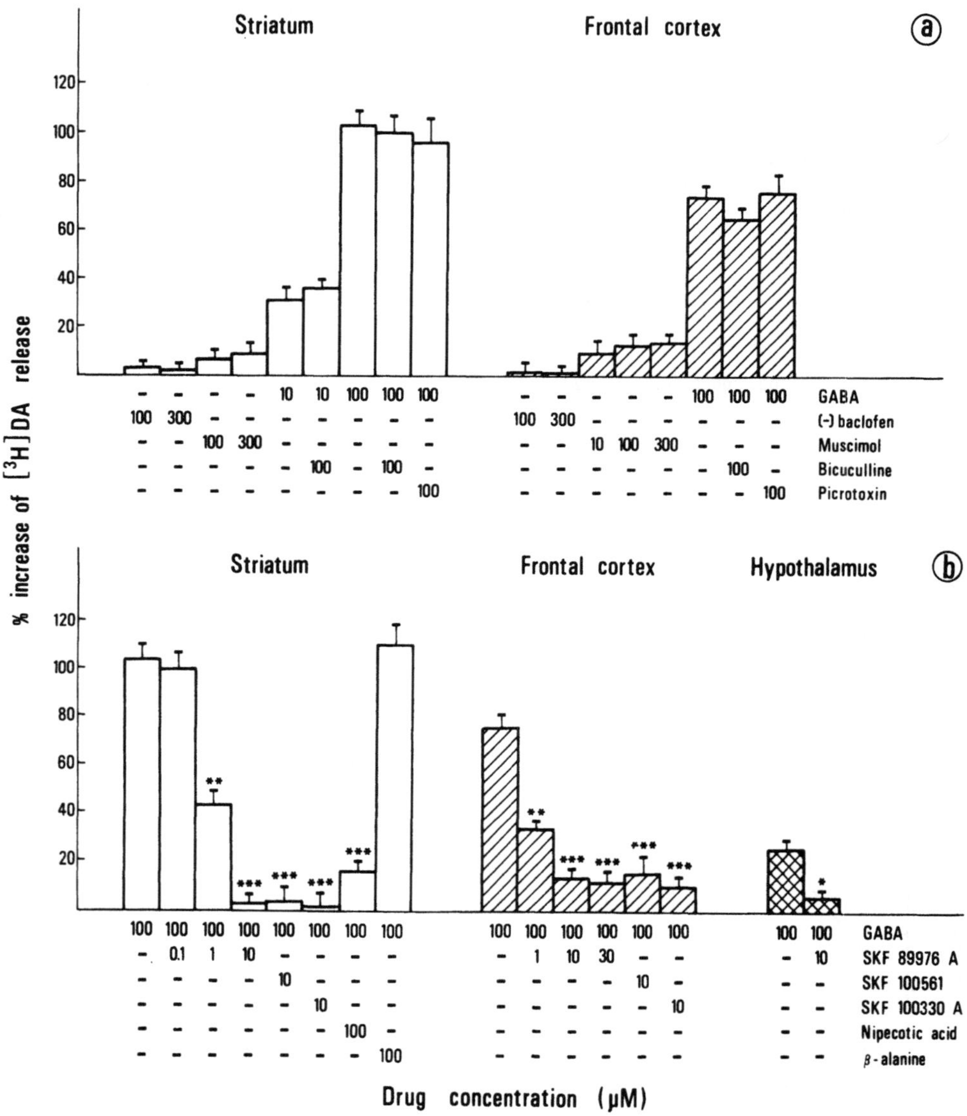

Fig. 6. Study of the GABA-induced ^{3}H-DA release using drugs selective for the GABA receptors (panel a) or GABA uptake inhibitors (panel b). Means ± S.E.M. of 4-6 experiments run in triplicate. * P < 0.05; ** P < 0.01; *** P < 0.001 using the two-tailed Student's t-test.

Effects of GABA on ^{3}H-DA basal release

Similarly to that observed with ^{3}H-ACh, GABA increased the spontaneous efflux of ^{3}H-DA in superfused synaptosomes prepared from rat corpus striatum and frontal cortex but much less from hypothalamus (Fig. 5). The effect of GABA did not seem mediated by $GABA_A$ and $GABA_B$ receptors (Fig. 6, panel **a**) and, also in this case, was sensitive to the GABA uptake inhibitors SK&F 89976A, SK&F 100330A, SK&F 100561 (Fig. 6, panel **b**).

Effects of GABA on ^{3}H-NA basal release

GABA increased also the basal release of ^{3}H-NA from superfused hippocampal synaptosomes prelabelled with ^{3}H-NA. The analysis of the

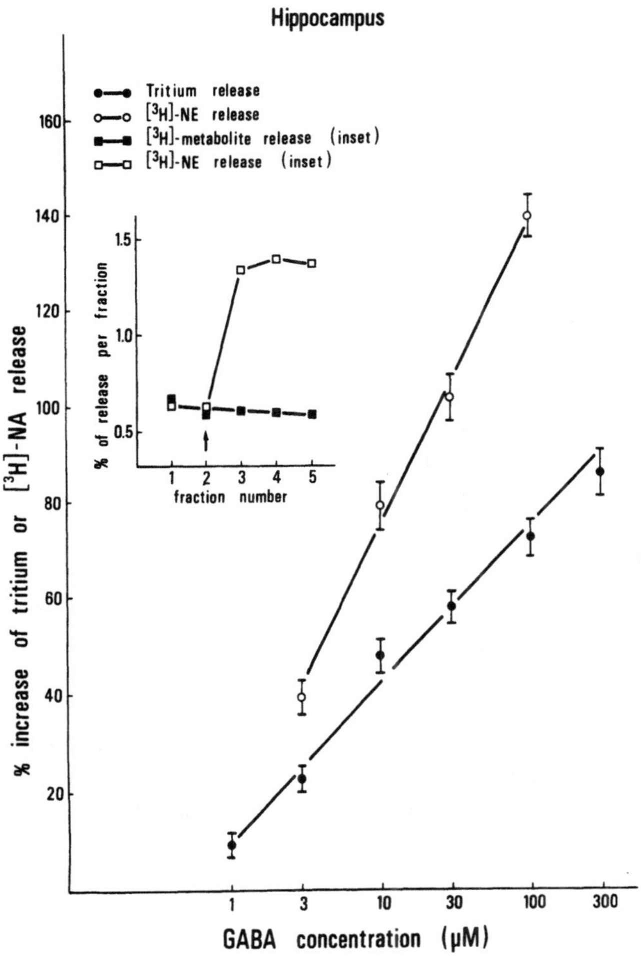

Fig. 7. Effect of GABA on the basal release of ^{3}H-NA or total tritium in rat hippocampus synaptosomes. For experimental details see the Legend to Fig. 1. Separation of ^{3}H-NA from ^{3}H-metabolites was carried out according to Smith et al. (1975). Means ± S.E.M. of 5-10 experiments in duplicate. Inset: effect of 100 μM GABA (added to the superfusion fluid when indicated by the arrow) on the basal efflux of ^{3}H-NA or ^{3}H-deaminated metabolites. Each curve represents the average of 3 experiments in triplicate.

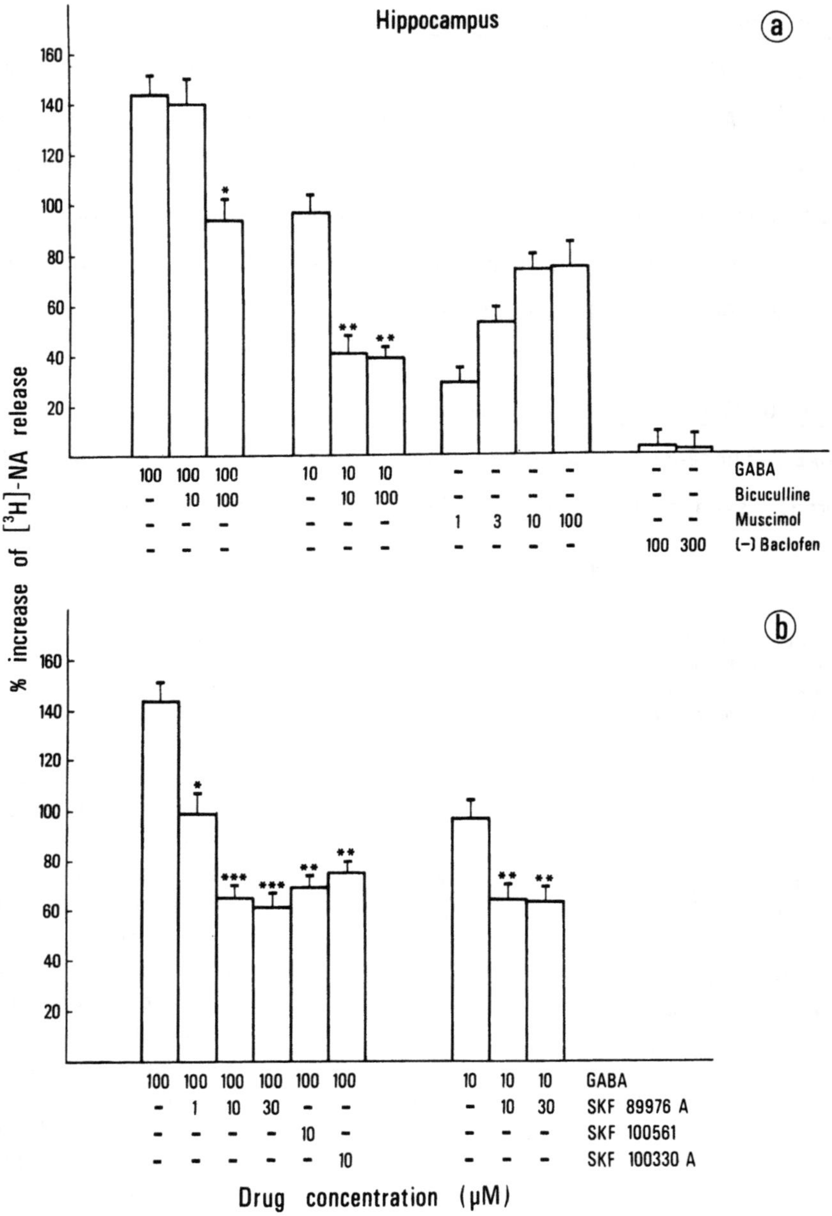

Fig. 8. Study of the GABA-induced ^{3}H-NA release in rat hippocampus synaptosomes using drugs selective for the GABA receptors (panel a) or GABA uptake inhibitors (panel b). Means ± S.E.M. of 3-6 experiments run in triplicate. * $P < 0.05$; ** $P < 0.005$; *** $P < 0.001$.

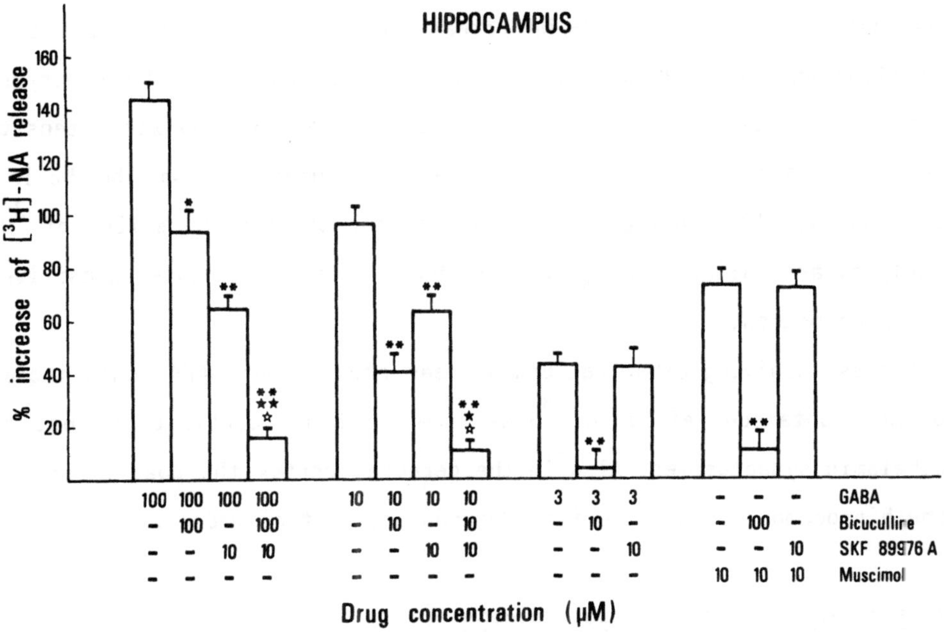

Fig. 9. Studies of the GABA-induced ^{3}H-NA release using drugs selective for the $GABA_A$ receptors and/or GABA uptake system. Means ± S.E.M. cf 3-6 experiments in triplicate. * $P < 0.01$; ** $P < 0.001$ compared to GABA alone. ★$P < 0.01$;★★$P < 0.001$ compared to GABA in presence of 100 μM bicuculline. ☆$P < 0.001$ compared to GABA in presence of 10 μM SK&F 89976A.

radioactivity released by the aminoacid showed it could be totally accounted for unmetabolized neurotransmitter (Fig. 7). The mechanisms by which GABA induced ^{3}H-NA release seem to be more complex than we have seen in the preceeding situations. Fig. 8, panel a shows that (-)baclofen was completely inactive while the effect of GABA was mimicked, but only in part, by muscimol. Bicuculline antagonized GABA but only a portion of the GABA effect was sensitive to the $GABA_A$ receptor antagonist. The enhancement of ^{3}H-NA release evoked by GABA was counteracted by the three above mentioned SK&F compounds (Fig. 8, panel b). Similarly to bicuculline, however, these compounds were unable to antagonize completely the effect of GABA. Fig. 9, shows that the GABA-induced ^{3}H-NA release was almost totally blocked when bicuculline and SK&F 89976A were present concomitantly in the superfusion medium. In contrast the effect of muscimol on ^{3}H-NA release was insensitive to SK&F 89976A but completely sensitive to bicuculline. On the basis of these results, GABA can apparently enhances the spontaneous efflux of ^{3}H-NA not only by activating a $GABA_A$ receptor but also by a mechanism sensitive to the SK&F compounds.

Studies involving other rat brain areas showed that GABA could increase also the spontaneous efflux of ^{3}H-NA from rat brain cortical but not from hypothalamic synaptosomes. Also in the cerebral cortex the dual effect seen in the hippocampus seems to exist (Bonanno and Raiteri 1987e).

Effects of GABA on ^{3}H-5HT basal release

When synaptosomes prepared from rat brain cortex, hypothalamus, hippocampus, midbrain or cerebellum and prelabelled with ^{3}H-5HT were exposed to exogenous GABA in superfusion (10-300 µM), the aminoacid did not affect significantly the basal efflux of radioactivity (data not shown).

DISCUSSION

The experiments described above have been carried out in rat brain synaptosomes prelabelled with nanomolar concentrations of ^{3}H-Ch, ^{3}H-DA, ^{3}H-NA, or ^{3}H-5-HT often in the presence of appropriate uptake inhibitors, in order to avoid the possibility that a given neurotransmitter could be taken up uncorrectly, i.e. ^{3}H-DA into noradrenergic terminals or vice versa. When the effects of Ch on GABA release were studied this problem has been overcome measuring directly the release of the endogenous neurotransmitter. Moreover, the release was studied by superfusing a thin layer of synaptosomes, a technique which minimizes the possibility of indirect effects often occurring using more intact brain preparations (see Raiteri and Levi 1978, for technical details). Finally ^{3}H-ACh was separated from

^{3}H-Ch and ^{3}H-DA or ^{3}H-NA from their ^{3}H-metabolites.Thus the first conclusion can be drawn is that GABA enhances the spontaneous efflux of ^{3}H-ACh, ^{3}H-DA and ^{3}H-NA (and Ch that of endogenous GABA) by acting directly and specifically on the respective nerve terminals.

The second consideration regards the nature of the mechanisms responsible for the effects observed. It is well known that GABA can activate two types of receptors, namely $GABA_A$ and $GABA_B$ (Enna 1984). The findings that the GABA-induced ^{3}H-ACh and ^{3}H-DA release was not mimicked by the $GABA_A$ agonist muscimol or by the $GABA_B$ agonist (-)baclofen and was insensitive to the $GABA_A$ antagonists bicuculline and picrotoxin tend to exclude a possible involvement of the above mentioned receptors. On the contrary, part of the GABA-induced ^{3}H-NA release seems to be mediated by a $GABA_A$ receptor sensitive mechanism, according with the data in literature (Fung and Fillenz 1983). But in spite of this, a still large portion of ^{3}H-NA release (varying with the different concentrations of GABA used: larger at the highest, negligible at the lowest) remained insensitive to bicuculline and picrotoxin. In addition, the effect of GABA was mimicked only partly by muscimol. All together these data support the possibility of the presence of a second unknown mechanism also in this latter case.

In this context a surprising occurrence was to discover that the effects of GABA were potently counteracted by three new GABA uptake inhibitors, SK&F 89976A, SK&F 100330A and SK&F 100561 (Yunger et al 1984). Such an observation may induce to different considerations. For example it could be suggested that GABA increased the spontaneous release of the considered neurotransmitters by activating a new GABA receptor subtype, not $GABA_A$ or $GABA_B$, where the three above mentioned substances would act as potent antagonists. However, if this hypothesis could not be excluded, several observations seem to weaken it. In fact, the three SK&F compounds are till now known to be only potent and selective GABA uptake inhibitors (Yunger et al 1984; Bonanno and Raiteri 1987c) without behaving as substrates for the GABA transport system (Bonanno and Raiteri unpublished results). These observations are in keeping with the inability by these substances to displace _in vitro_ ^{3}H-muscimol (Yunger et al 1984) and ^{3}H-GABA binding (Pittaluga and Raiteri unpublished results) from their specific sites. Furthermore, they show _in vivo_ a potent anticonvulsant activity in rats and mice (Yunger et al 1984), a finding suggestive of a facilitating action on the GABA transmission, more consistent with a GABA reuptake inhibition rather than with a receptor blocking activity. On the basis of the preceeding considerations it could be tempted to propose that GABA penetrates into the cholinergic, dopaminergic and noradrenergic nerve terminals through a transport system selective for the aminoacid and, in such a manner, stimulates neurotransmitter release.

The specificity of the phenomena observed is supported by some experimental evidences. The effect of GABA on ^{3}H-ACh release was not blocked by the Ch uptake inhibitor hemicholinium-3 (Fig. 3 and Bonanno and Raiteri 1987b) and the effect on ^{3}H-DA and ^{3}H-NA release was unaffected by the specific uptake inhibitors GBR 12909 and fluoxetine, respectively (Bonanno and Raiteri unpublished results). This fact tends to exclude that GABA could be transported into the nerve terminals by Ch, DA, or NA transport systems. The possibility that a non-selective low affinity amino acid transport site should be involved is unlikely since a number of neutral amino acids did not evoke neurotransmitter release (Bonanno and Raiteri 1987a,c,d). It seems also improbable that GABA penentrates into "gliosomes" which may be present in our synaptosomal preparations, since the effect of GABA was not prevented by β-alanine (Fig. 6 and Bonanno and Raiteri 1987a,e), an inhibitor of GABA glial uptake. Moreover, 2,4-diaminobutyric acid, considered to be a substrate of GABA transport, and nipecotic acid, a substrate-inhibitor of GABA uptake without a known receptor activity, mimicked, although less potently, the effect of GABA (Bonanno and Raiteri 1987a,c). Finally, the phenomenon of the carrier coexistence occurs specifically in discrete rat brain areas (see in Fig. 1,5 and Bonanno and Raiteri 1987e the lack of effect in the corpus striatum for ^{3}H-ACh and in the hypothalamus for ^{3}H-DA and ^{3}H-NA release) and seems to be transmitter selective. In fact, no effect of GABA on the spontaneous efflux of ^{3}H-5-HT could be observed in all the rat brain regions investigated (Bonanno and Raiteri 1987e).

The hypothesis of a carrier coexistence on a same nerve terminal is very strongly supported by the finding that, in rat hippocampal synaptosomes, not only GABA increases the spontaneous ^{3}H-ACh release through a putative carrier-mediated mechanism but, conversely, Ch increases the spontaneous efflux of endogenous GABA. Also Ch operates by a mechanism insensitive to muscarinic or nicotinic receptor agonists or antagonists. And also in this case the effect was completely blocked by a transport system blocker: the Ch uptake inhibitor hemicholinium-3 (Fig. 4).

Why a carrier for GABA, which has been thought of as a characteristic of GABAergic nerve endings, should also be present on cholinergic, dopaminergic and noradrenergic nerve terminals and a carrier of Ch is reciprocally identifiable on GABAergic synaptosomes can be only matter of speculation. Carriers for a given neurotransmitter are thought to be present on the nerve terminals releasing that transmitter in order to remove it from the synaptic gap. Receptors sensitive to the neurone's own transmitter (autoreceptors) are also present on the terminal to regulate release. Some nerve endings possess heteroreceptors through which modulators can regulate the release of the transmitter. Similarly, the presence of a heterocarrier on a given nerve terminal might be evidence for an hitherto unknown route of presynaptic modulation of neurotransmitter release, via penetration of the modulator into the nerve ending rather than activation of presynaptic heteroreceptors.

The coexistence of heterologous carriers on the same nerve terminal, as hypothesized, raises another interesting possibility. It is now well accepted that some neurones contain more than one substance though to act as a neurotransmitter. In particular, several cases of neuronal coexistence of two peptides or of a classical neurotransmitter with a peptide have been reported (see for example, Hökfelt et al 1980; Cuello 1982). On the basis of the results presented it is tempting also to propose the coexistence of two classical neurotransmitters. If this were true it should be the first functional evidence of such an occurrence. Thus, the evidentiation of the carrier coexistence could be useful as a support for other analytical techniques, i.e. using immunocyto- and immunoisto-chemical probes, and vice versa.

Supported by grants from the Italian Ministry of Education, from the Italian C.N.R. and from Regione Liguria.

REFERENCES

Arbilla S, Langer SZ (1979) Facilitation by GABA of the potassium-evoked release of ^{3}H-noradrenaline from rat occipital cortex. Naunyn-Schmiedeb Arch Pharmacol 306: 161-168

Balfour DJK (1980) Effects of GABA and diazepam on ^{3}H-serotonin release from hippocampal synaptosomes. Eur J Pharmacol 68: 11-16

Bianchi C, Tanganelli S, Marzola G, Beani L (1982) GABA induced changes in acetylcholine release from slices in guinea-pig brain. Naunyn-Schmiedeb Arch Pharmacol 318: 253-258

Bonanno G, Raiteri M (1987a) Presence of γ-aminobutyric acid (GABA) uptake system on cholinergic terminals of rat hippocampus: evidence for neuronal coexistence of acetilcholine and GABA? J Pharmacol Exp Ther 240: 294-297

Bonanno G, Raiteri M (1987b) Regional selectivity of a γ-aminobutyric acid induced [^{3}H]acetilcholine release sensitive to inhibitors of γ-aminobutyric acid uptake. J Neurochem 48: 1454-1458

Bonanno G, Raiteri M (1987c) Coexistence of carriers for dopamine and GABA uptake on a same nerve terminal in the rat brain. Br J Pharmacol 91: 237-243

Bonanno G, Raiteri M (1987d) Carriers for GABA and noradrenaline uptake coexist on a same nerve terminal in rat hippocampus. Eur J Pharmacol 136: 303-310

Bonanno G, Raiteri M (1987e) A carrier for GABA uptake exists on noradrenaline nerve endings in selective rat brain areas but not on serotonin terminals. J Neural Transmission 69: 59-70

Bowery NG, Hill DR, Hudson AL, Doble A, Middlemiss DN, Shaw J, Turnbull M, (1980) (-)Baclofen decreases neurotransmitter release in the mammalian CNS by an action at a novel GABA receptor. Nature 283: 92-94

Cuello AC (Ed) (1982) Co-transmission. Mac Millan Press 1982 New York

Enna SJ (Ed) (1984) The GABA receptors. Umana Press 1984 Clifton, NY

Ennis C, Cox B (1981) GABA enhancement of ^{3}H dopamine release from slices of rat striatum: dependence on slice size. Eur J Pharmacol 70: 417-420

Fung SC, Fillenz M (1983) The role of pre-synaptic GABA and benzodiazepine receptors in the control of noradrenaline release in rat hippocampus. Neurosci Lett 42: 61-66

Giorguieff MF, Kemel ML, Glowinsky J, Besson MJ (1978) Stimulation of dopamine release by GABA in rat striatal slices. Brain Res 139: 115-130

Gray EG and Whittaker VP (1962) The isolation of nerve endings from brain:

an electron microscope study of cell fragments derived by homogenization and centrifugation. J Anat 96: 79-87

Hökfelt T, Johansson O, Ljungdahl Å, Lundberg JM, Schultzberg M (1980) Peptidergic neurons. Nature 284: 515-521

Marchi M, Hoffman DW, Giacobini E, Volle R (1981) Development and aging of cholinergic synapses. VI. Mechanisms of acetylcholine biosynthesis in the chick iris. Dev Neurosci 4: 442-450

Pittaluga A, Raiteri M (1987) Choline increases endogenous GABA release in rat hippocampus by a mechanism sensitive to hemicholinium-3. Naunyn-Schmiedeb Arch Pharmacol 336: 327-331

Raiteri M, Angelini F, Levi G (1974) A simple apparatus for studying the release of neurotransmitters from synaptosomes. Eur J Pharmacol 25: 411-414

Raiteri M, Levi G (1978) Release mechanisms for catecholamines and serotonin in synaptosomes. In: Ehrenpreis S, Kopin I (Eds). Reviews of Neuroscience, vol 3. Raven Press 1978 New York: 77-130

Reimann W, Zumstein A, Starke K (1982) γ-aminobutyric acid can both inhibit and facilitate dopamine release in the caudate nucleus of the rabbit. J Neurochem 39: 961-969

Reubi JC, Emson PC, Jesell TM, Iversen LL (1978) Effects of GABA, dopamine, and substance P on the release of newly synthesized ^{3}H-5-hydroxytryptamine from rat substantia nigra in vitro. Naunyn-Schmiedeb Arch Pharmacol 304: 271-275

Scatton B, Bartholini G (1981) GABA-Acetylcholine interaction in the rat striatum. In: Pepeu G, Ladinsky H (Eds). Cholinergic mechanisms. Plenum Press 1981 New York: 771-780

Schlicker E, Classen K, Göthert M (1984) $GABA_B$ receptor-mediated inhibition of serotonin release in the rat brain. Naunyn-Schmiedeb Arch Pharmacol 326:99-105

Smith JE, Lane JD, Shea PA, McBride WJ, Aprison MH (1975) A method for concurrent measurement of picomole quantities of acetylcholine, choline, dopamine, norepinephrine, serotonin, 5-hydroxytryptophan, 5-hydroxyindoleacetic acid, tryptophan, tyrosine, glycine, aspartate, glutamate, alanine and gamma-aminobutyric acid in single tissue samples from different areas of rat central nervous system. Anal Biochem 64: 149-169

Starr MS (1978) GABA potentiates potassium-stimulated ^{3}H-dopamine release from slices of rat substantia nigra and corpus striatum. Eur J Pharmacol 48: 325-328

Starr MS (1979) GABA-mediated potentiation of amine release from nigrostriatal dopamine neurones in vitro . Eur J Pharmacol 53: 215-226

Stoof JC, Den Breejen EJS, Mulder AH (1979) GABA modulates the release of dopamine and acetylcholine from rat caudate nucleus slices. Eur J Pharmacol 57: 35-42

Yunger LM, Fowler PJ, Zarevics P, Setler PE (1984): Novel inhibitors of γ-aminobutyric acid (GABA) uptake: anticonvulsant actions in rats and mice. J Pharmacol Exp Ther 228: 109-115

SPONTANEOUS RELEASE OF ACETYLCHOLINE FROM *TORPEDO* ELECTROMOTOR NEURONS

Roxana Licht and Daniel M. Michaelson
Department of Biochemistry, Tel Aviv University, Ramat Aviv Israel 69978.

INTRODUCTION

Acetylcholine (ACh) is released from cholinergic nerve terminals both spontaneously and by presynaptic stimulation. Evoked ACh release is Ca^{2+} dependent and quantal (DelCastillo & Katz, 1954), whereas basal ACh release is Ca^{2+} independent and is distinct from the randomly occurring miniature end plate potentials (MEPPs) which account for only a few percent of the spontaneously released ACh (Katz & Miledi, 1977).

The function and mechanims responsible for spontaneous ACh release, whose origin seems to be cytoplasmic (Carroll, 1983), are not known. Recent studies revealed that 2-(4-phenylpiperidino) cyclohexanol (AH5183), a specific inhibitor of ACh uptake into isolated cholinergic synaptic vesicles, blocks basal ACh release from neuromuscular junctions (Edwards et al., 1985). This led to the suggestion that basal release is mediated by the vesicular transporter which is incorporated into the presynaptic plasma membrane during exocytosis. Since the density of neuromuscular junctions is relatively low, the kinetics of basal release from this preparation can be monitored only indirectly by electrophysiological measurements of the effects of the released ACh on the post synaptic membrane potential. This restriction can be overcome by utilizing the

NATO ASI Series, Vol. H21
Cellular and Molecular Basis of Synaptic Transmission
Edited by H. Zimmermann

Torpedo electric organ which is richly innervated with cholinergic neurons and therefore is most suitable for the study of presynaptic processes (Whittaker et al., 1972, 1975; Israel et al., 1970, 1976; Dowdall & Zimmerman 1977; Dunant et al, 1980; Michaelson & Sokolovsky, 1976).

In the present work we studied the kinetics of ACh release from *Torpedo* electric organ prisms and the effects thereon of AH5183 and other cholinergic ligands.

Experimental

Measurements of spontaneous ACh release: Single prisms were separated from freshly excised *Torpedo* electric organs and suspended in *Torpedo* buffer (250 mM NaCl, 4.8 mM KCl,2.4 mM $MgCl_2$, 10 mM glucose, 260 mM sucrose, 1.2 mM phosphate buffer, pH 7.0) for 30 min. The prisms were then transferred to fresh *Torpedo* buffer in which they were kept for up to 2 hours prior to use. At t=o the blocks were transferred into *Torpedo* buffer which contained 2 mM Ca^{2+} after which samples were taken at the indicated times. Unless otherwise indicated, experiments were performed at 25°C. Additions to the medium were done either 30 min. (α-bungarotoxin, atropine, phospholine iodide, paraoxon) or 15 min. (hemicholinium-3, carbamylcholine, oxotremorine and indomethacin) prior to initiation of the experiment. AH5183 was added at t=o.

Measurements of ACh: The released ACh was assayed by the chemiluminiscent method developed by Israel & Lesbats (1981). When the tissue acetylcholinesterase (AChE) was inhibited (100 μM phospholine iodide or 150 μM paraoxon) the released ACh was measured directly whereas when release was performed in the absence of AChE inhibitors ACh was assayed by measuring the levels of choline liberated following its release and hydrolysis. In the latter case chemiluminiscence due to molecules other than choline was eliminated by pretreating the samples with luminol and peroxidase. Results presented are average ± S.D. of the indicated number of experiments.

Control experiments in which trace amounts of [^{3}H]ACh were added to the reaction mixture revealed that in the absence of AChE inhibitors all the released ACh was hydrolyzed within less than 1 min whereas in the presence of phospholine iodide (100 μM) or paraoxon (150 μM) less than 10% were hydrolyzed in 30 min.

RESULTS

A. Spontaneous release in the presence and absence of acetylcholinesterase inhibitors

Incubation of *Torpedo* electric organ prisms at 25°C in modified *Torpedo* buffer in the presence of the acetylcholinesterase (AChE) inhibitor phospholine iodide (100 µm) and Ca^{2+} (2mM) resulted in spontaneous ACh release (Fig. 1). Release was linear for about 10 min (0.27 ± 0.1 nmol/gr tissue/min; n=11) after which it levelled off to a plateau (2.75 ± 1.0 nmol/gr tissue; n=11) which was unchanged for at least 60 min. The amount of ACh thus released corresponds to 0.4 ± 0.03 percent (n=3) of the total tissue ACh. Basal ACh release was not affected either by omitting Ca^{2+} from the incubation mixture or by the addition of EGTA (1 mM) (respectively 2.6 ±0.4 and 2.9 ± 0.3 nmol/gr tissue; n=3). Similar results were obtained when the AChE was inhibited by paraoxon (150 µm) except that under

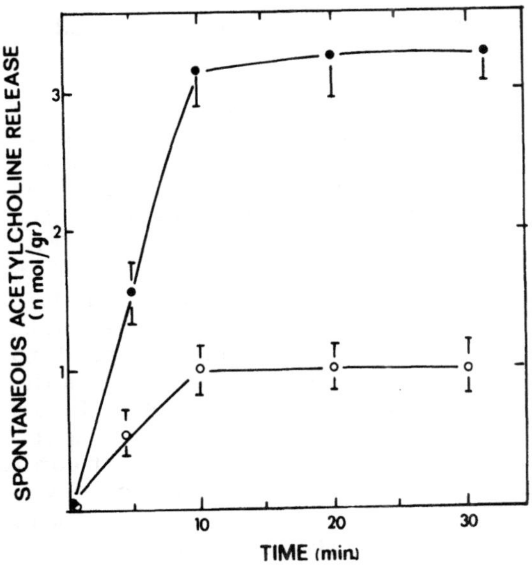

Fig.1. Spontaneous ACh release from the Torpedo electric organ. Single electric organ prisms were incubated at 25°C (●) and 4°C (○) in *Torpedo* buffer + Ca^{2+} (2 mM) which contained phospholine iodine (100 µM). Ach release was measured by withdrawing samples at the indicated times, as described in Experimental.

these conditions the plateau level was lower (1.7 ± 0.2 nmol/gr tissue; n = 2). The temperature dependence of ACh release is depicted in Fig. 1. As can be seen at 4°C the initial rate (0.1 ± 0.01 nmol/gr tissue/min; n = 3) and the extent of release at the plateau (1.0 ± 0.1 nmol/gr tissue; n = 3) were about three-fold lower than at 25°C.

The inhibition of ACh release was completely lost when electric organ prisms were transferred to fresh media every 10 min whereas when the tissue was transferred to media which had been preincubated with electric organ prisms, release was completely blocked (Fig. 2).The inhibition of ACh release by conditioned medium was completely abolished by heating the media to 90°C for 10 min. (Fig. 2). Since ACh was not hydrolyzed by this treatment this suggests that the electric organ prisms release a factor which inhibits release and that this factor is not ACh. The cholinergic agonists oxotremorine (5 μM) and carbamylcholine (0.1 μM) had no effect on release even when added

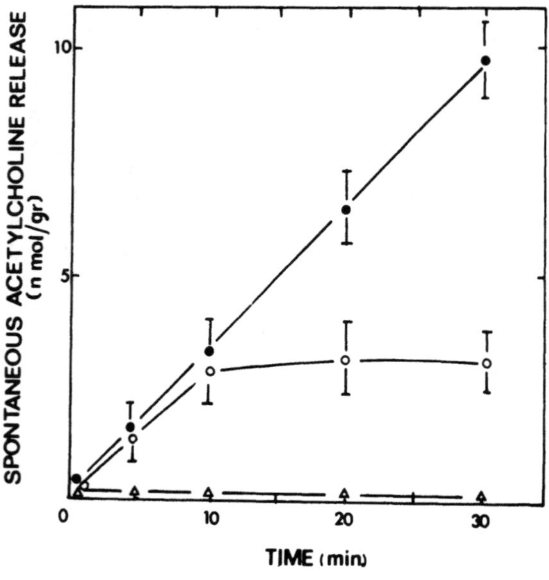

Fig.2. The effect of conditioned and fresh media on ACh release. Single electric organ prisms were incubated in Torpedo buffer + Ca^{2+} (2 mM) in the presence of phospholine iodine (100 μM) either continuously for 30 min. (○)or were transferred to fresh medium every 10 min (●). Release from preparations which were incubated in conditioned medium is indicated by (△). The medium was conditioned by preincubation with electric organ prisms (0.1gr/ml for 30 min). ACh release was measured as described in Experimental.

under conditions of maximal release, i.e. tissue transfers to fresh media every 10 min (respectively 107 ± 7 and 100 ± 3 percent of control; n = 2). The nicotinic antagonist α-bungarotoxin(0.1 μM) and the muscarinic antagonist atropine (10 μM) also did not affect ACh release (respectively 107 ± 7 and 99 ± 4 percent of control; n = 2) nor did indomethacin (25 μM) which in the Torpedo interferes with the synthesis of the muscarinic ACh receptor second messenger (Pinchasi et al., 1984)(110 ± 8 percent of control; n = 2).

In order to determine whether secretion of the inhibitory factor is affected by ACh it is pertinent that release be studied under conditions which ensure rapid ACh hydrolysis. This was done by characterizing basal release in the absence of AChE inhibitors. Incubation of electric organ prisms at 25°C in the absence of AChE inhibitors resulted in spontaneous ACh release which proceeded linearly for up to at least 30 min Fig.3). The initial rates (0 - 10 min) of ACh release in the absence and presence of phospholine iodide were similar (respectively 0.28 ± 0.15, and 0.27 ± 0.1 nmol/gr tissue/min; n = 11). However, at longer incubation times (10 - 30 min) ACh release in the absence of AChE inhibitors proceeded at the initial rate (Fig. 3) whereas it was completely blocked when phospholine iodide was present (Fig. 1, Fig. 2). Similar findings were obtained at 4°C (Fig. 1, Fig. 3). Hemicholinium-3 (5 μM) had no effect on ACh release either in the presence or absence of phospholine iodide. Carbamylcholine (0.1 μM) reduced the rate of ACh release in the absence of AChE inhibitors by 35 ± 10 percent (Fig. 4) but had no effect on ACh release in the presence of phospholine iodide.

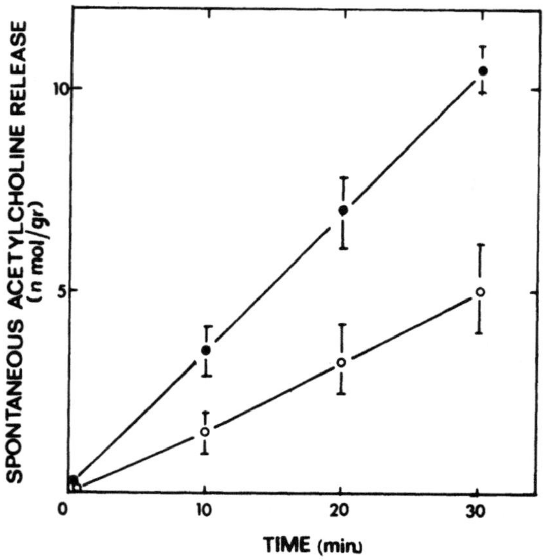

Fig.3. Spontaneous ACh release in the absence of acetylcholinesterase inhibitors. Single electric organ prisms were incubated at 25°C (●) and 4°C (o) in *Torpedo* buffer + Ca^{2+} (2mM). Release was measured as described in Experimental.

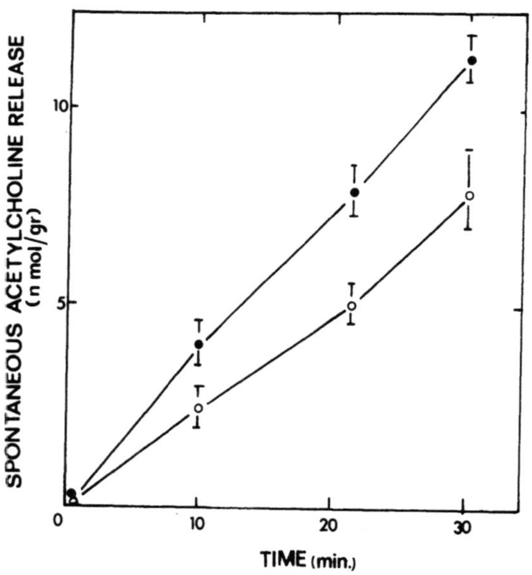

Fig.4. The effect of carbamylcholine on spontaneous ACh release in the absence of acetylcholinesterase inhibitors. Single electric organ prisms were incubated in *Torpedo* buffer + Ca^{2+} (2 mM) in the presence (o) or absence of carbamylcholine (0.1 μM) (●). Ach release was measured as described in Experimental.

B. The effect of AH5183 and pH on spontaneous acetylcholine release

Incubation of electric organ prisms with AH5183 (0.01 - 1 μM) in the absence of AChE inhibitors resulted in repressed basal release (Fig. 5). Maximal inhibition (35 ± 10 percent ; n = 12) was obtained at 0.1 μM with EC_{50} 0.01 μM. By contrast AH5183 (0.01 - 10 μM) had no effect on ACh release from electric tissue whose AChE was inhibited by either phospholine iodide or paraoxon (not shown). The finding that in the presence of AChE inhibitors AH5183 was ineffective raises the possibility that the effects of this compound on release are blocked by the secreted ACh. This was examined by measurements of the effects of AH5183 on basal ACh release in the presence of carbamylcholine. Indeed, as can be seen in Fig. 6 basal ACh release in the presence of carbamylcholine (0.1 μM) and no AChE inhibitors was not inhibited by AH5183 (0.1 μM).

The dependence on pH of ACh release and of the effects thereon of AH5183 were determined. Short incubations (10 min) of electric organ prisms in acidic (pH 6.0) and alkaline (pH 9.0) media had no effect on ACh release either in the presence of AChE inhibitors (respectively 98±15 and 91±17 percent of the release at pH7.0; n=2) or in their absence (Fig. 7). By contrast prolonged pre-incubation (1 hr) at pH 6.0 enhanced release whereas pre-incubation at pH 9.0 inhibited release (Fig. 7). These effects were reversible and were not affected by either phospholine iodide or paraoxon (not shown). During the phospholine iodide and paraoxon experiments the tissue was transferred to fresh medium every 10 min after the preincubation period to ensure that release was not self inhibited (Fig. 2).

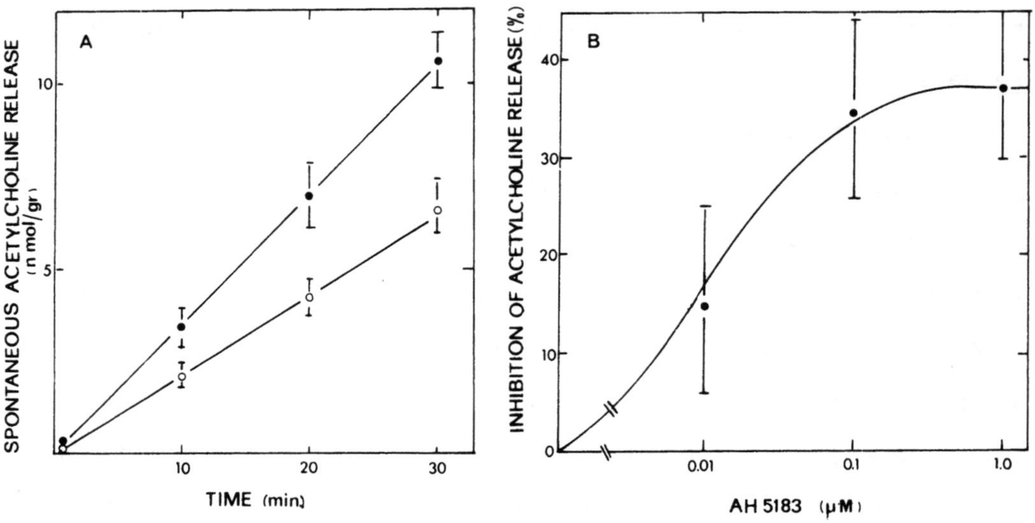

Fig. 5. The time dependency (A) and dose response (B) of the effects of AH5183 on ACh release in the absence of acetylcholinesterase inhibitors. In A(○) and (●)correspond respectively to Ach release in the presence and absence of AH5183 (0.1 μM) whereas in B (●) represents the percent inhibition of ACh release induced by the indicated AH5183 concentrations at t=10 min. ACh release was measured as in Experimental.

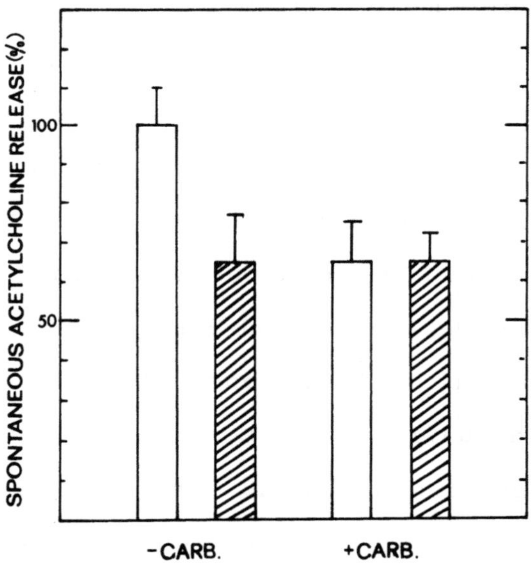

Fig. 6. The effect of carbamylcholine on the sensitivity of ACh release to AH5183. Single electric organ prisms were incubated in *Torpedo* buffer + Ca^{2+} (2 mM) in the absence of acetylcholinesterase inhibitors and in the presence or absence of carbamylcholine (0.1 μM) and AH5183 (0.1 μM). ACh release was measured as described in Experimental. Results presented are percent of control (100% = 0.26 ±0.1 nmol/gr/min; n=3). The dashed bars correspond to release in the presence of AH5183.

The effects of AH5183 on ACh release in the absence of AChE inhibitors were also pH dependent. At pH 6.0 AH5183 (0.1 μM) did not affect release whereas at pH 7.0 and pH 9.0 release was inhibited by respectively 35±10 and 45±10 precent. This pH sensitivity was evident within the first 10 min following the pH change. Since during this time basal ACh release was not affected by the external pH (Fig. 6) this finding suggests that the pH dependency of the effects of AH5183 on release is not due to changes in the extent of release.

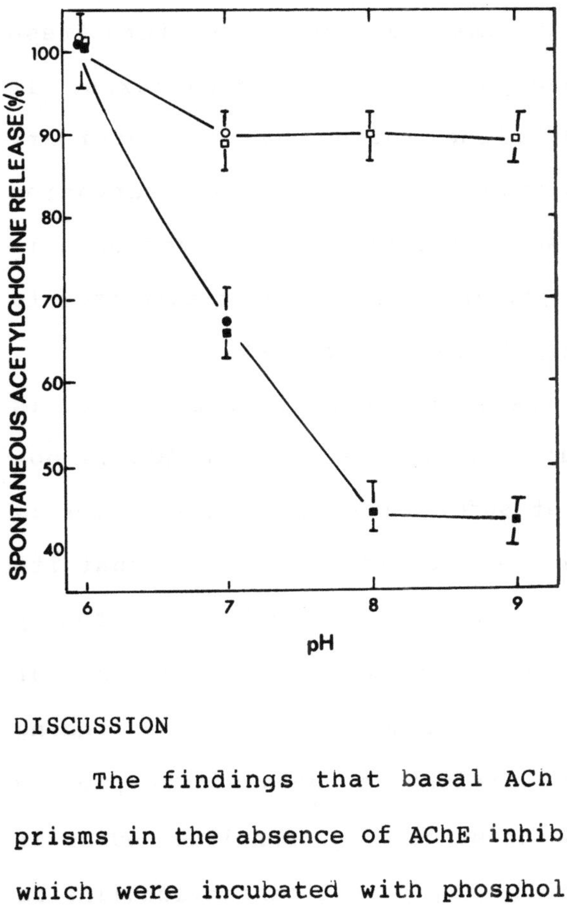

Fig. 7. The effects of pH on spontaneous ACh release in the absence of acetylcholinesterase inhibitors. ACh released after 10 min and 60 min preincubation periods at the indicated pH values are denoted by empty and filled symbols respectively. At neutral and acidic pH values release was performed in Torpedo buffer + Ca^{2+} (2 mM) in the presence (■,□) and absence of Tris (10 mM) (●,○) whereas at basic pH values it was performed in Torpedo buffer + Ca^{2+} (2mM) + Tris (10 mM)(■,□). ACh release was measured as described in Experimental.

DISCUSSION

The findings that basal ACh release from electric organ prisms in the absence of AChE inhibitors (Fig. 3) and from prisms which were incubated with phospholine iodide in non conditioned media are similar (Figs. 2 and 3) suggest that the secreted choline which is measured in the absence of AChE inhibitors is indeed the hydrolytic breakdown of the released ACh and is not due to spontaneous release of choline. This assertion is supported by the observation that hemicholinium-3 does not affect ACh release and that the pH dependencies of ACh release in the absence and presence of phospholine iodide are similar.

Incubation of electric organ prisms in the presence of AChE inhibitors (phospholine iodide or paraoxon) results in secretion of a heat labile factor which inhibits basal ACh release (Fig.2). This factor is not the AChE inhibitors themselves since release is restored when the prisms are transferred into fresh media which contain either phospholine iodide or paraoxon. It should be noted that phospholine iodide has a quaternary nitrogen whereas paraoxon is lipophilic. Thus, the observation that the extent of ACh release in the presence of paraoxon is about 40% lower than with phospholine iodide may result from to the permeation of paraoxon into the nerve terminals.

The molecular nature of the secreted inhibitory factor is not known. It is not ACh since unlike the factor ACh is not hydrolyzed by heating (10 min at 90°C). However as the factor is released only when the AChE is blocked it is possible that its secretion is triggered by ACh. This assertion is supported by the finding that in the absence of AChE inhibitors the rate of ACh release is reduced by carbamylcholine (Fig. 4). The finding that under these conditions carbamylcholine does not completely inhibit release and that in the presence of phospholine iodide it does not affect ACh release at all, imply that carbamylcholine is less effective than ACh in triggering the secretion of the inhibitory factor. Muscarinic and nicotinic antagonists do not disinhibit ACh release. Thus, the presumed effects of ACh on the secretion of the inhibitory factor are not mediated by either the postsynaptic nicotinic or the presynaptic muscarinic ACh receptors (Michaelson et al., 1979).

The marked temperature dependence of basal ACh release (Fig 1, Fig. 3) suggests that release is mediated by a transport system. The inhibition of ACh release in the absence of AChE inhibitors by low levels of AH5183 (EC_{50} ~ 10 nM) which are similar to those which block ACh uptake into synaptic vesicles (Anderson et al., 1983) suggests that at least part of the release is mediated by a transport system similar to the vesicular ACh carrier (Parsons et al., 1982). The pH dependency of ACh release (Fig. 7) is consistent with this interpretation since the accumulation of ACh in cholinergic synaptic vesicles requires a pH gradient in which the vesicle's interior is acidic compared to its external medium (Anderson et al., 1982). However, since the effects of external pH on basal ACh release are not immediate (Fig. 7) we cannot exclude the possibility that the pH dependency is due to indirect effects of pH on release. The findings that the sensitivity of ACh release to AH5183 follows immediately changes in external pH implies that the mechanisms underlying the effects of pH on release and on its inhibition by AH5183 are different. It is not known if both processes are regulated by pH gradients or whether protonation of the transporter results in enhanced release and in the concomitant resistivity to AH5183.

At present we do not understand why in the Torpedo ACh release in the presence of AChE inhibitors (phospholine iodide or paraoxon) is insensitive to AH5183 whereas spontaneous ACh release from rodent myoneural junctions in the presence of paraoxon is blocked by AH5183 (Edwards et al., 1985). This could be due to the presently reported inhibitory factor which is released from the electric tissue under these conditions and whose levels are expected to be much lower in the sparsely innervated myoneural preparation.

In summary, the present findings show that spontaneous ACh release from *Torpedo* nerve terminals is a complex biochemical process which is regulated by endogenous factors and is mediated by a transport system similar to the vesicular ACh carrier.

ACKNOWLEDGEMENTS

We thank Dr. S.M. Parsons for the gift of AH5183. This work was supported in part by grants from the U.S. National Institutes of Mental Health (MH 40294) and from the U.S.-Israel Binational Science Foundation (grant no. 85-00097/1).

REFERENCES

Anderson, D.C., King, S.C. and Parsons, S.M. (1982) Proton gradient linkage to active uptake of [^{3}H]acetylcholine by *Torpedo* electric organ synaptic vesicles. Biochem. 21: 3037-3043.

Anderson, D.C., King, S.C. and Parsons, S.M. (1983) Pharmacological characterization of the acetylcholine transport system in purified *Torpedo* electric organ synaptic vesicles. Mol Pharm 24: 48-54.

Carrol, P.T. (1983) Spontaneous release of acetylcholine and acetylhomocholine from mouse forebrain minces: cytoplasmic or vesicular origin. Neurochem Res 8: 1271-1283.

DelCastillo, J. and Katz, B. (1954) Quantal components of the endplate potential. J Physiol (London) 124: 560-573.

Dowdall, M.J. and Zimmermann, H. (1977) The isolation of pure cholinergic nerve terminal sacs (T-sacs) from the electric organ of juvenile *Torpedo*. Neuroscience 2: 405-421.

Dunant, Y., Eder, L . and Servetiadis-Hirt,L. (1980) Acetylcholine release evoked by single or a few nerve impulses in the electric organ of *Torpedo*. J Physiology Lond 298: 185-203.

Edwards, C., Dolczal, V., Tucek, S. Zemkova, H. and Vyskoci, F. (1985) Is an acetylcholine transport system responsible for nonquantal release of acetylcholine at the rodent myoneural junction. Proc Nat Acad Aci USA 82: 3504-3518.

Israel, M. and Lesbats, B. (1981) Chemiluminiscent determination of acetylcholine and continuous detection of its release from Torpedo electric organ synapses and synaptosomes. Neurochem Int 3: 87-90.
Israel, M., Gautron, J. and Lesbats, B. (1970) Fractionement de l'organe electrique de la Torpille: localisation subcellulaire de l'acetylcholine. J Neurochem 17: 1441-1450.
Israel, M., Manaranche, R., Mastour-Frachon, P. and Morel, N. (1976) Isolation of pure cholinergic nerve endings from the electric organ of Torpedo marmorata. Biochem J 160:113-115.
Katz, B., Sec, R.S. and Miledi, R. (1977) Transmitter leakage from motor nerve endings. Proc R Soc London B 196: 59-72.
Michaelson, D.M. and Sokolovsky, M. (1976) Neurotransmitter release from viable purely cholinergic Torpedo synaptosomes. Biochem Biophys Res Commun 73: 25-31.
Michaelson, D.M., Avissar, S., Kloog, Y. and Sokolovsky, M. (1979) Mechanisms of acetylcholine release: Possible involvement of presynaptic muscarinic receptors in regulation of acetylcholine release and protein phosphorylation. Proc Natl Acad Sci USA 176: 6336-6340.
Parsons, S.M., Carpenter, R.S., Koenigsberger, R. and Rothlin, J.E. (1982) Transport in the cholinergic synaptic vesicle. Fed Proc 41: 2765-2768.
Pinchasi, I., Burstein, M. and Michaelson, D.M. (1984) Metabolism of arachidonic acid and prostaglandins in the Torpedo electric organ: modulation by the presynaptic muscarinic acetylcholine receptor. Neuroscience 13: 1359-1364.
Whittaker, V.P., Essman, W.B. and Dowe, G.H.C. (1972) The isolation of pure synaptic vesicles from the electric organ of elasmobranch fish of the family Torpedinidae. Biochem J 128: 844-846.
Whittaker, V.P., Essmann, W.B. and Dowdall, M.J. (1975) The biochemistry of cholinergic synapses as exemplified by the electric organ of Torpedo. J Neural Trans Suppl XII: 36-60.

THE MECHANISM OF ACETYLCHOLINE RELEASE : ITS ESSENTIAL COMPONENT, THE MEDIATOPHORE

Maurice Israël
Centre National de la Recherche Scientifique
Laboratoire de Neurobiologie cellulaire et moléculaire,
1, avenue de la Terrasse,
91190 Gif-sur-Yvette,
France

INTRODUCTION

In a dialogue on virtue the greek philosopher Platon opposes his master Socrate to Menon. Socrate could paralyse his contradictors with a ferocious critique, this leads Menon to compare him to the sea water fish the Torpedo which numbs its preys. Socrate followed the joke but points out that like the paralysed prey the Torpedo is itself numbed after the attack.(ref. 14). Such an essential aspect of the Torpedo physiology (the fatigue after the discharge) had therefore been observed some 400 years B.C.

The fatigue that we shall consider corresponds to a blockade of synaptic transmission resulting from a reduction of acetylcholine (ACh) release without exhaustion of transmitter stores. The phenomenon was more recently observed by Katz and Miledi 1969, who found a fatigue of synaptic transfer at the squid giant synapse, they attributed it to the accumulation of calcium on the inside of the membrane. Similarly, Adams et al 1985 obtained a decline of transmission when metabolic inhibitors were used to increase cytoplasmic calcium liberated from intracellular stores where it is sequestered by energy linked processes. The fatigue was reversed by a calcium chelator injection in the nerve terminal.

Since it is also well established that the influx of calcium in the nerve terminal elicits transmitter release (see for example Llinas and Steinberg 1977), it appears paradoxical that calcium should also induce the fatigue. Probably acti-

NATO ASI Series, Vol. H21
Cellular and Molecular Basis of Synaptic Transmission
Edited by H. Zimmermann

vation and fatigue of the release mechanism take place at different calcium concentrations. This dual effect of calcium on release appears to be an essential characteristic which will help to identify the presynaptic membrane structure which translocates ACh. Another way to recognize it was to find a pharmacololgical test. The compound cetiedil recently used by Morot-Gaudry Talarmain et al 1987 to block the release of ACh from synaptosomes provided such a test since it acted on release after the calcium entry step.

In previous reports we showed that it was possible to endow artificial membranes with an ACh releasing system activated by calcium and inherited from the membrane of Torpedo electric organ synaptosomes. The reconstituted proteoliposomes showed release properties (kinetics, amounts of ACh transfered, calcium dependency) very close to those of parent synaptosomes. The presynaptic membrane element which translocated ACh upon calcium action was called mediatophore and purified. (see ref.2, 6, 8). We have shown that like for intact synaptosomes the compound cetiedil blocked the release of ACh performed by the mediatophore (ref. 11). Moreover we found a calcium dependent desensitization of the mediatophore very similar to the desensitization or fati gue of ACh release from synaptosomes.

Taken together these observations support the view that the mediatophore might well be the structural element which normally transfers the transmitter across the nerve terminal membrane. In relation to the properties of the mediatophore an interpretation of the quantal mode of ACh release is proposed.

ACETYLCHOLINE RELEASE FROM SYNAPTOSOMES AND FROM PROTEOLIPOSOMES DERIVED FROM NERVE TERMINAL PLASMA MEMBRANES

It is now a routine technique to monitor the release of ACh from Torpedo electric organ synaptosomes using the cholineoxidase procedure described by Israël and Lesbats (1981 a, b, 1982). This method is based on the following principle : Choline is oxidized by cholineoxidase giving betain and H_2O_2. The hydrogen peroxide is then easily measured, we have prefered to use the luminol-peroxidase reaction which is chemiluminescent. When an ACh assay is performed it is not necessary to separate choline and acetylcholine since they are measured in succession. After the decay of the light emission due to choline oxidation, acetylcholinesterase is added, the choline resulting from ACh hydrolysis gives a new light emission proportional to ACh. When one monitors transmitter release from cholinergic synaptosomes, most if not all the recorded light emission is due to ACh release as shown by the fact that the little or no light is measured after acetylcholinesterase inhibition. The figure 1 shows Torpedo electric organ synaptosomes and two ACh release curves obtained after KCl depolarization in the presence of calcium, or after inducing an influx of calcium with the calcium ionophore A23187. Notice that in both cases the rising slope of release curves is less steep than for ACh standards injected after the decay of release, showing that the enzymatic reactions were not rate limiting.

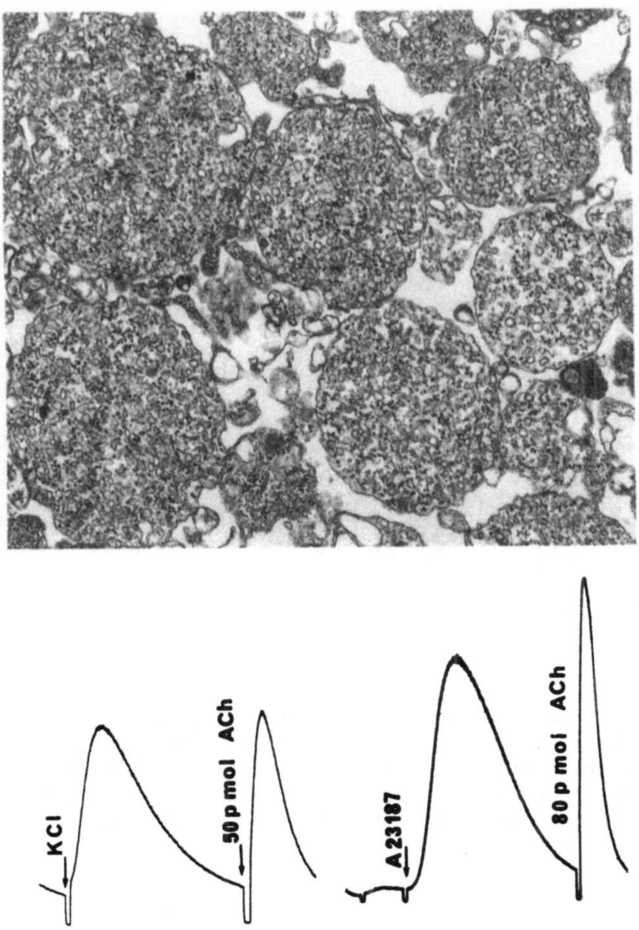

Fig. 1. Synaptosomes purified from Torpedo electric organ and two ACh release curves monitored with the cholineoxidase chemiluminescent procedure.

Proteoliposomes were made with synthetic lecithin mixed to lyophilized synaptosomal plasma membranes or to purified membrane fractions. They were filled with ACh (25 to 50 mM) and ions and were able to release the transmitter in response to the influx of calcium induced by the ionophore A23187. The figure 2 shows proteoliposomes made with total synaptosomal membrane components and a typical ACh release curve. It is clear that the artificial membrane has inherited a number or proteins from the synaptosomal membrane and that it acquires the capacity of releasing ACh. In previous reports Birman et al 1985 have compared the release from protéoliposomes and synaptosomes for a range of calcium and ionophore concentrations a striking similarity was observed.

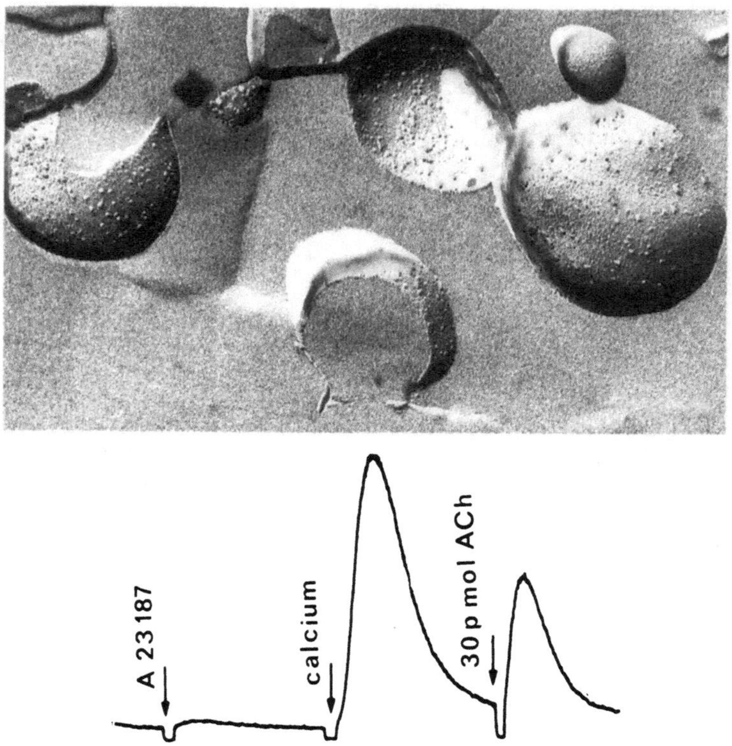

Fig. 2. Proteoliposomes observed after freeze-fracture methods. They were made with synaptosomal membranes mixed to synthetic lecithin. The proteoliposomes have inherited from the native membrane the release mechanism as shown by the fact that the influx of Calcium triggered by A23187 elicits the efflux of ACh.

PURIFICATION OF THE MEDIATOPHORE

The basic idea was to find out if it was possible to extract from synaptosomal membranes the element which confered to the artificial membrane its calcium dependent ACh release properties. We therefore compared the mediatophore content of the different extracts of synaptosomal membranes, after incorporating them to the liposomal membranes.

The mediatophore activity was expressed in p mol of ACh released per second from protéoliposomes containing 1nmol of occluded ACh, when stimulated with A23187 plus calcium. The activity of each fraction was normalized for 1mg protein. In all cases the proteoliposomes were made by mixing the fraction (400 to 15 µg protein) with 4 mg lecithin, the procedure has been described in several previous reports. (réf. 2,6,8). In a first attempt the membranes were solubilized in detergents a partial purification was achieved. After several purification steps however, the activity decreased which limited further work. (see Birman et al, 1986). We later found that it was possible to extract the mediatophore in strong alkaline conditions in the form of a protein-lipid micelle which could be purified by gel-filtration or sedimented by centrifugation. We then noticed that if the micelle was treated with organic solvents, that the mediatophore activity followed the lipid phase while most proteins were precipited. The mediatophore was then precipitated from the organic phase with ether (as proteolipids, Soto et al 1969). The lipid free mediatophore is shown fig.3. It has a doughnut shape and a diameter of 70-80 A. The structure is most probably pentameric. The lipid free mediatophore is inactive but becomes water soluble like it is often the case for proteolipids. A stokes radius of 53 Å was determined with a sedimentation coefficient of 9.8S the molecular weight should be close to 200 000 d but a marked tendency to form aggregates is sucrose gradients limited a more accurate determination. After SDS acrylamide gel electrophtoresis a band situated between, 14000 and 17000 d, gave the molecular weight of a single mediatophore subunit. Hence each monomer would be composed of 2 perhaps 3 chains.

The exact stochiometry will have to wait for further informations.

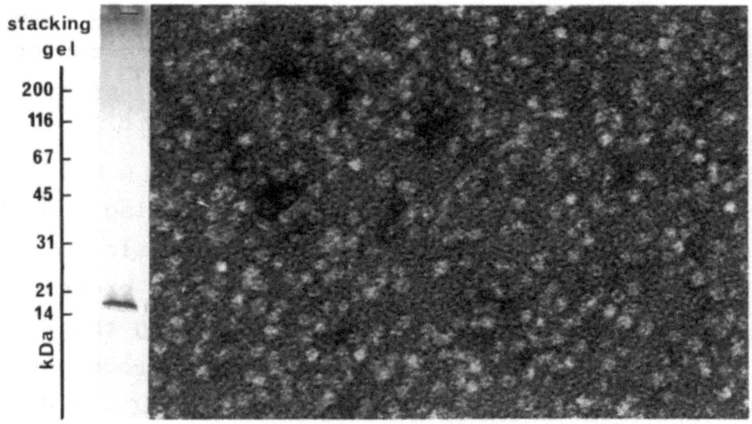

Fig. 3. Negatively stained purified mediatophore after removal of lipids. After SDS polycrylamide gel electrophoresis a single subunit was characterized.

EVIDENCE THAT THE MEDIATOPHORE REPRESENTS THE MECHANISM OF TRANSMITTER RELEASE :

Several essential properties were implicit in the functional test used to purifie the mediatophore : the calcium dependent release, or the localization in the presynaptic membrane. It was not inplicit however to find that the localization was specific, or that other substances were not released. These necessary properties of the release mechanism, are not however sufficient to demonstrate that the isolated mediatophore is indeed the nerve terminal membrane structure supporting the release mechanism. An essential argument in favour of such a view was brought by a pharmacoloigical inhibition test with the compound cetiedil. The release of ACh supported by the mediatophore was blocked similarly to the transmitter release from synaptosomes or tissue. The figure 4 shows on the left the inhibition by citiedil of ACh release from proteoliposomes made with mediatophore, release was elicited with calcium following the addition of the ionophore A23187. On the right the release from synaptosomes is inhibited at similar cetiedil concentrations for a variety of stimulating agents (KCl, gramicidin A23187, glycera neurotoxin) showing that the drug acted after the calcium entry step on the release mechanism. On top of each curve we show control and inhibited release at two cetiedil concentrations for proteoliposomes and synaptosomes. The pharmacological test would have been a definitive argument if the drug had been fully specific. Since other processes such as the choline pump or the uptake of ACh by synaptic vesicles were also inhibited, we did not consider that the pharmacological test was sufficient by itself to indentify the mechanism of release. The drug ressembles to the ACh molecule but with its large end groups it might affect all ACh or choline fluxes.

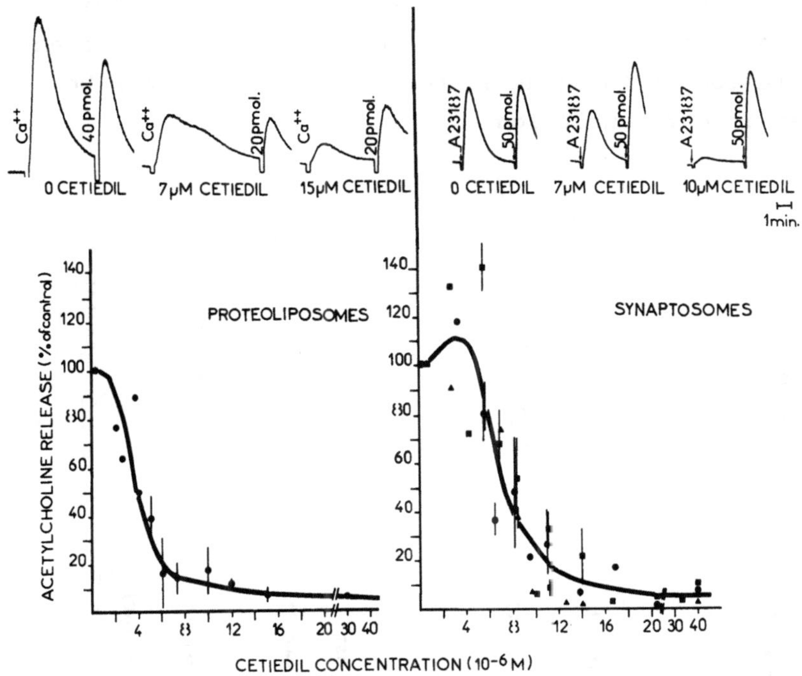

Fig. 4. Inhibition by the drug cetiedil of ACh release from synaptosomes or from proteoliposomes made with mediatophore inserts on top of each curve show control and inhibited release of ACh for two cetiedil concentrations (see Morot-Gaudry Talarmain et al 1987).

A typical characteristic of the release mechanism, is the dual effect of calcium on its activation and desensitization. The phenomenon is demonstrated in fig 5. A test stimulation of synaptosomes with the ionophore A23187 plus calcium is delivered alone, or a couple of minutes after a small conditioning stimulation at low calcium. The conditioning stimulus releases very small ACh amounts but blocks the release of the second test stimulation. Since this same phenomenon was observed on proteoliposomes made with mediatophore containing fractions or with pure mediatophore we may consider that the mediatophore represents most probably a fundamental structure of the release mechanism. The activation of ACh release by calcium took place in the millimolar range of calcium concentration while desensitization or fatigue was found for micromolar calcium concentrations. The consequence of this observation is that mediatophore molecules close to the calcium channel where the calcium concentration rises rapidly will be activated while mediatophore molecules situated at some distance will be desensitized. An evident interpretation of the quantal release of ACh may now be envisaged in our conclusion.

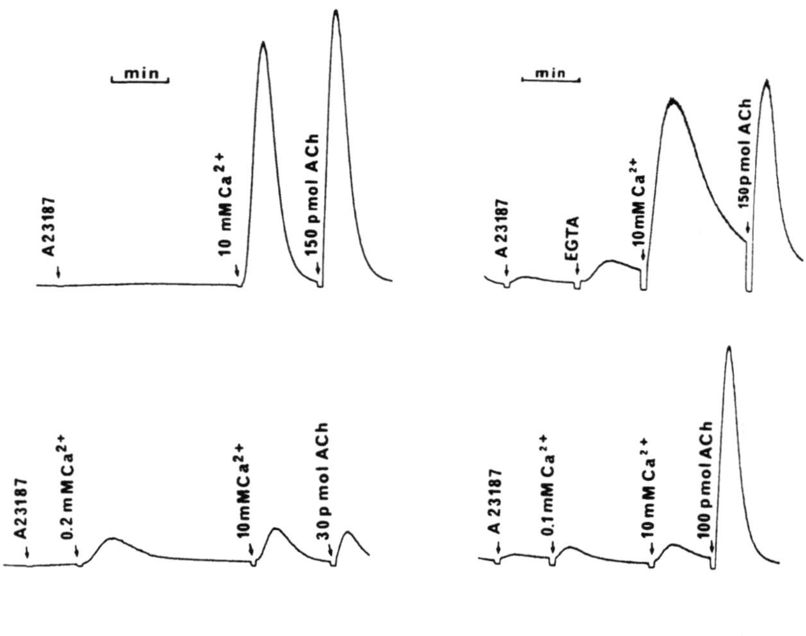

Fig. 5. Calcium induced activation and desensitization of ACh release from synaptosomes (right) or from proteoliposomes made with pure mediatophore (left). The top traces show the test release triggered by A23187 at 10mM Calcium. The bottom traces show that if a conditioning calcium influx 02mM in the presence of A23187 is first given that a subsequent 10mM Calcium addition fails to release ACh. (0.2 mM EGTA were added in the left trace to remove traces of external calcium).

CONCLUSION : An interpretation of Quantal acetylcholine release

If some 20 mediatophore molecules close to the calcium channel are simultaneously activated by the sudden rise of calcium at the cytoplasmic opening of the calcium channel, then a quantum of transmitter will be released giving a miniature end-plate potential similar to the one originally observed by Fatt and Katz 1952. The mediatophore molecules are functionally synchronized by the calcium diffusion gradient. The mediatophore molecules situated at some distance of the calcium channel in a diluted region of the calcium gradient are desensitized, and do not release transmitter. Hence the mediatophore molecules which are within a given range of calcium diffusion will open together giving the normal miniature end-plate potential. If stimulation synchronizes the opening of calcium channels then an additive response: the end-plate potential will necessarily represent a sum of miniatures end-plate potentials. Now if some experimental condition (heat - calcium challenge) disorganizes the calcium concentration gradient found at the opening of the calcium channel then one might record smallest events due to the activation of single mediatophore molecules giving sub-miniature end-plate potentials as those found by Kriebel and Gross 1974. This also explains that the normal miniature is itself the necessary addition of sub-miniatures. It should also be said that in some experimental conditions the boundary between activated and desensitized mediatophores can be changed giving giant miniature end-plate potentials. Some of these very big miniature might also result from an occasional exocytotic event similar to the exocytosis found for some glandular tissues. But here this phenomenon is not frequent. The view proposed explains most experimental results on transmitter release, but it is not my opinion that synaptic vesicles should only have a secondary storage role. They have more than the task of concentrating ACh, ATP, calcium peptides and protons. They should be considered in terms of populations of vesicles changing their contents in relation to synaptic work. In these conditions the histogram of vesicle populations appears as the recorded history of synaptic activity some sort of local memory. The stored information is probably averaged, distorted by the exchanges with the cytoplasmic compartment, the synaptic response would therefore be conditionned to the stored information. If verified, this working hypothesis would give to synaptic vesicles a major role in synaptic function.

REFERENCES

Adams D.J., Takeda K. and Umbach J.A.(1985) Inhibition of calcium buffering depress evoked transmitter release at the squid giant synapse. J.Physiol.**369**,145-159.

Birman S., Israël M., Lesbats B. and Morel N.(1986) Solubilization and partial purification of a presynaptic membrane protein ensuring calcium dependent acetylcholine release from proteoliposomes. J.Neurochem.**47**,433-444.

Israël M. and Lesbats B. (1981 a) Chemiluminescent determination of acetylcholine and continuous detection of its release from Torpedo electric organ and synaptosomes. Neurochem.Int.3,81-90.

Israël M. and Lesbats B. (1981 b) Continuous determination by a chemiluminescent method of acetylcholine release and compartmentation in Torpedo electric organ synaptosomes. J.Neurochem. **37**, 1475-1483.

Israël M. and Lesbats B. (1982) Application to mammalian tissues of the chemiluminescent method for detecting acetylcholine. J. Neurochem. **39**, 248-250.

Israël M., Lesbats B., Morel N., Manaranche R., Gulik-Krzywicki T. and Dedieu J.C (1984) Reconstitution of a functional synaptosomal membrane possessing the protein constituents involved in acetylcholine translocation. Proc. Natl. Acad. Sci.USA **81**, 277-281.

Israël M., Manaranche R., Mastour-Frachon P. and Morel N. (1976) Isolation of pure cholinergic nerve endings from the electric organ of Torpedo marmorata. Biochem. J. **160**,113-115.

Israël M., Morel N., Lesbats B., Birman S. and Manaranche R. (1986) Purification of a presynaptic membrane protein which mediates a calcium-dependent translocation of acetylcholine. Proc. Natl. Acad. Sci. USA, **83**, 9226-9230.

Katz B. and Miledi R. (1969) Tetrodotoxin-resistant electric activity in pre-synaptic terminals. J. Physiol. **203**, 459-487.

Kriebel M.E. and Gross C.E. (1974) Multimodal distribution of frog miniature end-plate potentials in adult and denervatice leg muscle. J. Gen. Physiol. **64**, 85-103

Llinas R.R. and Steinberg I.Z. (1977) The place of a calcium hypothesis in synaptic transmission. Neurosci. Res. Prog. Bull. **15**, 565-574.

Morot Gaudry-Talarmain Y. Israël M., Lesbats B. and Morel N. (1987) Cetiedil, a drug to inhibit acetylcholine release in Torpedo electric organ, Vol **49** in press.

Morel N., Israël M., Manaranche R. and Mastour-Franchon P. (1977) Isolation of pure cholinergic nerve endings from Torpedo electric organ. J. Cell. Biol.**75**, 43-55.

Platon. In the Menon. Edited by Flammarion p 341.

Soto E.F., Pasquini J.M., Placido R., Latorre J.L. (1969) Fractionation of lipids and proteolipids from cat grey and white matter by chromatography on an organophilic dextran gel. J. Chromatog. **41**,400-409.

POTASSIUM CHANNELS IN RAT BRAIN SYNAPTOSOMES: PHARMACOLOGY AND TOXICOLOGY

Mordecai P. Blaustein[1], Dieter K. Bartschat[1], Christina G. Benishin[1,3], William E. Brown[2], Kathryn A. Colby[1], Bruce K. Krueger[1], Mary J. Schneider[1] and Roger G. Sorensen[1]

[1]University of Maryland School of Medicine, Department of Physiology, 655 West Baltimore Street, Baltimore, Maryland 21201, USA

[2]Carnegie-Mellon University, Department of Biological Sciences, 4400 Fifth Avenue, Pittsburgh, Pennsylvania 15213, USA

[3]Present address: University of Alberta Faculty of Medicine, Department of Physiology, Edmonton, Alberta, Canada T6G 2H7

INTRODUCTION

Potassium channels appear to be the most diverse group of ion channels in biological systems (Hille, 1984; Yellen, 1987). Neuronal K channels play key roles in the control of membrane potential, action potential repolarization, repetitive firing, and higher functions such as learning and memory. However, relatively little is known about the properties of K channels in presynaptic nerve terminals because these channels are difficult to study with traditional electrophysiological methods. This is a significant gap in our knowledge because these nerve terminal K channels may play a critical role in the control of synaptic transmission.

Radioactive tracer flux studies have demonstrated that rat brain synaptosomes are physiologically competent nerve endings containing a large variety of functional ion channels. They have Na channels that are opened by veratridine, batrachotoxin, and certain scorpion and sea anemone toxins and blocked by tetrodotoxin (TTX) and saxitoxin (Krueger and Blaustein, 1980; Tamkun and Catterall, 1981). They also have voltage-gated Ca channels that are opened by depolarization and blocked by La and Cd (Nachshen and Blaustein, 1980 and 1982; Nachshen, 1985). More recently, we have employed ^{86}Rb and ^{42}K efflux methods to show that synaptosomes have a variety of K channels including voltage-gated K channels (Bartschat and Blaustein, 1985a), Ca-activated K channels (Bartschat and Blaustein, 1985b) and opiate-operated K channels (Bartschat and Blaustein, 1988).

A number of pharmacological agents and toxins are known to be K channel

NATO ASI Series, Vol. H21
Cellular and Molecular Basis of Synaptic Transmission
Edited by H. Zimmermann

blockers with well-established specificity for certain K channels. Several of these substances were used to distinguish the components of the ^{86}Rb efflux in synaptosomes that correspond to various types of K channels. We also identified some new K channel blockers that appear to be selective for specific types of K channels. In addition, we examined the role of certain second-messenger lipids in the modulation of K channel function. Our observations indicate that synaptosomes are a very valuable and convenient preparation for assaying agents that affect K channel function.

METHODS

Synaptosome Rb efflux. ^{86}Rb efflux from either crude (P_2) or purified synaptosomes was measured as described (Bartschat and Blaustein, 1985a).

Rat forebrain synaptosomes were prepared according to the procedure of Hajos (1975), as modified by Krueger et al. (1979). In some experiments the crude synaptosome preparation (the pellet, P_2, resulting from the second 10,000 g spin) was used without further purification. In other instances the synaptosomes were further purified in a one-step sucrose density gradient (Hajos, 1975).

Synaptosomes (either P_2 or the purified preparation) were slowly equilibrated at 0°C with a physiological salt solution (5K, containing, in mM: NaCl, 145; KCl, 5; RbCl, 0.1; $MgCl_2$, 2; glucose, 10; NaH_2PO_4, 0.5; and HEPES, 10, titrated to pH 7.4 with NaOH). The synaptosomes (approximately 40 mg protein/ml) were then incubated in 5K containing ^{86}Rb (10-20 μCi/ml) for 30 min at 30°C. An aliquot of tracer-loaded synaptosomes was pipetted onto a 13 mm diameter glass fiber filter (Schleicher and Schuell #25) and washed 5 times under vacuum filtration with a 30°C Wash Solution (5K containing 0.1% bovine serum albumin) to remove extracellular tracer. The Wash Solution also contained toxin, where appropriate, in order to expose the synaptosomes to the toxin for 12 to 15 sec before initiating the efflux measurement. The synaptosomes were then exposed for 1-6 sec to the following Efflux Solutions at 30°C, all with or without toxin: 5K; 50K or 100K (similar to 5K, but containing 50 or 100 mM KCl and 100 or 50 mM NaCl, respectively); or 50K/Ca or 100K/Ca (similar to 50K and 100K, but containing 1 mM $CaCl_2$ and only 1 mM $MgCl_2$, respectively).

Efflux incubations were terminated with a Stop Solution containing (in mM): tetraethylammonium chloride (TEA), 145; tetrabutylammonium chloride (TBA), 5; RbCl, 0.1; $NiCl_2$, 10; $MgCl_2$, 10; and HEPES, 10, titrated to pH 7.4 with NaOH. The synaptosome suspensions were rapidly filtered, and the filtrate (Efflux Solution plus Stop Solution) was collected in a scintillation vial. The ^{86}Rb in the filtrate and on the filters (containing the trapped synaptosomes) were both determined by liquid scintillation spectrometry. The percent of total radioactivity content released during the incubation was then calculated:

$$^{86}\text{Rb Efflux (\% of content)} = \frac{\text{(cpm in filtrate)}}{\text{(cpm in filtrate) + (cpm on filter)}} \times 100$$

In most experiments the efflux for each condition was measured as a function of efflux time (1-6 sec) in 3-6 replicate samples. The time course line for each condition was then calculated by linear regression analysis.

<u>"Resting" Rb efflux.</u> The representative time course experiment in Fig. 1 shows the various components of the ^{86}Rb efflux measured in the absence of toxins or other blocking agents (Bartschat and Blaustein, 1985a). These data, which were obtained with purified synaptosomes, were virtually identical to those obtained with the P_2 preparation (see Results). The slope of the 5K line corresponds to the resting ^{86}Rb efflux (component "R"). The average efflux into 5K, about 0.4% of the ^{86}Rb load per sec, is equivalent to a resting K permeability of about 2.4×10^{-7} cm/sec (Bartschat and Blaustein, 1985a).

The "residual efflux" in the 5K solution at "zero" time (usually 6-10% of total ^{86}Rb efflux) includes the efflux between the end of the wash period and the introduction of the Efflux Solution (at "zero" time), as well as incompletely washed extracellular tracer that was washed off with the Efflux and Stop Solutions.

<u>Component "T"</u> is the increment in the ordinate-intercept of the linear regression line produced by elevation of the extracellular K concentration (i.e., depolarization). This transient component probably represents ^{86}Rb efflux through voltage-gated, K channels that inactivate within 1 sec (Bartschat and Blaustein, 1985a).

<u>Component "S"</u> is the steady increase in the efflux indicated by the increase in the slope of the linear regression line upon elevation of extracellular K (i.e., the difference between the dotted line and the 5K

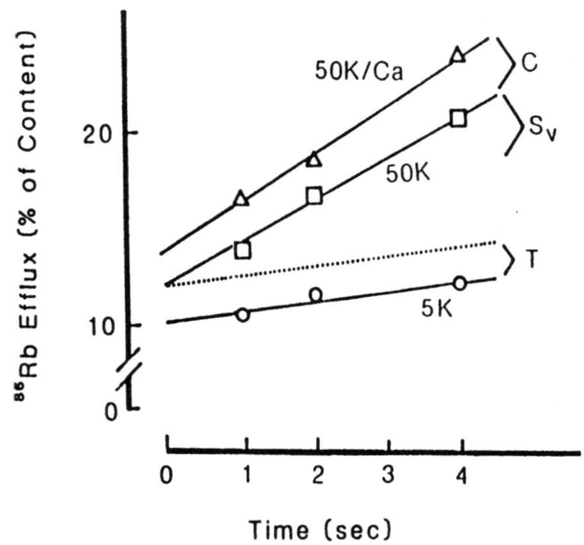

Fig. 1. Time course of ^{86}Rb efflux from synaptosomes. As described in Methods, purified synaptosomes were loaded with ^{86}Rb, and efflux was carried out in 5K solution or in depolarizing 50K solutions without or with Ca for 1-4 sec, as indicated. Each symbol indicates the mean of 4 determinations; the solid lines were determined by linear regression analysis. The efflux components labelled "T", "S_V" and "C" are described in the text.

solid line in Fig.1). At the K concentrations used in these experiments, approximately 1/2 to 2/3 of this increment is attributable to increased ^{86}Rb efflux through the ohmic "resting K channels" as a result of the increase in electrical driving force during depolarization (component "S_R"; see Bartschat and Blaustein, 1985a and 1986). The remaining 1/3-1/2 of component "S" (labelled "S_V" in Fig. 1) cannot be accounted for by electrodiffusion, and appears to be due to ^{86}Rb efflux through a voltage-gated, non-inactivating K channel.

Component "C" is the Ca-dependent increase in ^{86}Rb efflux that occurs when Ca is added to K-rich solutions. This component corresponds to Rb efflux through Ca-activated K channels (Bartschat and Blaustein, 1985b).

Purification of Toxins. The K channel blocking components of *Dendroaspis angusticeps* (Da) venom (Sigma Chemical Co., St. Louis, MO) were purified by a modification of the method of Harvey and Karlsson (1980). The crude venom was initially separated into five major peaks of UV absorbing material by chromatography on Sephadex G-50 (fine) eluted with 0.1 M ammonium acetate at pH 6.8. Only the third peak contained components that inhibited ^{86}Rb efflux from synaptosomes. This material was lyophilized, redissolved in 0.1 M ammonium acetate at pH 6.8, and loaded onto an HPLC ion exchange column (Brownlee Labs, Santa Clara, CA). Adsorbed material was eluted with an ascending gradient of ammonium acetate (0.1-0.8 M). Of the numerous peaks eluted, only four inhibited any of the components of ^{86}Rb efflux from synaptosomes; these are denoted as, α-, β-, γ- and δ-DaTX, respectively, in the order of their elution from the column. Ion exchange elution characteristics and amino acid composition (not shown) indicated that α-DaTX was identical to the polypeptide called "dendrotoxin" by Harvey and Karlsson (1980). Amino acid composition and partial sequencing indicated that δ-DaTX was identical to D. angusticeps toxin $C_{13}S_1C_3$ described by Joubert and Taljaard (1980). β- and γ-DaTX have not previously been identified as neurotoxins; they were comparable in size to the other two toxins, with molecular weights of about 7,000.

The following agents were also tested on synaptosome ^{86}Rb efflux: venoms from the scorpions, Leiurus quinquestriatus (from Latoxan, Rosans, France), Centruroides sculpturatus, and Tityus serrulatus (both from Sigma Chemical Co., St. Louis., MO); dexoxadrol and levoxadrol (gifts from Dr. P. Von Voightlander, The Upjohn Company, Kalamazoo, MI); the diacyglycerol analogs, OAG (1-oleoyl-2-acetylglycerol) and diC8 (1,2-sn-dioctanoyl-glycerol) (from Avanti Polar Lipids, Birmingham, AL); deoxy-diC8 (a gift from Dr. C. Loomis, Duke University).

Covalent labeling of the PCP and DaTx receptors. m-Azido[^{3}H]PCP was prepared and synaptic membranes were covalently labelled as described (Sorensen and Blaustein, 1986).

To label the DaTx receptors, α-DaTx and γ-DaTx were first radioiodinated in the presence of iodogen (Pierce Chemical Co., Rockford IL) using published methods (Salacinski et al., 1981). Synaptic membranes (1 mg/ml) were incubated at 37°C in binding buffer: 145 mM NaCl, 5 mM KCl, 2 mM $MgCl_2$ and 20 mM HEPES, pH 7.4, containing 2 nM radioiodinated toxin in the absence or presence of a 500-fold excess of the respective unlabeled toxin. After 30 min, bound ligand was separated from free ligand by centrifugation. The pellets were washed once with binding buffer and once with 250 mM triethanolamine (TE) buffer, pH 8.5. The pellets were suspended in TE buffer containing 1 or 5 mg/ml dimethyl suberimidate and kept at room temperature for 3 hr. The pellets collected after centrifugation of the crosslinking mixtures were suspended in sodium dodecyl sulfate (SDS)-dissociation buffer and kept overnight at room temperature. The membrane polypeptides were separated by SDS-polyacrylamide gel electrophoresis. The gels were stained and dried, and autoradiograms were prepared in order to locate attached radioiodinated toxin.

RESULTS AND DISCUSSION

Rb efflux component "T" corresponds to the A current. Component "T" is activated by K-rich media or other methods of depolarization; it inactivates spontaneously, within about 1 sec (Bartschat and Blaustein, 1985a). The pharmacology of this component of the ^{86}Rb efflux is similar to that of the A current that has been measured with electrophysiological methods in many types of neurons (Hille, 1984; Ragowski, 1985). This efflux component is selectively blocked by low concentrations of 4-aminopyridine, 4-AP (IC_{50} = 0.1-0.2 mM) and by the tetraalkylammonium ions, TEA and TBA (Bartschat and Blaustein, 1985a). It is also blocked by relatively high concentrations of phencyclidine (PCP, >10 μM; Bartschat and Blaustein, 1986).

Component "T" is also selectively blocked by two components of Da venom, α-DaTX (Fig. 2A) and δ-DaTX (not shown). These neurotoxins greatly reduce the ordinate intercept of the ^{86}Rb efflux time course curve in K-rich media, with negligible effect on the slope of the line (component "S"). Although not illustrated here, neither α-DaTX nor δ-DaTX affects component "C". In voltage clamp experiments on other neuronal preparations, α-DaTX has been observed to block an inactivating K current, the A current; it also promotes neurotransmitter release from nerve terminals (Harvey and Karlsson, 1980 and 1982; Dolly et al., 1984; Halliwell et al., 1986). This supports the view that component "T" corresponds to the A current.

Rb efflux component "S_V" corresponds to the current carried by a non-inactivating K channel. Like component "T", ^{86}Rb efflux component "S_V" is activated by depolarization. However, "S_V" is much longer lasting, and does not inactivate within at least 6 sec. This component is blocked by high concentrations of 4-AP (IC_{50} = 1-2 mM) and by tetraalkylamines (Bartschat and Blaustein, 1985a).

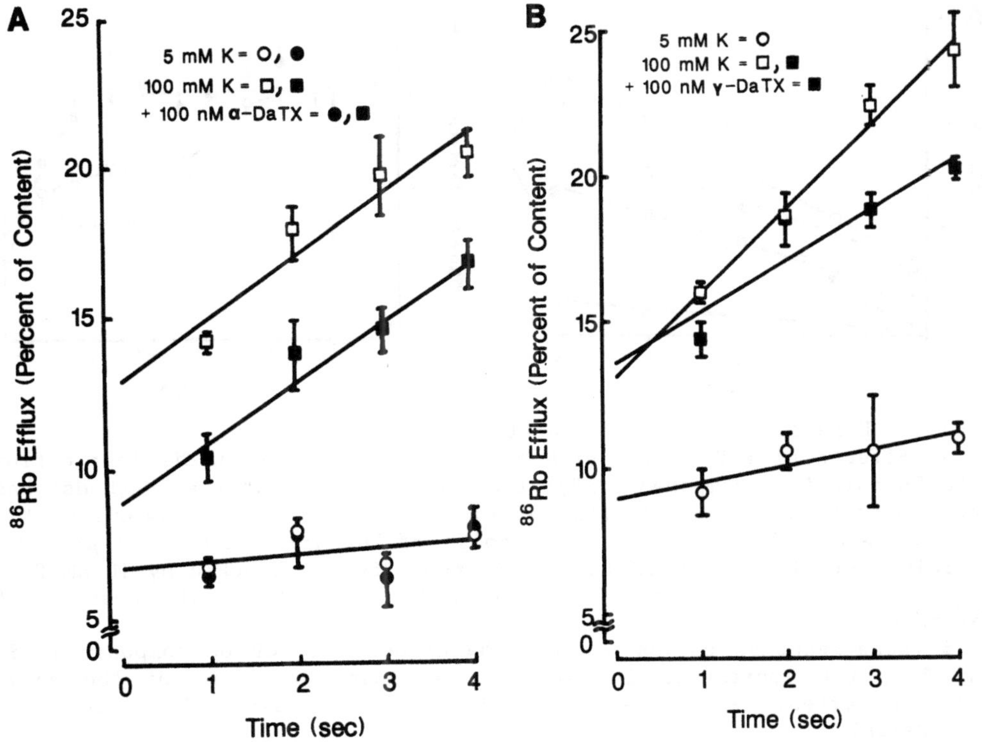

Fig. 2. Effects of 100 nM α-DaTX (A) and 100 nM γ-DaTX (B) on the time course of ^{86}Rb efflux from P_2 synaptosomes incubated in 5K (circles) or 100K (squares) solution. The toxin was present (solid symbols) only during the 15 sec wash and 1-4 sec incubation. See Methods for further details. Symbols represent the means of 4 or 5 determinations ± SE.

A: Note that α-DaTX reduced the increment in ordinate intercept (component "T"; see Fig. 1), but had no effect on the slope of the line in 100K solution.

B: Note that γ-DaTX reduced the slope of the regression line in the 100K solution, but did not affect the ordinate intercept. The reduction in slope corresponds to about 40% of the increment in slope produced when K was increased from 5 mM to 100 mM in the absence of toxin (see text).

Component "S_V" is selectively blocked by low concentrations of phencyclidine (IC_{50} <1 μM; Bartschat and Blaustein, 1986) as illustrated in Fig. 3. This is particularly noteworthy because phencyclidine [PCP; 1-(1-phenylcyclohexyl)piperidine; "angel dust"] is a drug of abuse; it induces a bizzare toxic psychosis in man that is often indistinguishable from acute primary schizophrenia (Domino, 1981). The block of component "S_V" by analogs of PCP parallels their potency in behavioral studies: e.g., TCP (the thienyl analog) > m-NH_3-PCP > PCP > m-NO_3-PCP (Albuquerque et al., 1981; Bartschat and Blaustein, 1986). This has led to the

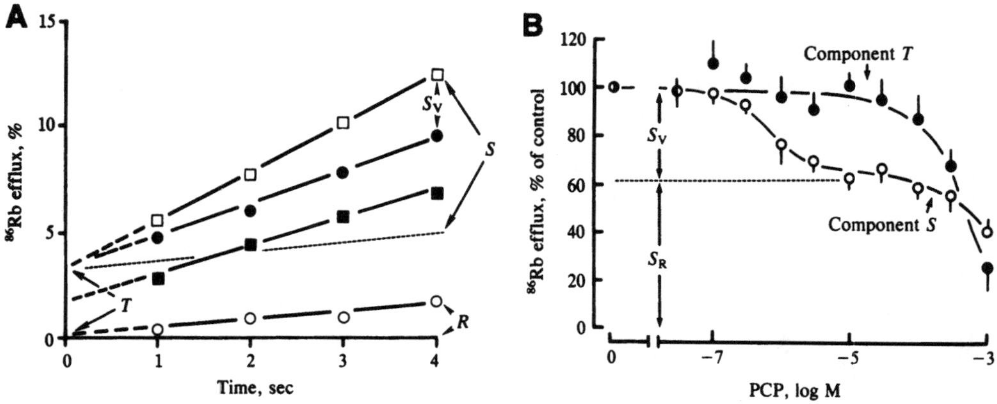

Fig. 3. Effects of PCP on ^{86}Rb efflux from synaptosomes.

A: Effects of PCP on the time course of efflux. Open circles: efflux into 5K solution (component "R"); the "residual efflux" (see Methods) has been subtracted. Open squares: efflux into 100K; efflux components "T" and "S" are indicated. Solid symbols: efflux into 100K + 10 μM PCP (circles) or 100K + 100 μM PCP. Component "S_V" is blocked by 10 μM PCP; "S_R" is the PCP-insensitive portion of "S". Data are the means of 4 determinations ±SE.

B: Dose-response curves showing the effects of PCP on components "S" and "T", the increments in slope and ordinate intercept of the dose-response curves, respectively (from Bartschat and Blaustein, 1986, with permission).

suggestion that the behavioral effects of PCP may be due to its block of a non-inactivating (delayed rectifier) K channel in presynaptic terminals (Albuquerque et al., 1981; Bartschat and Blaustein, 1986): this action would be expected to prolong action potentials and thereby enhance Ca entry and neurotransmitter release at the terminals. The bizzare behavior observed with PCP intoxication may then be a manifestation of the altered synaptic transmission.

Dioxadrol and "sigma" ligands interact with PCP receptors and generalize to PCP in behavioral paradigms. These substances are optically active, and their actions are stereoselective. For example, the "sigma" ligand (+)N-allylnormetazocine (NANM; SKF 10,047), which is more effective than (-)NANM in displacing PCP from its receptors, is also more potent in inducing PCP-like behavior. Likewise, the dioxolane, dexoxadrol, is very potent both in displacing PCP from its receptor and in inducing a PCP-like behavioral syndrome; its enantiomer, levoxadrol, is virtually inactive in both respects (Cone et al., 1984; Sorensen and Blaustein, 1987). The "sigma" ligands and dioxadrol display similar stereoselective effects with respect to their block of synaptosome ^{86}Rb efflux component "S_V" and their

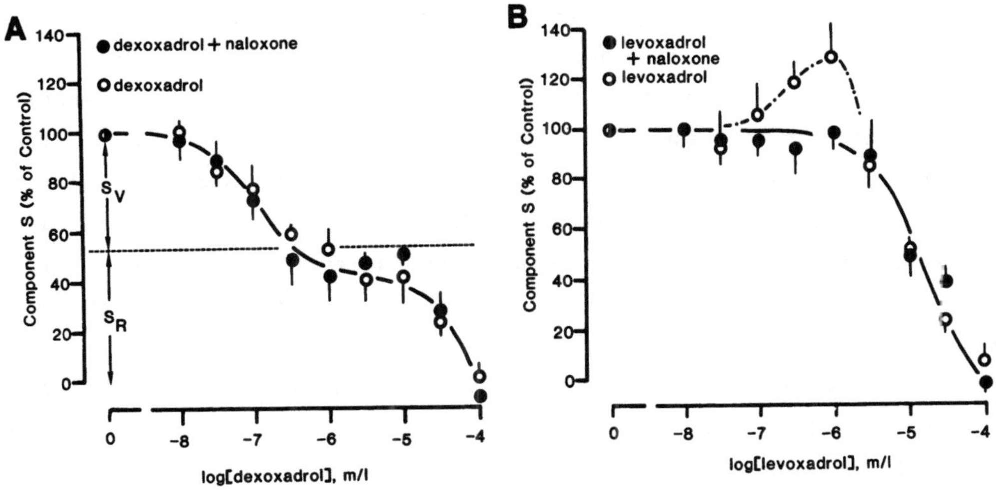

Fig. 4. Dose dependence for the effects of dexoxadrol (A) and levoxadrol (B) on ^{86}Rb efflux component "S" (see Fig. 1) in the presence of 1 mM 4-AP (to block component "T"; see Bartschat and Blaustein, 1985a). The label, "S_V", indicates the portion of component "S" that was blocked by 10 μM PCP. "S_R" is the PCP-insensitive portion of component "S" that presumably corresponds to ^{86}Rb efflux through the resting K conductance (see text). Data were obtained in the absence (open circles) and presence (closed circles) of 10 μM naloxone. Values are means of 12 determinations ± SE.

ability to displace PCP from its receptor sites and to discriminate PCP-like behavior (see Sorensen and Blaustein, 1987). (+)NANM is about 5-fold more potent than (-)NANM in blocking component "S_V", and dexoxadrol is very potent (K_I = 50 nM), while levoxadrol is ineffective except at high concentrations (> 1 μM) where it blocks all components of the ^{86}Rb efflux non-selectively (Bartschat and Blaustein, 1988). The effects of dexoxadrol and levoxadrol (with 10 μM naloxone present to block possible opiate-like effects of these agents; see below) on components "S_V", the PCP-sensitive portion of component "S", and "S_R", the PCP-insensitive portion, are illustrated in Fig. 4. Other ligand binding data also suggest that the PCP receptor is part of, or is associated with a K channel: typical K channel blockers (tetraalkylamines and aminopyridines) also displace ^3H-PCP from this site. TEA is a competitive inhibitor of PCP binding, while 4-AP interacts allosterically with the receptor (Sorensen and Blaustein, 1987). These results are all consistent with the hypothesis that the behavioral effects of PCP intoxication are related to its blockage of the voltage-gated, non-inactivating K channel (Albuquerque et al., 1981; Bartschat and Blaustein, 1987).

Levoxadrol opens a K channel that is insensitive to 4-AP. Certain opiates open K channels in resting neurons (e.g., Williams et al., 1982); this action should reduce excitability, and may thus account for the sedative effects of these agents. Dioxadrol, in addition to its PCP-like (psychotomimetic) action, also induces morphine-like (sedative) effects (Cone et al., 1984). The morphine-like effects of dioxadrol are also stereoselective, but the stereoselectivity is reversed: levoxadrol has a potent morphine-like action, while dexoxadrol has virtually none (Cone et al., 1984).

As already noted, levoxadrol did not reduce ^{86}Rb efflux component "S_V". However, 0.1-1.0 μM levoxadrol did increase ^{86}Rb efflux into both 100K (Fig. 4B) and 5K (not shown); these effects, which were not observed with dexoxadrol (Fig. 4A), were completely prevented by 10 μM naloxone (Fig. 4B). This (new) levoxadrol-activated, naloxone-sensitive (and apparently voltage-independent) component of ^{86}Rb efflux ("Σ") was not dependent on external Ca and was not blocked by 1 mM 4-AP or 10 μM PCP.

These data imply that the PCP-like and the morphine-like effects of dioxadrol are due to two different actions: 1) The PCP-like effect is the result of block of the voltage-gated, non-inactivating (delayed rectifier) K channel. And 2) The morphine-like effect is probably the result of interaction with an opiate receptor (cf. Cone et al., 1984) that opens a K channel that is different from the three channels characterized above. Whether or not the opiate receptor activated by levoxadrol is the same as either the "mu" or the "delta" opiate receptors, or is indeed a separate receptor, needs to be determined.

Effects of snake and scorpion toxins on Rb efflux component "S_V". The actions of several polypeptide toxins provide additional evidence that synaptosome ^{86}Rb efflux component "S_V" corresponds to a non-inactivating (or very slowly inactivating) K channel. We have isolated two previously unidentified polypeptides from D. angusticeps venom, β- and γ-DaTX, that preferentially block this component of the synaptosome ^{86}Rb efflux; the block of "S_V" by γ-DaTX is illustrated in Fig. 2B. High concentrations of β-DaTX also block component "T" (not shown). This cross-reactivity raises the possibility that these two types of voltage-gated K channels, the non-inactivating channel and the A channel, may have some structural homologies.

Venoms from several New World scorpions contain polypeptide toxins that

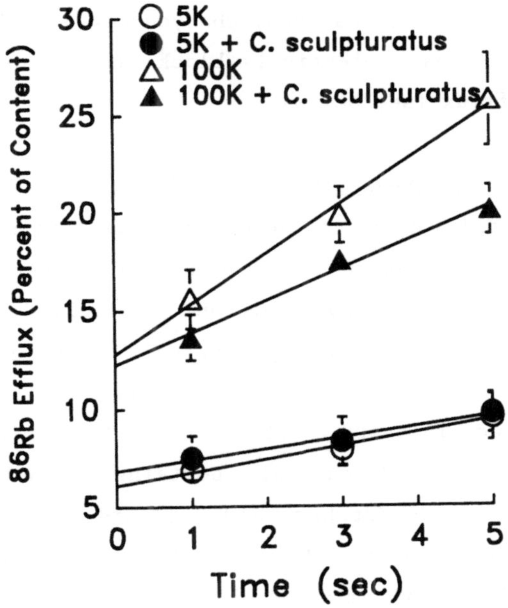

Fig. 5. Block of Component "S_V" by C. sculpturatus venom. Efflux was measured after a 16 min preincubation of P_2 synaptosomes with (solid symbols) or without (open symbols) 5 μg/ml C. sculpturatus venom, following a 15 min loading of the synaptosomes with ^{86}Rb. 1 μM TTX was present during preincubation to block any Na channel toxin activity present in the venom. The wash and efflux solutions contained 0.1 μM TTX. Symbols represent the means of 3 or 4 determinations ± SE.

selectively inhibit a non-inactivating (delayed rectifier) K channel (Possani et al., 1982; Carbone et al., 1987). Fig. 5 shows that unfractionated C. sculpturatus venom selectively blocks ^{86}Rb efflux component "S_V" in synaptosomes: it reduces the steady efflux in Ca-free, K-rich media, but does not affect the "resting" efflux or the ordinate intercept (component "T"). Identical results are obtained with T. serrulatus venom (not shown) as well as with γ-DaTX (Fig. 2B) and β-DaTX (not shown). None of these toxins or venoms affect component "C" (not shown).

The effects of the venom from the Old World scorpion, L. quinquestriatus, are more complex than those of the New World scorpion venoms. In Ca-free media, L. quinquestriatus venom, like that from C. sculpturatus and T. serrulatus, selectively blocks ^{86}Rb efflux component "S_V" (not shown). However, unlike the venoms from the New World scorpions or from D. angusticeps, L. quinquestriatus venom also contains a basic protein of about 7,000-10,000 daltons, charybdotoxin (Miller et al., 1985; Smith et al., 1986), that selectively blocks the Ca-dependent component ("C") of the ^{86}Rb efflux (see below).

Polypeptide toxin structure-activity relationships. We have determined the amino acid composition of the 4 polypeptide toxins isolated from D. angusticeps venom. These toxins all have molecular weights of about 7,000 daltons and contain 57-59 amino acids and 3 disulfide bridges. α-DaTX consists of 59 amino acids and is identical to the polypeptide called "dendrotoxin" (Harvey and Karlsson, 1982; Dolly et al., 1984) and toxin $C_{13}S_2C_3$ (Joubert and Taljaard, 1980). δ-DaTX, which, like α-DaTX, preferentially blocks the A channel, contains 57 amino acids and is identical to toxin $C_{13}S_1C_3$ (Joubert and Taljaard, 1980). α- and δ-DaTX have marked sequence homology at their C-terminal ends (beginning at position 30): 25 of the 30 amino acids are identical. They also show considerable homology in the N-terminal region: 9 of the 15 amino acids between positions 12 and 26 are identical. β- and γ-DaTX have been partially sequenced. Their sequences from position 30 to 52 are identical. Moreover, their structures are very similar to α- and δ-DaTX in this region near the C-terminal: 20 of the 23 amino acids in this region (the "core") are conserved in all four Da toxins.

A number of other snake toxins, as well as some peptidase inhibitors with no K channel blocking activity, contain a homologous amino acid core (Joubert and Taljaard, 1980; Harvey and Karlsson, 1982; Dufton, 1985). The significance of this core is uncertain, but it is not likely to be responsible for the selective effects of the 4 snake toxins we studied.

The sequences or partial sequences of three scorpion toxins that block a non-inactivating (delayed rectifier) K channel have been published: toxins II-11 (noxiustoxin, NTX) and II-10.2 from C. noxius, and toxin II-9A from T. serrulatus (Possani et al, 1982; Possani, 1984). Like the snake toxins, NTX has three disulfide bridges, but it consists of only 39 amino acids and has a molecular weight of about 4,000 daltons. Toxins II-10.2 and II-9A have sequence homology with NTX, and toxin II-9A has approximately the same molecular weight as NTX (Carbone et al, 1987). Toxin II-9A may be responsible for the non-inaactivating K channel blocking activity we observed with T. serrulatus venom.

There is no obvious sequence homology between these scorpion toxins and the 4 polypeptide toxins we isolated from D. angusticeps venom. Thus, the determinants of the structure-activity relationships of these snake and scorpion toxins are unclear at present.

Rb efflux component "C" corresponds to the current carried by a Ca-activated K channel. When Ca is added to the K-rich efflux media, a further stimulation of ^{86}Rb efflux is observed (Fig. 1). Most of the Ca-dependent component ("C") appears to inactivate within 1 sec: the linear extrapolation of the time course curve intersects the ordinate axis at a point above the Ca-free, high K curve (Fig. 1). However, other experiments indicate that this is a consequence of inactivation of the Ca entry mechanism (the voltage-gated Ca channels) and not of the ^{86}Rb efflux pathway (Bartschat and Blaustein, 1985b). Activation of this Ca-dependent component of ^{86}Rb efflux hyperpolarizes synaptosomes (Bartschat and Blaustein, 1985b), this implies that component "C" is due to an increased K conductance.

Several pharmacological agents selectively block this Ca-dependent ^{86}Rb efflux with IC_{50}'s on the order of 1 μM: quinine sulfate (Bartschat and Blaustein, 1985b), haloperidol, and the phenothiazines, fluphenazine and trifluoperazine (Benishin et al., 1986). These agents block Ca-activated K channels in other preparations (Amando-Hardy et al., 1975; McCann and Welsh, 1987; Dinnan et al., 1987). The Ca-dependent ^{86}Rb efflux from depolarized synaptosomes is also inhibited by tetraalkylamines but not by 4-AP (which enhances component "C" slightly; Bartschat and Blaustein, 1985b).

Two polypeptide toxins from invertebrate venoms, the bee toxin, apamin, and the scorpion (L. quinquestriatus) toxin, charybdotoxin, selectively block some Ca-activated K channels. Apamin blocks a small conductance (10-20 pS) TEA-insensitive channel (Hugues et al., 1982; Romey and Lazdunski, 1984; Blatz and Magleby, 1986), whereas charybdotoxin blocks a large conductance channel ("maxi Ca-activated K channel"; 180-240 pS) that is also blocked by TEA (Miller et al., 1984). In synaptosomes, unfractionated L. quinquestriatus venom (Fig. 6) as well as purified charybdotoxin (not shown) block ^{86}Rb efflux component "C" (= ΔC). The fact that this component is blocked by TEA and charybdotoxin, but is unaffected by apamin, suggests that it corresponds to current flow through the "maxi Ca-activated K channel".

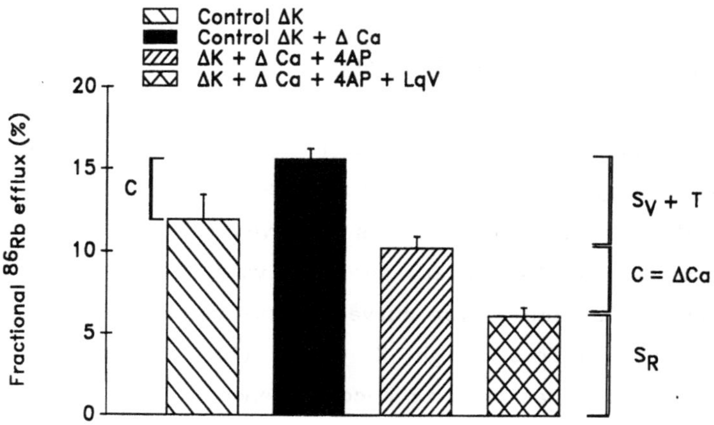

Fig. 6. Block of the Ca-dependent component of ^{86}Rb efflux (ΔC = component "C") by L. quinquestriatus venom (LqV). Preincubation of the synaptosomes with or without 3.3 μg/ml LqV was performed as described in the Fig. 5 legend. ΔK = [(efflux in 100K) minus (efflux in 5K)]; ΔCa = [(efflux in 100K/Ca) minus (efflux in 100K)]; (ΔK + ΔCa) = [(efflux in 100K/Ca) minus (efflux in 5K)]. Where indicated, 5 mM 4-AP was present in both the wash and efflux solutions to block efflux due to components "T" and "S_V"; thus, (ΔK + ΔCa) in the presence of 4-AP corresponds to components "S_R" + "C". LqV has no effect on "S_R" (not shown). Bars indicate means of 3 or 4 determinations ± S.E. of ^{86}Rb efflux at 5 sec.

Diacylglycerol analogs modulate K channels in synaptosomes. K channels are key regulators of neuronal excitability and signalling; they are therefore important targets for cellular control systems. Neuronal K channel activity can be modulated directly by cyclic nucleotide-dependent and calcium/phospholipid-dependent protein kinases (protein kinases A and C, respectively), as well as indirectly by neurotransmitters which act to stimulate the formation of second messengers (Levitan, 1985). Modulation of K channels via second messenger stimulation of protein phosphorylation is believed to underlie simple forms of learning and short-term memory in invertebrates (Castellucci et al., 1980; Farley and Auerbach, 1986).

Synaptosomes offer a convenient mammalian system in which both modulation of K channel activity and the biochemical mechanisms responsible for the modulation (e.g., phosphorylation) can be studied. With this in mind, we tested the effects of analogs of diacylglycerol, a second messenger that activates protein kinase C (Takai et al., 1979), on the various components of ^{86}Rb efflux from synaptosomes prepared from rat hippocampus. Incubation of hippocampal synaptosomes with diC8, an

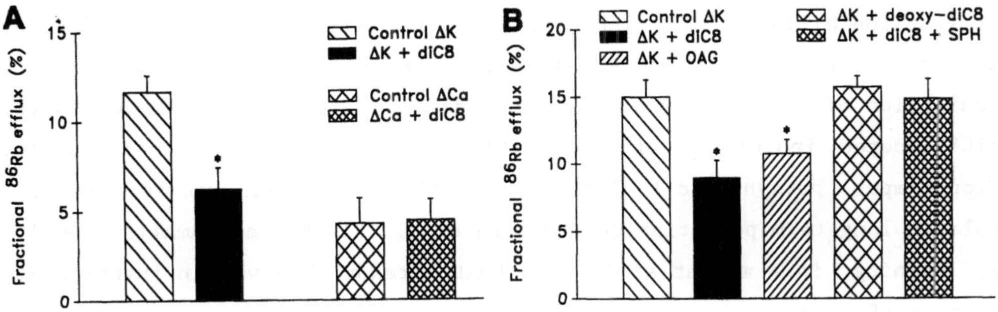

Fig. 7. A: Effects of diC8 on K-stimulated (ΔK) and Ca-dependent (ΔCa) ^{86}Rb efflux from hippocampal synaptosomes. Hippocampal P_2 synaptosomes were loaded with ^{86}Rb and then pre-incubated ± 150 μM diC8 at 37° C. for 6-9 min (see Methods). Extracellular ^{86}Rb was washed away and efflux was measured for 5 sec. ΔK = [(efflux into 50K) minus (efflux into 5K)]. ΔCa = [(efflux into 50K/Ca) minus (efflux into 50K)]. SE (ΔK) = $\sqrt{[SE(5K)]^2 + [SE(50K)]^2}$; SE (ΔCa) = $\sqrt{[SE(50K)]^2 + [SE(50K + Ca)]^2}$. The data are means of 4 determinations ± SE. *significantly different from control (no diC8) at $p < 0.05$.

B: Pharmacology of the diC8 effect. Hippocampal P_2 synaptosomes were pre-incubated ± 150 μM diC8, OAG or deoxy-diC8 or diC8 plus 100 μM sphingosine (SPH) for 6-9 min before ^{86}Rb efflux was measured for 5 sec. None of the drugs significantly affected resting efflux; SPH alone had no effect on ΔK. The depolarizing efflux solution contained 100 (rather than 50) mM K. Data from 3 representative experiments (one with each drug) are shown. N for ΔK and ΔK + diC8 is 15-18; N for each drug is 4-6. Similar results for each drug were obtained in at least 2 other experiments. *significantly different from control ΔK (no drug) at $p < 0.05$.

activator of protein kinase C (Conn et al., 1985), inhibited K-stimulated ^{86}Rb efflux, ΔK, but did not affect the Ca-dependent component, "C" or ΔCa (Fig. 7A). The effects of diC8 were time-dependent: a minimum of 15 sec of exposure was required to demonstrate inhibition; half-maximal inhibition occurred after approximately 30 sec of exposure to 150 μM diC8. Concentrations of diC8 from 15 μM to 450 μM inhibited ΔK; approximately 50 μM produced about a 25% decrease in ΔK, which was about half the maximal inhibition (50% of control ΔK) observed at the highest concentrations of diC8. Time course experiments demonstrated that diC8 treatment decreased both component "T" and component "S_V". This indicates that both the inactivating and non-inactivating voltage-gated K channels can be modulated by activation of protein kinase C (not shown).

Pharmacological evidence supports the idea that the diC8 effect resulted from activation of protein kinase C (Fig. 7B). OAG, another C-kinase-activating diacylglycerol analog (Conn et al., 1985), produced a similar inhibition of ΔK. Inhibition was not observed with deoxy-diC8, an analog that does not activate C-kinase (Conn et al., 1985). In addition, inhibition of C-kinase by sphingosine (Hannun et al., 1986) prevented the diC8-induced inhibition of ^{86}Rb efflux. It is possible that, in the hippocampus, presynaptic K channel modulation by protein kinase C plays a role in long-term potentiation (Malenka et al., 1986), a mammalian model of learning, in a manner similar to that already observed in invertebrate systems (Farley and Auerbach, 1986).

Identification of polypeptide subunits of voltage-gated K channels. The receptor polypeptides for PCP and two of the *D. angusticeps* neurotoxins have been tentatively identified. m-Azido-[^{3}H]PCP, a photoaffinity analog of PCP that binds to the PCP-receptor in synaptic membranes, was covalently attached to synaptic membranes (Sorensen and Blaustein, 1986). m-azido[^{3}H]PCP specifically labels two polypeptides of molecular weight (M_r) 95,000 and 80,000 (Fig. 8A). Some smaller peptides, including one of M_r = 30-35,000, also appear to be specifically labelled. This label is displaced by excess unlabelled PCP and TCP (the thienyl analog of PCP) and by TEA and 4-AP; the label in some other polypeptides (especially one with M_r = 56,000) is not displaced. Because PCP is a selective blocker of component "S_V", the polypeptides labelled by m-azido-[^{3}H]PCP may be subunits of the non-inactivating K channel.

Two polypeptide toxins from *D. angusticeps*, α- and γ-DaTX, have been radioiodinated and crosslinked to their receptor polypeptides. [^{125}I]α-DaTX specifically labels a polypeptide of M_r = 80,000 (uncorrected for bound toxin); this is similar to the polypeptide labelled by "dendrotoxin" (Mehraban et al. 1984). This polypeptide may be a subunit of the A channel in rat brain because α-DaTX preferentially blocks this channel (see above). In contrast, [^{125}I]γ-DaTX specifically labels polypeptides of M_r = 80,000 and 35,000 daltons (Fig. 8B), comparable to two of the polypeptides labelled by m-azido-[3H]PCP. This supports the view that at least some of the peptides labelled by m-azido-[^{3}H]PCP may be subunits of the non-inactivating K channel, because γ-DaTX, like PCP, preferentially blocks this channel. This represents the first identification of some of the subunits that may be part of the non-inactivating K channel. Purification and reconstitution of these polypeptides will be required to establish this definitively.

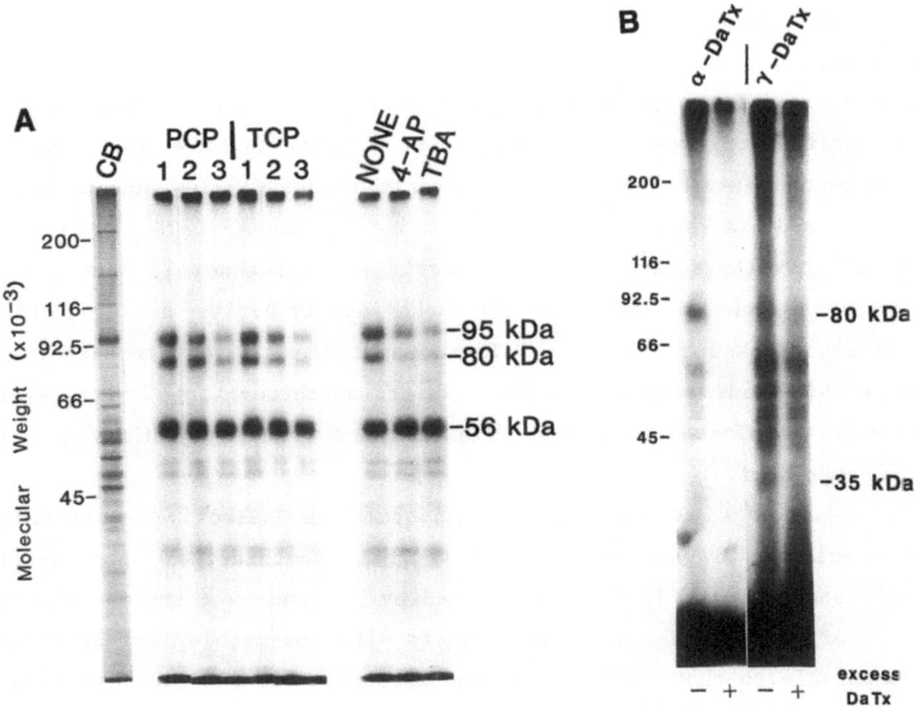

Fig. 8. Labeling of membrane receptor polypeptides.

A: m-Azido-[^{3}H]PCP was attached to synaptic membranes (see Methods and Sorensen and Blaustein, 1986) in binding mixtures with, respectively: NONE (no additions); PCP and TCP (lanes 1, 2 and 3 contained 1, 100 and 1000 μM of respective unlabeled ligand); 4-AP and TBA (5 mM compound). The resulting fluorogram is shown. The Coomassie blue (CB) staining pattern of a gel run in parallel is also shown; migration distances of standard proteins are indicated (from Sorensen and Blaustein, 1986, with permission).

B. [^{125}I]α-DaTx and [^{125}I]γ-DaTx were crosslinked to synaptic membrane polypeptides (see Methods) in the absence (-) and presence (+) of the respective unlabeled toxin. Molecular weights of polypeptides specifically labeled by the radioiodinated toxins are indicated on the right; migration distances of standard proteins are indcated on the left.

SUMMARY

Our results show that synaptosomes are an excellent preparation in which to study the physiological, pharmacological and biochemical properties of nerve terminal ion channels. Using ^{86}Rb efflux methods we have identified 5 different K channels in rat forebrain synaptosomes. One, which appears to be responsible for the resting K conductance, is blocked only by high concentrations of 4-AP, TEA, PCP, dexoxadrol and levoxadrol.

We have identified two voltage-gated channels, a Ca-activated K channel and an opiate-activated K channel. The rapidly-inactivating, voltage-gated K channel (A channel) is sensitive to relatively low concentrations of 4-AP, and is selectively blocked by α- and δ-DaTX. The non-inactivating voltage-gated (delayed rectifier) K channel is sensitive to low concentrations of PCP and its behaviorally-active analogs; it is selectively blocked by β- and γ-DaTX and by toxins present in certain Old World and New World scorpion venoms. Rb efflux through both voltage-gated K channels is blocked by tetraalkylamines and reduced by diacylglycerol analogs that activate protein kinase C.

The Ca-activated K channel in synaptosomes is selectively blocked by quinine sulfate, by haloperidol and by the phenothiazines, fluphenazine and trifluoperazine. It is also blocked by TEA and by charybdotoxin, but not by 4-AP or apamin. In these respects, its pharmacological properties are comparable to those of the large conductance (180-240 pS) Ca-activated K channel observed in other neuronal preparations (Hille, 1984; Yellen, 1987).

The opiate ligand-activated channel that we identified is opened by "sigma" ligands with morphine-like activity; it is insensitive to 4-AP or PCP and is not dependent upon external Ca. The activation of this channel by the morphine-like "sigma" ligands such as levoxadrol is prevented by naloxone.

Our results provide pharmacological profiles ("fingerprints") of these presynaptic terminal K channels. These findings may assist electrophysiologists who require selective pharmacological tools to dissect out the roles of each type of channel in more intact preparations. Furthermore, we have used some of the polypeptide toxins and other channel blockers as selective molecular probes to help identify the channel molecules. This represents a first step in the effort to purify and characterize the K channel proteins.

ACKNOWLEDGEMENTS

Supported by NIH grants NS-16106 (to MPB), NS-21758 (to DKB), NS-16285 and NS-20106 (to BKK), by a NRSA (to RGS) and a MDA Fellowship (to CGB), and by the Medical Biotechnology Center of the University of Maryland Biotechnology Institute. Protein sequencing was performed in the Central Protein Sequencing Facility of Carnegie-Mellon University directed by WEB.

REFERENCES

Albuquerque EX, Aguayo LG, Warnick JE, Weinstein H, Glick SD, Maayani S, Ickowicz R and Blaustein MP (1981) The behavioral effects of phencyclidines may be due to their blockade of potassium channels. Proc Natl Acad Sci USA 78: 7792-7796

Amando-Hardy M, Ellory JC, Ferreira MG, Fleminger S and Lew VL (1975) Inhibition of the calcium-induced increase in the potassium permeability of human red blood cells by quinine. J Physiol (Lond) 250: 32-33P

Bartschat DK and Blaustein MP (1985a) Potassium channels in isolated presynaptic nerve terminals from rat brain. J Physiol (Lond) 361: 419-440

Bartschat DK and Blaustein MP (1985b) Calcium-activated potassium channels in presynaptic nerve terminals from rat brain. J Physiol (Lond) 361: 441-457

Bartschat DK and Blaustein MP (1986) Phencyclidine in low doses selectively blocks a presynaptic voltage-regulated potassium channel in rat brain. Proc Natl Acad Sci USA 83: 189-192

Bartschat DK and Blaustein MP (1988) Psychotomimetic sigma ligands, dexoxadrol and phencyclidine block the same presynaptic potassium channel in rat brain. J Physiol (Lond) in press

Benishin CG, Sorensen RG, Krueger BK and Blaustein MP (1987) Four toxin components of green mamba (Dendroaspis angusticeps) venom with different specificities for voltage-gated K channels in rat brain synaptosomes. Fed Proc 46: 504

Blatz AL and Magleby KL (1986) Single apamin-blocked Ca-activated K^+ channels of small conductance in cultured rat skeletal muscle. Nature 323: 718-720

Carbone E, Prestipino G, Spadavecchia L, Franciolini F and Possani LD (1987) Blocking of the squid axon K^+ channel by noxiustoxin: a toxin from the venom of the scorpion Centruroides noxius. Pflugers Arch 408: 423-431

Castellucci VF, Kandel ER, Schwartz JH, Wilson FD, Nairn AC and Greengard P (1980) Intracellular injection of the catalytic subunit of cyclic AMP-dependent protein kinase simulates facilitation of transmitter release underlying behavioral sensitization in Aplysia. Proc Natl Acad Sci USA 77: 7492-7496

Cone EJ, McQuinn RL and Shannon HE (1984) Structure-activity relationship studies of phencyclidine derivatives in rats. J Pharmacol Exp Therap 228: 147-153

Conn PM, Ganong BR, Ebeling J, Staley D, Neidel JE and Bell RM (1985) Diacylglycerols release LH: Structure-activity relations reveal a role for protein kinase C. Biochem Biophys Res Comm 126: 532-539

Dinnan TG, Crunelli V and Kelly JS (1987) Neuroleptics decrease calcium-activated potassium conductance in hippocampal pyramidal cells. Brain Research 407: 159-162

Dolly JO, Halliwell JV, Black JD, Williams RS, Pelchen-Matthews A, Breeze AL, Mehraban F, Othman IB and Black AR (1984) Botulinum neurotoxin and dendrotoxin as probes for studies on transmitter release. J Physiol (Paris) 74: 280-303

Domino EF (ed) (1981) PCP (Phencyclidine): Historical and Current Perspectives. NPP Books, Ann Arbor, MI.

Dufton M.J. Protease inhibitors and dendrotoxins. Sequence classification, structural prediction and structure/activity. Eur J Biochem 153: 647-654 (1985).

Farley J and Auerbach S (1986) Protein kinase C activation induces conductance changes in Hermissenda photoreceptors like those seen in associative learning. Nature **319**: 220-223

Hajos F (1975) An improved method for the preparation of synaptosomal fractions in high purity. Brain Res 93: 485-489

Halliwell JV, Othman IB, Pelchan-Matthews A and Dolly JO (1986) Central action of dendrotoxin: selective reduction of a transient K conductance in hippocampus and binding to localized acceptors. Proc Natl Acad Sci USA 83: 493-497

Hampton RY, Mdzihradsky F, Woods JH and Dahlstrom PJ (1982) Stereospecific binding of ^{3}H-phencyclidine in brain membranes. Life Sci 30: 2147-2154

Hannun YA, Loomis CR, Merrill AH Jr. and Bell RM (1986) Sphingosine inhibition of protein kinase C activity and of phorbol dibutyrate binding in vitro and in human platelets. J Biol Chem **261**: 12604-12609

Harvey AL and Karlsson E (1980) Dendrotoxin from the venom of the green mamba, Dendroaspis angusticeps. A neurotoxin that enhances acetylcholine release at neuromuscular junctions. Naunyn-Schmeideburg's Arch Pharmacol 312: 1-6

Harvey AL and Karlsson E (1982) Protease inhibitor homologues from mamba venoms: facilitation of acetylcholine release and interactions with prejunctional blocking toxins. Br J Pharmacol 77: 153-161

Hille B (1984) Ionic channels of excitable membranes. Sinauer Associates, Inc. Sunderland, MA

Hugues M, Romey G, Duval D, Vincent JP and Lazdunski M (1982) Apamin as a selective blocker of the calcium dependent potassium channel in neuroblastoma cells: voltage clamp and biochemical characterization of the toxin receptor. Proc Natl Acad Sci USA 79: 1308-1312

Joubert FJ and Taljaard N (1980) The amino acid sequence of two proteinase inhibitor homologues from Dendroaspis angusticeps venom. Hoppe-Seyler's Z. Physiol. Chem 361: 661-674

Krueger BK and Blaustein MP (1980) Sodium channels in presynaptic nerve terminals. Regulation by neurotoxins. J Gen Physiol 76: 287-313

Krueger BK, Ratzlaff RW, Strichartz GR and Blaustein MP (1979) Saxitoxin binding to synaptosomes, membranes, and solubilized binding sites from rat brain. J Membrane Biol 50: 287-310

Levitan IB (1985) Phosphorylation of ion channels. J Membrane Biol 87: 177-190

Malenka RC, Madison DV and Nicholl RA (1986) Potentiation of synaptic transmission in the hippocampus by phorbol esters. Nature 321: 695-697

McCann JD and Welsh MJ (1987) Neuroleptics antagonize a calcium-activated potassium channel in airway smooth muscle. J Gen Physiol 89:339-352

Mehraban F, Breeze AL and Dolly JO (1984) Identification by cross-linking of a neuronal acceptor protein for dendrotoxion, a convulsant peptide. FEBS Lett. 174: 116-122

Miller C, Moczydlowski E, Latorre R and Phillips M (1985) Charybdotoxin, a protein inhibitor of single Ca^{2+}-activated K^+ channels from mammalian skeletal muscle. Nature (Lond) 313: 316-318

Nachshen DA (1985) The early time course of potassium-stimulated calcium influx into presynaptic terminals from rat brains. J Physiol (Lond)

361: 251-268
Nachshen DA and Blaustein MP (1980) Some properties of potassium-stimulated calcium influx in presynaptic nerve endings. J Gen Physiol 76: 709-728
Nachshen DA and Blaustein MP (1982) Influx of calcium, strontium and barium in presynaptic nerve endings. J Gen Physiol 79: 1065-1087
Possani LD (1982) The primary structure of noxiustoxin: a K^+ channel blocking peptide, purified from the venom of the scorpion Centruroides noxius Hoffmann. Carlsberg Res Commun 47: 285-289
Possani LD (1984) Structure of scorpion toxins. In: Tu AT (ed). Handbook of natural toxins. Vol. 2. Insect poisons, allergens and other invertebrate venoms. Marcel Dekker 1984 New York: 513-550
Ragowski MA (1985) The A-current: How ubiquitous a feature of excitable cells is it? Trends Neurosci 8: 214-219
Romey G and Lazdunski M (1984) The coexistence in rat muscle of two distinct classes of Ca^{2+}-dependent K^+ channels with different pharmacological properties and different physiological functions. Biochem Biophys Research Commun 118: 669-674
Salacinski PRP, McLean C, Sykes JEC, Clement-Jones VV and Lowry PJ (1981) Iodinationof proteins, glycoproteins, and peptides using a solid-phase oxidizing agent, 1,3,4,6-tetrachloro-3α,6α-diphenyl glycoluril (iodogen). Analyt Biochem 117: 136-146
Smith C, Phillips M and Miller C (1986) Purification of charybdotoxin, a specific inhibitor of the high-conductance Ca^{2+}-activated K^+ channel. J Biol Chem 261: 14607-14613
Sorensen RG, and Blaustein MP (1986) m-Azido-Phencyclidine covalently labels the rat brain PCP receptor, a putative K channel. J Neurosci 6: 3676-3681
Sorensen RG and Blaustein MP (1987) The rat brain phencyclidine (PCP) receptor. A putative K channel. Biochem Pharmacol, in press.
Takai Y, Kishimoto A, Iwasa Y, Kawahara Y, Mori T and Nishizuka Y (1979) Calcium-dependent activation of a multifunctional protein kinase by membrane phospholipids. J Biol Chem 254: 3692-3695
Tamkun MM and Catterall WA (1981) Ion flux studies of voltage-sensitive sodium channels in synaptic nerve-ending particles. Molec Pharmacol 19: 78-86
Williams JT, Egan TM and North RA (1982) Enkephalin opens potassium channels on mammalian central neurones. Nature 299: 74-77
Yellen G (1987) Permeation in potassium channels: implications for channel structure. Ann Rev Biophys Biophys Chem 16: 227-246

CALCIUM CHANNELS AND Na^+/Ca^{2+} EXCHANGE IN SYNAPTOSOMES

A. P. Carvalho, M. S. Santos, A. O. Henriques, P. Tavares and C. M. Carvalho
Centro de Biologia Celular, Departamento de Zoologia, Universidade de Coimbra,
3049 Coimbra Codex
Portugal

INTRODUCTION

Synaptosomes, the subcellular fraction of brain homogenates rich in presynaptic nerve terminals (Gray and Whittaker 1962; De Robertis 1962; Whittaker et al 1964; Whittaker 1965; Bradford 1969), possess the principal functional features of intact neurons: maintenance of membrane potential (Bradford 1971; Blaustein and Goldring 1975; Coutinho et al 1984), neurotransmitter uptake and release (Blaustein 1975; Raiteri and Levi 1978), depolarization dependent Ca^{2+} entry (Nachshen and Blaustein 1980; Blaustein 1975) and Na^+/Ca^{2+} exchange (Blaustein 1974; Blaustein and Osborn 1975; Blaustein and Ector 1976; Carvalho 1982; Coutinho et al 1983, 1984).

The use of the synaptosome as a model for the nerve terminal has constituted a useful biochemical approach to studying Ca^{2+} channels and Na^+/Ca^{2+} exchange in the mammalian central nervous system. However, the contribution of each of these Ca^{2+} translocating mechanisms during depolarization of the synaptosomal membrane is not clear, since there is no specific inhibitor for either mechanism in the nervous system. Even the organic Ca^{2+} channel blockers, which selectively block Ca^{2+} fluxes through the Ca^{2+} channels in smooth and cardiac muscles, do not seem to work in synaptosomes at concentrations which saturate their binding sites (Nachshen and Blaustein 1979; Daniell et al 1983; Carvalho et al 1986 a; Carvalho and Carvalho 1983; Rampe et al 1984; Miller and Freedman 1984; Chin 1986; Miller 1987).

Turner and Goldin (1985) have indicate that, under certain conditions, the organic Ca^{2+} channel blockers, at nM concentrations, also inhibit $^{45}Ca^{2+}$ influx into synaptosomes mediated by K^+-depolarization. Moreover, White and Bradford (1986) showed that the agonist BAY K8644 increases the Ca^{2+} uptake into submaximally stimulated synaptosomes, and Middlemiss and Spedding (1985) showed that BAY K8644 increases the Ca^{2+}-dependent neurotransmitter release in brain slices. More important, there are several reports that neuronal cell lines (Toll 1982; Takahashi and Ogura 1983; Abus et al 1984; Freedman et al 1984) and primary cell cultures (Miller 1987; Thayer et al 1986; Gähwiler and Brown 1987) display Ca^{2+} channels sensitive to Ca^{2+} blockers. These results indicate that the Ca^{2+} antagonist

NATO ASI Series, Vol. H21
Cellular and Molecular Basis of Synaptic Transmission
Edited by H. Zimmermann

binding sites in brain may have a functional correlate, but there is difficulty in observing it consistently, especially in isolated brain fractions.

We reasoned that some of the inherent difficulty in showing the effect of organic Ca^{2+} channel blockers on $^{45}Ca^{2+}$ fluxes through the Ca^{2+} channels in synaptosomes might be, in part, related to the experimental limitation imposed by the fact that K^+ depolarization-stimulated Ca^{2+} uptake includes a variable quantity of Ca^{2+} influx through the Na^+/Ca^{2+} exchange mechanism (Coutinho et al 1984; Mullins et al 1985; Carvalho et al 1986 a, b). Therefore, we explored experimental conditions which permitted us to distinguish the effect of the Ca^{2+} channel blockers on Ca^{2+} entry through channels from the effect on Ca^{2+} entry through the Na^+/Ca^{2+} exchange mechanism, and studied the effect of membrane depolarization on the [^{3}H]nitrendipine binding by synaptosomes and synaptic plasma membrane vesicles. In addition, we studied the effect of the blockers on Ca^{2+}-dependent and Ca^{2+}-independent release of [^{3}H]GABA.

MATERIAL AND METHODS

Preparation of Synaptosomal Fractions from Brain

Synaptosomes were prepared from sheep brain cortex homogenates according to the method described by Hajós (1975), with some modifications (Carvalho and Carvalho 1979). The final synaptosomal pellets were either suspended in 0.32 M sucrose buffered with 10 mM HEPES-Tris, pH 7.4, or pre-incubated for 30 min at 30°C in Na^+-rich medium containing (in mM): 128 NaCl, 5 KCl, 1 $MgCl_2$, 10 glucose, 10 HEPES-Tris, pH 7.4. After the incubation, the suspension was centrifuged for 15 min at 20,000 x g and the pellet was resuspended in choline medium containing (in mM): 128 choline chloride (ChCl), 5 KCl, 1 $MgCl_2$, 10 glucose; 10 HEPES-Tris, pH 7.4, at 20 mg protein/ml.

Synaptic plasma membrane (SPM) vesicles were prepared by osmotic lysis of synaptosomal pellets in 5 mM HEPES-Tris, pH 8.6, and partially purified from the lysate, as described previously (Coutinho 1983). The SPM vesicles utilized for Na^+/Ca^{2+} exchange or ATP-dependent Ca^{2+} uptake studies were washed and resuspended in Na^+ medium or K^+ medium containing (in mM): 150 NaCl or KCl, respectively, and 1 $MgCl_2$, 10 glucose, 10 HEPES-Tris, pH 7.4, at 20 mg protein/ml.

Calcium Uptake Studies

Ca^{2+} uptake due to K^+- depolarization was performed by diluting the synaptosomal suspension (20-40 fold) into Ca^{2+} uptake media containing KCl (0-128 mM) substituted for NaCl or ChCl, as referred in the individual figures, and 1 mM

$MgCl_2$ 10 mM glucose, 20 µM or 1 mM $CaCl_2$ (plus $^{45}CaCl_2$, 2.5 µCi/µmol), 10 mM HEPES-Tris, pH 7.4. The reaction was conducted at 30°C, and was terminated by rapid mixing with 0.1 mM $LaCl_3$ prepared in NaCl medium (128 mM NaCl, 5 mM KCl), and the suspension was filtered through Whatman GB/F filters, as described previously (Carvalho et al 1986 a).

Na^+/Ca^{2+} exchange and ATP-dependent Ca^{2+} uptake by SPM fractions were studied as described previously (Coutinho et al 1983; Carvalho et al 1986 a), by diluting membrane vesicles preloaded with NaCl (Na_i^+) or KCl (K_i^+) into Ca^{2+} uptake media containing (in mM): 150 NaCl or KCl, 1 $MgCl_2$, 10 glucose, 10 glucose, 10 HEPES-Tris, pH 7.4, and 20 µM $CaCl_2$+$^{45}Ca^{2+}$ (2.5 µCi/µmol), at a final protein concentration of 0.5 mg/ml). In the case of ATP-dependent Ca^{2+} uptake, the reaction was initiated by adding 1 mM ATP-Mg. All reactions were terminated as described for synaptosomal Ca^{2+} uptake.

Measurement of Synaptosomal Membrane Potential

The membrane potentials of synaptosomes and SPM vesicles were determined from the accumulation of tetraphenylphosphonium (TPP^+) by using a TPP^+ selective electrode, as described previously (Coutinho et al 1984).

[^{3}H] Nitrendipine Binding

The [^{3}H]nitrendipine binding to synaptosomal and SPM fractions was performed as described previously for [^{3}H]nimodipine (Carvalho et al 1986 a).

Materials

All chemicals used were of analytical grade. [^{3}H]nitrendipine (87 Ci/mmol) were purchased from New England Nuclear. [^{3}H]GABA was purchased from Amersham International. Nimodipine was provided by Prof. Hoffmeister, from Bayer AG, Wuppertal, FRG. Nifedipine and d-cis-diltiazem were provided by Prof. T. Macedo, Faculty of Medicine, Coimbra, Portugal. Verapamil was obtained from Knoll Lusitana, Portugal. PY 108-86 was provided by Dr. Hof, from Sandoz, Basel, Switzerland. The 1,4-dihydropyridines were first dissolved in absolute ethanol at 1 mM and then were diluted to the appropriate concentrations with buffer. Due to the lability of the 1,4-dihydropyridines to light, all solutions were stored light-protected at -20°C.

RESULTS AND DISCUSSION

Pre-incubation of Synaptosomes in Na^+-Rich Media Increases the "K^+-depolarization"-dependent $^{45}Ca^{2+}$ Uptake

Storage of synaptosomes in Na^+-rich medium (in mM: 128 NaCl, 5 KCl, 10 glucose, 1.2 MgCl, 10 HEPES-Tris, pH 7.4), at 30°C, for 30 min, about triples the Na^+ content and decreases the K^+ by 30-40%, as compared to synaptosomes which are kept in 0.32 M sucrose, buffered with 10 mM HEPES-Tris, at pH 7.4 (Fig. 1 and Coutinho et al 1984). Incubation of isolated synaptosomes in Na^+-rich media is current procedure in most studies published on the various aspects of K^+-depolarization Ca^{2+} uptake (Blaustein 1975; Blaustein and Ector 1976; Nachshen and Blaustein 1980).

Comparision of the results of studies of the $^{45}Ca^{2+}$ uptake due to K^+-depolarization in Na^+-rich and sucrose-synaptosomes shows that the Na^+-rich synaptosomes take up larger amounts of $^{45}Ca^{2+}$ than do the sucrose synaptosomes, when depolarized under identical conditions. Furthermore, depolarization with 60 mM KCl in which osmolarity is maintained by choline (Ch^+), instead of Na^+, further increases the $^{45}Ca^{2+}$ uptake in the Na^+-rich synaptosomes, but not in the sucrose synaptosomes (Fig. 1 A, B).

The initial rate of $^{45}Ca^{2+}$ uptake in Na^+-rich synaptosomes, due to the addition of K^+, is at least twice as high as the rate in sucrose synaptosomes, and the maximal uptake is particularly increased when choline, instead of Na^+, is used to maintain the osmolarity. In the sucrose synaptosomes the total uptake at 120 sec, in K^+, Ch^+ medium (60 mM K^+; 73 mM Ch^+) is about 3 nmol/mg protein, whereas the uptake by the Na^+-rich synaptosomes, under identical conditions, is about 9 nmol/mg protein, and is not yet approaching saturation (Fig. 1 A, B).

We interpret the results as follows: The high intrasynaptosomal Na^+ content of synaptosomes pre-incubated in Na^+-rich medium increases the level of Na^+/Ca^{2+} exchange, and this effect is particularly visible when the uptake media is free of Na^+, i.e., when Ch^+ substitutes Na^+ during K^+-depolarization. This substitution of Ch^+ for Na^+ has no significant effect if the internal Na^+ content of synaptosomes is low, as is the case for sucrose synaptosomes that have not been pre-incubated in Na^+-rich medium (Fig. 1 A). In any case, the pre-incubation of synaptosomes in Na^+-rich media promotes increase in $^{45}Ca^{2+}$ uptake during subsequent K^+-depolarization studies, whether the uptake medium has Na^+ or Ch^+, and this increase is particularly significant for initial rates of uptake (Fig. 1 A, B).

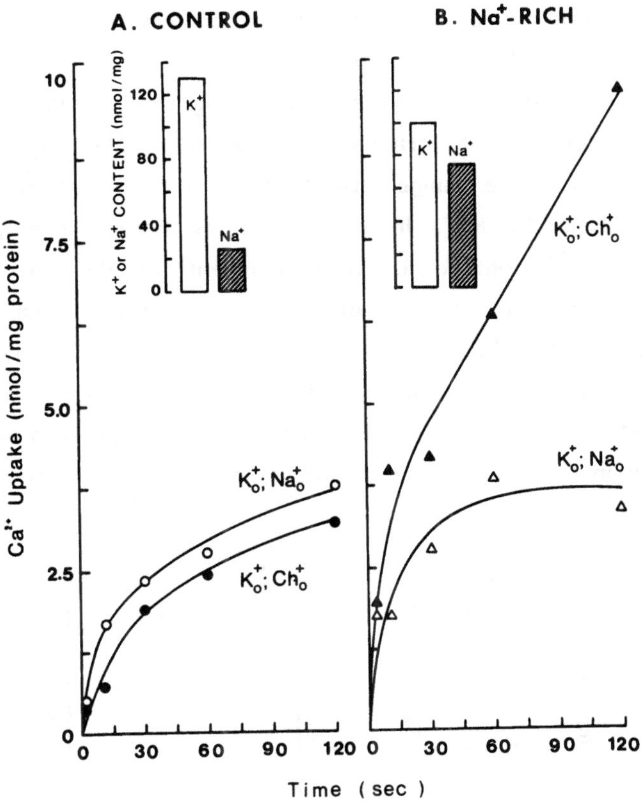

Fig. 1. Time course of Ca^{2+} uptake by synaptosomes pre-incubated (30 min, at 30°C) in: **A.** Sucrose medium (control) or **B** Na^+-rich medium containing (in mM): 128 NaCl, 5 KCl, 10 glucose, 1.2 $MgCl_2$, 10 HEPES-Tris, pH 7.4. $^{45}Ca^{2+}$ uptake represents the difference between the uptake in basal (128 Na^+; 5 K^-) and depolarizing media, in the presence of external Na^+ (60 K^+; 73 Na^+), or in the absence of external Na^+ (60 K^+; 73 Ch^+), all containing 1 mM $^{45}Ca^{2+}$. The inserted bars represent the intrasynaptosomal K^+ and Na^+ content of control (A) and of synaptosomes pre-incubated in Na^+-rich medium (B).

The Relative Importance of Na^+ Gradient and Membrane Depolarization on Ca^{2+} Uptake

Synaptosomes which have been pre-incubated in Na^+-rich medium have a membrane potential of about -70 mV when placed in a choline chloride medium, as measured by a TPP^+ electrode (Fig. 2). The potential decreases with increasing concentrations of external K^+ (K_o^+) according to the Nernst equation, and, thus, when the (K_o^+) is about equal to the value that we measure for the internal K^+ concentration (K_i^+), we obtain a membrane potential of 0-10 mV (Fig. 2 A). The initial membrane potential is lower by 10-20 mV if the membrane potential measurements are made in 150 mM Na_o^+, rather than in 150 mM Ch^+. This partial depolarization induced by Na^+ probably reflects the contribution of the Na^+ diffusion potential which opposes the K^+ diffusion potential.

In Fig. 2A we present the results of changing the membrane potential on $^{45}Ca^{2+}$ uptake at a constant Na^+ gradient, by substituting the Ch_o^+ in the outside medium by K_o^+. In Fig. 2B, we change both membrane potential and the Na^+ gradient by substituting the Na_o^+ for K_o^+. In both conditions there is a voltage drop, but the pattern of $^{45}Ca^{2+}$ uptake is very different in the two cases. The K^+-depolarization in the Ch_o^+ medium occurs under a constant high Na^+ gradient directed outward, and the $^{45}Ca^{2+}$ uptake is initially about 2,5 nmol/mg protein, as compared to less than 1 nmol/mg protein in the Na_o^+ medium (Fig. 2 A, B). Addition of 10 mM K_o^+ causes an increase of $^{45}Ca^{2+}$ uptake to a value of about 4 nmol/mg protein in the Ch_o^+ medium, but has much smaller effect in the Na_o^+ medium.

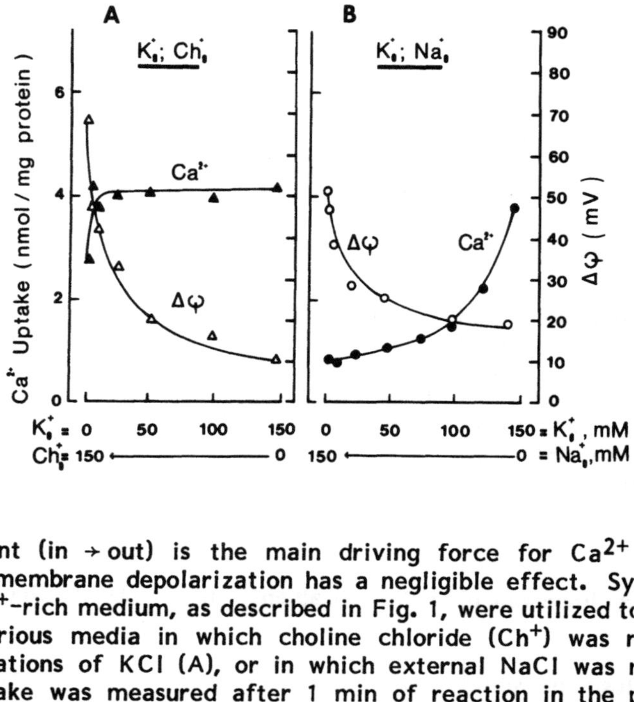

Fig. 2. Na^+ gradient (in → out) is the main driving force for Ca^{2+} uptake by synaptosomes, and membrane depolarization has a negligible effect. Synaptosomes pre-incubated in Na^+-rich medium, as described in Fig. 1, were utilized to determine Ca^{2+} uptake in various media in which choline chloride (Ch^+) was replaced by increasing concentrations of KCl (A), or in which external NaCl was replaced by KCl (B). Ca^{2+} uptake was measured after 1 min of reaction in the presence of 20 µM $^{45}CaCl_2$. Membrane potential ($\Delta\Psi$) was determined in the same conditions, by following the uptake of TPP^+ by synaptosomes at a TPP^+ concentration of 25 µM, using a TPP^+ selective electrode as described previously (Modified from Coutinho et al 1984).

Therefore, we conclude that under our experimental conditions, the $^{45}Ca^{2+}$ uptake induced by K^+ is not mainly due to membrane depolarization. Also, the substitution of Na_o^+ for K_o^+ in Fig. 2B should contribute to some Na^+/Ca^{2+}

exchange, whereas this component does not exist in Fig. 2 A, in which the Na^+ gradient is maintained constant. Thus, it seems that the larger abrupt increase in $^{45}Ca^{2+}$ uptake when K^+ is added to the Ch^+ medium, as compared to Na^+ medium, reflects an effect of K_o^+ on the Na^+/Ca^{2+} exchange, independently of alterations on the Na^+ gradient. The fact that the K^+ effect is observed only when a large outward Na^+ gradient exists can be explained if we assume that the Na^+/Ca^{2+} exchange, which is electrogenic, is promoted by membrane depolarization by K_o^+ and/or that K^+ functions as a counter ion in Na^+/Ca^{2+} exchange.

Other studies have shown that K^+ stimulates Na^+/Ca^{2+} exchange in mitochondria (Crompton et al 1980). Therefore, we designed an experiment to test unequivocally whether K^+ stimulates Na^+/Ca^{2+} exchange across synaptosomal membranes. We used synaptic plasma membranes vesicles (SPM), prepared from lysed synaptosomes, loaded with 150 mM Na_i^+. When suspensions of these SPM are diluted 20-fold in Ca^{2+} uptake medium containing 150 mM Na_o^+, there is a $^{45}Ca^{2+}$ uptake of only about 1 nmol/mg protein, but when a Na^+ gradient is created by sustituting all external Na_o^+ for Ch_o^+, the $^{45}Ca^{2+}$ uptake increases to about 5 nmol/mg protein (Fig. 3). If we now maintain the outward gradient of Na^+ constant, but also add K^+ to the medium, the Na^+/Ca^{2+} exchange increases to about 7 nmol/mg protein when the external medium contains 75 mM Ch_o^+ plus 75 mM K_o^+ and an even larger value for $^{45}Ca^{2+}$ uptake is obtained (10 nmol/mg protein) if all Ch_o^+ is substituted for K_o (K_o = 150 mM).

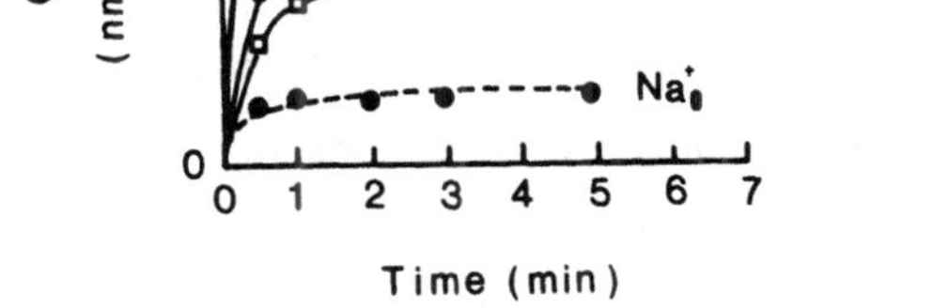

Fig. 3. Stimulation by K^+ of Na^+/Ca^{2+} exchange by synaptic membrane vesicles (SPM) loaded with Na^+. SPM vesicles are preloaded with 150 mM NaCl medium (Na_i^+) and are diluted 20-fold in Ca^{2+} uptake medium containing the following external ions: 150 mM NaCl (Na_o^+), 150 mM choline chloride (Ch_o^+), 75 mM KCl plus 75 mM choline chloride (K_o^+; Ch_o^+) or 150 mM KCl (K_o^+), and 20 μM $^{45}CaCl_2$, as described in Methods.

These results show, unequivocally, that K^+ in the external medium stimulates Na^+/Ca^{2+} exchange, and that this effect is particularly pronounced when synpatosomes have been preincubated in Na^+-rich medium and the uptake medium is Na^+ free, i.e., contains Ch^+ instead of Na^+ (Fig. 2 and 3). These results also are relevant to the K^+-depolarization studies which have shown that pre-incubation of the synaptosomes with K^+ reduces the subsequent $^{45}Ca^{2+}$ uptake from a K^+-depolarizing medium (Nachshen and Blaustein 1980; Drapeau and Blaustein 1983; Turner and Goldin 1985). Thus, the presence of K^+ in the external medium during the pre-depolarization would obviate the effect of K^+ on Na^+/Ca^{2+} exchange during subsequent K^+-depolarization. This would also explain that the effect of pre-depolarization is not instantaneous, as it would more likely be the case if it were mediated through inactivation of the Ca^{2+} channels. The fact that the effect is observed maximally only after several seconds (Nachshen 1985) may reflect a discharge of the Na^+ gradient accelerated by K_o^+, in addition to the effect of K_o^+ on the Na^+/Ca^{2+} exchange through the depolarizing and counter ion effect of K^+.

Ca^{2+} Channel Blockers vs. Ca^{2+} Fluxes and Transport in Synaptosomes

In an attempt to determine the effect of Ca^{2+} channel blockers on the fraction of Ca^{2+} which enters synaptosomes through the Ca^{2+} channels, we first designed experimental conditions which eliminate the Na^+/Ca^{2+} exchange component (Carvalho et al 1986 a), which may be occluding detection of the effect of the Ca^{2+} blockers on $^{45}Ca^{2+}$ fluxes through the channels. The simplest way to eliminate the Na^+/Ca^{2+} exchange is to use synaptosomes isolated in sucrose and which have not been pre-incubated in Na^+/rich medium, so that the internal Na^+ concentration is low, i.e., about 10-20 mM.

If $^{45}Ca^{2+}$ uptake is measured in synaptosomes isolated in sucrose and then placed in a medium containing 60 mM KCl and 73 mM KCl in the presence of 1 mM Ca^{2+} supplemented with $^{45}Ca^{2+}$ (Fig. 1), the $^{45}Ca^{2+}$ is taken up predominantly through the channels since about the same uptake occurs whether Na^+ or Ch^+ are used to maintain osmolarity. Under these conditions the internal Na^+ concentration (Na_i^+) is not sufficient to promote Na^+/Ca^{2+} exchange, since the (Na_i^+) is about 10-20 mM and the K_D value for Na^+ of the Na^+/Ca^{2+} exchanger is of the order of 25-50 mM (Blaustein 1974; Blaustein and Oborn, 1975).

Our results reported earlier (Carvalho et al 1986 a) show that (+)-verapamil, nifedipine and nimodipine are much less potent in inhibiting $^{45}Ca^{2+}$ influx than in inhibiting the binding of [^{3}H]nimodipine to purified SPM. Thus, (+)-verapamil has a IC_{50} for inhibiting Ca^{2+} influx due to depolarization of about 500 μM, whereas, under similar conditions, it inhibits 50% of the [^{3}H]nimodipine binding at concentration of 0.7 μM. The discrepancy between the affinities determined from binding studies and the effectiveness in blocking $^{45}Ca^{2+}$ influx are even larger for nimodipine and nifedipine (Carvalho et al 1986 a). Therefore, there is no correlation between the affinities of these Ca^{2+} blockers for SPM binding sites and their potency for blocking Ca^{2+} influx due to K^+ depolarization of synaptosomes, under conditions of minimal Na^+/Ca^{2+} exchange. These results are at variance with those reported by Turner and Goldin (1985), but are in accord with those of other workers who, however, performed the experiments under conditions in which there occurs considerable Na^+/Ca^{2+} exchange (Nachshen and Blaustein 1979; Rampe et al 1984; Miller et al 1984; Daniell et al 1983).

We re-examined the problem by studying the effect of the Ca^{2+} channel blockers on Ca^{2+} fluxes in Na^+-rich synaptosomes, but in experiments designed to distinguish in the same preparation, between Ca^{2+} entry through Ca^{2+} channels and through the Na^+/Ca^{2+} exchange mechanism (Fig. 4). In synaptosomal

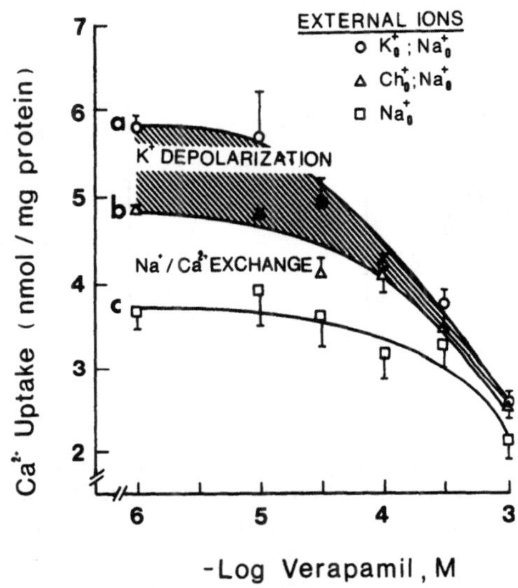

Fig. 4. Verapamil inhibits both Ca^{2+} uptake by synaptosomes due to K^+ depolarization and due to Na^+/Ca^{2+} exchange. Ca^{2+} uptake was initiated by transferring synaptosomes pre-incubated in Na^+-rich medium to each of three different media: (curve c) 128 mM NaCl plus 5 mM KCl (basal medium); (curve b) 60 mM ChCl plus 73 mM NaCl; or (curve a) 60 mM KCl plus 73 mM NaCl, all containing, additionally, 10 mM glucose, 1 mM $MgCl_2$, 10 mM HEPES-Tris, pH 7.4, 1 mM $^{45}CaCl_2$ and increasing concentrations of verapamil. Ca^{2+} uptake was terminated after 2 min of reaction by filtration. Ca^{2+} uptake due to Na^+/Ca^{2+} exchange is the difference in Ca^{2+} uptake between the uptake in $Ch^+ + Na^+$ medium (curve b) minus the uptake in Na^+ medium (curve c). Ca^{2+} uptake due to K^+ depolarization (shaded area) is the difference in Ca^{2+} uptake in $K^+ + Na^+$ medium (curve a) minus the uptake in $Ch^+ + Na^+$ medium (curve b).

preparations which have been pre-incubated in Na^+/rich media, it is not possible to measure the influx of $^{45}Ca^{2+}$ due to K^+ depolarization in the absence of Na^+/Ca^{2+} exchange, since the addition of 60 mM K^+ to depolarize the synaptosomes is accompanied by an equivalent reduction of the external Na^+ to maintain osmolarity, and Na^+/Ca^{2+} exchange also accurs under these conditions (Fig. 4, curve _a_). However, when the Na_o^+ concentration is reduced to the same extent in substitution for Ch^+ rather than for K^+, which avoids depolarization, $^{45}Ca^{2+}$ enters the synaptosomes only by Na^+/Ca^{2+} exchange (Fig. 4, curve _b_). The $^{45}Ca^{2+}$ influx due to K^+ depolarization was calculated indirectly as the difference between curves _a_ of Fig. 4, representing total $^{45}Ca^{2+}$ influx (due to K^+ depolarization + due to Na^+/Ca^{2+} exchange) and curve _b_, representing Na^+/Ca^{2+} exchange. Curve _c_ of Fig. 4 represents the control uptake in a high Na^+ medium (128 mM NaCl and 5 mM KCl).

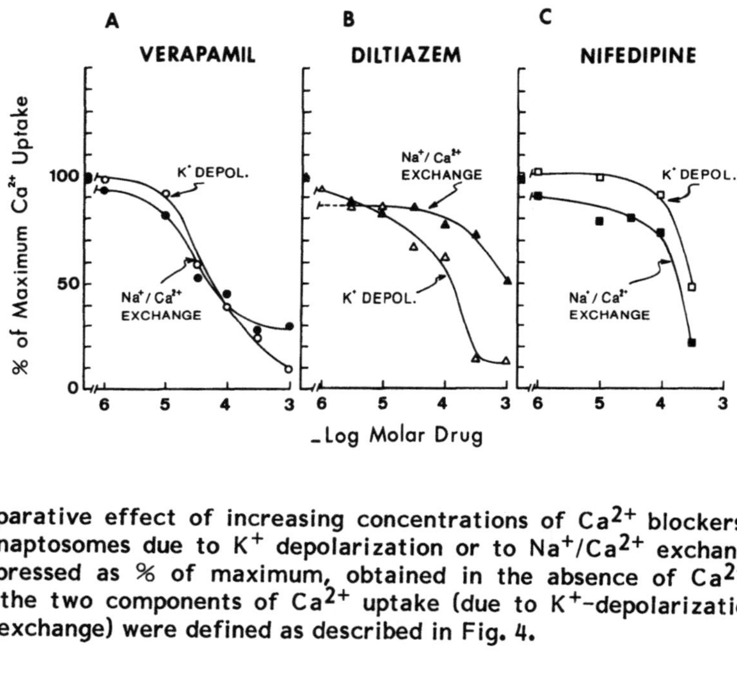

Fig. 5. Comparative effect of increasing concentrations of Ca^{2+} blockers on Ca^{2+} uptake by synaptosomes due to K^+ depolarization or to Na^+/Ca^{2+} exchange. Ca^{2+} uptake is expressed as % of maximum, obtained in the absence of Ca^{2+} channel blocker, and the two components of Ca^{2+} uptake (due to K^+-depolarization or due to Na^+/Ca^{2+} exchange) were defined as described in Fig. 4.

Verapamil for concentrations higher than 10^{-5} M inhibits $^{45}Ca^{2+}$ uptake by synaptosomes due to both K^+ depolarization and to Na^+/Ca^{2+} exchange, and the potency of the drug is about equal in both cases (Fig. 5). Therefore, in the concentration range of the drug studied there is no selective effect of verapamil on Ca^{2+} channels. Similar results were obtained for the Ca^{2+} blockers, _d-cis_-diltiazem and nifedipine (Fig. 5), as reported earlier (Carvalho et al 1986 a, b). There is some indication from Fig. 5B that diltiazem may be less efective in inhibiting Na^+/Ca^{2+} exchange than it is in inhibiting Ca^{2+} influx due to

K^+-depolarization.

Verapamil, nifedipine and diltiazem also inhibit Na^+/Ca^{2+} exchange and ATP-dependent Ca^{2+} uptake by SPM. The concentrations required to inhibit 50% of the Ca^{2+} uptake by Na^+/Ca^{2+} exchange are $4.6x10^{-4}$ M, $3x10^{-4}$M and 10^{-3} M for verapamil, nifedipine and d-cis-diltiazem, respectively, and similar values are obtained for inhibiting 50% of the ATP-dependent Ca^{2+} uptake ($8.7x10^{-4}$M, $8.7x10^{-4}$M and 10^{-3}M) for verapamil, nifedipine and d-cis-diltiazem (Table 1). The inhibition of Na^+/Ca^{2+} exchange has also been reported earlier for cardiac membrane vesicles (Erdreich and Rahamimoff 1984) and for nerve membrane systems (Erdreich et al 1983; Carvalho et al 1986 a, b).

TABLE 1. Comparative effects of Ca^{2+} channel blockers on the Ca^{2+} transport systems in SPM

Calcium blocker	IC_{50} (M)	
	Na^+/Ca^{2+} Exchange	ATP-dependent Ca^{2+} Uptake
Verapamil	4.6×10^{-4}	8.7×10^{-4}
Nifedipine	3.0×10^{-4}	8.7×10^{-4}
d-cis-diltiazem	10^{-3}	10^{-3}

IC_{50} are concentrations of the drugs producing 50% inhibition of the Ca^{2+} uptake by: Na^+/Ca^{2+} exchange or ATP-dependent Ca^{2+} uptake

Ca^{2+}-Channel Blockers Inhibit Ca^{2+}-Dependent and Ca^{2+}-Independent [^{3}H]GABA Release

Exogenous tritiated γ-aminobutyric acid, [^{3}H]GABA, is taken up into synaptosomes by a high affinity system which is Na^+ dependent (Martin, 1973; Iversen and Kelly 1975; Blaustein and King 1976). This accumulated [^{3}H]GABA can be released in response to depolarization, and two mechanisms have been proposed for the release; one mechanism requires Ca^{2+}, whereas the other is Ca^{2+} independent, and may represent a reversal of the Na^+-dependent GABA carrier (Blaustein and King 1976; Levi et al 1976; Cunningham et al 1981; Santos et al 1987). It is not clear whether the [^{3}H]GABA released in the presence or absence of Ca^{2+} come from different pools, although there is some evidence that they are both cytoplasmatic (De Belleroche and Bradford 1971; Santos et al 1987).

In Fig. 6, we report the $^{45}Ca^{2+}$ uptake and the Ca^{2+}-independent and the Ca^{2+}-dependent [^{3}H]GABA release from synaptosomes in various ionic media. We observe that Ca^{2+}-independent release of [^{3}H]-GABA due to K^+ depolarization

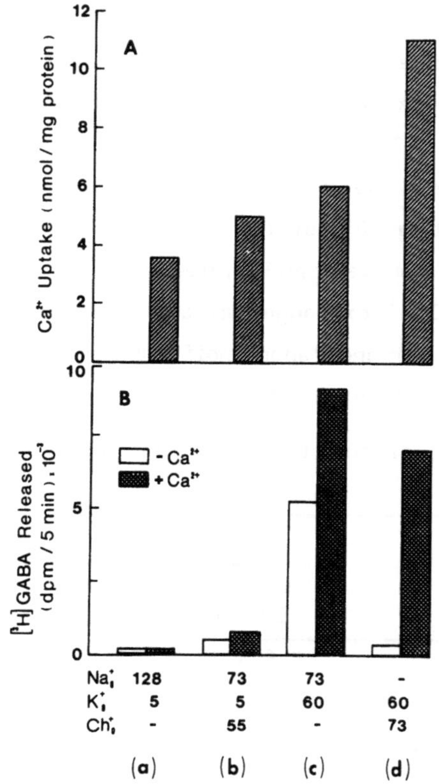

Fig. 6. Relationship between Ca^{2+} entry and [^{3}H]GABA release by synaptosomes. Synaptosomes pre-incubated in Na^+-rich medium either in the absence or in the presence of 0.5 µM [^{3}H]GABA, were subsequently utilized to study Ca^{2+} uptake (A) or [^{3}H]GABA release (B) in similar ionic media, as described in the abcissa. Ca^{2+} uptake by synaptosomes was studied in basal Na^+ medium (a), in Na^+ + Ch^+ medium (b), in Na^+ + K^+ medium (c) and in K^+ + Ch^+ medium (d), all containing 1 mM $^{45}CaCl_2$. The reaction was conducted for 2 min, at 35°C and was terminated by filtration. The release of [^{3}H]GABA by preloaded synaptosomes was conduted by superfusion with external media of the composition indicated, (a) through (d), in the absence or in the presence of 1 mM $CaCl_2$.

(60 mM K^+) requires Na^+ in the medium (Fig. 6, column c) and that Ca^{2+} further increases the release by about 40%. Depolarization by K^+ (60 mM) in a choline medium, in the absence of Na^+, does not release [^{3}H]GABA unless Ca^{2+} is present in the medium (Fig. 6, column d).

We studied whether Ca^{2+} blockers inhibit [^{3}H]GABA release selectively through interference with Ca^{2+} entry into the synaptosomes. Thus, we determined the effect of verapamil and PY 108-68 on the two fractions of [^{3}H]GABA release, and the results show that the Ca^{2+}-independent release of [^{3}H]GABA was partially inhibited by verapamil and PY 108-68 for concentrations even lower than those at which the K^++Ca^{2+}-evoked release was inhibited (> 10^{-5}M), although a larger fraction of the release is inhibited in the latter case (Fig. 7). Therefore, although the Ca^{2+} channel blockers studied seem to inhibit the Ca^{2+}-dependent [^{3}H]GABA release more drastically, they also inhibit the Ca^{2+}-independent [^{3}H]GABA release. Thus, it appears that the effect of the blockers is non-specific and occurs at concentrations which are at least 1000 times higher than those required to saturate their binding sites, as determined from saturation binding studies (Bellemann et al 1981; Glossmann et al 1982; Murphy and Snyder 1982; Carvalho et al 1986 a, b).

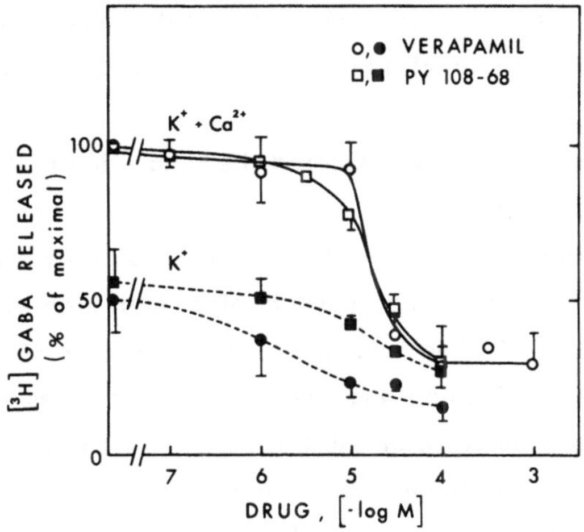

Fig. 7. Ca^{2+} channel blockers (verapamil and PY 108-68) inhibit both Ca^{2+}-dependent and Ca^{2+}-independent [^{3}H]GABA release evoked by K^+-depolarization in synaptosomes. Experimental conditions are similar to those described in Fig. 6, except that the relaese of [^{3}H]GABA from preloaded synaptosomes was tested in K^+ medium (Fig. 6B, (c)), in the presence or in the absence of 1 mM $CaCl_2$. The amount of GABA released was expressed as % of maximal GABA release in K^+ medium plus Ca^{2+}, without drugs present. Taken from Carvalho et al 1986 b.

Effect of Membrane Potential on [^{3}H]Nitrendipine Binding by Synaptosomes and SPM Vesicles

The effect of Ca^{2+} channel blockers on $^{45}Ca^{2+}$ fluxes in synaptosomes at concentrations of the blockers much higher than those reported for saturation of their high affinity binding sites raises the question of whether the receptors are linked to the voltage dependent Ca^{2+} channels. However, this discrepancy is not observed in certain clonal cell lines, such as NG108-15 (Freedman et al 1984; Creba and Karobath 1986) and PC12 cells (Toll 1982; Takahashi and Ogara 1983; Freedman et al 1984), where binding affinities correlate well with the blockade of $^{45}Ca^{2+}$ influx (Freedman et al 1984) induced by K^+ depolarization.

Furthermore, the agonist BAY K8644 enhances [^{3}H]noradrenaline release in PC12 cells (Albus et al 1984), $^{45}Ca^{2+}$ uptake in synaptosomes (White and Bradford 1986) and serotonin release from brain slices (Middlemiss and Spedding 1985), and these effects are counteracted by nanomolar concentrations of Ca^{2+} blockers. It has also been reported by Turner and Goldin (1985) that nanomolar concentrations of Ca^{2+} channel blockers inhibit $^{45}Ca^{2+}$ uptake and [^{3}H]norepinephrine in rat brain synaptosomes, when the uptake was measured in a choline medium, but most reports with synaptosomes do not show correlation between the concentrations of the blockers required to saturate their binding sites and those required to inhibit $^{45}Ca^{2+}$ fluxes (Nachshen and Blaustein 1979; Daniell et al 1983; Rampe et al 1984; Miller and Freedman 1984; Carvalho et al 1986 a, b; Chin 1986; Miller 1987; Suskiw et al 1986).

In this part of the study we have explored the suggestion that the Ca^{2+} channel proteins may have different sensitivities to the organic Ca^{2+} antagonists depending on the membrane potential (Hess et al 1984; Green 1985; Porzig and Becker 1985; Schwartz 1985). Thus, the inactivated channels bind, with high affinity, verapamil, diltiazem (Kanaya et al 1983; Lee and Tsien 1983), nitrendipine and nisoldipine (Bean 1984; Kass et al 1984; Shilling and Drewe, 1986), whereas the Ca^{2+} agonists bind preferentially to the open state of the Ca^{2+} channels and stabilize them (Albus et al 1984). Therefore, we have studied the effect of depolarizing the synaptosomal membrane by K^+ or veratridine on the [^{3}H]nitrendipine binding, since, to our knowledge, this type of study has not been attempted in cell fractions. Two types of nerve terminal preparations were used: synaptosomes with the intrinsic ionic content and synaptic plasma membrane (SPM) vesicles whose internal ionic content was experimentally controlled.

TABLE 2. Effect of $[K^+]_o$ depolarization on the equilibrium binding parameters of [^{3}H]nitrendipine specific binding to sheep brain synaptosomes

External $[K^+]$	K_D (nM)	B_{max} (fmol/mg protein)
0 mM	0.36 ± 0.04	127 ± 5
60 mM	0.33 ± 0.03	152 ± 5

The results are means ± SD of 3-5 different experiments performed in triplicate. K_D and B_{max} values were obtained by Scatchard analysis of equilibrium saturation binding data performed in the concentration range of 0.05 to 1.5 nM [^{3}H]nitrendipine. Specific [^{3}H]nitrendipine binding was defined as the difference in binding in the absence (total binding) and in the presence (non-specific binding) of 5 μM unlabelled nifedipine. Incubation was conduted at 25° C, for 45 min, and membrane bound and free [^{3}H]nitrendipine were separated by filtration through Whatman GF/B filters, as described previously (Carvalho et al 1986 a).

Equilibrium binding studies were performed in synaptosomes in 0 mM and 60 mM K_o buffers using [^{3}H]nitrendipine concentrations in the range of 0.05 to 1.5 nM, and Scatchard analysis of these saturation binding data permitted calculating the binding constants (K_D) and maximal binding capacity (B_{max}) for the fully polarized and depolarized membranes. Table 2 summarizes these parameters obtained in 3-5 different experiments. The K_D values, for synaptosomes are 0.36 ± 0.04 nM at 0 mM [K_o^+] and 0.33 ± 0.03 nM at 60 mM [K_o^+]. Thus, depolarization has no significant effect on the affinity of the membranes for [^{3}H]nitrendipine. The B_{max} values also are not very different (Table 2). Thus, the B_{max} value for polarized synaptomes is 127 ± 5 fmol/mg protein at 0 mM [K_o^+] and 152 ± 5 fmol/mg protein under K^+ depolarization (60 mM K_o^+).

In another series of experiments, we utilized SPM vesicles loaded with 150 mM KCl, 10 mM HEPES-Tris, pH 7.4, to study the association and dissociation kinetics of the binding of [^{3}H]nitrendipine to this membrane preparation in 0 mM and 60 mM K^+ buffers, in the presence of valinomycin (Fig. 8). It is observed that the equilibrium specific binding of [^{3}H]nitrendipine is increased by about 30% in 60 mM K^+ relative to 0 mM K^+. Both association and dissociation were complete within 10 min, at 25°C, in the presence of valinomycin. The K_D values derived from the calculated rate constants are 0.41 ± 0.05 nM and 0.34 ± 0.09 nM for 0 mM

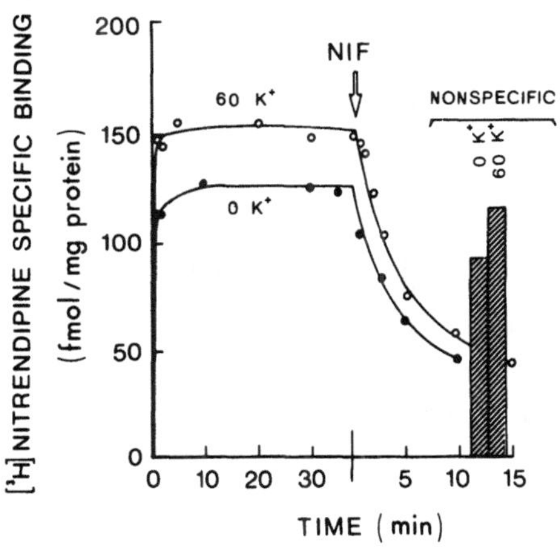

Fig. 8. Effect of membrane depolarization by K^+ (60 mM) on the association and dissociation rates of [^{3}H]nitrendipine binding to SPM vesicles preloaded with 150 mM KCl, 10 mM HEPES-Tris, pH 7.4. The binding of 0.35 nM [^{3}H]nitrendipine to SPM membranes was performed in two different external ionic media: 0 K^+ (133 mM NaCl) and 60 K^+(60 mM KCl, 73 mM NaCl), in the presence of 8 μg/ml valinomycin. At various time periods samples were taken from the incubation medium and were filtered. At the time indicated by the arrow, an excess of cold nifedipine (5 μM) was added to initiate the dissociation of [^{3}H]nitrendipine from the membrane binding sites. Non-specific binding was determined as defined in Table 2. The non-specific binding has been subtracted from the total binding to obtain the specific binding.

K^+ and 60 mM K^+ buffers, respectively. These results are in good agreement with the results obtained for synaptosomes (Table 2). It is also evident from Fig. 8 that the specific binding of [^{3}H]nitrendipine is increased by depolarization. Again the difference is relatively small, but we consistently obtain a clear increase in the specific [^{3}H]nitrendipine binding induced by depolarization. Taken together, the results suggest that depolarization does not cause a significant increase in binding affinity, but increases the total binding sites. Others have reported a clear increase in binding affinities for Ca^{2+} channel blockers for PC12 cells (Greenberg et al 1986) and for myocardial cells (Porzig and Becker 1985). However, there are also reports of enhancement on the number of binding sites due to depolarization in cardiac myocytes (Green et al 1984) and in frog sartorious muscle cells (Schwartz et al 1985), while Gredal et al (1987) report that neither the B_{max} nor the K_D values are influenced by depolarization in synaptosomes.

Veratridine, another depolarizing agent, was tested for its effect on [^{3}H]nitrendipine binding in a Na^+-rich medium. At concentrations ranging from 10 to 100 μM, veratridine has an inhibitory effect (Fig. 9) which may indicate that veratridine binds to Ca^{2+} channels and interfers with the binding of the Ca^{2+} channel blockers, in accord with previous observations by Reuter (1983) that veratridine at 2 x 10^{-4} M blocks Ca^{2+} currents in neuroblastoma cells. Therefore, veratridine may not be as specific for Na^+ channels as usually assumed.

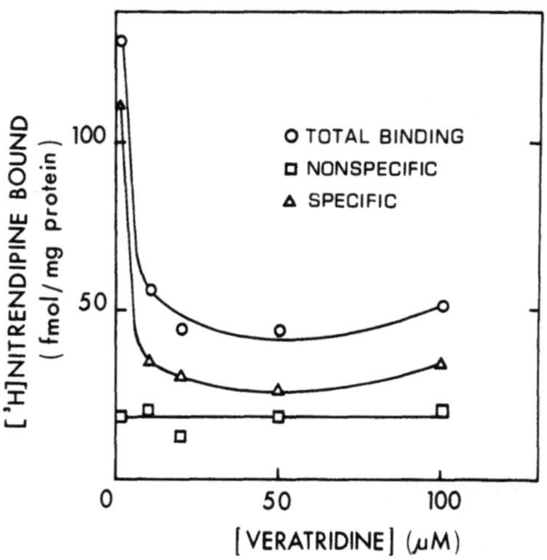

Fig. 9. Effect of veratridine on [^{3}H]nitrendipine binding to synaptosomes. Synaptosomes (0.138 mg protein/ml) were incubated for 45 min at 25°C in a medium containing (in mM): 133 NaCl, 1 $MgCl_2$, 1 $CaCl_2$, 10 HEPES-Tris, pH 7.4, 0.5 nM [^{3}H]nitrendipine, and veratridine (0-100 μM), in 1.8 ml final volume. Total, specific and non-specific binding were defined as described in Table 2.

CONCLUSIONS

The influx of $^{45}Ca^{2+}$ due to K^+-depolarization into synaptosomes which have been pre-incubated in a Na^+-rich medium is composed of one component which enters through the Ca^{2+} channels and of another which enters through the Na^+/Ca^{2+} exchange mechanism. The relative contribution by each component, at constant K^+-depolarization, depends on the Na^+ load of the synaptosomes, but, under most experimental conditions utilized, the Na^+/Ca^{2+} exchange accounts for about 50% of the total $^{45}Ca^{2+}$ influx occuring during K^+-depolarization. Complete substitution of the external Na_o^+ for choline to ensure a constant Na^+ electrochemical gradient during K^+ depolarization, does not eliminate the Na^+/Ca^{2+} exchange component because: 1) complete replacement of external Na^+ increases the Na^+/Ca^{2+} exchange due to a larger Na^+ gradient in → out; 2) the reduction in membrane potential due to K^+ depolarization stimulates Na^+/Ca^{2+} exchange, which is electrogenic; and 3) addition of K^+ to the external medium apparently has a specific stimulating effect on Na^+/Ca^{2+} exchange which can be observed even in SPM vesicles loaded with Na^+ (Fig. 3). For all these reasons, the relationship between membrane depolarization and the $^{45}Ca^{2+}$ influx into synaptosomes is very different in a Na_o^+ or choline medium, and there is no consistent correlation between membrane potential and Ca^{2+} influx (Fig. 3). Therefore, it is concluded that $^{45}Ca^{2+}$ influx due to K^+-depolarization of synaptosomes which have been pre-incubated in Na^+-rich media includes, under most experimental conditions utilized, $^{45}Ca^{2+}$ influx through the Na^+/Ca^{2+} exchange mechanism, in addition to Ca^{2+} influx through voltage dependent Ca^{2+} channels.

We reasoned that this could have occluded detection of the specific effect of Ca^{2+} channel blockers on $^{45}Ca^{2+}$ influx in synaptosomes in previous studies (Nachshen and Blaustein 1979; Daniell et al 1983; Carvalho and Carvalho, 1983; Rampe et al 1984). However, in experiments designed to distinguish, in the same synaptosomal preparation, between Ca^{2+} entry through channels and through the Na^+/Ca^{2+} exchange mechanism (Fig. 4), we studied the effect of three Ca^{2+} channel blockers (verapamil, nifedipine and d-cis-diltiazem) on each of the two mechanisms of Ca^{2+} entry. Both mechanisms show about equal sensitivity to inhibition by the Ca^{2+} channel blockers, and concentration above 10 μM are required for effective inhibition of $^{45}Ca^{2+}$ influx (Fig. 5), whereas the binding studies give much higher affinities of the membranes for the blockers (K_D ≃ nM). Furthermore, verapamil and PY 108-68 inhibit both the Ca^{2+}-dependent and Ca^{2+}-independent release of [^{3}H]GABA from synaptosomes, due to K^+-depolarization, with K_D values of

2.2 x 10^{-5}M and 3.3 x 10^{-5}M, respectively. Thus, the inhibition of [^{3}H]GABA release by Ca^{2+} blockers also occurs only at relatively high concentrations and it is not mediated exclusively through their effect on Ca^{2+} entry.

The effect of Ca^{2+} channel blockers in synaptosomes at concentrations much higher than those reported for saturation of the high affinity binding sites raises the question of whether the receptors are linked to the voltage dependent Ca^{2+} channels. Therefore, we explored in synaptosomes the suggestion that the Ca^{2+} channel proteins may change their affinity for the channel proteins with change in membrane potential, as has been found for muscle cells (Hess et al 1984; Bean 1984; Greenberg et al 1985; Schwartz et al 1985). Our results show only a small increase in the binding capacity of the synaptosomal membranes for [^{3}H]nitrendipine due to depolarization. It is unlikely that the differences found explain the large difference that exists between the affinity of the Ca^{2+} channel for their binding sites in the brain membranes and the effectiveness of the drugs in inhibiting $^{45}Ca^{2+}$ influx and neurotransmitter release in cell culture. Depolarization of the synaptosomes by veratridine drastically decreases the binding of [^{3}H]nitrendipine at concentrations above 10 μM, which may indicate that veratridine binds to the Ca^{2+} channel proteins in competition with the Ca^{2+} channel blockers (Reuter 1983).

The lack of effect of Ca^{2+} channel blockers on $^{45}Ca^{2+}$ fluxes and neurotransmitter release in brain fractions, at the concentrations which saturate the channel protein binding sites, probably reflects that the brain fractions loose some factor(s) essential for coupling the binding of the blockers to the physiological effects, since $^{45}Ca^{2+}$ fluxes and neurotransmitter release in neuronal cells in culture shows high sensitivity to the Ca^{2+} channel agonists and antagonists (Toll 1982; Takahashi and Ogura 1983; Abus et al 1984; Freedman et al 1984; Miller 1987).

AKNOWLEDGEMENTS

This work was supported by grants in aid from INIC and JNICT (Portuguese Ministry of Education) and by the Calouste Gulbenkian Fundation.

Albus U, Habermann E, Ferry DR and Glossmann H (1984) Novel 1,4-dihydropyridine (Bay K 8644) facilitates calcium dependent [^{3}H]noradrenaline release from PC12 cells. J Neurochem 42:1186-1189

Bean BP (1984) Nitrendipine block of cardiac calcium channels: high affinity binding to the inactivated state. Proc Natl Acad Sci 81:6388-6392

Bellemann P, Ferry DR, Lübbecke F and Glossmann H (1981) [^{3}H]Nitrendipine, a potent calcium antagonist, binds with high affinity to cardiac membranes. Arzneim Forch/Drug Res 31:2064-2067

Blaustein MP (1974) The interrelationship between sodium and calcium fluxes across cell membranes. Rev Physil Biochem Pharmacol 70:33-82

Blaustein MP (1975) Effect of potassium, veratridine and scorpion venom on calcium accumulation and transmitter release by nerve terminals in vitro. J Physiol (Lond) 247:617-655

Blaustein MP and Ector AC (1976) Carrier-mediated sodium-dependent and calcium-dependent calcium efflux from pinched-off nerve terminals (synaptosomes) in vitro. Biochim Biophys Acta 419:295-308

Blaustein MP and Goldring JM (1975) Membrane potentials in pinched-off nerve terminals monitored with a fluorescent probe: evidence that synaptosomes have potassium difusion potentials. J Physiol 247:589-615

Blaustein MP and King AC (1976) Influence of membrane potential on sodium-dependent uptake of GABA by pre-synaptic nerve terminals. J. Memb Biol 30:153-173

Blaustein MP and Oborn CJ (1975) The influence of sodium on calcium fluxes in pinched-off nerve terminals in vitro. J Physiol (Lond) 247:657-686

Bradford HF (1975) Isolated nerve terminals as an in vitro preparation for the study of dynamic aspects of transmitter metabolism and release. In: Iversen SD and Snyder SH (eds) Handbook of Psycopharmacology, Vol 1. Plenum Press, New York, p 191

Bradford HF (1969) Respiration in vitro of synaptosomes from mammalian cerebral cortex. J Neurochem 16:675-684

Bradford HF (1971) Membrane potential and metabolic performance in mammalian synaptosomes. In: Benson PF (ed), Cellular Organelles and Membranes in Mental Retardation, Churchill-Livingstone, London

Carvalho AP (1982) Calcium in the nerve cell. In: Lajtha A (ed) Handbook of Neurochemistry, Vol 1. Plenum Press, New York, p 69

Carvalho CAM and Carvalho AP (1979) Effect of temperature and ionophores on the permeability of synaptosomes. J Neurochem 33:309-317

Carvalho CAM and Carvalho AP (1984) Identification of calcium channels with radiolabelled calcium blockers. In: Burton RM and Guerra FC (eds) Biomembranes. Dynamics and Biology, Plenum Press, New York, p 317

Carvalho AP, Coutinho OP, Madeira VMC and Carvalho CAM (1984) Calcium transport in synaptosomes and synaptic plasma membrane vesicles. In: Burton RM and Guerra FC (eds) Biomembranes. Dynamics and Biology, Plenum Press, New York, p291

Carvalho CAM, Coutinho OP and Carvalho AP (1986 a) Effect of Ca^{2+} channel blockers on Ca^{2+} translocation across synaptosomal membranes. J Neurochem 47:1774-1784

Carvalho CAM, Santos SV and Carvalho AP (1986 b) γ-Aminobutyric acid release from synaptosomes as influenced by Ca^{2+} and Ca^{2+} channel blockers. Eur J Pharmacol 131:1-12

Chin JH (1986) Differential sensivity of calcium channels to dihydropyridines. The modulated receptor hypotesis. Biochem Pharmacol 35:4115-4120

Coutinho OP, Carvalho AP and Carvalho CAM (1983) Effect of monovalent cations on Na^{+}/Ca^{2+} exchange and ATP-dependent Ca^{2+} transport in synaptic membranes. J Neurochem 41:670-676

Coutinho OP, Carvalho CAM and Carvalho AP (1984) Calcium uptake related to K^{+}-depolarization and Na^{+}/Ca^{2+} exchange in sheep brain synaptosomes. Brain Res 290:261-271

Creba JA and Karobath M (1986) The effect of dihydropyridine calcium agonists and antagonist on neuronal voltage-sensitive calcium channels. Biochem Biophys Res Commun 134:1038-1047

Crompton M, Heid I and Carafoli E (1980) The activation by potassium of the sodium-calcium carrier of cardiac mitochondria. FEBS Lett 115:257-259

Cunningham JO and Neal MJ (1981) On the mechanism by which veratridine causes a calcium-independent release of γ-aminobutyric acid from brain slices. Brit J Pharmacol 73:655-667

Daniell LC, Barr EM and Leslie SW (1983) $^{45}Ca^{2+}$ Uptake into Rat whole brain synaptosomes unaltered by dihydropyridine calcium antagonists. J Neurochem 41:1455-1459

De Belleroche JS and Bradford HF (1977) On the site of origin of transmitter amino acids released by depolarization of nerve terminals in vitro. J Neurochem 29:335-343

De Robertis E, Pellegrino De Iraldi A, Rodriguez de Lores Arnaiz G and Salganicoff L (1962) Cholinergic and non-cholinergic nerve endings in rat brain. J Neurochem 9:23-25

Drapeau P and Blaustein MP (1983) Initial release of [^{3}H]dopamine from rat stiatal synaptosomes: correlation with calcium entry. J Neurosci 3:703-713

Erdreich A and Rahamimoff H (1984) The inhibition of Ca uptake in cardiac membrane vesicles by verapamil. Biochem Pharmacol 33:2315-2323

Erdreich A, Spanier R and Rahamimoff H (1983) The inhibition of Na-dependent Ca uptake by verapamil in synaptic plasma membrane vesicles. Eur J Pharamcol 90:193-202

Freedman SB, Dawson G, Villereal ML and Miller RJ (1984) Identification and characterization of voltage sensitive calcium channels in neuronal cell lines. J Neurosc 4:1453-1467

Freedman SB and Miller RJ (1984) Calcium channel activation: a different type of drug action. Proc Natl Acad Sci USA 81:5580-5583

Gähwiler BH and Brown DA (1987) Effects of dihydropyridines on calcium currents in CA3 pyramidal cells in slice cultures of rat hippocampus. Neuroscience 20:731-738

Glossmann H, Ferry DR, Lübbecke F., Mewes R and Hoffmann F (1982) Calcium channels: direct identification with radioligand binding studies. Trends Pharmacol Sci 3:431-437

Gray EG and Whittaker VP (1962) The isolation of nerve endings from brain: an electron-microscopic study of the cell fragments derived by homogeneization and centrifugation. J Anat 96:79-88

Green FJ, Farmer BB, Wiseman GL, Jose MJL and Watanabe AM (1985) Effect of membrane depolarization on binding of |^{3}H|nitrendipine to rat cardiac myocytes. Circ Res 56:576-585

Greenberg DA, Carpenter CL and Messing RO (1986) Depolarization-dependent binding of the calcium channel antagonist, (+)-|^{3}H|PN 200-110, to intact cultured PC 12 cells. J Pharmacol Expl Ther 238:1021-1027

Gredal O, Drejer J and Honoré T (1987) Different target sizes of the voltage-dependent Ca^{2+} channel and the |^{3}H|nitrendipine binding site in brain. Eur J Pharmacol 136:75-80

Hajós F (1975) An improved method for the preparation of synaptosomal fraction in high purity. Brain Res 93:485-489

Hess P, Lansman JB and Tsien RW (1984) Different modes of calcium channel gating by dihydropyridine Ca agonists and antagonists. Nature 311:538-544

Kass RS, Sanguinetti MC, Bennet PB, Coplin BE and Krafte DS (1984) Voltage-dependent modulation of cardiac calcium channels by dihydropyridines. In: Fleckenstein A, Van Breemen C, Gross R and Hoffmeister F (eds) Cardiovascular Effects of Dihydropyridines Type Calcium Antagonists and Agonists, Springer-Verlag, New York, p198

Iversen LL and Kelly JS (1975) Uptake and metabolism of γ-aminobutyric acid by neurons and glial cell. Biochem Pharmacol 24:933-938

Lee KS and Tsien RW (1983) Mechanism of calcium channel blockade by verapamil, D 600, diltiazem and nitrendipine in single dialysed heart cells. Nature 302:790-

Levi G, Rusca G and Raiteri M (1976) Diaminobutyric acid: a tool for discriminating between carrier-mediated and non-carrier-mediated release of GABA from ·synaptosomes? Neurochem Res 1:581-598

Martin DL (1973) Kinetics of the sodium dependent transport of gamma-aminobutyric acid by synaptosomes. J Neurochem 21:341-356

Midlemiss DN and Spedding M (1985) A functional correlate for the dihydropyridine binding site in rat brain. Nature 314:94-96

Miller RJ (1987) Multiple calcium channels and neuronal function. Science 235:46-52

Miller R and Freedman SB (1984) Are dihydropyridine binding sites voltage sensitive calcium channels? Life Sci 34:1205-1221

Mullins LJ, Requena J and Whittembury J (1985) Ca^{2+} entry in squid axons during voltage-clamp pulses is mainly Na^{+}/Ca^{2+} exchange. Proc Natl Acad Sci USA 82:1847-1851

Murphy KMM and Snyder SH (1982) Calcium antagonist receptor binding sites labeled with $|^3H|$-nitrendipine. Eur J Pharmacol 77:201-202

Nachshen DA (1985) The early time course of potassium-stimulated calcium uptake in presynaptic nerve terminals isolated from rat brain. J Physiol 361:251-268

Nachshen DA and Balustein MP (1979) The effects of some calcium antagonists on calcium influx in presynaptic nerve terminals. Mol Pharmacol 16:579-586

Nachshen DA and Blaustein MP (1980) Some properties of potassium stimulated calcium influx in presynaptic nerve ending. J Gen Physiol 76:709-728

Noris PJ, Dhaliwal DK, Druce DP and Bradford HF (1983) The supression of stimulus-evoked release of aminoacid neurotransmitters from synaptosomes by verapamil. J Neurochem 40:514-521

Porzig H and Becker C (1985) Binding of dihydropyridine Ca-channel ligands to living cardiac cells at different membrane potentials. Naunyn-Schmiedeberg's Arch Pharmacol 329:R47

Raiteri M, Cerrito F, Cervoni AM, Carmine R, Ribera MT and Levi G (1978) Studies on dopamine uptake and release in synaptosomes. In: Roberts PJ et al (eds) Advances in Biochemical Psycopharmacology, Vol. 19. Raven Press, New York, p35

Raiteri M and Levi G (1978) Release mechanisms for catecholamines and serotonin in synaptosomes. In: Ehrenpreis S and Kopin IJ (eds) Reviews in Neuroscience, Vol. 3. Raven Press, New York, p77

Rampe D, Janis RA and Triggle DJ (1984) BAY K8644, a 1,4-dihydropyridine Ca^{2+} channel activator: dissociation of binding and functional effects in brain synaptosomes. J Neurochem 43:1688-1692

Reuter H (1983) Calcium channel modulation by neurotransmitters, enzymes and drugs. Nature 301:569-574

Santos MS and Carvalho AP (1982) A simple superfusion technique for studying neurotransmitter release. Effect of monovalent ions on $|^3H|$GABA release from synaptosomes. Cienc Biol (Portugal) 7:95-113

Santos MS, Gonçalves PP and Carvalho AP (1987) Compartmentation and release of exogenous GABA in sheep brain synaptosomes. Neurochem Res 12:297-304

Schilling WP and Drewer JA (1986) Effect of membrane potential on $|^3H|$nitrendipine (NIT) binding determined in an isolated cardiac sarcolemma preparation. Biophys J 49: abst.)

Sihra TS, Scott IG and Nicholss DG (1984) Ionophore A23187, verapamil, protonionophores, and veratridine influence the release of γ-aminobutyric acid from synaptosomes by modulation of the plasma membrane potential rather than the cytosolic calcium. J Neurochem 43:1624-1630

Shwartz LM, McCleskey EW and Almers W (1985) Dihydropyridine receptors in muscle are voltage-dependent but most are not functional calcium channels. Nature 314:747-751

Suskiw JB, O'Leary ME, Murawsky MM and Wang T (1986) Presynaptic Calcium channels in rat cortical synaptosomes: fast kinetics of phasis calcium influx, channel inactivation and relationship to nitrendipine receptors. J Neurosci 6:1349-1357

Takahashi M and Ogura A (1983) Dihydropyridines are potent calcium channel blockers in neuronal cells. FEBS Lett 152:191-194

Thayer SA, Murphy SN and Miller RJ (1986) Widespread distribution of dihydropyridines sensitive calcium channels in central nervous system. Mol Pharmacol 30:505-509

Toll L (1982) Calcium antagonists. High affinity binding and inhibition of calcium transport in a clonal cell line. J Biol Chem 257:13189-13192

Turner TY and Goldin SM (1985) Calcium channels in rat brain synaptosomes: identification and pharmacological characterization. J. Neurosc 5: 842-849

White EJ and Bradford HF (1986) Enhancement of depolarization-induced synaptosomal Ca^{2+} uptake and neurotransmitter release by BAY K8644. Biochem Pharmacol 35:2193-2197

Whittaker VP (1965) The application of subcellular fractionation techniques to the study of brain function. Progr Biophys Mol Biol 15:39-96

Whittaker VP, Michaelson IA and Kirland RJA (1964) The separation of synaptic vesicles from nerve ending particles ("synaptosomes"). Biochem J 90:293-303

PROPERTIES OF PRESYNAPTIC VOLTAGE-SENSITIVE CALCIUM CHANNELS IN RAT SYNAPTOSOMES

J.B. Suszkiw
Department of Physiology and Biophysics
University of Cincinnati School of Medicine
Cincinnati, OH 45267-0576, U.S.A.

INTRODUCTION

Voltage-sensitive calcium channels (VSCC) in presynaptic nerve terminals control influx of Ca^{2+} and thereby determine the quantity of transmitter released by presynaptic action potentials. Biochemical ^{45}Ca tracer flux experiments with isolated nerve endings (synaptosomes) indicate that K^{+}-depolarization dependent ^{45}Ca entry consists of fast transient and slow sustained components (Nachshen & Blaustein, 1980). This article summarizes recent studies from this laboratory, which indicate that the transient Ca influx represents a channel mediated process, whereas the sustained component reflects Ca entry via reversed Na/Ca exchange. The kinetic and pharmacological characteristics of the channel-mediated Ca influx (J_{Ca}) in rat cortical synaptosomes are described.

VOLTAGE AND TIME DEPENDENCE OF THE CHANNEL-MEDIATED CALCIUM INFLUX IN SYNAPTOSOMES

In order to characterize the transient component of Ca entry, Ca uptake in synaptosomes must be measured at the subsecond time range. This has been accomplished through the employment of quench-flow apparatus (Fig. 1) which enables measurements of ^{45}Ca uptake in synaptosomes at times as short as 50 msec (Nachshen 1985, Suszkiw et al 1986).

When synaptosomes, suspended in Na-rich saline Hepes-buffered Krebs-Ringer, HKR (Suszkiw et al 1986), are depolarized by high K, the time course of Ca entry is described by sum of exponential and linear terms (Fig. 2 curves A, A', and A"). In contrast, K-depolarization dependent Ca entry in synaptosomes suspended in Na-deficient (choline substitution) media is a single exponential process (Fig. 2-B). Since synaptosomes suspended in standard saline have high Na_i ($\simeq$ 90 mM), whereas synaptosomes in Ch-substituted media, have low Na_i ($\simeq$ 5 mM) (Suszkiw et al 1986), it is concluded that the slow (linear) component of Ca entry is associated with high Na_i and represents reversed,

NATO ASI Series, Vol. H21
Cellular and Molecular Basis of Synaptic Transmission
Edited by H. Zimmermann

membrane potential sensitive Na/Ca exchange. The channel-mediated, exponential influx in synaptosomes in choline media, is described by the expression: $J_{Ca(50K,\ 1\ Ca)} = 3.0\ (1-e^{-1.3t})$, i.e., approaches plateau with time constant tau ≃ 800 msec.

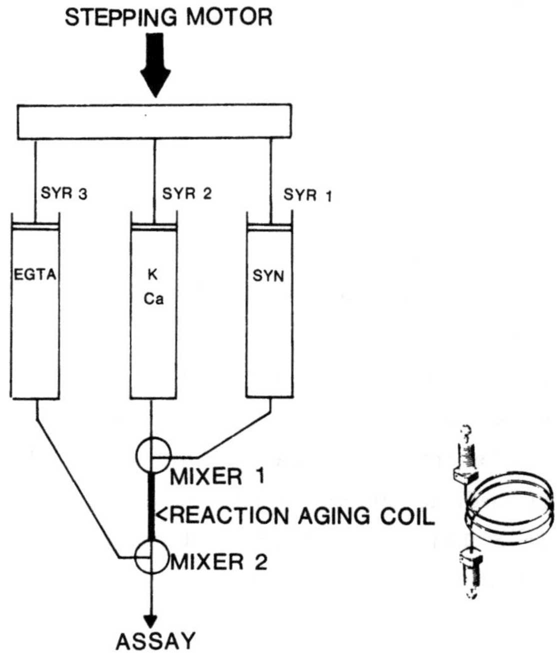

Fig. 1. Schematic diagram of quench-flow (QF) procedure. Syringes contain suspension of synaptosomes in basal saline (syr 1), basal or high K saline containing ^{45}Ca tracer (syr 2), and EGTA-quench solution (syr 3). Activation of stepping motor drives syringe pistons and initiates flow and mixing of solutions in Berger mixers (mixing time < 5 msec). The uptake time is determined by the transit time of reaction mixture through the aging coil between mixers 1 and 2. Quenched reaction mixture is collected and synaptosomes are immediately separated from the medium by filtration. ^{45}Ca retained with synaptosomes on the filters is measured by scintillation spectrometry.

J_{Ca} is a saturable function of Ca_o (Fig. 3). The limiting value of Ca influx, J_{Ca}^{max}, is ≃ 6 nmol Ca/sec/mg protein, corresponding to Ca current of about 0.06 pA/μm^2 synaptosomal surface area. Relative Ca influx as function of K_o (Fig. 4-A) suggests that J_{Ca} channels in synaptosomes activates upon 20-30 mV depolarization. Predepolarization of synaptosomes in high K Ca-free media, results in voltage and time dependent reduction of channel activity (Fig. 4-A). The voltage-dependent channel inactivation is a slow process.

Three types of calcium currents (channels) have been distinguished in electrophysiological recordings from cell bodies of vertebrate neurons. The T type currents

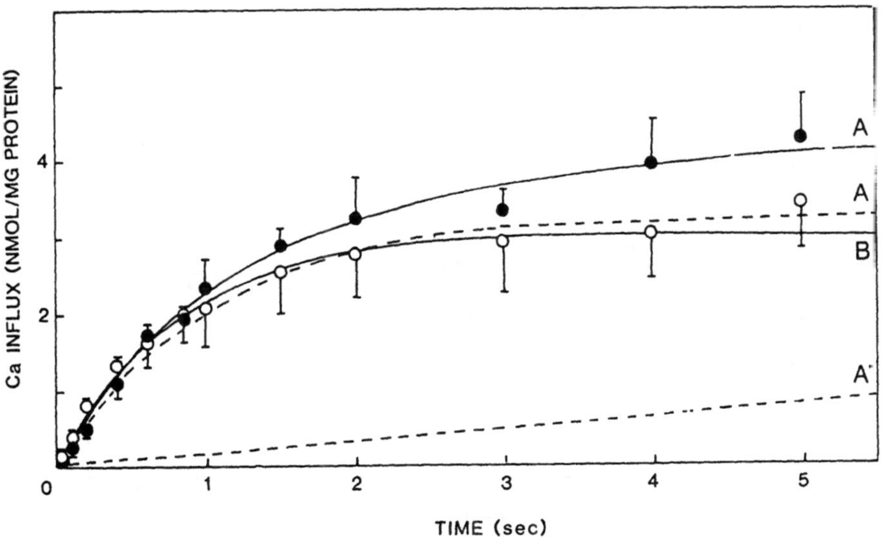

Fig. 2. Time course of K-depolarization-dependent Ca entry in synaptosomes. Ca influx is ^{45}Ca uptake in 52.5 mM K minus uptake in 5 mM K solutions; Ca_o = 1 mM. Curve A, obtained with synaptosomes suspended in standard Hepes-buffered Krebs-Ringer (HKR) solution is a composite of exponential (A') and linear (A") terms. Curve B represents influx of Ca in synaptosomes suspended in Na-deficient, choline-substituted HKR. The curves were obtained by non-linear regression. Individual data points (open and filled circles) are means ± SD.

have low threshold of activation and inactivate moderately rapidly at the range of resting membrane potentials.The N and L type channels have high threshold of activation (-20 to -10 mV), but are distinguished from each other in that the former inactivate relatively rapidly whereas the later inactivate slowly (Tsien 1987). Even though J_{Ca} reflects at best some sort of time and population average of macroscopic Ca fluxes in individual nerve ending particles, and for these reason does not allow direct comparison with biophysical characteristics of Ca currents recorded in electrophysiological experiments, some tentative conclusions can be reached with regard to the nature of the channels involved. Voltage-dependence of J_{Ca} in rat cortical synaptosomes suggests presence of channels which have moderate activation threshold ($\simeq$ -50 to -40 mV) and inactivate only very slowly during sustained depolarization. Thus, it is unlikely that the low threshold, inactivating T type or the high threshold,inactivating N type channels are predominantly responsible for the Ca influx, as measured in synaptosomes.

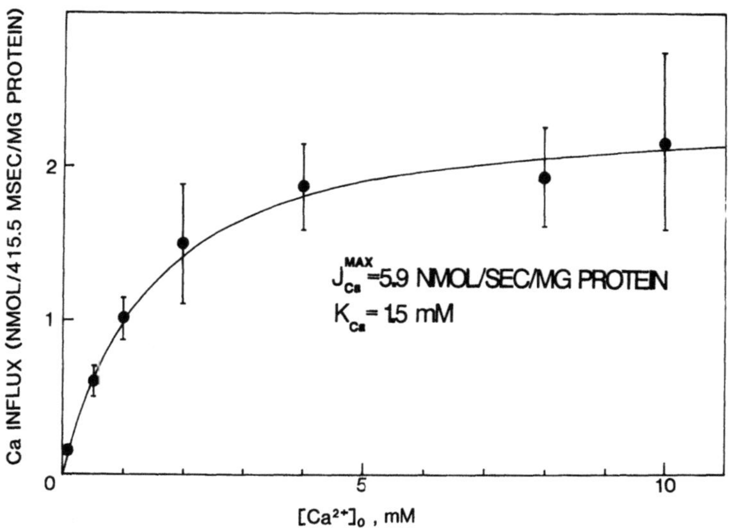

Fig. 3. Dependence of J_{Ca} on $[Ca]_o$. Initial rates of Ca influx were measured during 415.5 msec depolarizations in 50 mM K, and corrected for Ca uptake in 5 mM K solutions. Best fit curve to Michaelis-Menten equation: $J_{Ca} = J_{Ca}^{max}[Ca]_0/(K_{Ca} + [Ca]_o)$, was obtained using nonlinear regression analysis.

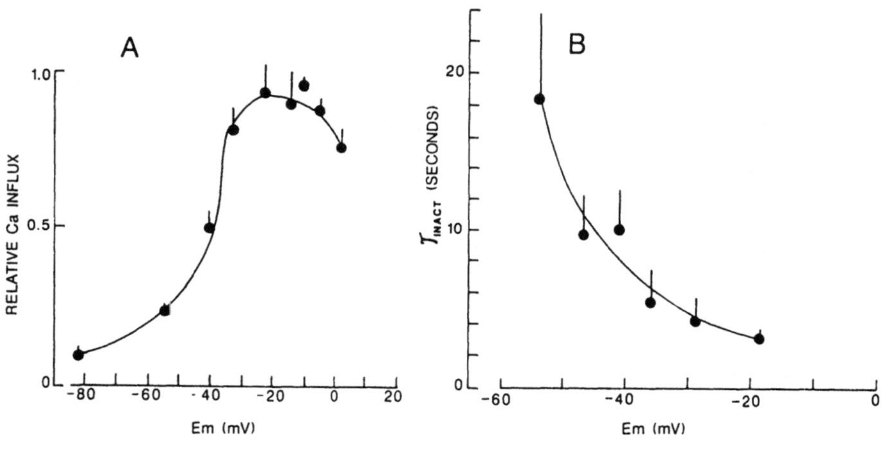

Fig. 4. Voltage-dependence of Ca influx (A) and channel inactivation (B). A: Ca influx (1 mM Ca_o, Ch-media) was measured during 1 sec exposures of synaptosomes to 5-130 mM K_o, and was expressed relative to the maximal value observed. B: Channel inactivation was measured by predepolarizing synaptosomes in Ca free, Ch-media with various K_o concentrations, and assaying for the channel activity during 1 sec exposures of synaptosomes to 0.05 mM ^{45}Ca after various times of predepolarization. The inactivation time constants (tau) were obtained from plots of ln (Ca influx) vs time (t) of predepolarization. The membrane potential of synaptosomes was estimated from 60 mV log $[K]_o/[K]_i$. Internal concentration of potassium (120 mM) was calculated from the K content of synaptosomes, measured by atomic absorption spectrometry (0.287 ± 0.064 μmol/mg protein) and the ^3H_2O space of synaptosomes (2.4 μl/mg protein).

PHARMACOLOGY OF J_{Ca} IN SYNAPTOSOMES

In addition to the differences in biophysical characteristics, various neuronal Ca channel types also exhibit different pharmacological properties, i.e., susceptibility to inorganic Ca channel blockers, dihydropyridine drugs, and the recently discovered neurotoxin, ω-conotoxin GVIA (ω-CgTx) (Tsien 1987, Miller 1987).

TABLE 1. Pharmacology of J_{Ca} in Synaptosomes.

DRUG	Ca Influx, % of control
Cd^{2+}, 10 μM	16.1 ± 3.6
Co^{2+}, 10 μM	102.1 ± 1.0
Bay K 8644, 1 μM	103.8 ± 12.3
Nifedipine, 10 μM	90.5 ± 2.8
ω-CgTx GVIA, 1 μM	89.2 ± 4.5
10 μM	63.4 ± 10
20 μM	57.4 ± 2.2
Forskolin, 1 μM	87.5 ± 11.4
PMA, 1 μM	88.6 ± 8.6

Synaptosomes were preincubated without and with the indicated concentrations of metal ions or drugs for 5 min. In the case of ω-CgTx, 25 min preincubations were employed. Following preincubations, Ca influx was assayed (1 sec, 50K, 0.5Ca) and ^{45}Ca uptake in the treated prepations was expresssed as percent of uptake in control preparations. Results are means ± SD.

Effects of various pharmacological agents on Ca influx in synaptosomes are summarized in Table 1. Similar to the N and L type currents (Tsien, 1987), J_{Ca} in synaptosomes is more susceptible to block by Cd^{2+} than Co^{2+} ions. However, the synaptosomal J_{Ca} is not significantly altered by L-type channel agonist Bay K 8644, or the antagonist nifedipine.This suggests that L type channels are not responsible for the measured J_{Ca}. The inhibition of J_{Ca} by ω-CgTx is weak and partial, with only about 50% of Ca being blocked by 20 μM toxin. This contrasts with the report by Reynolds et al

(1987) that J_{Ca} in synaptosomes is strongly blocked by ω-CgTx, and is inconsistent with their conclusion that Ca channels in synaptosomes are predominantly the N type.

Since regulation of presynaptic Ca channels by phosphorylation/dephosphorylation provides one potential mechanism of transmitter release modulation, it was of interest to determine whether J_{Ca} in synaptosomes is altered by adenylate cyclase activator, forskolin, or kinase C activator phorbol myristate acetate ester (PMA). Pretreatment of synaptosomes with either forskolin or PMA had no significant effect on the channel activity in synaptosomes (Table 1). If it is assumed that similar to cardiac Ca currents, neuronal L currents are enhanced by a c-AMP dependent phosphorylation (Hofmann et al 1987), then the lack of effect of forskolin treatment on J_{Ca} is consistent with absence of functional L type channels in synaptosomes. The lack of effect of phorbol ester treatment suggests that synaptosomal channels are also not regulated by kinase C mediated phosphorylation. Since both forskolin and phorbol esters have been reported to alter transmitter release (e.g. see Miller 1986), it appears that these effects are exerted at sites other than through the phosphorylation of Ca channels in nerve terminals.

CONCLUSIONS

The properties of J_{Ca} do not fully coincide with those expected of macroscopic T,N or L type currents. Although single channel recordings will be needed to unambigously define presynaptic Ca channels in synaptosomes,the currently available evidence suggests that these channels may be different from the Ca channel subtypes present in neuronal cell bodies.

ACKNOWLEDGEMENTS

The work summarised in this article was supported by National Institutes of Health grant NS 20786. Figures 1-3 have been reproduced from Suszkiw et al 1986. I thank Mr. M. Murawsky for his assistance with the experiments.

REFERENCES

Hofmann F, Nastainczyk W, Rohrkasten A, Schneider T and Sieber M (1987) TIPS 8:393-398.

Miller RJ (1985) Second messengers, phosphorylation and neurotransmitter release. TINS 8:463-465.

Miller RJ (1987) Multiple calcium channels and neuronal function. Science 235:46-52.

Nachshen DA (1985) The early time-course of potassium-stimulated calcium uptake in presynaptic nerve terminals isolated from rat brain. J. Physiol. (Lond.) 361:251-268.

Nachshen DA and Blaustein MP (1980) Some properties of potassium-stimulated calcium influx in presynaptic nerve endings. J. Gen. Physiol. 79:1065-1087.

Reynolds IJ, Wagner JA, Snyder SH, Thayer SA, Olivera BO and Miller RJ (1987) Brain voltage-sensitive calcium channel subtypes differentiated by ω-conotoxin GVIA. Proc. Natl. Acad. Sci. USA 83:8804-8807.

Suszkiw JB O'Leary ME, Murawsky MM and Wang T (1986) Presynaptic calcium channels in rat cortical synaptosomes: fast-kinetics of phasic calcium influx, channel inactivation, and relationship to nitrendipine receptors. J. Neurosci. 6:1349-1357.

Tsien RW (1987) Calcium currents in heart cells and neurons. In: Kaczmarek LK and Levitan IB (eds) Neuromodulation. Oxford University Press 1987 London New york: ch 11: 206-242.

MECHANISMS OF FAILURE OF SYNAPTIC TRANSMISSION DURING ANOXIA

K. Krnjević and J. Leblond
McGill University, 3655 Drummond St., Anaesthesia Research Department;
Montréal, Québec Canada H3G 1Y6.

SUMMARY

Evidence from intracellular recording and voltage-clamping in hippocampal slices indicates that brief anoxia (2-4 min) strongly depresses excitability and synaptic potentials by raising K conductance and inactivating Ca-currents. Both effects could be caused by a rise in intracellular free Ca^{2+}, but other explanations are not excluded.

INTRODUCTION

Numerous observations over many years (Verworn 1900; Porter 1912; van Harreveld 1944; Brooks and Eccles 1947; Andersen 1960; Lipton and Whittingham 1979) have demonstrated a rapid loss of synaptic transmission in the vertebrate CNS in the absence of oxygen. At the "highest" level, the cerebral cortex and the cerebellum, electrical activity disappears within a matter of seconds of the onset of anoxia or ischemia (Sugar and Gerard 1938); and synaptic responses are correspondingly quickly lost (Lipton and Whittingham 1984; Hansen 1985).

All this may not seem surprizing in view of the rapid loss of cortical function produced by anoxia. But the precise mechanisms are far from fully understood. It is well known that nerve cells have little in the way of energy reserves and that their immediate source of energy (ATP and especially creatine phosphate) soon diminish in the absence of O_2. But why this leads to such a rapid loss of electrical activity and synaptic transmission is still quite mysterious.

Some 12 years ago, after the discovery of Ca-activated K channels, it was suggested (Krnjević 1975) that because the maintenance of a very low intracellular concentration of free Ca^{2+} must require a constant supply of energy, anoxia could lead to an early rise in $[Ca]_i$, and

NATO ASI Series, Vol. H21
Cellular and Molecular Basis of Synaptic Transmission
Edited by H. Zimmermann

thus rapidly enhance the K permeability of cortical neurons. And indeed, clear evidence was already available that central neurons of cats are hyperpolarized during anoxia (Glötzner 1967; Speckmann et al 1970; Grossman and Williams 1971). This was confirmed by further experiments, mainly *in vitro*, which showed that the hyperpolarization is associated with a fall in input resistance, caused by G_K increase (Hansen et al 1982; Fujiwara et al 1987).

These observations were therefore in keeping with the proposed model of a $[Ca]_i$-triggered rise in G_K. But there were some other possibilities, for example that the depletion of ATP triggered an ATP-sensitive G_K (Noma 1983). It was also by no means clear that a rise in post-synaptic G_K would account for the early loss of synaptic potentials.

The present paper reviews some of our recent findings in experiments on hippocampal slices, which provide fresh insight into the mechanisms of transmission failure (Krnjević and Leblond 1987a,b).

METHODS

The experiments were all performed on conventional transverse slices from adult Sprague-Dawley rats, maintained in an interface (Haas)-type chamber at 33.5°. Anoxia was induced by substituting 95% N_2 and 5% CO_2 for the usual 95% O_2 and CO_2 mixture. Measurements of P_{O_2} at the upper surface of the slice in some experiments, with an O_2 electrode, showed that P_{O_2} drops to its minimum within 20-25 s - and that it returns to 90% of its normal level within 10 s of resuming oxygenation.

RESULTS

The typical sequence of events is illustrated in Fig. 1. After a 30-40 s latent period, there is a marked hyperpolarization, fall in input resistance (measured with hyperpolarizing current pulses) and disappearance of on-going firing. Though not obvious in Fig. 1, a brief phase of increased excitability and minor depolarization is often the earliest detectable change, shortly preceding the more prominent hyperpolarization.

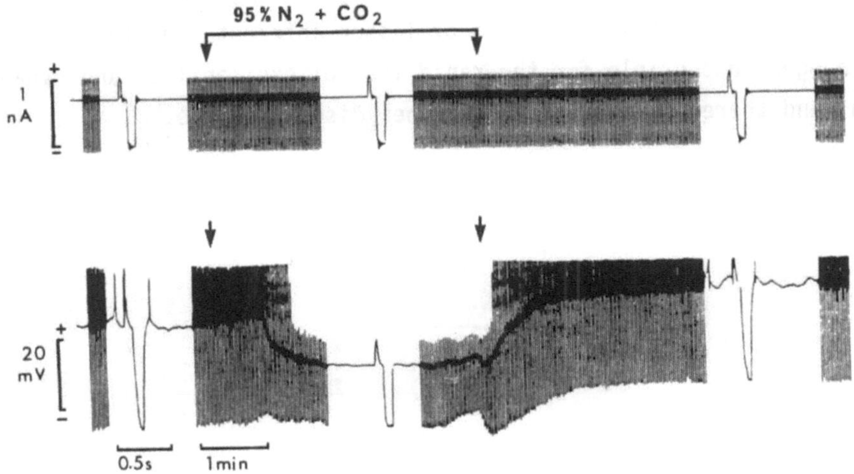

Fig. 1 Hyperpolarizing effect of anoxia on hippocampal neurons is associated with marked fall in resistance. Intracellular recording with 1.5 M KCl microelectrode was from a CA1 pyramidal layer cell, at depth of 200 μm. It was initally somewhat depolarized (V_m -55 mV; spike 72 mV) and there was ongoing firing. Spikes were also evoked at regular intervals by 0.14 nA depolarizing pulses; 0.8 nA hyperpolarizing pulses measured input resistance (initally 27.5 MΩ). Current pulses are monitored in upper trace. Records were obtained both on slow and fast paper speed. At first arrow, aerating gas mixture was changed from 95% O_2 and CO_2 to 95% N_2 and CO_2. At second arrow, 95% O_2 was reintroduced. During anoxia, note 12 mV hyperpolarization, 44% drop in R_N and disappearance of spikes. (From Fig 1 in Krnjević and Leblond 1987b).

Upon returning to O_2, after a latent period of 5-10 s, a variable short further hyperpolarization coincides with the start of the recovery of R_N and is followed by a slower return of potential to the initial level.

If EPSPs (and possibly spikes) are evoked by stimulation of afferent fibres (of the Schaffer collaterals), the EPSPs typically vanish within 2-4 min of the onset of anoxia.

This sequence of changes is readily reproducible if anoxia is not repeated at too short intervals (< 10-12 min) and for periods not exceeding 4-5 min. Excessive applications, however, may lead to a sudden very large depolarization, associated with a gross loss of R_N, having all the characteristics of spreading depression (including a very large increase in extracellular K and fall in Ca_o (Morris et al 1987)). Certain slices are unusually prone to this phenomenon, which may be triggered by only 1-2 min of exposure to N_2. Though not irrelevant, this paroxysmal event is of lesser interest than the more subtle changes responsible for the rapid loss of excitability and synaptic function, and therefore will not be further discussed here.

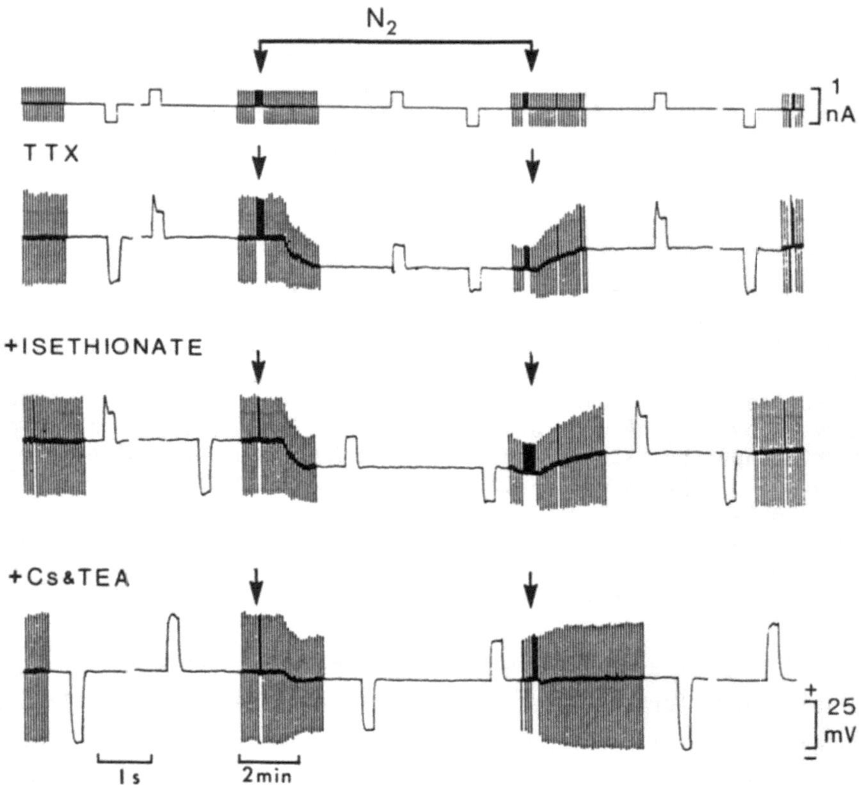

Fig. 2. Hyperpolarizing effect of anoxia is little changed after exposure to TTX and replacement of Cl^- with isethionate, but is much diminished by extracellular Cs^+ and TEA. Cell at depth of 120 µm in another slice, recorded with 3M K acetate microelectrode in presence of TTX (0.5 µM). Initial V_m -63 mV and R_N 42.5 MΩ; during anoxia V_m dropped -12 mV and R_N by 50%. After 90% replacement of $[Cl]_o$ with isethionate V_m fell -15 mV and R_N by 45%. After adding 3 mM Cs^+ and 10 mM TEA, V_m was -70 mV and R_N 65 MΩ; during anoxia V_m dropped by only -5 mV and R_N by 27%. (From Fig 1 in Krnjević and Leblond 1987b).

MECHANISM OF HYPERPOLARIZING CONDUCTANCE INCREASE

A variety of tests (Fig. 2) point to a rise in G_K as the principal underlying conductance change; for example its lack of sensitivity to tetrodotoxin (excluding a role for fast Na channels); a very negative reversal potential (near -90 mV), even when recording with Cl^--containing microelectrodes (which always cause reversal of IPSPs), or after replacement of the greater part of extracellular Cl^- with a non-permeant anion, like isethionate (Fig. 2); as well as clear sensitivity to changes in extracellular [K].

All these observations are in agreement with those of previous authors (Hansen et al 1982; Fujiwara et al 1987). There is no comparable consensus, however, about effective blockers of this phenomenon. According to Hansen et al 1982, the hyperpolarization can be blocked by 4-aminopyridine. But this was not the case in the experiments of Fujiwara et al 1987, nor in our own. On the other hand, Fujiwara et al 1987 reported substantial (though incomplete) antagonism by Ba^{2+}, while in our experiments, quinine - also a G_K blocker (Atwater et al 1979) - was at least partly effective.

ROLE OF INTRACELLULAR Ca^{2+}

If G_{KCa} is indeed the predominant conductance that is activated, one would expect that the anoxically-evoked changes should be either reduced or eliminated by intracellular injection of a strong chelator, such as EGTA (cf. Krnjević et al 1978). When we recorded with EGTA-containing microelectrodes, we observed a consistent trend towards smaller increases in conductance during anoxia (Krnjević and Leblond, 1987a). However, in spite of a significant correlation between the depression of the effects of anoxia on one hand and the intensity of currents applied to inject EGTA into the cells on the other, the change produced by EGTA was seldom dramatic. Perhaps for that reason, Fujiwara et al 1987, reported no reduction of anoxic effects by EGTA injected into three cells in their experiments.

ATP-SENSITIVE G_K

The other obvious possible mechanism of activation of G_K is via ATP-sensitive K channels. Such channels were first identified by Noma (1983), in cardiac cells. Their outstanding characteristic is that they are closed in the presence of "normal" intracellular ATP levels (typically 3-5 mM), but they open when $[ATP]_i$ falls much below 1 mM (Noma 1983; Kakei et al 1985). They have a relatively large unitary conductance, and they probably make a major contribution to the G_K increase produced by anoxia in cardiac ventricular fibres (Noma and Shibahi 1985), as well as in frog skeletal muscle (Spruce et al 1985; Castle and Haylett 1987). Similar channels control the membrane potential of pancreatic cells (Cook and Hales 1984; Trube et al 1986) but are much less prominent in rat myotubes (Spruce et al 1985) or in chromaffin cells (Trube et al 1986).

In view of the variable results of intracellular EGTA injections, activation of ATP-sensitive channels in hippocampal neurons is an attractive alternative explanation. Against it, however, is the lack of evidence so far that such channels exist in nerve cells; and more importantly, the marked discrepancy between loss of function and the corresponding changes in ATP levels. For example, according to Yamamoto and Kurokawa (1970), when synaptic transmission in cortical slices is reduced by 93% during anoxia, their ATP content has fallen by only 20%. Moreover, according to our own preliminary observations, tolbutamide, a specific blocker of ATP-dependent channels (Trube et al 1986; Castle and Haylett 1987) does not prevent the anoxic hyperpolarization and G_K increase in hippocampal slices.

ANOXIC BLOCK OF Ca-CURRENTS

When recording from voltage-clamped hippocampal pyramidal cells (with a single electrode), a variety of membrane currents can be demonstrated

(Jahnsen 1986), including large, relatively slow, Ca-inward currents when fast Na^+ channels are inactivated by tetrodotoxin and K^+ outward currents are depressed by TEA or intracellular Cs (Johnston et al 1980; Brown and Griffith 1983). According to our observations (Krnjević and Leblond 1987b), anoxia quite consistently either strongly depresses or eliminates such Ca-inward currents, within 1-2 min of changing from O_2 to N_2. This effect has been seen when evoking a low threshold, very rapidly inactivating I_{Ca} (Fig. 3, upper traces) or a higher threshold, more slowly inactivating I_{Ca} (Fig. 3, lower traces). On returning to

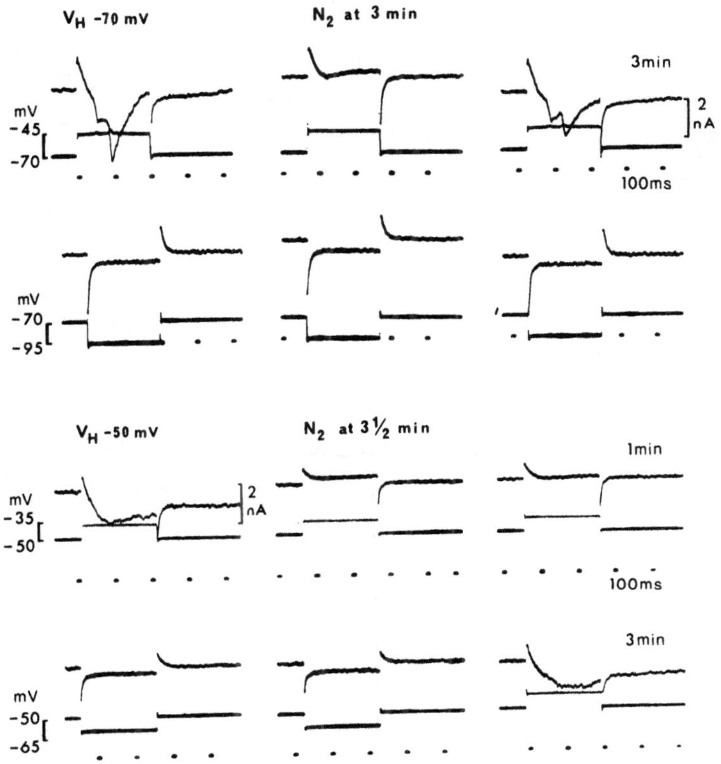

Fig. 3. Another cell, recorded with 3M KCl microelectrode at depth of 230 μm (initial V_m -80 mV and R_N 51 MΩ), shows marked reduction of Ca-inward currents during 4 min period of anoxia. Slice was treated with 0.5 μM TTX, 3 mM Cs^+ and 10 mM TEA. In upper traces (V_H - 70 mV), 25 mV depolarizing pulses evoke large, mainly transient inward current; corresponding leak currents, evoked by similar hyperpolarizing pulses, are immediately below. In lower traces, where V_H was -50 mV, a more sustained inward current (and tail) was evoked by 14 mV depolarizing pulses (leak current is shown below). Times when traces were obtained during anoxia and after subsequent reoxygenation are indicated. (From Fig 3 in Krnjević and Leblond 1987b).

normal oxygenation, there is a rapid initial return of I_{Ca}, though complete recovery may take several minutes.

In the final example, illustrated in Fig. 4, Ca-inward current is recorded with a Cs-containing microelectrode (intracellular Cs blocks K

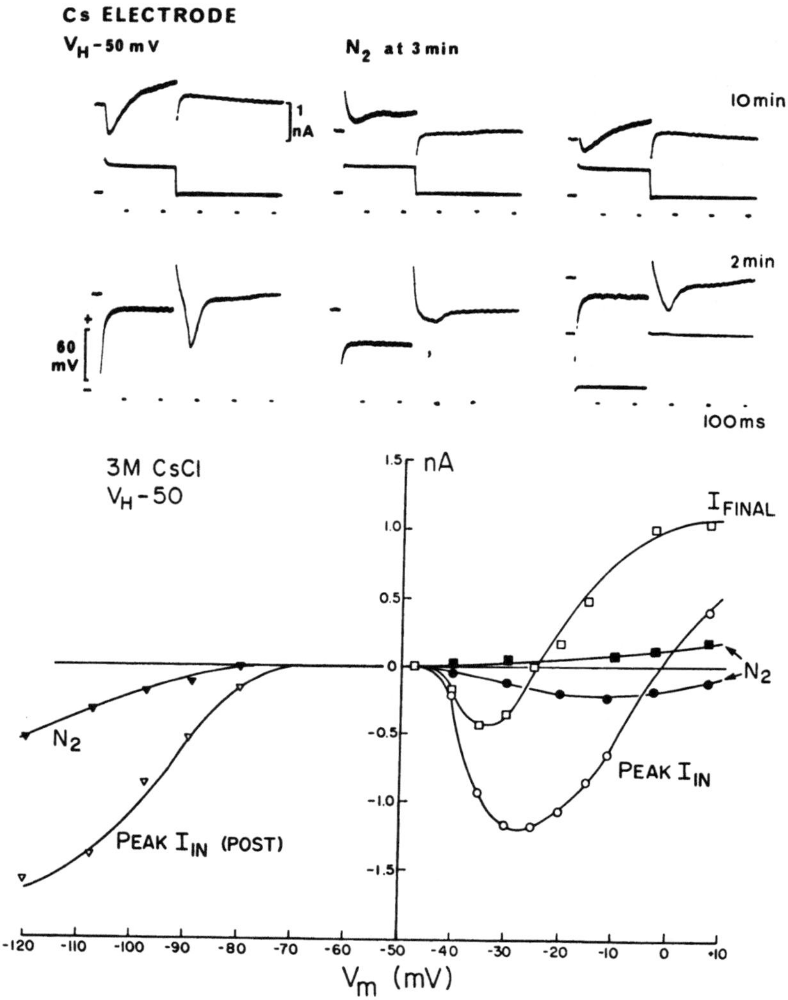

Fig. 4. Further example of depression of Ca inward currents by anoxia, in a cell recorded at depth of 210 μ m with 3 M CsCl electrode; superfusate contained 0.5 μM TTX and 3 mM Cs^+, and V_H was -50 mV. Brief inward currents were evoked either by 30 mV depolarizing pulses (upper series) or just after end of 60 mV hyperpolarizing pulses (lower series; equal pulses were applied throughout). Times when traces were recorded during 4.5 min period of anoxia or after its end are indicated. Graph below shows corresponding current-voltage plots for peak inward current (circles) and final outward current (rectangles), as well as for delayed inward current (after end of hyperpolarizing pulses, triangles). Open symbols are initial control values, closed symbols values recorded 2 min after onset of anoxia. Leak currents have been subtracted. (From Fig 4 in Krnjević and Leblond 1987b).

channels and so eliminates the outward currents that normally mask I_{Ca}). In Fig. 4 two types of Ca-current are evident in the same cell, at V_H - 50 mV: above, a relatively high-threshold inward current evoked directly by depolarizing steps; below a low threshold current that follows hyperpolarizing steps. Both these currents largely disappear after 3 min of anoxia, as shown by the middle traces, as well as by the graphs.

DISCUSSION

The highly reproducible block of I_{Ca} observed in most cells studied in these experiments - as well as in parallel experiments on slices from Wistar rats, maintained in a submersion-type chamber (Krnjević et al 1987) - may well explain why synaptic transmission is rapidly abolished by anoxia. If, as seems likely, presynaptic Ca currents are equally sensitive to anoxia, transmitter release would soon vanish following O_2 removal. Although post-synaptic G_K increase and hyperpolarization are usually sufficient to account for the failure of orthodromic spikes, they cannot alone explain the disappearance of EPSPs. Other explanations involving a failure of nerve terminal invasion, owing to a marked rise in presynaptic G_K or in extracellular K^+, are not consistent with the fact that the afferent volley recorded in the region of the dendritic synapses is not abolished at a time when EPSPs have disappeared; moreover, according to our own measurements in hippocampal slices (Morris et al 1987), increases of extracellular K^+ are commonly too small during 2-4 min periods of anoxia (only up to 1-2 mM) to block transmission (Balestrino et al 1986), except when spreading depression takes place.

It is well established that I_{Ca} is readily inactivated by a rise in free internal [Ca] (Eckert and Chad 1984). Therefore the same factor - a rise in $[Ca]_i$ caused by anoxia - could trigger both the higher G_K and the depression of I_{Ca}.

On the other hand, a good supply of ATP may also be essential for the activity of Ca-channels; so a fall in ATP consequent on anoxia could also be responsible not only for the rise in G_K (Noma and Shibasaki 1985) but also for the depression of I_{Ca}; moreover, ATP may be directly involved in the process of exocytotic transmitter release (Sanchez-Prieto et al 1987). Further experiments will be needed to decide whether anoxic depression is mediated principally by intracellular changes in Ca^{2+} or ATP (or both acting together).

ACKNOWLEDGEMENTS

This research was financially supported by the Medical Research Council of Canada.

REFERENCES

Andersen P (1960) Interhippocampal impulses. Acta Physiol Scand 48: 178-208

Atwater I, Dawson CM, Ribalet B and Rojas E (1979) Potassium permeability activated by intracellular calcium ion concentration in the pancreatic β-cell. J Physiol 288: 575-588

Balestrino M, Aitken PG and Somjen GG (1986) The effects of moderate changes of extracellular K^+ and Ca^{2+} on synaptic and neural function in the CA1 region of the hippocampal slice. Brain Res 377: 229-239

Brooks CMcC and Eccles JC (1947) A study of the effects of anaesthesia and asphyxia on the monosynaptic pathway through the spinal cord. J Neurophysiol 10: 349-360

Brown DA and Griffith WH (1983) Persistent slow inward calcium current in voltage-clamped hippocampal neurones of the guinea-pig. J Physiol 337: 303-320

Castle NA and Haylett DG (1987) Effect of channel blockers on potassium efflux from metabolically exhausted frog skeletal muscle. J Physiol 383: 31-43

Cook DL and Hales CN (1984) Intracellular ATP directly blocks K^+ channels in pancreatic β-cells. Nature 311: 271-273.

Eckert R and Chad JE (1984) Inactivation of Ca channels. Prog Biophys Mol Biol 44: 215-267

Fujiwara N, Higashi H, Shimoki K and Yoshimura M (1987) Effects of hypoxia on rat hippocampal neurones *in vitro*. J Physiol 384: 131-151

Glötzner F (1967) Intracelluläre Potentiale, EEG und corticale Gleichspannung an der sensomotorischen Rinde der Katze bei akuter Hypoxie. Archiv Psychiat Nervenkr 210: 274-296

Grossman RG and Williams VF (1971) Electrical activity and ultrastructure of cortical neurons and synapses in ischemia. In: Brierly JB and Meldrum BS (eds). Brain Hypoxia. JB Lippincott Co Philadelphia: 61-75

Hansen AJ (1985) Effect of anoxia on ion distribution in the brain. Physiol Rev 65: 101-148.

Hansen AJ, Hounsgaard J and Jahnsen H (1982) Anoxia increases potassium conductance in hippocampal nerve cells. Acta Physiol Scand 115: 301-310

Jahnsen H (1986) Responses to neurons in isolated preparations of the mammalian central nervous system. Prog Neurobiol 27: 351-372
Johnston D, Hablitz JJ and Wilson WA (1980) Voltage clamp discloses slow inward current in hippocampal burst-firing neurones. Nature 286: 391-393
Kakei M, Noma A and Shibasaki T (1985) Properties of adenosine-triphosphate-regulated potassium channels in guinea-pig ventricular cells. J Physiol 363: 441-462
Krnjević K (1975) Coupling of neuronal metabolism and electrical activity. In: Ingvar DH and Lassen NA (eds). Brain Work: The coupling of function, metabolism and blood flow in the brain. Alfred Benzon Symposium VIII/Copenhagen Munsgaard: 65-78
Krnjević K, Cherubini E and Ben-Ari Y (1987) Effects of anoxia on synaptic transmission and calcium currents in immature hippocampus. Soc Neurosci Abstr 13: 1355
Krnjević K and Leblond J (1987) Mechanism of hyperpolarizing response of hippocampal cells in isolated slices to anoxia. J Physiol 382: 79P
Krnjević K and Leblond J (1987b) Anoxia reversibly suppresses neuronal Ca-currents in rat hippocampal slices. Can J Physiol Pharmacol 65: 2157-2161
Krnjević K, Puil E and Werman R (1978a) EGTA and motoneuronal after potentials. J Physiol 275: 199-223
Lipton P and Whittingham TS (1979) The effect of hypoxia on evoked potentials in the *in vitro* hippocampus. J Physiol 287: 427-438
Lipton P and Whittingham TS (1984) Engery metabolism and brain slice function. In: Dingledine R (ed). Brain slices. Plenum Press New York: 113-153
Morris ME, Krnjević K and Leblond J (1987) Changes in extracellular $[K^+]$ and $[Ca^{2+}]$ evoked by anoxia in rat hippocampal slices. Soc Neurosci Abstr 13: 1355
Noma A (1983) ATP-regulated K^+ channels in cardiac muscle. Nature 305: 147-148
Noma A and Shibasaki T (1985) Membrane current through adenosine-triphophate-regulated potassium channels in guinea-pig ventricular cells. J Physiol 363: 463-480
Porter EL (1912) Variations in irritability of the reflex arc. Variations under asphyxial conditions, with blood-gas determinations. Amer. J Physiol 31: 223-244
Sanchez-Prieto J, Sihra TS and Nicholls DG (1987) Characterization of the exocytotic release of glutamate from guinea-pig cerebral cortical synaptosomes. J Neurochem 49: 58-64
Speckmann E-J, Caspers H and Sokolov H (1970) Aktivitätsänderungen spinaler Neurone während und nach einer Asphyxie. Pflügers Arch 319 122-138
Spruce AE, Standen NB and Stanfield PR (1985) Voltage-dependent ATP-sensitive potassium channels of skeletal muscle membrane. Nature 316: 736-738
Sugar O and Gerard RW (1938) Anoxia and brain potentials. J. Neurophysiol 1: 558-572
Trube G, Rorsman P and Ohno-Shosaku T (1986) Opposite effects of tolbutamide and diazoxide on the ATP-dependent K^+ channel in mouse pancreatic beta-cells. Pflügers Arch 407: 493-499
van Harreveld, A (1944) Survival of reflex contraction and inhibition during cord asphyxiation. Amer J Physiol 141: 97-101

Verworn M (1900) Ermüdung, Erschöpfung und Erholung der nervösen Centra des Rückenmarkes. Arch f (Anat) Physiol 1900 (Suppl):152-176

Yamamoto C and Kurokawa M (1970) Synaptic potentials recorded in brain slices and their modification by changes in the level of tissue ATP. Exp Brain Res 10: 159-170

ADRENERGIC AND CHOLINERGIC VESICLES: ARE THERE COMMON ANTIGENS AND COMMON PROPERTIES?

Hans Winkler, Reiner Fischer-Colbrie, Dieter Obendorf and Ulrike Schwarzenbrunner
University of Innsbruck, Department of Pharmacology, Peter-Mayr-Straße 1, A-6020 Innsbruck, Austria.

INTRODUCTION

In mammalian tissues both sympathetic and cholinergic terminals contain two types of vesicles: the large dense cored ones (LDV) and the small dense cored ones (SDV). In this short review we discuss the possibility that this resemblance is not only expressed at the morphological level but also by the biochemical and functional properties of these vesicles.

PROPERTIES OF LDV IN ADRENERGIC AND CHOLINERGIC NERVES

Adrenergic LDV: Most of the data have been obtained for adrenergic vesicles and therefore we will discuss these organelles first. Adrenergic LDV very closely resemble adrenal chromaffin granules (Klein & Thureson-Klein 1984) as far as function and biochemical composition is concerned. For chromaffin granules the uptake of catecholamines and nucleotides via specific carriers has been established (see Winkler et al 1986). The driving force for this mechanism is a chemiosmotic gradient which is provided by a proton pumping ATPase (Johnson 1987). The conversion of dopamine to noradrenaline within these vesicles requires the presence of the enzyme dopamine ß-hydroxylase, the co-factor ascorbic acid and the cytochrome b-561 which provides reducing equivalents by allowing passage of electrons into the vesicle interior (Fleming & Kent 1987; Njus et al 1987). Although for LDV these mechanisms have not been studied in comparable detail the available results on their functional and biochemical

NATO ASI Series, Vol. H21
Cellular and Molecular Basis of Synaptic Transmission
Edited by H. Zimmermann

properties indicate that they possess analogous properties (Klein & Thureson-Klein 1984). In agreement it appears that both chromaffin granules and LDV have a very similar biochemical composition (see Table 1). The only apparent exception is the low content of lysolecithin in LDV when compared with chromaffin granules.

In electron micrographs adrenergic LDV have an electron-dense content even after catecholamine depletion by reserpine (see Winkler & Carmichael 1982). It seems now obvious that this electron dense core is due to the presence of the chromogranins which are likely to be the major protein constituents of the soluble content of LDV, just as they are in chromaffin granules (see Winkler et al 1986).

Cholinergic LDV: Very little is known about these vesicles. By immunhistology it has been established that cholinergic LDV can contain VIP (Lundberg et al 1981; Johansson & Lundberg 1981). However just as in the case of adrenergic LDV it seems unlikely that these peptides are responsible for the electron dense content of these vesicles. It seems therefore significant that recent studies indicate that also cholinergic vesicles contain chromogranins. Immunohistological studies of chromogranins in the central nervous system revealed that the cholinergic moto-neurons in the spinal cord stained positively for these antigens (Somogyi et al 1984). Furthermore, this study also established that only the LDV but not SDV stained positively for these proteins. Finally chromogranin A immunoreactivity was found in cholinergic nerve terminals in the rat diaphragm (Volknandt et al 1987). Thus we would like to suggest that the dense core of cholinergic LDV as seen in electron micrographs is due to the presence of chromogranins.

We have just discussed that adrenergic LDV transport and store transmitter and nucleotides. Is this also true for cholinergic ones? An answer is still lacking. A prerequisite for biochemical uptake studies is the availability of purified cholinergic LDV. Apparently attempts to purify these vesicles from cholinergic nerves in the intestine have only been started recently (Agoston & Conlon 1987).

TABLE 1: Constituents of adrenal chromaffin granules and large (LDV) and small (SDV) dense cored vesicles of sympathetic nerve

SOLUBLE CONTENT

CHROMAFFIN GRANULES	LDV	SDV
Catecholamines (1)	Catecholamines (5)	Catecholamines (5)
Nucleotides (1)	Nucleotides (5)	Nucleotides (5)
Calcium (1)		
Ascorbic acid (1)		
Chromogranin A (1)	Chromogranin A (5)	absent (6)
PG-Chromogranin A (4)		
Chromogranin B (2)	Chromogranin B (7)	
Secretogranin II (2)	Secretogranin II (7)	
Enkephalins (2)	Enkephalins (5)	absent(6,8)
Carboxypeptidase H (2)		
Neuropeptide Y (2)	Neuropeptide Y (8)	absent (8)

MEMBRANES

CHROMAFFIN GRANULES	LDV	SDV
Phospholipids (1)	Phospholipids (5)	Phospholipids (5)
Cholesterol (1)	Cholesterol (5)	Cholesterol (5)
Gangliosides (3)		
DBH (1)	DBH (5)	DBH(6)
Cytochrome b-561 (1)	Cytochrome b-561 (5)	Cytochrome b-561 (5)
H^+-ATPase (2)	H^+-ATPase (5)	H^+-ATPase (5)
Amine carrier (2)	Amine carrier (5)	Amine carrier (5)
Nucleotide carrier (2)	Nucleotide carrier (9)	Nucleotide carrier (9)
Glycoprotein II (2)		
Glycoprotein III (2)		
Carboxypeptidase H (2)		

The numbers in parentheses after each constituent refer to the following references where the evidence for the presence of the various components can be found. In most cases the evidence is based on a direct determination of the constituent by biochemical, immunological or immunohistochemical analysis. For the presence (in LDV and SDV) of the amine and nucleotide carrier and the proton pumping ATPase the evidence rests on functional studies (Klein & Thureson-Klein 1984; Aberer et al 1978; Neuman et al 1984). Blank spaces indicate that the possible presence of a constituent was not yet determined. H^+ATPase, proton pumping ATPase; DBH, dopamine ß-hydroxylase.

(1) Winkler 1976; (2) Winkler et al 1986; (3) Winkler & Westhead 1980; (4) Eiden et al 1987; (5) Klein & Thureson-Klein 1984; (6) Neuman et al 1984; (7) Hagn et al 1986; (8) Fried et al 1986; (9) Aberer et al, 1978.

PROPERTIES OF SDV IN CHOLINERGIC AND ADRENERGIC VESICLES

In contrast to LDV much more data are available on cholinergic SDV than on adrenergic ones. Since the electric organ of the eel provides a suitable source for isolating pure cholinergic SDV the biochemical composition and functional properties of these vesicles is well established (Whittaker 1987). These vesicles can accumulate acetylcholine via a carrier which apparently is driven by a proton pumping ATPase providing an electrochemical gradient (Parsons et al 1987). Similarly the vesicles can accumulate nucleotides (Luqmani 1981). In accordance these vesicles contain a proton-pumping ATPase, carriers for acetylcholine and nucleotides although the molecular properties of these proteins are not yet established (see Whittaker 1987), in addition they contain phospholipid, cholesterol and a proteoglycan. Defined protein antigens present in these vesicles are SV 2 (Buckley & Kelly 1985) and p65 (Matthew et al 1981).

For adrenergic SDV data are limited, since only partly purified preparations are available (Fried et al 1978; see also Klein & Thureson-Klein 1984). These preparation were found to contain cytochrome b-561 (Fried 1978). The presence of dopamine ß-hydroxylase seems controversial (Bisby et al 1973; Klein et al 1979; Neuman et al 1984; see also Klein & Thureson-Klein 1985).

Functional studies in vivo, however indicate that these vesicles can synthesize noradrenaline from dopamine and therefore must contain DBH (Neuman et al 1984) and cytochrome b-561 which is functionally coupled to this enzyme (Fleming & Kent 1987). LDV also accumulates nucleotides in vivo (Aberer et al 1978). Thus they should contain carrier for amines, nucleotides and the proton pumping ATPase providing the driving force (see Table 1).

RELATIONSHIP OF LDV TO SDV

Fig. 1 gives a schematic presentation of the functional properties of LDV and SDV in adrenergic nerves: In cholinergic nerves SDV have an analogous function to adrenergic SDV, although they are not involved in the synthesis of the transmitter but take up the final product, i.e. acetylcholine. As already pointed out above we do not yet know whether

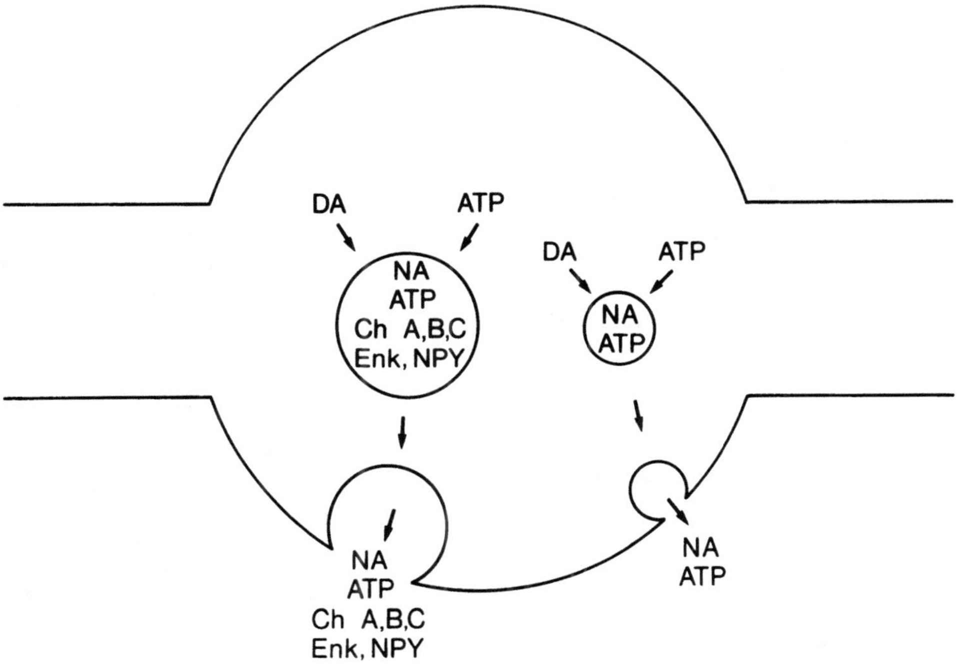

Fig. 1: Functional properties of LDV and SDV in sympathetic nerve. This figure compiles our present knowledge on the functional properties of sympathetic LDV and SDV.
ATP: adenosine triphosphate, Ch: chromogranin, DA: dopamine, Enk: enkephalin, NA: noradrenaline, NPY: neuropeptide Y.

cholinergic LDV are comparable to adrenergic LDV as far as transmitter uptake is concerned.

One crucial questions remains: What is the biogenesis of these two types of vesicles. Two main possibilities will be considered:

(i) As shown in Fig. 2 it has been proposed (Smith 1972) that SDV are formed from LDV after these vesicles have released their content by exocytosis. Several points are in agreement with this hypothesis: LDV are transported to terminals by axonal flow, they secrete by exocytosis (Smith & Winkler 1972; Klein & Thureson-Klein 1984) since their soluble proteins like chromogranin A and dopamine ß-hydroxylase are released concomitantly with the transmitters. Finally retrieval of membranes after exocytosis has at least been demonstrated for

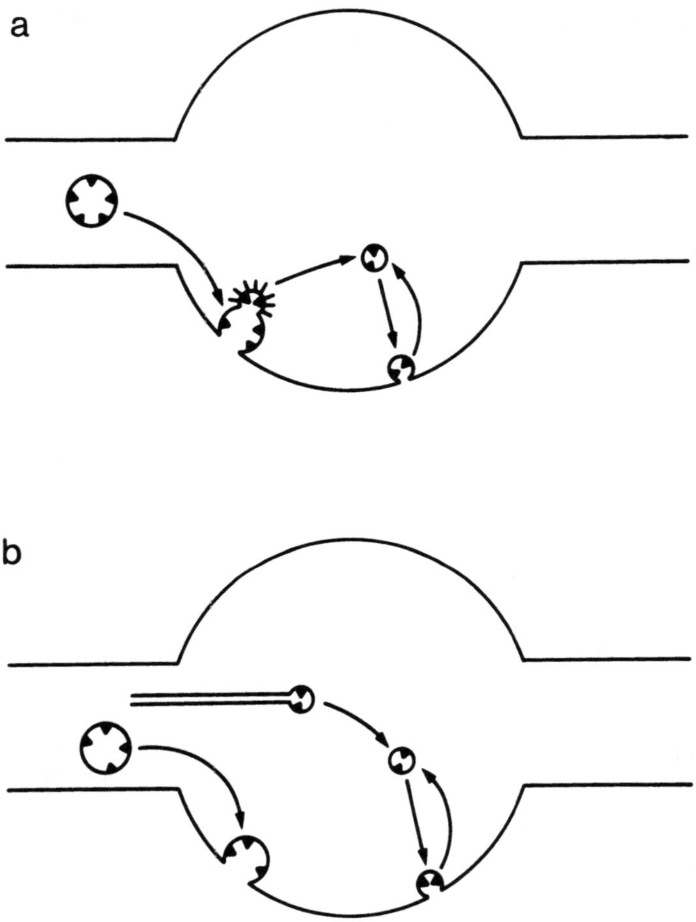

Fig. 2: Biogenesis of LDV and SDV: The figure illustrates two possible models: a) LDV release their content by exocytosis. The membranes are retrieved and become functional as SDV, which can be filled with transmitter and used for several secretion cycles (see Smith 1972) b) LDV and SDV represent separate vesicle populations. LDV which contain peptides have to be transported from the Golgi region to the terminal. SDV could be formed locally (as shown) or could also be transported from the cell body to the terminal.

the chromaffin granules of the adrenal medulla (Diner 1967; Lingg et al 1983; Patzak et al 1984; Patzak & Winkler 1986). If in nerve the retrieved vesicles are used as SDV the functional and biochemical properties of these two types of vesicles should be identical. As we have discussed above no difference has yet been found for adrenergic vesicles, but

such a concept of biogenesis should be valid for all nerve terminals and recent studies on synaptophysin (see below) indicate that at least one antigen is only present in SDV, but not in LDV of brain synaptic terminals. These crucial findings will be discussed below.
(ii) The second possibility is that SDV and LDV are separate vesicle populations. Since LDV have to transport peptides from the Golgi region to the plasma membrane they must arise from the Golgi complex in accordance with other vesicles belonging to the regulated pathway (Gumbiner & Kelly 1982). In this concept SDV represent a separate entity possibly also arising from the Golgi complex (constitutive pathway, Gumbiner & Kelly 1982) or from membranes of the endoplasmic reticulum. In the adrenergic system it is difficult to reconcile this concept with the similarity of the biochemical and functional properties of the membranes of the two types of vesicles. It seems that at least four proteins are present in both types of membranes, however as pointed out above if one antigen differs between these two membranes such a finding apparently contradicts a formation of SDV from LDV.
The crucial question therefore is: Does such a difference in protein composition exist? Recent studies on synaptophysin seem to support such a difference. Synaptophysin or p38 is a protein originally isolated from synaptic vesicles (SDV) of brain (Jahn et al 1985; Wiedenmann & Franke 1985). A careful analysis by immunohistological methods revealed its presence only in SDV, but not in LDV (Navone et al 1986). In adrenal medulla the same authors found that synaptophysin was present in a special population of small vesicles but not in chromaffin granules (Navone et al 1986). Furthermore it was also reported that by immunoblotting synaptophysin could not be detected in chromaffin granules (Wiedenmann & Franke 1985). These studies apparently establish that only SDV contain this protein but not LDV and not the analogous vesicles in adrenal medulla, the chromaffin granules.
On the other hand, in preliminary biochemical studies (D.Obendorf, U.Schwarzenbrunner, R.Fischer-Colbrie & H.Winkler) we obtained evidence that chromaffin granules do

contain small amounts of synaptophysin. These results were based on immunoblotting of fractions recovered after density gradient centrifugation of a large granule fraction of bovine adrenal medulla. Synaptophysin was found in two peaks, one on top of the gradient (microsomal membrane) and one exactly coinciding with chromaffin granules. If further data exclude the possibility that synaptophysin vesicles became adsorbed to chromaffin granules during homogenization and thus co-sediment with chromaffin granules one has to accept the possibility that chromaffin granule do indeed contain synaptophysin. Does this then also apply to LDV of nerve? Obviously an answer to this question is crucial for deciding whether the membrane of SDV can be derived from LDV or not.

CONCLUSIONS

The questions raised in the title: "Are there common antigens and common properties of adrenergic and cholinergic vesicles" has been answered in the affirmative for both types of vesicles, i.e. LDV and SDV. One crucial question remains: Are SDV formed from LDV after exocytosis or do they represent a completely separate vesicle population?

ACKNOWLEDGEMENTS

The original work by the authors quoted in this review was supported by the Dr. Legerlotz-Stiftung and by the Fonds zur Förderung der wissenschaftlichen Forschung (Austria).

REFERENCES

Aberer W, Kostron H, Huber E and Winkler H (1978) A characterization of the nucleotide uptake by chromaffin granules of bovine adrenal medulla. Biochem J 172: 353-360

Agoston DV and Conlon JM (1987) Presence of a neuropeptide in a model cholinergic system. Ann NY Acad Sci 493: 135-137

Bisby MA, Fillenz M and Smith AD (1973) Evidence for the presence of dopamine ß-hydroxylase in both populations of noradrenaline storage vesicles in sympathetic nerve terminals of the rat vas deferens. J Neurochem 20: 245-248

Buckley K and Kelly RB (1985) Identification of a transmembrane glycoprotein specific for secretory vesicles of neural and endocrine cells. J Cell Biol 100: 1284-1294

Diner O (1967) L'expulsion des granules de la medullo-surrénale chez le hamster. Compt Rend 265: 616-619

Eiden LE, Huttner WB, Mallet J, O'Connor DT, Winkler H and Zanini A (1987) A nomenclature proposal for the chromogranin/secretogranin proteins. Neuroscience 21: 1019-1021

Fleming PJ and Kent UM (1987) Secretory vesicle cytochrome b_{561}: a transmembrane electron transporter. Ann NY Acad Sci 493: 101-107

Fried G (1978) Cytochrome b-561 in sympathetic nerve terminal vesicles from rat vas deferens. Biochim Biophys Acta 507: 175-177

Fried G, Lagercrantz H and Hökfelt T (1978) Improved isolation of small noradrenergic vesicles from rat seminal ducts following castration. A density gradient centrifugation and morphological study. Neuroscience 3: 1271-1291

Fried G, Terenius L, Brodin E, Efendic S, Dockray G, Fahrenkrug J, Goldstein M and Hökfelt T (1986) Neuropeptide Y, enkephalin and noradrenaline coexist in sympathetic neurons innervating the bovine spleen. Cell Tiss Res 243: 495-508

Gumbiner B and Kelly RB (1982) Two distinct intracellular pathways transport secretory and membrane glycoproteins to the surface of pituitary tumor cells. Cell 28: 51-59

Jahn R, Schiebler W, Ouimet C and Greengard P (1985) A 38,000-dalton membrane protein (p38) present in synaptic vesicles. Proc Natl Acad Sci USA 82: 4137-4141

Johansson O and Lundberg JM (1981) Ultrastructural localization of VIP-like immunoreactivity in large dense-core vesicles of "cholinergic-type" nerve terminals in cat exocrine glands. Neuroscience 6: 847-862

Johnson RG (1987) Proton pumps and chemiosmotic coupling as a generalized mechanism for neurotransmitter and hormone transport. Ann NY Acad Sci 493: 162-177

Klein RL and Thureson-Klein ÅK (1984) Noradrenergic Vesicles. Molecular organization and function. In: Lajtha A (ed.) Handbook of Neurochemistry Vol 7. Plenum Publishing Corporation: 71-109

Klein RL, Thureson-Klein ÅK, Chen Yen S-H, Baggett JMacC, Gasparis MS and Kirksey DF (1979) Dopamine ß-hydroxylase distribution in density gradients: physiological and artifactual implications. J Neurobiol 10: 291-307

Lingg G, Fischer-Colbrie R, Schmidt W and Winkler H (1983) Exposure of an antigen of chromaffin granules on cell surface during exocytosis. Nature 301: 610-611

Lundberg JM, Fried G, Fahrenkrug J, Holmstedt B, Hökfelt T, Lagercrantz H, Lundgren G and Änggård A (1981) Subcellular fractionation of cat submandibular gland: comparative studies on the distribution of acetylcholine and vasoactive intestinal polypeptide (VIP). Neuroscience 6: 1001-1010

Luqmani YA (1981) Nucleotide uptake by isolated cholinergic synaptic vesicles: evidence for a carrier of adenosine 5'-triphosphate. Neuroscience 6: 1011-1021

Matthew WD, Tsavaler L and Reichardt LF (1981) Identification of a synaptic vesicle-specific membrane protein with a wide distribution in neuronal and neurosecretory tissue. J Cell Biol 91: 257-269

Navone F, Jahn R, Di Gioia G, Stukenbrok H, Greengard P and De Camilli P (1986) Protein p38: an integral membrane protein specific for small vesicles of neurons and neuroendocrine cells. J Cell Biol 103: 2511-2527

Neuman B, Wiedermann CJ, Fischer-Colbrie R, Schober M, Sperk G and Winkler H (1984) Biochemical and functional properties of large and small dense-core vesicles in sympathetic nerves of rat and ox vas deferens. Neuroscience 13: 921-931

Njus D, Kelley PM, Harnadek GJ and Pacquing YV (1987) Mechanism of ascorbic acid regeneration mediated by cytochrome b_{561}. Ann NY Acad Sci 493: 108-119

Parsons SM, Bahn BA, Gracz LM, Kaufman R, Kornreich WD, Nilsson L and Rogers GA (1987) Acetylcholine transport: fundamental properties and effects of pharmacologic agents. Ann NY Acad Sci 493: 220-233

Patzak A and Winkler H (1986) Exocytotic exposure and recycling of membrane antigens of chromaffin granules: ultrastructural evaluation after immunolabeling. J Cell Biol 102: 510-515

Patzak A, Böck G, Fischer-Colbrie R, Schauenstein K, Schmidt W, Lingg G and Winkler H (1984) Exocytotic exposure and retrieval of membrane antigens of chromaffin granules: quantitative evaluation of immunofluorescence on the surface of chromaffin cells. J Cell Biol 98: 1817-1824

Smith AD (1972) Mechanisms involved in the release of noradrenaline from sympathetic nerve. Br med Bull 29: 123-129

Smith AD and Winkler H (1972) Fundamental mechanisms of the release of catecholamines. In: Blaschko H and Muscholl E (eds.) Catecholamines, Handbook of Experimental Pharmacology, vol.33, Springer, Berlin: pp. 538-617

Somogyi P, Hodgson AJ, DePotter RW, Fischer-Colbrie R, Schober M, Winkler H and Chubb IW (1984) Chromogranin immunoreactivity in the central nervous system. Immunochemical characterization, distribution and relationship to catecholamine and enkephalin pathways. Brain Res Rev 8: 193-230

Volknandt W, Schober M, Fischer-Colbrie R, Zimmermann H and Winkler H (1987) Cholinergic nerve terminals in the rat diaphragm are chromogranin A immunoreactive. Neurosci Lett, submitted

Whittaker VP (1987) Cholinergic synaptic vesicles from the electromotor nerve terminals of torpedo. Composition and life cycle. Ann NY Acad Sci 493: 77-91

Wiedenmann B and Franke WW (1985) Identification and localization of synaptophysin, and integral membrane glycoprotein of M_r 38,000 characteristic of presynaptic vesicles. Cell 41: 1017-1028

Winkler H (1976) The composition of adrenal chromaffin granules: an assessment of controversial results. Neuroscience 1: 65-80

Winkler H, Apps DK and Fischer-Colbrie R (1986) The molecular function of adrenal chromaffin granules: established facts and unresolved topics. Neuroscience 18: 261-290

Winkler H and Carmichael SW (1982) The chromaffin granule. In: The secretory granule, eds. Poisner and Trifaró, Elsevier Biomedical Press, pp. 3-79

Winkler H and Westhead E (1980) The molecular organization of adrenal chromaffin granules. Neuroscience 5: 1803-1823

STRUCTURE AND FUNCTION OF SYNAPTIC VESICLES:NEW ASPECTS

Herbert Stadler, Edilio Borroni, Martina Ploghöft and Marie-Luise Kiene

Abteilung Neurochemie, Max-Planck-Institut für biophysikalische Chemie, D-3400 Göttingen, FRG

INTRODUCTION

Synaptic vesicles play a central role in neurotransmission as the transmitter storing and releasing organelles of the nerve terminal. In recent years progress has been made in functional identification of vesicle specific components. Most of this work has been done on synaptic vesicles isolated from electric organs of Torpedo marmorata (Stadler et al., 1985), a purely cholinergic model system. More recently a few proteins specific to brain synaptic vesicles have been isolated and characterized as well. Integration of the findings leads to a first model of the vesicle structure including aspects of uptake and storage of the solutes within these organelles. The major core protein of cholinergic vesicles, a heparan-sulfate proteoglycan (Stadler and Dowe, 1982), is a secretory protein. This protein can be labelled in vivo with ^{35}S-sulfate and this labelling technique enabled us to study the heterogeneity of these vesicles and their life cycle in the nerve terminal in more detail than previously. These findings provide new aspects towards understanding quantal release and synapse formation in this system. Furthermore we present evidence that mammalian brain synaptic vesicles may contain a proteoglycan-like component suggesting that it is a secretory protein as well.

NATO ASI Series, Vol. H21
Cellular and Molecular Basis of Synaptic Transmission
Edited by H. Zimmermann

RESULTS

Structure and life cycle of cholinergic synaptic vesicles in Torpedo electric organ

Our results on the structure of electric organ vesicles is summarized in Fig. 1.

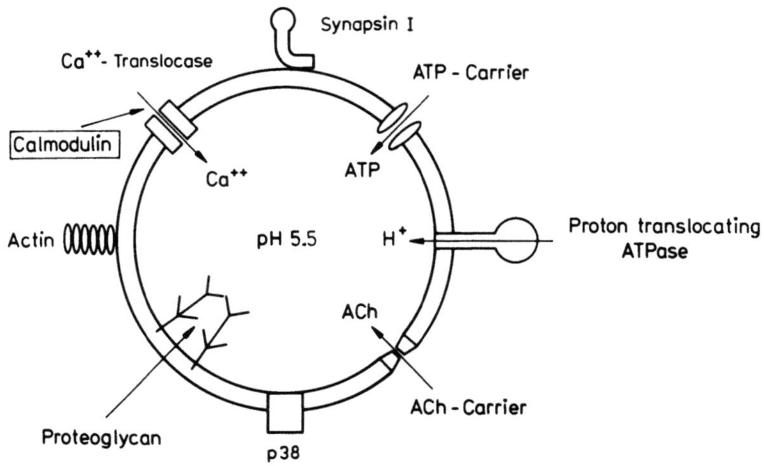

Figure 1. Structure of cholinergic synaptic vesicles from Torpedo marmorata. Explanation see text.

The core is filled with a concentrated solution of acetylcholine and ATP at a pH around 5.5 (Stadler and Füldner,1980; Füldner and Stadler,1982).Uptake of acetylcholine is electrogenic (Anderson et al.,1982) and driven by an ATP dependent proton pump (Harlos et al.,1984),ATP uptake is mediated by a translocase resembling the nucleotide carrier of mitochondria (Stadler and Fenwick,1983;Lee and Witzemann,1983) .

In addition an uptake system for Ca^{++} is present (Israël et al.,1980).More recently other components originally iden - tified in brain synaptic vesicles have been shown to be present in the Torpedo vesicle.One of them is the peri - pherally associated protein synapsin I (compare chapter by Volknandt et al.,this book) and the other is synaptophysin (p38) a probably integral membrane protein,that has a slightly larger molecular weight in Torpedo (40 kD) than in brain (Stadler,unpublished).The major core component is a heparansulfate proteoglycan.It can be labelled in vivo by injection of ^{35}S-sulfate into the electric lobe of the fishbrain,the site where the cellbodies of the electromotoneurones are.

Using this labelling technique,we have carried out a study of axonal transport of vesicles and followed their maturation and turnover in the nerve terminal.We found that the synaptic vesicle proteoglycan moves anterogradely in electromotoneurones with a speed of around 110 mm/day at 17°C which corresponds to the fast transport class in this system.Isolation of synaptic vesicles from ^{35}S-labelled axons showed that the proteoglycan is incorporated in the vesicles.These axonal vesicles have a protein composition identical to the vesicles isolated from nerve terminals, but they are not filled up with acetylcholine and ATP. These "empty" axonal vesicles (VP_o vesicles) accumulate after arrival in the nerve terminal acetylcholine and ATP and constitute then the fully mature and exocytosis competent population VP_1.After a further period of time VP_1 changes to a third population (VP_2).VP_2 obviously has undergone exocytosis since its proteoglycan content is much lower as compared to VP_1.The three vesicle populations can be separated on the basis of density dif - ferences on shallow sucrose gradients.They differ furthermore in uptake properties.VP_o and VP_2 take up acetyl - choline and ATP in contrary to VP_1 which exhibits Ca^{++}

uptake properties not observed in VP_o and VP_2.Probably Ca^{++} uptake into VP_1 is an important step in processing this secretory organelle. One possibility would be that Ca^{++} entry into the organelle activates a protease that splits the membrane anchored proteoglycan creating its secretory form.

Analysis of the acetylcholine content of VP_1 and VP_2 showed that VP_2 contains approximately 5-10 times acetylcholine less than VP_1.Since the two populations are present in the nerve terminal in approximately equally amounts,the question arises, which population is the source of acetylcholine quantal release? A recent study in electric organ suggests that acetylcholine quanta in this tissue are in the range of 7000-10000 molecules per quanta (Dunant and Muller, 1986). This is in relatively good agreement with the VP_2 population on the basis that VP_1 has been estimated to contain around 100,000 molecules acetylcholine per vesicle.

From these results a new aspect towards understanding the vesicle hypothesis and the molecular basis of acetylcholine release is emerging: Cholinergic synaptic vesicles although seen in the electronmicroscope as a rather homogeneous population, are heterogeneous in terms of acetylcholine content and most probably only one population is involved in quantal release.Beside the reasonable match between acetylcholine content found in VP_2 and electrophysiological measurements on the number of acetylcholine molecules per quanta, another fact is in favour of VP_2 beeing the major source of nerve terminal acetylcholine release. VP_2 have been described as "recycling" vesicles (Zimmermann and Whittaker, 1977) reloaded with transmitter in agreement with our findings that they possess a comparatively longer halflife than VP_1. Therefore exocytosis of VP_1 vesicles may represent, as compared to exocytosis of VP_2 a rare event, the electrophysiological consequences of which have not been recognized yet.

Exocytosis of VP_1 is not only accompanied by release of acetylcholine but as well by secretion of a large part of the core proteoglycan. This view is especially supported by the finding that a smaller form of the proteoglycan is actually found in the extracellular matrix of the electric organ. Most interestingly, a factor in the extracellular matrix of the electric organ described to cluster acetylcholine receptors (Fallon et al.,1985) has been identified as a heparansulfate proteoglycan. In the case that further experiments can show that the vesicular proteoglycan found in the extracellular matrix is identical with the acetylcholine receptor clustering factor an attractive hypothesis towards a transsynaptic signalling hypothesis evolves. The cholinergic vesicle is not only involved in neurotransmission by releasing acetylcholine but in formation and maintenance of the synapse by depositing a secretory protein into the synaptic extracellular matrix guiding receptor formation.

Wether vesicle heterogeneity is related to the phenomenon of subminiature endplate potentials described by Kriebel (see chapter by Kriebel and Fox, this book) remains unanswered yet. A detailed account of axonal transport of vesicles and their life cycle in electromotoneurones is given in Kiene and Stadler (1987) and Stadler and Kiene (1987). The life cycle of cholinergic synaptic vesicles is summarized in Figure 2.

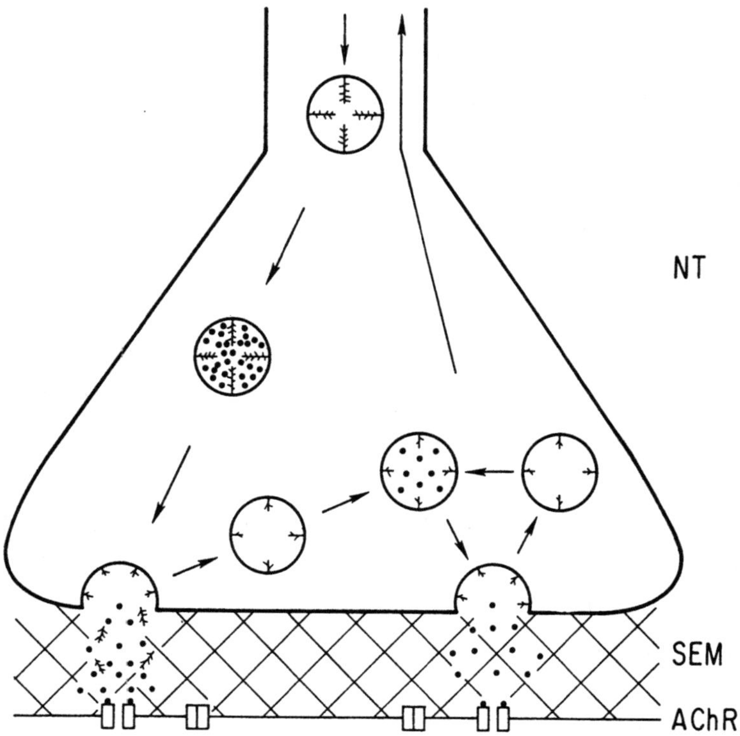

Figure 2. Life cycle of synaptic vesicles in Torpedo electromotor nerve terminals (NT). Empty vesicles arrive in the terminal and are filled up with acetylcholine (dots). Exocytosis releases acetylcholine and part of the membrane anchored core proteoglycan which binds to the synaptic extracellular matrix (SEM). Vesicles are retrieved by endocytosis, refill with acetylcholine and can undergoe further exo- endocytotic cycles ("recycling"). These recycled vesicles are probably the major source of quantal acetylcholine release acting on the acetylcholine receptor (AChR).

A proteoglycan like component of brain synaptic vesicles

We have isolated synaptic vesicles from pig brain and analysed them for the presence of proteoglycans. By gel filtration of vesicle preparations solubilized in deter - gent, we found that a high molecular weight fraction (>350 kD) is present that shows proteoglycan like features, e.g. the presence of uronic acid and glucosamine. Against this fraction we have raised a library of monoclonals and chosen one of the clones (C98) for function studies. Immunblotting against synaptic vesicles using C98 anti - bodies shows a broad band in the range above 200 kD (Fig. 3).

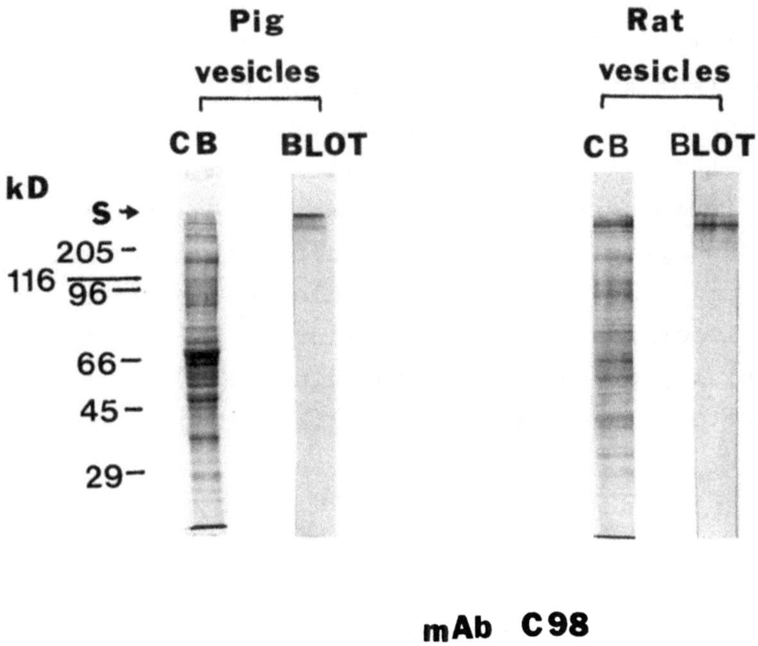

Fig. 3. Pig (left) and (right) rat brain vesicle preparations after SDS gel electrophoresis stained with Coomassie blue (CB) and their immunblots incubated with the monoclonal antibody C98. C98 recognizes a high molecular weight antigen (> 200 kD) migrating as a smear at the top of the gel.

Along subcellular fractionation of pig or rat brain the antigen is found in the synaptic vesicle fraction, but as well in a second soluble nonmembrane bound pool suggesting that this represents a second form of the proteoglycan. Highly purified vesicles, obtained by chromatography on Sephacryl S 1000, were found to contain the C98 antigen together with synaptophysin (p38) used as a vesicle marker, again underlining the probable vesicle specific origin of the antigen. Nevertheless sucrose density gradient centrifugation of synaptic vesicles indicated that immunreac - tivity of synaptophysin and C98 did not exactly coincide , suggesting that C98 might be present only in a sub-population of synaptic vesicles. It cannot be ruled out at the moment that C98 is present in transport vesicles of the constitutive secretory pathway. Immunocytochemistry in rat brain areas suggests that C98 is enriched mainly in synapse rich areas like the molecular layer in the cere - bellum and that it is not a constituent of glial cell. This view is supported by results on cell cultures derived from chicken optic tectum where exclusively neurones and not glial cells are stained (A.Michler,unpublished). The cell free supernatant of these cultures contains C98 immunreactivity probably corresponding to the soluble secreted form of the proteoglycan like antigen.

CONCLUSION

Synaptic vesicles from Torpedo electromotor nerve terminals are heterogeneous in nature and consist of three populations. Two of them contain acetylcholine in different amounts and only a population that has undergone exo-endo-cytotic cycles seems to be involved in quantal acetylcholine release responsible for the creation of miniature endplate potentials. The vesicles release beside acetylcholine a core heparansulfate proteoglycan that binds to the extracellular matrix and might play a functional role in cholinergic transmission.

Evidence has been obtained that synaptic vesicles from mammalian brain might contain a proteoglycan like antigen as well. Probably this antigen is also secreted. The classical view of synaptic vesicles as a purely transmitter containing and releasing organelle may have to be extended. The function of these secretory proteins is as yet unknown.

REFERENCES

Anderson, D.C.,King, S.C. and Parsons, S.M. (1982). Proton gradient linkage to active uptake of [^{3}H]acetylcholine by Torpedo electric organ synaptic vesicles. Biochemistry 21, 3037-3043

Dunant, Y. and Muller, D. (1986). Quantal release of acetylcholine evoked by focal depolarization at the Torpedo nerve-electroplaque junction. J. Physiol. 379, 461-478

Fallon, J.R., Nitkin, R.M., Reist, N.E., Wallace, B.G. and McMahan, U.J. (1985). Acetylcholine receptor-aggregating factor is similar to molecules concentrated at neuro - muscular junctions. Nature 315, 571-574

Füldner, H.H. and Stadler, H. (1982). ^{31}P-NMR analysis of synaptic vesicles. Status of ATP and internal pH. Eur. J. Biochem. 121 519-524

Harlos,P., Lee, D.A. and Stadler, H. (1984). Characterization of a Mg^{++}-ATPase and a proton pump in cholinergic synaptic vesicles from the electric organ of Torpedo marmorata.

Israël, M.,Manaranche, R., Marsal, J., Meunier, F.M., Morel, N., Frachon, P. and Lesbats, B. (1980).ATP-dependent calcium uptake by cholinergic synaptic vesicles isolated from Torpedo electric organ. J. Membrane Biol. 54, 115-126

Kiene, M.L. and Stadler, H. (1987). Synaptic vesicles in electromotoneurones.I. Axonal transport, site of transmitter uptake and processing of a core proteoglycan during maturation. EMBO J. 6, 2209-2215

Lee, D.A. and Witzemann, V. (1983). Photoaffinity labeling of a synaptic vesicle specific nucleotide transport system from Torpedo marmorata . Biochemistry 22, 6123-6130

Stadler, H. and Füldner, H.H. (1980). Proton NMR detection of acetylcholine status in synaptic vesicles. Nature 286, 293-294

Stadler, H. and Dowe, G.H.C. (1982). Identification of heparansulfate proteoglycan in the core of synaptic vesicles of Torpedo marmorata. EMBO J. 1, 1381-1384

Stadler, H. and Fenwick, E.M. (1983). Cholinergic synaptic vesicles from Torpedo marmorata contain an atractyloside-binding protein related to the mitochondrial ADP/ATP carrier. Eur. J. Biochem. 136, 377-382

Stadler, H., Kiene, M.L., Harlos, P. and Welscher, U. (1985). Structure and function of cholinergic synaptic vesicles. In Hamprecht, B. and Neuhoff, V. (eds.),Neurobiochemistry, Springer Verlag Berlin, p. 55-65

Stadler, H. and Kiene, M.L. (1987). Synaptic vesicles in electromotoneurones.II. Heterogeneity of populations is expressed in uptake properties; exocytosis and insertion of a core proteoglycan into the extracellular matrix. Embo J. 6, 2217-2221

Zimmermann, H. and Whittaker, V.P. (1977). Morphological and biochemical heterogeneity of cholinergic synaptic vesicles. Nature 267, 633-635

COMPLEXITY AND REGULATION IN THE ACETYLCHOLINE STORAGE SYSTEM OF SYNAPTIC VESICLES

Stanley M. Parsons, Krystyna Noremberg, Gary A. Rogers, Lawrence M. Gracz, Wayne D. Kornreich, Ben A. Bahr and Rose Kaufman
Department of Chemistry and the Neurosciences Research Program
Institute of Environmental Stress
University of California
Santa Barbara, CA 93106, USA

INTRODUCTION

Synaptic vesicles isolated from the electric organ of Torpedo exhibit active transport of acetylcholine (AcCh). The transport system is composed of at least three components; namely, an ATPase thought to pump protons into the vesicle, an AcCh transporter which draws on the proton-motive-force to drive secondary active transport of AcCh, and a receptor for the drug l-2-(4-phenylpiperidino)cyclohexanol (vesamicol, formerly known as AH5183) (Bahr and Parsons, 1986a). When the vesamicol receptor is occupied, AcCh active transport is blocked noncompetitively with no effect on the ATPase activity (Anderson et al., 1983; Bahr and Parsons, 1986b).

Vesamicol originally was discovered as a neuromuscular blocking agent with unusual characteristics of action (Marshall, 1970). The drug inhibits evoked quantal release of AcCh from every cholinergic nerve terminal preparation which has been examined, presumably as a secondary effect of storage block (reviewed in Marshall and Parsons, 1987). The drug also inhibits nonquantal release of AcCh from the motoneural terminal, which suggests that nonquantal release is mediated by synaptic vesicle membrane incorporated into the cytoplasmic membrane as a result of exocytosis (Edwards et. al., 1985). Quantitative autoradiography has shown that the vesamicol receptor is distributed heterogeneously in brain in a manner highly correlated with other cholinergic terminal markers (Marien et al., 1987). It now is clear that

NATO ASI Series, Vol. H21
Cellular and Molecular Basis of Synaptic Transmission
Edited by H. Zimmermann

the vesamicol receptor is a new major component of the cholinergic terminal, but its function *in vivo* is uncertain. We report here experiments designed to explore the linkage between occupation of the receptor and the occurrance of AcCh transport inhibition, and whether regulatory phenomena are present.

AcCh TRANSPORT AND VESAMICOL BINDING BY VP_1 and VP_2 SYNAPTIC VESICLES

VP_2 (actively recycling) vesicles are much more active *in vivo* in AcCh uptake than VP_1 (resting or depot) vesicles are (Zimmermann and Denston, 1977). We asked whether this also was the case *in vitro*. A crude membrane pellet obtained from a homogenate of *Torpedo californica* electric organ was resuspended and centrifuged to equilibrium in an isosmotic sucrose density gradient. After fractionation, the amount of $[^3H]$AcCh actively transported or $[^3H]$vesamicol specifically bound by each fraction was determined using glass fiber filter assays. As shown in Figure 1, two peaks of vesamicol-sensitive transport activity were found. One around fraction 6 at lower density occurred coincident with VP_1 vesicles. Another around fraction 17 at higher density occurred coincident with VP_2 vesicles. A control demonstrated that both peaks of transport activity were sensitive also to the uncoupler nigericin, as expected.

Each fraction was assayed for binding of $[^3H]$vesamicol at an essentially saturating concentration of the drug (see below). The receptor for vesamicol was present in all of the fractions of the gradient where $[^3H]$AcCh transport activity was present. More important is the observation that in the VP_2 fractions the ratio of $[^3H]$AcCh transport activity to the concentration of the vesamicol receptor was about 4 times higher than in the VP_1 fractions. This increase in ratio suggests that in the VP_2 vesicles either the AcCh transporter is more active, or there is less apparent vesamicol receptor per equally active transporter, or both changes have occurred as compared to VP_1 vesicles.

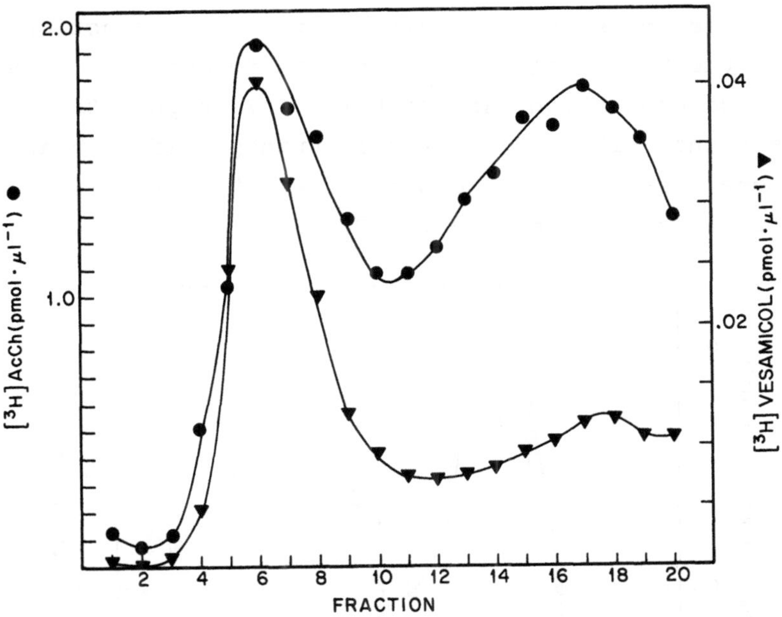

Fig.1 AcCh transport and vesamicol binding in VP_1 and VP_2 vesicles. Crude _Torpedo_ synaptic vesicles were centrifuged for 3.5 hrs. at 4° and 196,200 x g in the Beckman VTi50 rotor in 800 milliosmolar glycine-sucrose which varied from about 1.03 to 1.085 g/ml. MgATP-dependent active transport of 50 µM [^{3}H]AcCh was carried out with 60 µl of each fraction for 30 min. at 23° as described, (Bahr and Parsons, 1986b). Equilibrium binding of [^{3}H]vesamicol at 22° and 500 nM total concentration was determined with 100 µl of each fraction similarly as described (Bahr and Parsons, 1986a) except that Whatman 2.4 cm GF/F filters coated 2 hrs. with 0.5% polyethyleneimine were substituted. Filters were soaked in 0.5 ml water after which radioactivity was determined by liquid scintillation spectroscopy in 10 ml of aqueous counting cocktail.

SV2 ANTIGEN LEVELS RELATIVE TO THE VESAMICOL RECEPTOR

We cannot tell from the above data which alternative is correct because the gradient fractions are impure. To make the decision we quantitated the SV2 antigen, which is recognized by a monoclonal antibody raised against electric organ synaptic vesicles by Buckley and Kelley (1985). As shown in Figure 2, the SV2 antigen was present in all of the fractions of the gradient where specific [^{3}H]vesamicol binding was present. However, in the VP_2 fractions, the ratio of the SV2 antigen concentration to apparent vesamicol receptor concentration was about 2.2-fold higher than in the VP_1 fractions. Thus, VP_2 vesicles transported AcCh about 2 - fold more actively but expressed about one-half as much vesamicol receptor compared to VP_1 vesicles.

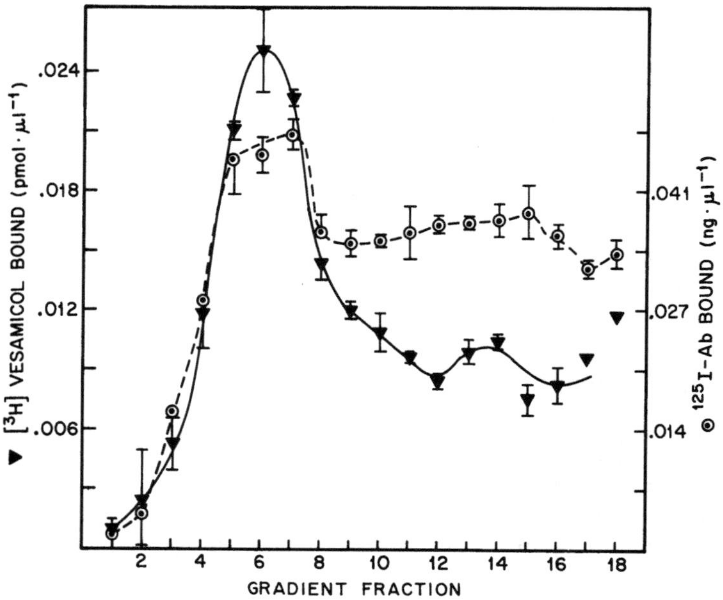

Fig.2 SV2 antigen levels and vesamicol binding in VP_1 and VP_2 vesicles. A standard curve for 10H3 anti-SV2 monoclonal antibody (gift of R. B. Kelley) binding to highly purified VP_1 vesicles was generated with a quantitative dot-immunobinding assay on nitrocellulose paper (Jahn et al., 1984) using ^{125}I-labelled sheep anti-mouse Ig as the secondary antibody and gamma counting. The amount of ^{125}I-antibody bound by 10 µl of each fraction of a density gradient was determined. Also shown are the amounts of [^{3}H]vesamicol specifically bound in the same fractions. One standard error and the mean of triplicate determinations are shown.

REGULATION OF THE RECEPTOR BY AN INTERNAL FACTOR WHICH MIGHT BE AcCh

Different preparations of highly purified VP_1 vesicles contain widely different apparent amounts of the vesamicol receptor (Bahr and Parsons, 1986a). In a chance observation, we noted that the amount of receptor which could be titrated increased slowly with aging of freshly isolated vesicles or

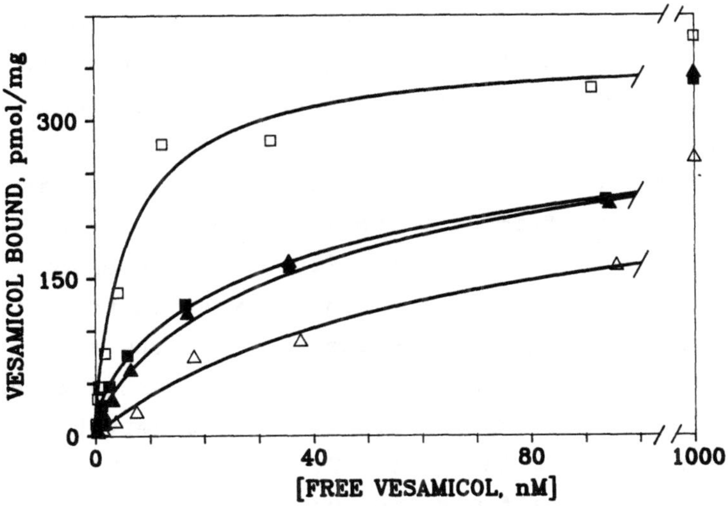

Fig.3. Effects of exogenous AcCh on the titration of [^{3}H]vesamicol binding to highly purified intact VP_1 vesicles or ghosts. Concentrated vesicles (0.3 mg protein/ml) were diluted with a 10-fold volume of isosmotic 0.70 M glycine, 0.10 M Hepes, 1 mM EDTA,1 mM EGTA, 0.02% NaN_3, adjusted to pH 7.4 with 0.80 M NaOH, or with a 10-fold volume of 0.020 M Hepes adjusted to pH 7.4 with NaOH and incubated 5 min. after which was added one volume of 4.0 M NaCl, 1.0 M Hepes, 10 mM EDTA, 10 M EGTA, 0.2% NaN_3, adjusted to pH 7.4 with NaOH to make resealed ghosts. Vesicles to be exposed to exogenous AcCh were incubated with 0.4 mM paraoxon for 30 min. Intact (■ , ▲) or lysed vesicles (□ , △) in the absence (■, □,) or presence (▲ , △) of 50 mM exogenous AcCh chloride and different concentrations of [^{3}H]vesamicol were incubated 60 min. at 23° afterwhich the amount of bound vesamicol was determined by the glass fiber filter assay and liquid scintillation spectroscopy. Regression analysis using the Hill equation give the curve parameters quoted in the text. Three hundred pmol receptor/mg protein corresponds to 6.0 receptors/synaptic vesicle (Anderson et al, 1986).

immediately upon hyposmotic lysis. It was realized that this could be due to loss of AcCh, and this possibility was investigated. Figure 3 demonstrates that [^{3}H]vesamicol binding to the receptor in intact vesicles is esssentially unaffected by the presence of 50 mM exogenous AcCh. Hyposmotic lysis of the vesicles led to a nearly 10-fold decrease in the vesamicol dissociation constant and a small increase in the total amount of receptor which could be titrated. Other preparations of vesicles gave larger effects. In contrast to the lack of response in intact vesicles, in ghosts 50 mM exogenous AcCh increased the vesamicol dissociation constant 12-fold and decreased the total amount of receptor which could be titrated by about 20 percent. Other experiments demonstrated that exogenous AcCh can completely block binding of moderate concentrations of [^{3}H]vesamicol to vesicle ghosts. Thus, release of an internal factor from the vesicles by hyposmotic lysis dramatically altered the affinity of the receptor and its sensitivity to AcCh. In other experiments, we also demonstrated that after lysis the vesicles reseal right-side-out, but they become rapidly and fully permeable to exogenous AcCh. The simplest explanation of these observations is that the internal factor in intact VP_1 vesicles which decreases vesamicol receptor affinity is endogenous AcCh.

VESAMICOL BINDING AND AcCh TRANSPORT INHIBITION ARE NOT COINCIDENT

The relationship between receptor occupancy and transport inhibition in intact VP_1 vesicles was studied using a double label approach where [^{14}C]AcCh active transport was carried out in the presence of variable levels of [^{3}H]vesamicol, and the amount of each isotope bound to the vesicles was determined. Also, the effect of the vesicle concentration was studied. As shown in Figure 4, complex results were obtained. At low vesicle concentration the transport inhibition and binding curves were noncooperative and noncoincident with the binding curve being shifted to about 3-fold higher free vesamicol concentrations relative to the transport inhibition curve. At high vesicle concentration the two curves were highly positively cooperative, more nearly coincident, and shifted to a somewhat higher free vesamicol concentration range. Also, a biphasic effect of the vesicle concentration on the AcCh uptake specific activity in the absence of vesamicol was seen. These results suggest that some endogenous dissociable factor which controls the vesamicol receptor affinity and cooperativity copurifies with the vesicles. Furthermore, it appears that at low vesicle concentration, only a fraction of the receptors need be occupied to get full transport inhibition.

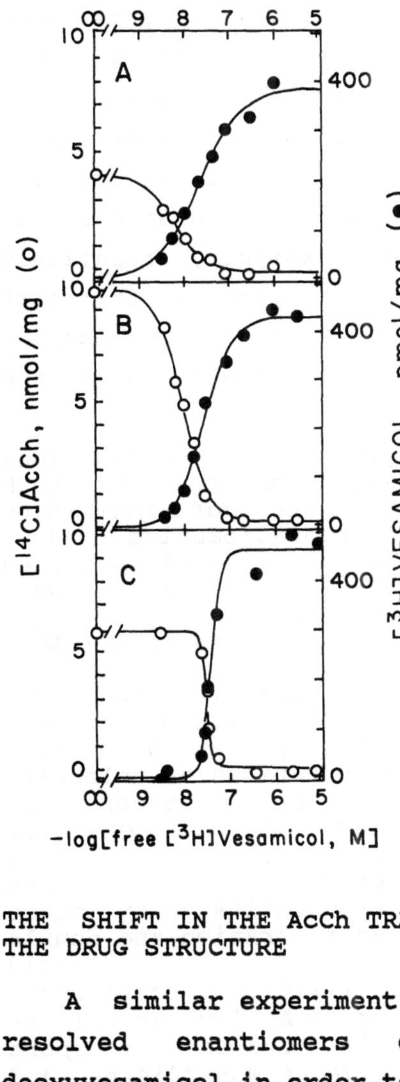

Fig. 4. Vesamicol binding and AcCh transport inhibition at different vesicle concentrations. Intact highly purified VP_1 vesicles were allowed to actively transport 50 µM [^{14}C]AcCh (0) for 30 min. at 23° in the presence of different concentrations of [^{3}H]vesamicol (●) at 0.050 (frame A), 0.275 (frame B) or 1.5 (frame C) mg protein/ml. The amount of each bound isotope was assayed by centrifugation-gel filtration and double channel liquid scintillation spectroscopy as described (Anderson et al., 1982). The Hill equation was fit to each data file. For A the IC_{50} value for transport inhibition was 6.5±0.8 nM with a Hill coefficient n of 1.3±0.2. The K_d for receptor occupancy was 22±4 nM with n of 0.9±0.1. For B the IC_{50} was 9.0 ±0.8 nM with n of 1.5±0.2 and the K_d was 25±3 nM with n of 1.4 ±0.2. For C the IC_{50} was 29±1 nM with n of 3.8±0.8 and K_d was 38±3 nM with n of 7±2.

THE SHIFT IN THE AcCh TRANSPORT INHIBITION CURVE DEPENDS ON THE DRUG STRUCTURE

A similar experiment as above was carried out with the resolved enantiomers of vesamicol and the analogue deoxyvesamicol in order to determine whether the relationship between receptor occupancy and transport inhibition depends on the drug structure. Because not all of these compounds are available in radiolabeled form, a low concentration of [^{3}H]vesamicol (not enough to inhibit AcCh transport) was equilibrated with the vesicles, and [^{14}C]AcCh active transport was carried out in the presence of different concentrations of each unlabeled drug. Figure 5 shows the results. AcCh transport was inhibited at lower unlabeled drug concentrations than required to fully occupy the receptor as monitored by displacement of [^{3}H]vesamicol. This behavior is similar to that seen in the direct experiment of Fig. 4. The different drugs exhibited different potencies, as expected, but very interestingly the ratio of midpoint concentrations for the ^{3}H and ^{14}C curves clearly depended on the identity of the drug. It was 3.9-fold for l-vesamicol (the potent enantiomer), 8.4-fold for d-vesamicol, and 24-fold for deoxyvesamicol. Thus, whereas deoxyvesamicol was the least potent at blocking active transport, relatively fewer occupied receptors were required to produce transport inhibition. This result suggests that the drug-receptor interaction involves an induced fit which depends upon the

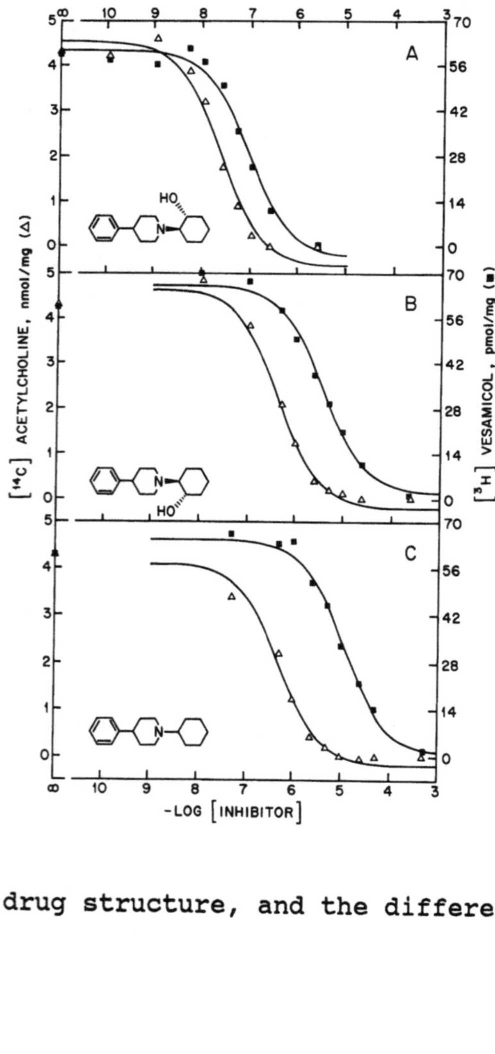

Fig.5. [^{3}H]Vesamicol displacement and [^{14}C]AcCh active transport inhibition by nonradioactive drug analogues. Highly purified VP_1 vesicles were equilibrated at 23° with enough [^{3}H]vesamicol (■) to saturate about 10% of the available receptor, which gave little transport inhibition, and 0 to high concentrations of unlabeled drugs. Active transport of 50 µM [^{14}C]AcCh (Δ) was carried out for 30 min. after which bound isotopes were quantitated by glass fiber filtration and double channel liquid scintillation spectroscopy. In frame A l-vesamicol was studied, and the transport IC_{50} value was 22±4 nM and the [^{3}H]vesamicol displacement K_i value was 86±16 nM. In frame B d-vesamicol gave IC_{50} 430±60 nM and K_i 3,600±700 nM. In frame C deoxyvesamicol gave IC_{50} 510±80 nM and K_i 12,000±2,000 nM.

drug structure, and the different resulting conformations of the drug-receptor complex exhibit different efficacies for transport inhibition.

DISCUSSION

The properties of the AcCh transport-vesamicol receptor system *in vitro* appear to depend on the original physiological status of the vesicles. The observation that

VP_2 vesicles transport AcCh better than VP_1 vesicles is consistent with their hypothesized *in vivo* role. Since this behavior was retained after homogenization of the electric organ and partial purification of the vesicles, the VP_2 vesicles *in vivo* are not more active merely because they lie closer to the high affinity choline uptake system and the site of AcCh synthesis. Rather, some long-lived presumably regulatory phenomena probably exists.

Close study of highly purified VP_1 vesicles has revealed a level of biochemical complexity which is consistent with the presence of regulatory phenomena. It is clear that some endogenous factor releasable from the inside of vesicles by hyposmotic lysis acts to decrease the binding of vesamicol, and in the extreme this factor can make some of the receptor cryptic. Exogenous AcCh has the same effect on vesicle ghosts, and since it rapidly equilibrates across the ghost membrane the straightforward hypothesis is that internal AcCh controls the conformation of the vesamicol receptor on the outside. Such a signal would be very useful to a nerve terminal by reporting which vesicles were fully loaded with AcCh and ready for release. The hypothesized AcCh signal, however, can not explain the lower level of vesamicol receptor expressed in VP_2 vesicles since these vesicles contain much less endogenous AcCh than VP_1 vesicles. This suggests that another "negative" signal controlling the receptor also exists.

A strong indication for the presence of an endogenous

dissociable factor came from study of vesicle concentration effects. The behavior induced by this hypothesized factor was to increase the positive cooperativity of vesamicol binding while decreasing the affinity at higher vesicle concentrations. This is reminiscent of the effect of 2, 3-diphosphoglycerate on oxygen binding to hemoglobin. The nature of the putative external factor is completely unknown.

Since AcCh can completely inhibit binding of vesamicol to ghosts, it seems likely that each vesamicol receptor is coupled to an internal AcCh binding site. As a minimal model not invoking any new AcCh binding sites, we hypothesize that the internal site could be the inwardly oriented AcCh transporter. This model suggests that AcCh transporters and vesamicol receptors are related stoichiometrically, but not necessarily one-to-one or within the same polypeptide. However, other observations show that transport inhibition requires occupation of only a fraction of the receptors (as little as 1/25 at the IC_{50} value of deoxyvesamicol) and involves an induced conformational change. These observations suggest that the coupling between the occupied receptor and the transporter is loose and involves an amplification mechanism. Thus, we are as yet unable to offer a comprehensive model of the mechanism of vesamicol-dependent inhibition of AcCh storage. It is clear though, that the system is complex and probably involves regulatory phenomena.

ACKNOWLEDGEMENTS

This work was supported by grant NS15047 from the National Institute of Neurological and Communicative Disorders and Stroke of the United States Public Health Service and a grant from the American Muscular Dystrophy Association.

REFERENCES

Anderson DC, Bahr BA, Parsons SM(1986)Stoichiometries of acetylcholine uptake, release and drug inhibition in Torpedo synaptic vesicles:Heterogeneity in acetylcholine transport and storage.J Neurochem 46:1207-1213

Anderson DC, King SC, Parsons SM(1982)Proton gradient linkage to active uptake of [^{3}H]acetylcholine by Torpedo electric organ synaptic vesicles.Biochemistry 21:3037-3043

Anderson DC, King SC, Parsons SM(1983)Pharmacological characterization of the acetylcholine transport system in purified Torpedo electric organ synaptic vesicles.Molec Pharmacol 24:48-54

Bahr BA, Parsons SM(1986a)Demonstration of a receptor in Torpedo synaptic vesicles for acetylcholine storage blocker l-trans-2-(4-phenyl[3,4-3H]piperidino)cyclohexanol.Proc Nat Acad Sci USA 83:2267-2270

Bahr BA, Parsons SM(1986b)Acetylcholine transport and drug inhibition kinetics in Torpedo synaptic vesicles.J Neurochem 46:1214-1218

Buckley D, Kelly RB(1985)Identification of a transmembrane glycoprotein specific for secretory vesicles of neural and endocrine cells.J Cell Biol 100:1284-1294

Edwards C, Dolezal V, Tucek S, Zemkova H, Vyskocil, F(1985)Is an acetylcholine transport system responsible for nonquantal release of acetylcholine at the rodent myoneural junction?Proc Natl Acad Sci USA 82:3514-3518

Jahn R, Schiebler W, Greengard P(1984)A quantitative dot-immunobinding assay for proteins using nitrocellulose membrane filters.Proc Natl Acad Sci USA 81:1684-1687

Marien MR, Parsons SM, Altar CA(1987)Quantitative autoradiography of brain binding sites for the vesicular acetylcholine transport blocker 2-(4-phenylpiperidino)cyclohexanol (AH5183). Pro Acad Sci USA 84:876-880

Marshall IG(1970)Studies on the blocking action of 2-(4-phenylpiperidino)cyclohexanol(AH5183).Br J Pharmac 38:503-516

Marshall IG, Parsons SM(1987)The vesicular acetylcholine transport system and its pharmacology.Trends in Neurosci 10:174-177

Zimmermann H, Denston DR(1977)Recycling of synaptic vesicles in the cholinergic synapses of the Torpedo electric organ during induced transmitter release.Neuroscience 2:695-714

SYNAPTIC VESICLE PROTEINS

Walter Volknandt, Andreas Henkel and Herbert Zimmermann
AK Neurochemie, Zoologisches Institut
der J.W. Goethe-Universität
6000 Frankfurt am Main
Federal Republic of Germany

INTRODUCTION

Most nerve terminals contain various types of vesicles which differ with regard to their appearance in the electron-microscope and presumably also their molecular properties. Profiles of cholinergic nerve terminals from e.g. motor neurons or parasympathetic fibres contain in addition to the numerous electron-lucent vesicles with a diameter of about 45 nm also few dense-cored vesicles with a diameter of about 80 - 150 nm. In addition, such nerve terminals may contain coated vesicles.

Electron-lucent cholinergic synaptic vesicles, particularly those derived from electric ray electric organs have been characterized in some molecular detail: Besides acetylcholine (ACh) they contain ATP as well as a proteoglycan with glycosaminoglycan side chains of the heparan sulfate type. Among functional properties they possess a proton-pumping ATPase which acidifies the vesicle interior, and uptake systems for ACh, ATP and Ca^{2+} (ref. Volknandt and Zimmermann 1986; Stelzl et al 1987). The general molecular composition of dense-cored cholinergic vesicles is not known. However, subcellular fractionation studies on the feline submandibular gland and the guinea pig myenteric plexus suggest that cholinergic dense-cored vesicles contain the peptide VIP, possibly together with the neurotransmitter ACh (ref. Ágoston et al 1985). Coated vesicles as characterized by their coat of clathrin molecules have been characterized from brain tissue (ref. Wileman et al 1985). They are involved in synaptic vesicle recycling following exocytosis and also in receptor mediated endocytosis and uptake of extracellular molecules for retrograde transport to the cell body.

NATO ASI Series, Vol. H21
Cellular and Molecular Basis of Synaptic Transmission
Edited by H. Zimmermann

This diversity of synaptic vesicle types within one and the same nerve terminal raises a number of questions: What are the contents and membrane properties of the various vesicle types? Are there common molecular properties between the different vesicle types, within the same nerve terminal or between different transmitter types? Are the molecular properties of synaptic vesicles conserved during evolution? Do, e.g. vesicles from mammalian cholinergic nerve terminals possess the same characteristics as those described for the Torpedo electric organ?

CHROMOGRANIN A IS ASSOCIATED WITH CHOLINERGIC NERVE TERMINALS

Chromogranin A is the most abundant soluble protein of chromaffin granules. This acidic glycoprotein has recently been demonstrated also in dense-cored vesicles of adrenergic axons and in various endocrine tissues like the anterior pituitary, parathyroid gland, thyroid gland or endocrine pancreas (ref. Volknandt et al 1987b). Using polyclonal antibodies against chromogranin A isolated from either bovine or rat adrenal medulla we could label motor endplates in the rat diaphragm by means of FITC-immunofluorescence (Fig. 1) (Volknandt et al 1987b). This

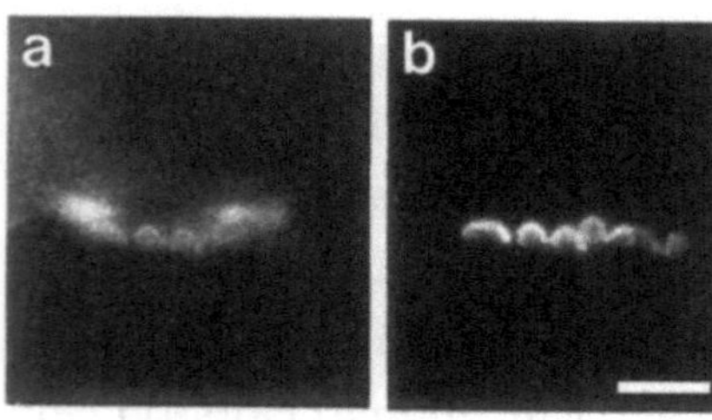

Fig. 1. Staining of the motor endplate in the rat diaphragm using an antiserum raised against chromogranin A isolated from bovine adrenal medulla. Immunofluorescence using FITC-labelled second antibody (a) or binding of rhodamine-conjugated α-bungarotoxin (b). Bar = 10 μm. (Methods in Volknandt et al 1987b).

fluorescence coincides with that of rhodamine labelled α-bungarotoxin (postsynaptic nicotinic ACh-receptors) suggesting that only the nerve terminals are labelled. Our observation is in accord with that of Somogyi et al (1984) that cholinergic motor neurons in the rat spinal cord are chromogranin A immunoreactive. Since cholinergic motor nerve terminals in the rat diaphragm contain a low percentage (5 - 10%) of dense-cored vesicles, our finding would be conceivable with the idea that these vesicles contain chromogranin A. Whether this protein may also be released on stimulation of the nerve terminals (as it is from chromaffin cells or adrenergic nerve terminals) needs to be elucidated.

COMMON MOLECULAR PROPERTIES OF SYNAPTIC VESICLES FROM TORPEDO TO RAT

Cholinergic synaptic vesicles of the electron-lucent type are abundant in the electric organ of the electric rays (cartilagenous fish) and the basic knowledge on the functional and molecular properties of this vesicle type is derived from this source only. We have now isolated cholinergic vesicles from animals of higher evolutionary state like the electric eel (Electrophorus electricus), electric catfish (Malapterurus electricus) (both bony fish) as well as the rat (diaphragm). For vesicle isolation density gradient centrifugation was followed by chromatography on columns of Sephacryl-1000 (Volknandt and Zimmermann 1986). Synaptic vesicles purified from all three tissues contain ATP in addition to ACh. The molar ratios of ACh/ATP are around 10 and thus somewhat higher than reported for the Torpedo electric organ (5). The notion of ATP as a general constituent of cholinergic synaptic vesicles is further supported by the detection of ATP in cholinergic vesicles isolated from immunoaffinity purified cholinergic nerve terminals of guinea pig brain (molar ratio 7) (Richardson and Brown 1987). The corelease of ACh and ATP observed at central and peripheral cholinergic synapses as well as the presence of ectonucleotidases at synaptic membranes suggest a neuroactive role for ATP at the cholinergic synapse (Grondal and Zimmermann this volume).

Another core constituent common to cholinergic synaptic vesicles from Torpedo to the rat diaphragm is a proteoglycan of the heparan sulfate type (Volknandt and Zimmermann 1986). The proteoglycan was identified by means of polyclonal antibodies (dot blotting) raised against vesicle-proteoglycan from either Torpedo electric organ or pig brain (gift of Dr. Herbert Stadler, Göttingen). In immunofluorescence studies on the electric organs of either Torpedo, Electrophorus, Malapterurus or the rat diaphragm antibody binding was restricted to the nerve terminal region. The functional role of the vesicular proteoglycan is not yet clear. Its function may be in the balance of charge on storage of the transmitter amine or of Ca^{2+}, but also in vesicle recycling (stabilization of the patch of vesicle membrane for exocytosis) or in transcellular signalling. Although under in vitro conditions the proteoglycan is tightly bound to the luminal side of the vesicle membrane (Kuhn et al 1988) it may be released under physiological conditions of vesicle exocytosis and thus become deposited in the extracellular matrix. There it may serve as a signal released from the nerve cell to specify the synaptic contact (Stadler and Kiene 1987).

SYNAPSIN I IN TORPEDO SYNAPTIC VESICLES

Vesicle-surface associated proteins may be involved in the interaction of vesicles with the cytoskeleton (e.g. kinesin, ref. Pratt 1987) or in the regulation of transmitter release. This latter mechanism has been proposed for the phosphoprotein synapsin I (Llinas et al 1985). Synapsin I is exclusively present in the nervous system and is associated with the surface of synaptic vesicles at the great majority of synapses (De Camilli et al 1983).

Using an affinity-purified monospecific polyclonal antibody against bovine brain synapsin I we could demonstrate that the protein is specifically associated with the nerve terminals in the electric organs of Torpedo, Electrophorus and Malapterurus (Volknandt et al 1987) (Fig. 2). On subcellular fractionation in media of low ionic strength (0.8 M glycine) synapsin I copurifies with synaptic vesicles from Torpedo electric organ on density gradient centrifugation and also on subsequent chromatography on a column of Sephacryl-1000. Immunotransfer blots of tissue homogenates of the electric organs of the three strongly electric fish reveal molecular weights (Mr 72 - 80,000) similar to that of synapsin I from the cerebral cortex or the diaphragm of the rat (72,000, 75,000) (Fig. 2). The synapsin I band of

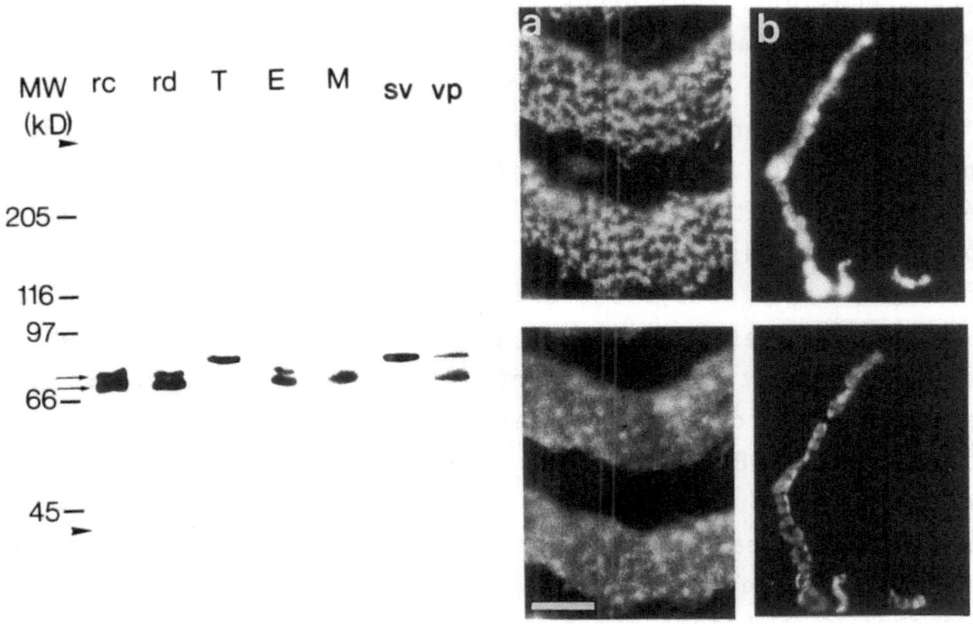

Fig. 2. Immunotransfer blots and staining of nerve terminal regions with anti-synapsin I antibody. Immunotransfer blots of tissue homogenates of rat cortex (rc); rat diaphragm (rd); and electric organs of Torpedo marmorata (T), Electrophorus electricus (E), Malapterurus electricus (M) and of subcellular fractions obtained from Torpedo electric organ: sv, synaptic vesicle fraction purified by density gradient centrifugation; vp, further purification of vesicle fraction by chromatography on Sephacryl-1000. FITC-Immunofluroescence of nerve terminals in electric organ of Torpedo (a) and Electrophorus (b) (upper panel). Identical images with binding of rhodamine-conjugated α-bungarotoxin (lower panel). Bar = 10 μm. (Methods in Volknandt et al 1987a).

Mr 80,000 observed in homogenates of Torpedo electric organ is also associated with vesicles isolated by sucrose density gradient centrifugation and column chromatography. However, after chromatography, an additional band of Mr 72,000 becomes apparent.

Our results suggest that synapsin I is associated with a large variety of synaptic vesicles including the cholinergic synaptic vesicles from electric rays. The Torpedo electric organ may thus be an ideal source to further study the functional and molecular properties of synapsin I.

HETEROGENOUS DISTRIBUTION OF MEMBRANE PROTEINS IN VESICLE POLULATIONS OF RAT BRAIN

Using monoclonal antibodies against two identified brain synaptic vesicle membrane proteins we could reveal both, preferential association of these proteins with certain vesicle types as well as differences in their protein composition (Volknandt et al 1988). One antibody was directed against a rat brain synaptic vesicle membrane protein of Mr 65,000 (p65) which has been shown to bind to many types of nerve terminals in brain and also to endocrine cells (Matthew et al 1981; Bixby and Reichardt 1985). The other antibody was directed against a membrane-bound glycoprotein of 38,000 (p38, synaptophysin) which occurs in many types of nerve terminals but also in neuroendocrine and endocrine cells. To this protein which is likely to occur as a homotetramer, Ca^{2+}-binding activity has been ascribed (Wiedenmann et al 1986; Rehm et al 1986).

Synaptic vesicles were isolated from rat brain by sucrose density gradient centrifugation (harvested on top of the 0.4 M sucrose layer) and further separated by chromatography on Sephacryl-1000 which allows the separation of subcellular particles up to a size of 300 nm (Fig. 3). Whereas p65 is distributed parallel to ATP in a broad peak (size range of all vesicles), p38 is preferentially associated with vesicles of smaller size (peak at 104 ml). Synapsin I which has been in-

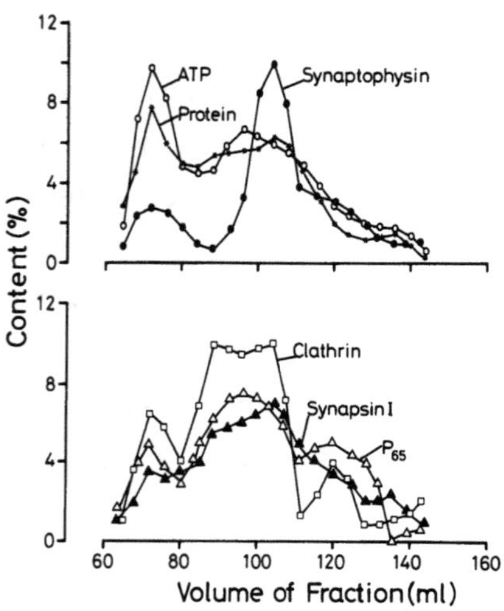

Fig. 3. Additional fractionation of a synaptic vesicle fraction derived from rat cerebral cortex using chromatography on Sephacryl-1000. Contents in protein and ATP as well as binding activities for clathrin, synapsin I, synaptophysin and protein 65 are expressed as percentage of recovered material (means of three independent experiments). (From Volknandt et al 1988).

cluded in the investigation has also a broad distribution but its main peak is with vesicles of the smaller size.

Using polyacrylamide beads for solid matrix immunoprecipitation of synaptic vesicles significant differences could be revealed in the protein composition of the vesicle population bound by the two antibodies (Fig. 4). Vesicles immunoprecipitated by the anti-p38-antibody were 3.2 fold enriched in ACh (molar ratio of ACh/ATP = 8.3). The protein pattern of vesicles

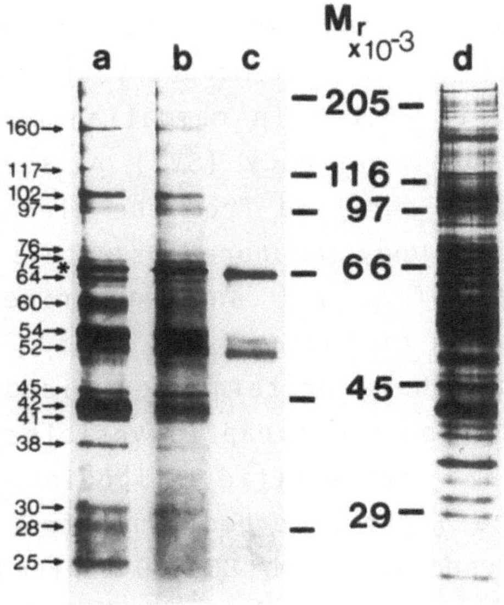

Fig. 4. Silver staining of synaptic vesicle polypeptides from rat cerebral cortex subjected to SDS PAGE. Vesicles immunoprecipitated by a) anti-synaptophysin antibody and b) asv48 antibody. c) material precipitated by control beads; d) polypeptide composition of parent vesicle fraction (purified by density gradient centrifugation). Molecular weights of standards are indicated by bars and the molecular mass of polypeptides by arrows. *, Mr of BSA. (From Volknandt et al 1988).

precipitated with anti-p65 differs from those precipitated with anti-p38. Proteins in the low molecular weight range (25,000 to 38,000) but also at higher molecular weight (60,000 and 64,000) are less apparent or absent in p65 vesicles. On immunoblot analysis synapsin I can be detected in synaptic vesicles precipitated by either antibody. p38 is higly enriched in anti-p38 immunoprecipitates but there is only a faint band with anti-p65 immunoprecipitated vesicles. The monoclonal antibody directed against p65 could not be used in immunoblotting.

Our results suggest that there are vesicle populations containing p65 but only little, if any, p38. These are likely to be of the larger vesicle type. p38 is associated with populations of smaller vesicles. Due to the failure of the immunoblot analysis for p65 we cannot decide whether p65 is also associated with the small p38 positive vesicles.

A V.ESICLE-SPECIFIC MEMBRANE-SPANNING GLYCOPROTEIN

Walker et al (1986) described a vesicle-specific polypeptide of Mr 86,000 in rat brain fractions which was recognized selectively by polyclonal antisera raised against isolated synaptic vesicles derived from the Torpedo electric organ. They showed that the antigen is associated with synaptic vesicles and demonstrated by immunocytochemical techniques the occurence of the antigen in the diaphragm, cerebellum, hippocampus and cerebral cortex of the rat. Similar findings were reported by Buckley and Kelly (1985) who identified a transmembrane vesicle-specific glycoprotein of Mr 75,000 to 85,000 in mammalian and avian brains by means of a monoclonal antibody (SV2) raised against cholinergic synaptic vesicles purified from the electric ray Discopyge ommata. The SV2 antibody immunoreacts with a glycoprotein of secretory vesicles of neural and endocrine cells indicating the wide distribution of the antigen.

In order to further characterize these proteins we raised two polyclonal antisera against cholinergic synaptic vesicles from Torpedo electric organ. To obtain monospecific antibodies, IgG fractions of the antisera were immunoadsorbed on nitrocellulose to a synaptic vesicle-polypeptide of Mr 86,000 derived from the bovine cerebral cortex. The monoclonal antibody SV2 was a gift of Dr. R.B. Kelly (San Francisco).

Comparison of our antibody and the SV2 antibody by immunocytochemistry revealed that the respective antigens are specifically associated with the electromotor nerve terminals of the electric organs of electric ray (Torpedo marmorata), electric eel (Electrophorus electricus) and electric catfish (Malapterurus electricus) (Fig. 5). Moreover, in indirect FITC-immunofluorescence both antibodies stain the peripheral motor nerve terminals in Torpedo dorsal muscle cells, the inner plexiform layer of the Torpedo retina and the adrenal medulla of rat.

On immunotransferblotting our antibody recognizes a polypeptide of about Mr 86,000 in a variety of tissues and species like rat and bovine cerebral cortex, rat cerebellum, locust brain (Locusta migratoria), electric organ of electric eel and catfish, rat diaphragm and rat adrenal medulla. Similar immuno-

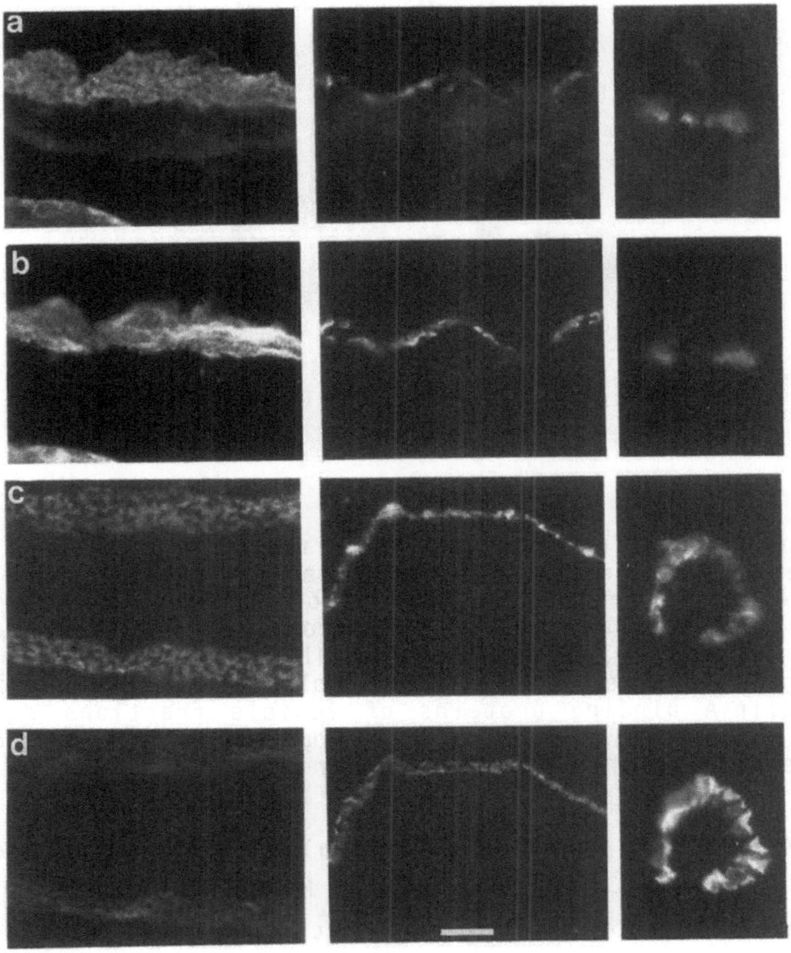

Fig. 5. Staining of nerve terminal regions in electric organs of Torpedo (left), Electrophorus (middle) and Malapterurus (right). a) Right and middle: indirect immunfluorescence using polyclonal IgG-fractions obtained from Torpedo vesicles purified on DEAE AFFI-GEL BLUE; left: indirect immunofluorescence using IgGs immuno-affinity purified with ox cortex vesicle protein. c) Indirect immunofluorescence using SV2-monoclonal antibody. b, d) Corresponding images revealing binding sites for rhodamine-conjugated α-bungarotoxin. Bar = 10 μm.

reactions were obtained with the SV2 antibody. Furthermore, both antibodies revealed an identical pattern on immunostaining of synaptic vesicle proteins purified from rat and bovine cerebral cortices (Fig. 6, a - d).

In Torpedo synaptic vesicles the SV2 monoclonal antibody preferentially recognizes a glycoprotein of higher molecular

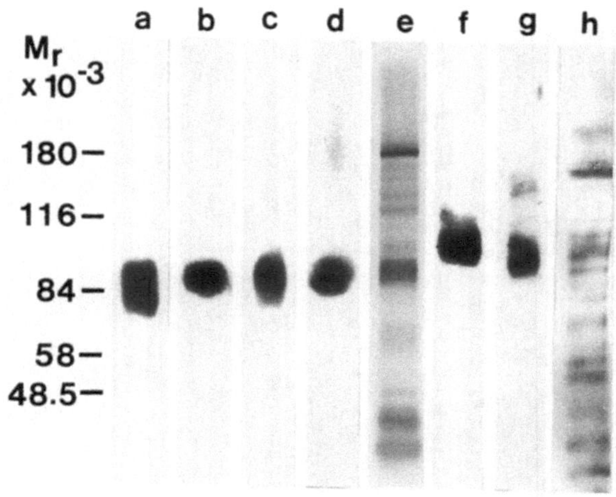

Fig. 6. Immunotransferblots and concanavalin A binding of synaptic vesicle proteins from mammalian cerebral cortex and Torpedo electric organ. Immunoblots of vesicles derived from rat (a, b) and bovine (c, d) cerebral cortices and the electric organ of Torpedo (f, g) using SV2 antibody (a, c, f) or the immunoaffinity-purified monospecific polyclonal IgGs (b, d, g). Concanavalin A binding proteins of vesicle fractions derived from bovine cerebral cortex (e) and from Torpedo electric organ (h). Molecular weights of standards are indicated by bars.

weight (Mr 100,000) than our polyclonal antibody which preferentially recognizes a glycoprotein of Mr 90,000 (Fig. 6, f - g). However, both antibodies also crossreact to a small extend with the respective higher and lower molecular weight component (compare Buckley and Kelly 1985). Thus, in Torpedo there might be two glycoproteins very similar in their protein core but different in the degree of glycosylation.

Concanavalin A binding to electrophoretically transfered proteins derived from brain vesicles and Torpedo vesicles revealed that the proteins recognized by the antibodies are glycoproteins (Fig. 6, e, h). Furthermore, after elution from a concanavalin A sepharose column by α-methylmannoside these proteins can be immunostained with the two antibodies. For both vesicle types the concanavalin A-peroxidase reaction product reveals a broad electrophoretic mobility of the glycoproteins similar to that obtained on immunostaining (compare Fig. 6). This is probably due to the higly glycosidic nature of these proteins.

Preliminary studies of the antigens derived either from brain or electric organ synaptic vesicles using 2-dimensional immunostaining with both antibodies revealed a very similar pI (around 5). On solubilisation and phase separation by nonionic detergents the two antigens display a similar behaviour. Moreover, the protein isolated from a synaptic vesicle fraction of bovine cerebral cortex using an IgG-fraction of our antisera coupled to a protein A column was also recognized by the SV2 antibody. This is strong evidence that in mammalian brain both antibodies recognize the same polypeptide.

TABLE 1. Location of SV2 and p86 antibody-binding sites in isolated vesicles

Source	Trypsin		Collagenase		Neuroaminidase ß-Endoglucosidase Mannosidase		Trypsin NP40	
	SV2	P86	SV2	P86	SV2	P86	SV2	P86
	(antibody binding activity, % of control)							
Torpedo electric organ	2.7 ±0.9	50.7 ±3.8	85.0 ±21.4	120.5 ±12.5	81.3 ±18.6	108.2 ±16.8	3.8 ±1.7	6.7 ±3.0
Ox cerebral cortex	5.2 ±2.5	86.4 ±3.2	128.1	100.3	96.6	90.6	0.9 ±0.2	14.2 ±7.1

Vesicles isolated by density gradient centrifugation were incubated at 4°C for 30 min and centrifuged at 220,000 g_{av} for 1 h. The pellet obtained was analyzed by immunodotting. Vesicles corresponding to 100 µg of protein were incubated in 2.0 ml of isotonic solution (control = 100%) to which the following substances were added: trypsin (0.7 U), collagenase (3.2 U), neuraminidase (3.6 U), ß-endoglucodiase H (0.1 U), mannosidase (0.1 U), NP40 (0.1%). Values represent means of ± S.E.M. of 4 experiments (Torpedo) or means of ± range (2 experiments) ox, or single experiments using triplicate determinations.

By immunodotting of synaptic vesicles incubated with proteases we found a major difference in the location of the antigenic determinants. Addition of trypsin to intact vesicles led to a complete loss of SV2 immunoreactivity suggesting that the epitope is accessible to the enzyme and must face the cytoplasmic side (compare Buckley and Kelly 1985). In contrast the antigenic determinants of our monospecific IgGs are likely to face both the cytoplasmic and the luminal side of vesicles (Table 1). Addition of collagenase or sialic acid and sugar cleaving enzymes to intact vesicles had no effect on the electrophoretic mobility of the respective antigens.

SUMMARY AND DISCUSSION

Whereas some molecular properties may be shared between synaptic vesicle types others may differ. Vesicle types may differ within the same transmitter type and nerve terminal (e.g. electron-lucent versus dense-cored) but also between transmitter types. Cholinergic vesicles of the electron-lucent type ubiquitously contain in their core ACh, ATP and a proteoglycan of the heparan sulfate type. Cholinergic dense cored vesicles may contain chromogranin A. Brain synaptic vesicles may differ with regard to their outfit with membrane proteins. Whereas a protein of Mr 38,000 may be generally associated with all vesicles of the small electron-lucent type it appears to be absent from large (dense-cored?) synaptic vesicles. Protein 65 is associated with synaptic vesicle populations of larger diameter and may in addition occur in the small vesicle type. Synapsin I as a surface associated vesicle protein appears to be bound to vesicles of a larger size range. A glycoprotein with a relative molecular mass of 90,000 - 100,000 is contained both in the membrane of Torpedo and bovine brain synaptic vesicles. It is recognized both by a monoclonal and a polyclonal antibody. In Torpedo the molecular weight range recognized after polyacrylamide gel electrophoreses differs between the two antibodies.

The application of antibodies directed against epitopes exposed to the surface of membrane associated vesicle proteins offers a possibility to separate brain synaptic vesicle populations and to study their individual functional and molecular properties. Further studies will reveal which proteins are specific to individual populations or groups of populations of brain vesicles and thus may be involved in a function specific to these vesicles (e.g. uptake of the particular neurotransmitter, electron- or proton-transport). Other proteins may be ubiquitous and thus are likely to be of relevance to the function of many or all vesicles.

REFERENCES

Ágoston DV, Ballmann M, Coulon JM, Dowe GCH, Whittaker VP (1985) Isolation of neuropeptide-containing vesicles from the guinea pig ileum. J Neurochem 45:398-406

Bixby JL, Reichardt LF (1985) The expression and localization of synaptic vesicle antigens at neuromuscular junctions in vitro. J Neurosci 5:3070-3080

Buckley K, Kelly RB (1985) Identification of a transmembrane glycoprotein for secretory vesicles of neural and endocrine cells. J Cell Biol 100:1284-1294

De Camilli P, Harris SM, Huttner WB, Greengard P (1983) Synapsin I (protein I), a nerve terminal-specific phosphoprotein. II. Its specific association with synaptic vesicles demonstrated by immunocytochemistry in agarose-embedded synaptosomes. J Cell Biol 96:1355-1373

Kuhn DM, Volknandt W, Stadler H, Zimmermann H (1987) Cholinergic vesicle specific proteoglycan. Stability in isolated vesicles and in synaptosomes during induced transmitter release. J Neurochem, in press

Llinás R, McGuiness TL, Leonard CS, Sugimori M, Greengard P (1985) Intraterminal injection of synapsin I or calcium/calmodulin-dependent protein kinase II alters neurotransmitter release at the squid giant synapse. Proc Natl Acad Sci USA 82:3035-3039

Matthew D, Tsavaler L, Reichardt LF (1981) Identification of a synaptic vesicle-specific membrane protein with a wide distribution in neuronal and secretory tissue. J Cell Biol 91: 257-269

Pratt MM (1986) Stable complexes of axoplasmic vesicles and microtubules: protein composition and ATPase activity. J Cell Biol 103:957-968

Rehm H, Wiedenmann B, Betz H (1986) Molecular characterization of synaptophysin, a major calcium-binding protein of the synaptic vesicle membrane. EMBO J 5:535-541

Richardson PJ, Brown SJ (1987) ATP release from affinity-purified rat cholinergic nerve terminals. J Neurochem 48:622-630

Somogyi P, Hodgson AJ, De Potter RW, Fischer-Colbrie R, Schober M, Winkler H, Chubb IW (1984) Chromogranin immunoreactivity in the central nervous system. Immunochemical characterization, distribution and relationship to catecholamine and enkephalin pathways. Brain Res Rev 8:192-230

Stadler H, Kiene M-L (1987) Synaptic vesicles in electromotoneurones. II. Heterogeneity of populations is expressed in uptake properties; exocytosis and insertion of a core proteoglycan into the extracellular matrix. EMBO J 6:2217-2221

Stelzl H, Grondal EJM, Zimmermann H (1987) Ca^{2+} dependency and substrate specificity of cholinergic synaptic vesicle ATPase. Neurochem Int 11:107-111

Volknandt W, Zimmermann H (1986) Acetylcholine, ATP, and proteoglycan are common to synaptic vesicles isolated from the electric organs of electric eel and electric catfish as well as from rat diaphragm. J Neurochem 47:1449-1462

Volknandt W, Naito S, Ueda T, Zimmermann H (1987a) Synapsin I is associated with cholinergic nerve terminals in the electric organs of Torpedo, Electrophorus, and Malapterurus and copurifies with Torpedo synaptic vesicles. J Neurochem 49: 342-347

Volknandt W, Schober M, Fischer-Colbrie R, Zimmermann H, Winkler H (1987b) Cholinergic nerve terminals in the rat diaphragm are chromogranin A immunoreactive. Neurosci Lett 81:241-244

Volknandt W, Henkel A, Zimmermann H (1988) Heterogenous distribution of synaptophysin and protein 65 in synaptic vesicles isolated from rat cerebral cortex. Neurochem Int, in press

Walker JH, Kristjansson GI, Stadler H (1986) Identification of a synaptic vesicle antigen (Mr 86,000) conserved between Torpedo and rat. J Neurochem 46:875-881

Wiedenmann B, Franke WW, Kuhn C, Moll R, Gould VE (1986) Synaptophysin: a marker protein for neuroendocrine cells and neoplasms. Proc Natl Acad Sci USA 83:3500-3504

Wileman T, Harding C, Stahl P (1985) Receptor-mediated endocytosis. Biochem J 232:1-14

SYNAPTOPHYSIN: A MAJOR CELL TYPE-SPECIFIC VESICLE PROTEIN OF NEUROENDOCRINE CELLS

Bertram Wiedenmann and Werner W. Franke
Med. Univ. Klinik, Department of Medicine, University of Heidelberg, Bergheimer Str. 58, 6900 Heidelberg, and Institute of Cell and Tumor Biology, German Cancer Research Center, Im Neuenheimer Feld 280, 6900 Heidelberg, FRG

In recent years, three integral membrane proteins, which appear to be specifically associated with small (40-80 nm) vesicles with electron microscopically translucent contents have been detected in various neuroendocrine (NE) cells, primarily by the use of monoclonal antibodies. On SDS-polyacrylamide gel electrophoresis (PAGE) of proteins from brain tissue homogenates, these molecules appear with relative mobilities corresponding to M_r 95,000 (protein termed SV2; Buckley and Kelly, 1985), M_r 65,000 (Matthew et al., 1981) and M_r 38,000 (termed synaptophysin; Jahn et al., 1985; Wiedenmann and Franke, 1985). Particular progress has recently been made in our understanding of the molecular and cellular topology and distribution of synaptophysin.

Synaptophysin is a major integral membrane protein of a certain type of small neuroendocrine (NE) vesicles, originally isolated from - and identified in - presynaptic vesicles of neurons (Jahn et al., 1985; Wiedenmann and Franke, 1985). This protein accounts for approximately 2 % of the total protein of synaptic vesicles purified from rat brain (Rehm et al., 1986). The protein also occurs in similar vesicles present in a broad spectrum of non-neuronal neuroendocrine cells (Wiedenmann and Franke, 1985; Navone et al., 1986; Wiedenmann et al., 1986). On SDS-PAGE, varying molecular weights, ranging from M_r 38,000 to 42,000, depending on the specific cell type, have been reported (Jahn et al., 1985; Wiedenmann and Franke, 1985; Rehm et al., 1986; Wiedenmann et al., 1986; Lee et al., 1987), such variations have been tentatively ascribed to different extents of glycosylation (Navone et al., 1986; Rehm et al., 1986) since in all cases examined chemical deglycosylation and inhibition of glycosylation reduced the M_r value of synaptophysin to approx. 34,000. Biochemical analyses have shown that synaptophysin can be readily extracted and solubilized with

NATO ASI Series, Vol. H21
Cellular and Molecular Basis of Synaptic Transmission
Edited by H. Zimmermann

detergents (Jahn et al., 1985; Wiedenmann and Franke, 1985), and the solubilized protein has been purified by immunoaffinity chromatography (Wiedenmann and Franke, 1985) and DEAE column chromatography (Rehm et al., 1986). Chromatographically purified synaptophysin has been identified as an acidic glycoprotein (pI ~ 4.8; Wiedenmann and Franke, 1985). Biochemical and biophysical studies on the purified molecule show further that synaptophysin forms an oligomeric disulfide bridges containing particles (Wiedenmann and Franke, 1985; Rehm et al., 1986). Recent cDNA sequencing data and previous immunological and proteolytic studies have shown that synaptophysin contains a relatively large cytoplasmic, i.e. extravesicular domain of approx. M_r 10,000 which contains the epitopes of the monoclonal antibodies characterized so far and the Ca^{++}-binding site. Furthermore, the cytoplasmic domain is also sensitive to proteolysis with various proteases, including collagenase (Jahn et al., 1985; Wiedenmann and Franke, 1985; Rehm et al., 1986; Buckley et al., 1985; Leube et al., 1985; Südhof et al., 1987).

Both the high immunological cross-reactivity between different species (e.g. man, rat, cow and frog; Jahn et al., 1985; Wiedenmann and Franke, 1985; Wiedenmann et al., 1986; Jahn et al., 1987) and the observed cross-hybridization of the approx. 2.5 kb synaptophysin mRNAs from different mammals (e.g. Buckley et al., 1987; Leube et al., 1987; Südhoff et al., 1987), indicate that synaptophysin is a protein highly conserved during evolution. Furthermore, synaptophsin has been identified in a wide variety of normal and neoplastic NE cells and tissues of both epithelial and neuronal origin, by immunohistochemistry, immunoblotting and "Northern blotting" of RNA, demonstrating that this protein is continually expressed during malignant transformation, which makes it a good candidate for monitoring the differentiated state of neuroendocrine tumor cells (Wiedenmann et al., 1986, 1987; Gould et al., 1987). The three groups which have sequenced synaptophysin have found one potential N-glycosylation site in a putatively glycosylation-accessible positions, i.e. at the vesicular lumen. Whether the additional N-glycosylation site emphasized by Buckley et al. (1987) is potentially glycosylated is questionable. The calculated total molecular weight of synaptophysin as derived from cDNA sequence analysis is 33,300 (Leube et al., 1987; Südhof et al., 1987) which corresponds well with the M_r value of 34,000 determined for the non-glycosylated synaptophysin by SDS-PAGE (Navone et al., 1986; Rehm et al., 1986). The hydrophobicity profile of the primary sequence reveals four major hydrophobic domains, the putative membrane-spanning regions, which are flanked by clusters of charged amino acids whereas the carboxyterminal domain of approximately 90 amino acids is hydrophilic and has a relatively high glycine and proline content (Buckley et al., 1987; Leube et al., 1987; Südhof et al., 1987). This domain also contains several successive collagen-resembling motifs (Leube et al., 1987; Südhof et al., 1987).

So far, only very few data exist on the synthesis processing and intracellular transport of synaptophysin. Recently, Tixier-Vidal et al. (1986) found synaptophysin in small (30-100 nm), electron microscopically clear-appearing vesicles near the trans-cisternae of the Golgi apparatus of undifferentiated neurons. After synapse formation, a similar trans-Golgi staining was only observed after drug-induced interruption of axonal transport. This has been interpreted to show that newly synthesized, glycosylated synaptophysin is inserted into a specific kind of vesicles. The localisation studies of Tixier-Vidal et al. (1988) suggest also that a subpopulation of the synaptophysin-containing vesicles contains a protein coat similar to that of coated vesicles, and some synaptophysin has also been reported in purified coated vesicles, suggesting that, after neurotransmitter release, retrieval of synaptic vesicle membrane components may be mediated by endocytosis in coated pits and vesicles (Pfeffer and Kelly, 1985; Wiedenmann et al., 1985). Future studies of vesicular and synaptophysin processing and transport will, hopefully, contribute to our understanding of the formation, function and turnover of synaptophysin and the vesicles that contain it.

Table 1: Synaptophysin-containig Cells and Tissues

Differentiation	Normal cell types	Tumors (incl. metastases and cell lines
Neuronal	Neurons of the central and peripheral nervous system[7,18]	Neuroblastoma[5,15,21] Medulloblastoma[5,15,21] Ganglioneuroblastoma[20] Ganglioneuroma[5] Paraganglioma[20]
	Pituitary gland[12]	Pituitary adenoma[5]
	Retina/inner and outer plexiform layers[18]	Retinoblastoma[ND]
	Chromaffin cells[18]	Pheochromocytoma[5,11,20] Pheochromocytoma (rat) cell line PC12[11,14]
Epithelial	Kuschiltzky cells of bronchial tract[10]	Bronchial carcinoids[5,20] NE carcinoma of the lung[8,20] a) well differentiated[5,10,14] b) intermediate[5,10,14] c) small cell type[2,5,10,14] Human bronchial carcinoma cell lines NCI-H69, NCI-H82, SCLC-22H, SCLC-16H[11]
	C-cells of thyroid[ND]	Medullary thyroid carcinoma[5,20,21]
	Enterochromaffin cells of small and large intestine[22]	Carcinoids NE carcinoma of small and large bowel[20-22]
	G-cells of the gastric antrum[22]	G-cell hyperplasia[22] Gastrinoma[22]
	Pancreatic islet cells[20]	Islet cell adenoma[20] Islet cell carcinoma[3,20-22]
	Parathyroid cells[ND]	Parathyroid adenoma[5] Parathyroid carcinoma[ND]
	Merkel cells of the skin[ND]	Merkel cell tumors[20]
	NE cells of the thymus[ND]	Thymoma[6] NE carcinoma of the thymus[6]

ND, not determined

References

1. Buckley, K., Kelly, R.B. (1985): Identification of a transmembrane glycoprotein specific for secretory vesicles of neural and endocrine cells. J. Cell Biol. 100:1284-1294.

2. Buckley, K.M., Floor, E., Kelly, R.B. (1987): Cloning and sequence analysis of cDNA encoding p38, a major synaptic vesicle protein. J. Cell Biol. 105:2447-2456.

3. Chejfec, G., Falkmer, S., Grimelius, L., Jacobsson, B., Rodensjö, M., Wiedenmann, B., Franke, W.W., Gould, V.E. (1987): Synaptophysin, a new marker for pancreatic neuroendocrine tumors. Am. J. Surg. Pathol. 11:241-247.

4. Franke, W.W., Grund, C., Achtstätter, T. (1986): Co-expression of cytokeratins and neurofilament proteins in a permanent cell line: cultured rat PC12 cells combine neuronal and epithelial features. J. Cell Biol. 103:1933-1943.

5. Gould, V.E., Wiedenmann, B., Lee, I., Schwechheimer, K., Dockhorn-Dworniczak, B., Radosevich, J.A., Moll, R., Franke, W.W. (1987): Synaptophysin expression in neuroendocrine neoplasms as determined by immunocytochemistry. Am. J. Pathol. 126:243-257.

6. Hofmann, W.J., Waldherr, R., Eberlein-Gonska, M., Wiedenmann, B., Otto, H.F.: Immunohistochemical examination of carcinoids of the thymus and thymomas with neuroendocrine markers. in preparation.

7. Jahn, B., Schiebler, W., Ouimet, C., Greengard, P. (1985): A 38 000 dalton membrane protein (p38) present in synaptic vesicles. Proc. Natl. Acad. Sci. USA 82:4137-4141.

8. Kayser, W., Wiedenmann, B. (1988): Neuroendocrine markers in bronchial carcinomas. A comparison of synaptophysin neuron-specific enolase and bombesin. Pathol. Res. Pract., in press.

9. Knaus, P., Betz, H., Rehm, H. (1986): Expression of synaptophysin during postnatal development of the mouse brain. J. Neurochem. 47:1302-1304.

10. Lee, I., Gould, V.E., Moll, R., Wiedenmann, B., Franke, W.W. (1987): Synaptophysin expressed in the broncho-pulmonary tract: Neuroendocrine cells, neuroepithelial bodies and neuroendocrine neoplasms. Differentiation 34:115-125.

11. Leube, R.E., Kaiser, P., Seiter, A., Zimbelmann, R., Franke, W.W., Rehm, H., Knaus, P., Prior, P., Betz, H., Reinke, H., Beyreuther, K., Wiedenmann, B. (1987): Synaptophysin: molecular organization and mRNA expression as determined from cloned cDNA. EMBO J. 6:3261-3268.

12. Navone, F., Jahn, R., Di Gioia, G., Stukenbrok, G., Greengard, P., DeCamilli, P. (1986): Protein p38: an integral

membrane protein specific for small vesicles of neurons and neuroendocrine cells. J. Cell Biol. 103:2511-2527.

13. Pfeffer, S.R., Kelly, R.B. (1985): The subpopulation of brain coated vesicles that carries synaptic vesicle proteins contains two unique polypeptides. Cell 40:949-957.

14. Rehm, H., Wiedenmann, B., Betz, H. (1986): Molecular characterization of synaptophysin, a major calcium-binding protein of the synaptic vesicle membrane. EMBO J. 5:535-541.

15. Schwechheimer, K., Wiedenmann, B., Lee, I., Franke, W.W. (1987): Synaptophysin: a reliable marker for medulloblastomas. Virchows Arch. A 411:53-59.

16. Südhof, T.C., Lottspeich, F., Greengard, P., Mehl, E., Jahn, R. (1987): A synaptic vesicle protein with a novel cytoplasmic domain and four transmembrane regions. Science 238:1142-1144.

17. Tixier-Vidal, A., Faivre-Bauman, A., Picart, R., Wiedenmann, B. (1988): Immunoelectron microscopic localization of synaptophysin in a Golgi subcompartment of developing hypothalamic neurons. Neuroscience, in press.

18. Wiedenmann, B., Franke, W.W. (1985): Identification and localization of synaptophysin, an integral membrane glycoprotein of M_r 38,000 characteristic of presynaptic vesicles. Cell 41:1017-1028.

19. Wiedenmann, B., Lawley, K., Grund, C., Branton, D. (1985): Solubilization of proteins from bovine brain coated vesicles by protein perturbants and Triton X-100. J. Cell Biol. 101:12-18.

20. Wiedenmann, B., Franke, W.W., Kuhn, C., Moll, R., Gould, V.E. (1986): Synaptophysin: a marker protein for neuroendocrine cells and neoplasms. Proc. Natl. Acad. Sci. USA 83:3500-3504.

21. Wiedenmann, B., Kuhn, C., Schwechheimer, K., Waldherr, R., Raue, F., Brandeis, W.E., Kommerell, B., Franke, W.W. (1987): Synaptophysin identified in metastases of neuroendocrine tumors by immunocytochemistry and immunoblotting. Am. J. Clin. Pathol. 87:560-569.

22. Wiedenmann, B., Rehm, H., Franke, W.W. (1987): Synaptophysin, an integral membrane protein of vesicles present in normal and neoplastic neuroendocrine cells. Ann. N. Y. Acad. Sci. 463:501-503.

23. Wiedenmann, B., Waldherr, R., Rosa, P., Huttner, W.: Gastroenteropancreatic neuroendocrine cells and tumors, identified by proteins specifically associated to dense and clear vesicles, using synaptophysin, chromogranin A and B, and secretogranin II. Gastroenterology, submitted.

THE NEW PARADIGM: SYMPATHETIC NEUROTRANSMISSION BY LATERAL INTERACTION BETWEEN SINGLE MIXED QUANTA ACTING IN TWO DIFFERENT BIOPHASES

Lennart Stjärne
Department of Physiology, Karolinska Institutet, S-10401 Stockholm, Sweden

INTRODUCTION

Numerous new findings challenge widely accepted views of the **microanatomical basis** and the **chemical complexity** of neuroeffector communication in many systems (Burnstock 1976, Dismukes 1979, Schmitt 1984,Vizi 1985, Changeux 1986, Eccles 1986, Hökfelt et al 1986, Iversen 1986). The aim of this mini-review is to discuss five aspects of an emerging **new paradigm** of sympathetic neurotransmission (Burnstock 1986, Lundberg and Hökfelt 1986, Stjärne 1986a,b, Stjärne and Lundberg 1986), namely its view on (1) the relative roles of the different sympathetic messengers, (2) the factors determining the composition of the 'transmitter cocktail' the nerve impulse releases from a varicosity, (3) the intermittent and monoquantal secretory activity of individual varicosities, (4) the heterogeneity of varicosities and the complementary roles of quanta released into the 'intra-' or 'extrajunctional' biophases, and (5) the conditions determining whether a released messenger will act directly and phasically (in 'transmitter mode') or indirectly and tonically (in 'modulator' mode).

DIFFERENT MESSENGERS AND THEIR ROLES IN NEUROTRANSMISSION

There is by now convincing evidence, contrary to Dale's Princple as commonly understood (Eccles 1986), that individual sympathetic neurons may use multiple messengers to communicate with effector cells (Burnstock 1986; Lundberg and Hökfelt 1986).

A. **Sympathetic nerves may utilize three classes of messengers**: Sympathetic nerve stimulation causes release of noradrenaline (NA), and many effector responses are mimicked by exogenous NA and depressed or abolished by NA antagonists or by depletion of the NA stores. Hence as earlier established, NA is the **catecholamine** messenger in sympathetic nerves (Euler 1956).

NATO ASI Series, Vol. H21
Cellular and Molecular Basis of Synaptic Transmission
Edited by H. Zimmermann

However, some effects of sympathetic nerve stimulation, in some systems, cannot be mimicked by NA, are resistant to NA antagonists and persist after depletion of the NA stores, i.e. do not behave as if they were caused by NA. There is growing evidence that these 'atypical' effects are caused by other messengers, released from sympathetic nerves along with NA. Sympathetic stem cells have open genes for transcription and formation of messenger RNAs coding for up to a dozen different putative transmitters (Potter et al 1986). In tissue culture, single sympathetic neurons show considerable plasticity; the culture conditions cause selective expression of some but not all of these genetic capabilities. The decisive factors are the secretory activity of the nerves (induced by depolarizing concentrations of K^+), and environmental factors (hormones, or chemical feedback from the effector cells). By using appropriate stimuli it is even possible to cause a single sympathetic neuron to switch from being mainly 'noradrenergic' to becoming mainly 'cholinergic' (Potter et al 1986).

In situ also the particular combination of messengers individual sympathetic neurons express in their mature state is determined during the establishment of their adult connectivity; again, the decisive factors are their afferent input (from the CNS), chemical feedback (from the effector cells they innervate) and hormonal and other factors in their microenvironment (Changeux 1986, Potter et al 1986). In the adult state it seems to be the rule rather than the exception that individual sympathetic nerves synthesize, store and release members of three classes of putative messengers, namely (1) a catecholamine, NA, (2) a nucleotide, adenosine 5'-triphosphate (ATP, Fedan et al 1981, Burnstock 1986, Stjärne and Åstrand 1984, 1985a, Stjärne and Lundberg 1986, Stjärne et al 1986), and (3) a neuropeptide, for example neuropeptide Y (NPY, Lundberg and Hökfelt 1986) or metenkephalin (ENK, Klein 1982).

A change of the established nomenclature thus seems to be called for: Sympathetic nerves which utilize several messengers should not be called 'noradrenergic' just because they **contain** NA (Stjärne 1986a,b, Stjärne and Lundberg 1986). Perhaps each neuron should be denoted accurately according to the chemical code it utilizes, for example as a 'NA'-, 'NA/ATP'-, 'NA/ATP/NPY'-, 'NA/ATP/ENK'- etc. neuron (Costa et al 1986).

B. **On the relative roles of different sympathetic messengers**: The fact that different sympathetic neurons utilize different messengers complicates the analysis of sympathetic neurotransmission, particularly since the varicosity can vary the composition of the 'transmitter cocktail' secreted (Lundberg and Hökfelt 1986) and the different messengers play different pre- and/or postjunctional roles in different tissues (see below).

Evaluation of the physiological importance of the different sympathetic messengers, i.e. NA, ATP and for example, NPY, requires specific pharmacological blocking agents or drugs selectively depleting the neuronal stores of one class of messengers. Numerous agents blocking NA receptors or depleting NA stores are now available. Two chemically different compounds, thought either to block competitively (Fedan et al 1981), or to desensitize, P_2 purine receptors (Meldrum and Burnstock 1983) are available for analysis of the messenger role of ATP. However, for most neuropeptides (including NPY) specific antagonists are not available; hence the importance of e.g., NPY as a sympathetic messenger is difficult to estimate.

The currently available evidence suggests that ATP is the only truly **ionotropic** (Eccles, 1986) messenger in sympathetic nerves, i.e. the only one which triggers fast depolarization of target cells (Stjärne and Åstrand 1984). NA and NPY seem to be mainly **metabotropic**, i.e., often act without depolarizing the target cells (Bolton and Large 1986, Stjärne and Lundberg 1986, Stjärne et al 1986). However, there are tissue differences in this regard: In some tissues the contractile response to NA (Cheung 1982, Suzuki 1984) or NPY (Neild 1987) is preceded/accompanied by a slow depolarization.

The role of the **'electrical link'** in sympathetic neurotransmission, e.g., the muscle action potential in smooth muscle, is more variable than earlier thought (Burnstock and Costa 1975). In some tissues in which the entire contractile response to sympathetic nerve stimulation is blocked by NA antagonists and hence mediated mainly or exclusively by NA, the contraction is not preceded by depolarization of the smooth muscle cells; here the 'electrical link' is lacking entirely (Bolton and Large 1986). In some other tissues (e.g., the mouse vas deferens), in which the contractile response to sympathetic nerve stimulation is normally preceded by an (ATP-mediated) fast depolarization of the smooth muscle cells and a muscle action potential, 'elimination' of these ATP-induced effects does not markedly alter the contractile response to nerve stimulation; here the 'electrical link' exists but is of minor importance (Stjärne and Åstrand 1985a). In some other tissues, e.g., the guinea-pig vas deferens (Fedan et al 1981,Sneddon and Burnstock 1984, Sneddon and Westfall 1984, Stjärne and Åstrand 1985a), or rabbit (Kügelgen and Starke 1985) or dog (Muramatsu 1987) mesenteric artery, pharmacological 'elimination' of ATP-induced EJPs and muscle action potentials depresses (particularly the twitch phase of) the contractile response to nerve stimulation; here the 'electrical link' is of major quantitative importance.

The relative importance of the pre- and postjunctional effects of different sympathetic messengers also differ in different tissues. As a sympathetic cotransmitter, ATP often exerts a potent excitatory postjunctional effect, e.g., triggering the EJPs in smooth muscle both in the guinea-pig and mouse vas deferens. In some tissues (e.g., the guinea-pig vas deferens) ATP in addition has (weak and inhibitory) prejunctional effects, but in most tissues (e.g., in the mouse vas deferens) ATP-mediated prejunctional effects seem to be lacking entirely (Stjärne and Åstrand 1985a).

Similarly, as a putative sympathetic cotransmitter NPY has different effects in different tissues: In the mouse vas deferens (exogenous) NPY has strong inhibitory prejunctional, and weak excitatory postjunctional effects; the net effect is **depression** of neuromuscular transmission, particularly on stimulation with short trains and/or at low frequency (Stjärne and Lundberg 1986, Stjärne et al 1986). In rat tail artery (exogenous) NPY has no or very weak inhibitory prejunctional effects, but contracts the smooth muscle and also sensitizes it to the transmitters released by nerve stimulation (and also to circulating NA, Neild 1987); the net effect is **potentiation** of neuromuscular transmission.

In conclusion: Many different lines of evidence indicate that sympathetic neurotransmission is much more complex than earlier thought, and also differently organized in different tissues (Burnstock 1986, Stjärne and Lundberg 1986).

VARIATION IN THE COMBINATION OF MESSENGERS SECRETED

There is growing evidence that sympathetic nerve varicosities are able to vary the composition of the 'transmitter cocktail' they secrete (Lundberg and Hökfelt 1986). The mechanisms involved are not fully understood, in part because the vesicular storage of the different messengers is incompletely known.

Sympathetic varicosities contain two classes of transmitter vesicles (Thuresson-Klein 1983), the 'small dense cored' vesicles (SDVs) and the 'large dense cored' vesicles (LDVs). Both vesicle fractions contain NA and ATP, and LDV fractions in addition may contain neuropeptides. However, with few exceptions (Pelletier et al 1981) firm proof is lacking as yet that messengers in the same vesicle **fraction** actually coexist in the same **individual** SDV or LDV.

The number of vesicles in different varicosities varies widely. For example, sympathetic nerve varicosities in rat iris have been reported to contain from 45 to 850 SDVs (Hökfelt 1969). In small animals (rat, mouse, guinea-pig) the proportion of LDVs is low (less than 5% of all vesicles); in larger animals and in man it is higher (up to 30%, Thuresson-Klein 1983). SDVs and LDVs differ quantitatively as well as qualitatively: LDVs are larger (internal diameter about twice as large as that of SDVs) and hence presumably contain at least 8-fold more NA and ATP than SDVs (Klein 1982). Further, LDVs but not SDVs contain the macromolecules synthesized in the perikaryon, i.e. neuropeptides and other specific proteins such as the enzyme dopamine β-hydroxylase. Hence NA is formed only in LDVs; SDVs obtain their complement of newly formed NA by uptake from the axoplasm of material 'leaking' from LDVs (Klein 1982).

The composition of the secreted 'transmitter cocktail' is regulated both by the nerve impulse pattern and by activation of prejunctional autoreceptors (Lundberg and Hökfelt 1986, Stjärne and Lundberg 1986). During low frequency stimulation, or under the influence of α_2-autoinhibition, primarily the contents of SDVs (NA and ATP) are released.

However, on stimulation at high frequency, or during block of α_2-autoreceptors, the contents of LDVs are preferentially released; the neuron has become in part **'peptidergic'** (Lundberg and Hökfelt 1986).

There is some reason to assume that SDVs and LDVs may secrete transmitter from different sites on the varicosity membrane, mainly by analogy with other systems, namely electronmicroscopical evidence from (non-sympathetic) boutons of central neurons in rat brain. Here signs of release from 'small' synaptic vesicles occurred almost exclusively at specialized membrane areas, i.e., in the 'presynaptic grids'. In contrast, evidence for exocytosis from 'large' vesicles occurred mainly at the outer margin of the presynaptic grids, or in unspecialized areas of the bouton. The conclusion was drawn that transmitter secretion from 'small' synaptic vesicles is 'directed' while that from large vesicles is not (Zhu et al 1986).

Sympathetic nerve varicosities lack a morphologically well-defined membrane presynaptic grid (Gabella 1981), but according to functional evidence may possess 'preferred release sites' (Cunnane and Stjärne 1984). By analogy with the abovementioned system, it is possible that SDVs in sympathetic nerve varicosities secrete their contents from 'preferred sites', while the contents of LDVs are secreted from 'random' sites on the varicosity membrane.

The mechanisms underlying selective translocation of SDVs and LDVs to their respective release sites are not known. An attractive possibility is that they involve Ca^{2+}-dependent phosphorylation of protein components of the cytoskeleton of the varicosity, for example Synapsin I, which in their unphosphorylated state restrict the mobility of the vesicles (Bähler and Greengard 1987).

ON SYMPATHETIC TRANSMITTER QUANTA

It is widely assumed that NA is secreted in 'quanta', i.e., in multimolecular packets of preset size (Burnstock and Holman 1966, Folkow et al 1967, Bevan et al 1969). However, the new evidence that sympathetic nerves contain at least two classes of transmitter vesicles, storing up to three classes of different messengers, implies that these nerves may utilize more than one 'quantal' and/or 'graded' release mechanism.

A. **Are all sympathetic messengers secreted in 'quanta'?** It is not known if the two classes of sympathetic transmitter vesicles are homogeneous; different SDVs or LDVs in a varicosity may differ both qualititatively (storing different messengers) and quantitatively (storing different amounts of each messenger). Nor is it known if SDVs or LDVs discharge their entire contents in each secretory cycle. SDVs, which do not seem to contain macromolecules may be able to afford all-or-none exocytosis, since they can be refilled by NA and ATP locally synthesized in LDVs and mitochondria, respectively (Wakade and Wakade 1984). However the macromolecules in LDVs (neuropeptides as well as specific proteins, such as the enzyme dopamine β-hydroxylase), which are synthesized in the perikaryon (Klein 1982, Lagercrantz and Fried 1982) may be too 'expensive' to be lost by all-or-none exocytosis (cf. Folkow et al 1967).

In conclusion: There is considerable evidence (see below) that nerve impulses release the (entire?) contents of individual SDVs, i.e. that the NA and ATP in an SDV constitutes the 'SDV quantum'. Whether LDVs release their contents of messengers (NA, ATP and neuropeptides) in quanta (either corresponding to their entire contents, or to a fixed subfraction), or in a graded fashion, is not known.

B. **Methodological limitations**: For any transmitter, proof that it is secreted in quanta is possible only if it is 'ionotropic', i.e., acts by opening fast ionic channels. According to most reports NA is not ionotropic in smooth muscle (Bolton and Large 1986). However, there is one exception: when applied iontophoretically close to the sympathetic nerve varicosities innervating smooth muscle in guinea-pig mesenteric arteriole, NA has been reported to cause a fast depolarization which mimics the nerve impulse induced EJPs (Hirst and Neild 1980). Both the EJP and the NA-induced depolarization were resistant to agents blocking α- and β-adrenoceptors; both responses therefore were hypothesized to be mediated by a separate receptor class, 'γ-adrenoceptors' (Hirst and Neild 1980). Definitive evaluation of these findings is not possible since specific antagonists for the hypothetical 'γ -adrenoceptors' are lacking; the finding therefore remains controversial (Bevan 1984, Neild and Hirst 1984).

In contrast to NA, its cotransmitter, ATP, exerts 'ionotropic' effects e.g., in the guinea-pig vas deferens, where it triggers two classes of excitatory junction potentials (EJPs) in smooth muscle cells (Burnstock and Holman 1966, Bennett 1973, Stjärne and Åstrand 1984). In these nerves SDVs make up more than 90% of all vesicles (Burnstock and Costa 1975), and during low frequency stimulation transmitter is secreted almost exclusively from SDVs (Basbaum and Heuser 1979). By analysis of spontaneous EJPs and 'fast' EJPs evoked by stimulation at low frequency it has been shown that the transmitter causing these responses (presumably the ATP contained in individual SDVs ,Cunnane and Stjärne 1984, Stjärne and Åstrand 1984). is secreted in quanta (Blakeley and Cunnane1979).

C. **The 'SDV quantum'**: As explained above our knowledge concerning sympathetic transmitter quanta is meagre; only the 'SDV quantum' is reasonably well documented. SDVs within a varicosity vary in electron density and hence in degree of filling (with NA, Fillenz 1977). This may imply that 'SDV quanta' vary in size, but it is possible that only 'filled' SDVs are translocated to the release site. In that case the amount of NA and ATP in **released** 'SDV quanta' might be relatively constant.

The NA/ATP ratio in SDVs has been reported to be about 50/1 (Fredholm et al 1982; Lagercrantz and Fried 1982), and current estimates of the number of NA molecules in SDVs, and hence in the 'SDV quantum', vary from 1000 (Fredholm et al 1982) to at least 60000 (calculated from Dahlström et al 1966, Griffith et al 1982). Hence the 'SDV quantum' may contain as few as 20 or as many as 1200 ATP molecules. Interestingly, the estimated number of channels opened by a single sympathetic transmitter quantum (in the guinea-pig vas deferens) varies within the same range, from 40 up to 1000 (Finkel et al 1984).

NERVE VARICOSITIES BASICALLY OPERATE IN DIGITAL MODE

Current models do not consider, or underestimate, the **microanatomical** complexity of the physiology of sympathetic neurotransmission. They have no answer to the following crucial questions: How and to what extent does a nerve impulse in a sympathetic parent axon induce transmitter secretion in each of the perhaps 20000 varicosities (Dahlström and Häggendal 1966) in its terminal arborization? Do most or all nerve impulses activate every varicosity (Folkow et al 1967), or do they often fail to release transmitter (Bevan et al 1969, Blakeley and Cunnane 1979)? When successful, do nerve impulses release single quanta (Bevan et al 1969, Cunnane and Stjärne 1984) or do they release a variable amount of transmitter, i.e. up to ten quanta (Blakeley and Cunnane 1979, Blakeley et al 1984)? Do factors modulating sympathetic transmitter secretion from a varicosity alter the number of quanta released per pulse, or the probability of monoquantal release (Stjärne 1985)?

A. **Study of NA overflow has provided the constraints**: Two decades ago studies of the stimulus-induced overflow of NA from the sympathetic vasomotor innervation in cat skeletal muscle led Folkow and his coworkers to define the constraints (Folkow et al 1967, Folkow and Häggendal 1970). Based (1) on the observation that the NA release per nerve impulse from the whole tissue corresponded to about 10^{-5} of its NA content, and (2) the estimate that the average varicosity contains about 1000 vesicles, they argued as follows: If essentially all NA is stored in vesicles and the contents of a vesicle represents the 'NA quantum', and if most varicosities are potentially secretory and all the secreted NA is recovered in the effluent, then the overflow results imply that each nerve impulse releases a quantum from only (about) 1% of the varicosities.Neurotransmission based on exceedingly high local concentrations of NA, but only in 1% of the neuro-effector gaps, appeared to them both wasteful (why should the local NA concentration have to be several orders of magnitude higher than the concentration of exogenous NA required to cause maximal contraction?) and inefficient (is effective neurotransmission possible in a system with 99% failure in each release site?). It would be "almost like shooting at a flock of birds with occasional artillery grenades instead of using ... a shotgun" (Folkow and Häggendal 1970). On these grounds they proposed the opposite alternative (the **'non-intermittent model'**), namely that each nerve impulse activates all varicosities, but releases from each of them 'NA quanta' corresponding to a small fraction (1% or less) of the contents of individual vesicles (Folkow et al 1967, Folkow and Häggendal 1970).

However, based on study of the NA secretion from sympathetic nerves in rabbit pulmonary artery, Bevan and his group arrived at the opposite conclusion (the **'intermittent model'**), namely that the individual varicosity is intermittently activated, by 1 out of every 7-8 nerve impulses and each time releases the entire NA contents of a single vesicle (Bevan et al 1969). Their experiments were carried out in the presence of the α-adrenoceptor antagonist phenoxybenzamine, which has later been found out to increase the nerve stimulation-induced NA secretion by at least 5-fold. After correction of their estimate for the effects of the α-blocker their view would imply that the average individual varicosity 'normally' (i.e., in the absence of the drug) releases the contents of a vesicle in response to about 2-3% of the nerve impulses (Stjärne 1985).

The question whether every nerve impulse activates every varicosity, releasing (about) 1% of the contents of a single vesicle, as proposed by Folkow et al (1967), or whether only (about) 1% of the nerve impulses activate the individual varicosity, each time releasing the total contents of a single vesicle, as proposed by Bevan et al (1969; after correction as explained above, Stjärne 1985), could not be answered only by study of NA overflow.

B. **Electrophysiological methods have provided resolution**: By combination of data by recently introduced electrophysiological methods and by study of NA overflow, it seems now possible in suitable model tissues (the guinea-pig or mouse vas deferens) to determine the probability that the individual varicosity will be activated by a nerve impulse in the parent axon, and the amount of transmitter(s) it releases. The nerve stimulation-induced overflow of ^{3}H-NA is used as a marker for the NA component, and the **'discrete event'**, i.e., the transient peak in the first time differential of the rising phase of intracellularly recorded 'fast' EJPs (Blakeley and Cunnane 1979), or the **excitatory junction current (EJC)** recorded by an extracellular microelectrode (Brock and Cunnane 1987), as markers for the ATP component, of the secreted 'mixed' SDV quanta (Stjärne and Åstrand 1984).

The interpretation of data by these methods is controversial (note for example the differences between Cunnane and Stjärne 1984, and Blakeley et al 1984). The following evidence seems to be difficult to reconcile either with the historical 'nonintermittent' model described above, according to which every nerve impulse activates every varicosity, releasing 'small' quanta (1-3% of the contents of individual vesicles, Folkow et al 1967, Folkow and Häggendal 1970), or with the hypothesis of graded release from a varicosity, implying that each 'successful' nerve impulse releases a variable number (1 - 10) of quanta equal to the contents of individual vesicles (Blakeley and Cunnane 1979, Blakeley et al 1984): (1) In the guinea-pig and mouse vas deferens, spontaneous EJPs which are tetrodotoxin-resistant and hence are thought to be caused by random release of **single** quanta, not by synchronous release of quanta from several sites, due to spontaneous nerve spikes, may be very large, up to 20 mV or more (Burnstock and Holman 1966); hence sympathetic transmitter quanta are likely to be 'big', i.e. not small fractions of the contents of a vesicle (Bennett 1973). (2) When resolution is optimal the overwhelming majority of nerve impulses can be seen to fail to evoke a discrete event (Blakeley and Cunnane 1979, Cunnane and Stjärne 1984) or EJC (Brock and Cunnane 1987); hence transmitter secretion from the individual sympathetic nerve varicosity is highly intermittent. (3) The amplitude distribution of spontaneous and nerve impulse-induced discrete events (Blakeley and Cunnane 1979, Cunnane and Stjärne 1984) or EJCs (Brock and Cunnane 1987) in a cell (or patch) often is similar, and in many cases individual spontaneous responses can be found which accurately match individual evoked responses; hence presumably nerve impulses basically release **single** quanta from a varicosity. (4) Comparison of the quantal content of EJPs and/or EJCs with the histologically determined number of varicosities in the sympathetic nerves of guinea-pig submucous arterioles suggests that each nerve impulse in the parent axon of this tissue activates only 1-2% of the varicosities (Hirst and Neild 1980, Finkel et al 1984). (5) In many other systems transmitter release from a varicosity ('bouton') basically is monoquantal, except when drugs such as 4-aminopyridine (4AP) have been added. In boutons (i.e., varicosities) in mammalian primary afferent fibre terminals in spinal cord, or in interneurons innervating Mauthner cells in goldfish, or in afferent fibres to auditory nuclei in avian brain stem, single nerve impulses release single quanta from each 'presynaptic grid'; in motor nerve terminals in striated muscle nerve impulses release single quanta from each 'active zone' (see references and discussion in Korn 1984).

In conclusion: Most data indicate that the average individual sympathetic nerve varicosity responds to about 1% (range: 0.2-3%) of the nerve impulses in the parent axon (during stimulation at low frequency), 'normally' secreting a single 'mixed' transmitter quantum, equal to the NA and ATP content of a SDV (Cunnane and Stjärne 1984; for a different view see Blakeley et al 1984). Hence under these conditions each impulse may basically release single quanta from 200 of the 20000 varicosities in the terminals of each neuron (Stjärne 1985); the varicosities act in a one-or none manner, i.e. as **digital** units.

However, stimulation with bursts at high frequency may bring out a latent **analogue** feature, namely selective increase in the probability that the transmitters in a LDV (i.e. NPY together with NA and ATP) will be secreted (Lundberg and Hökfelt 1986, Stjärne and Lundberg 1986, Stjärne et al 1986).

C. **Is the intermittency due to failure of invasion?** The possibility has been discussed that each nerve impulse in the parent axon may invade only a small part of the terminal arborization of sympathetic nerves (Haefely 1972). In that case intermittent invasion by the nerve impulse may be a major cause of the low probability of transmitter secretion from individual varicosities (Cunnane and Stjärne 1984). However, recently direct experimental evidence against this possibility has been obtained by electrophysiological techniques: In the guinea-pig vas deferens the extracellularly recorded nerve terminal spike was 'non-intermittent' even though the evoked release of a quantum (as reflected in the EJC) was highly intermittent. Thus most or all nerve impulses may invade all varicosities (Brock and Cunnane 1987; for similar conclusions from studies of locus coeruleus neurons in rat brain, see Ryan et al 1985). Hence sympathetic nerve varicosities may fail to release transmitter mainly because the secretory mechanisms are refractory to the triggering effect of depolarization by the nerve impulse.

D. **Why do nerve impulses release single quanta?** The mechanisms underlying the **monoquantal** secretion from sympathetic nerve varicosites are not understood. In nerve endings in which transmitter is secreted from a 'presynaptic grid', the explanation of monoquantal release may be purely mechanical: a vesicle in the process of exocytosis deforms the grid enough to 'laterally' inhibit exocytosis from other parts (Korn 1984).

This explanation cannot apply to monoquantal secretion from sympathetic nerve varicosities, since they seem to lack a morphologically distinct, preformed presynaptic grid. However, evidence has been reported that they may possess **functional** 'preferred release sites', so small (and therefore difficult to detect by morphological methods) that they accomodate only a few (perhaps maximally two) vesicles at a time (Cunnane and Stjärne 1984). Such a (still hypothetical) mechanism does not explain why only **single** quanta are released but it may help to explain why the secretory mechanisms of a varicosity often appear to be refractory to a nerve impulse: The secretory responsiveness would be maximal when both 'preferred release sites' are charged with a SDV, and virtually zero when both are empty (Cunnane and Stjärne 1984, Stjärne 1985, Stjärne and Åstrand 1985b).

BASIC RESTRICTION AND MODULATION OF TRANSMITTER SECRETION

The low probability of activation of the secretory mechanisms of sympathetic nerve varicosities is 'desirable' for economical reasons, since each varicosity possesses only some 500 SDVs (i.e., quanta). To what extent is it subject to physiological control?

A. **Basic, physiologically unregulated restrictive mechanisms**: The secretory responsiveness of sympathetic nerve varicosities to nerve impulses may be restricted by two classes of 'intrinsic' mechanism, each apparently only to a limited extent subject to physiological regulation:

(1) Restriction due to the **'chemo-mechanical'** properties of the axoplasm: As mentioned above, SDVs and LDVs may not be freely mobile; they may be 'trapped' by the cytoskeleton of the varicosity (Baines 1987, Burgoyne and Cheek 1987). It seems possible that this restriction is relieved by phosphorylation of critical proteins, triggered for example by nerve action potential-induced influx of Ca^{2+} ions. Hence factors modulating transmitter secretion by changing the regional concentration of Ca^{2+}, i.e. the facilitation induced by trains of nerve impulses, as well as the inhibition by activation of prejunctional α_2-adrenoceptors, may operate in part by influencing the degree of phosphorylation of such specific proteins, e.g., Synapsin I (Bähler and Greengard 1987, Starke 1987).

(2) Restriction due to properties of **ionic channels** in the varicosity membrane: Activation of voltage-gated K^+ channels in the nerve varicosities may suppress the secretory responsiveness by shortening the duration of the nerve action potential and hence restricting Ca^{2+} influx. There is evidence that this mechanism may be of major importance for the basic restrictiveness of transmitter secretion from sympathetic nerve varicosities: Block of voltage-gated K^+ channels by tetraethylammonium (TEA) and/or 4-aminopyridine (4AP) dramatically increases the secretion of NA evoked by electrical stimulation of sympathetic nerves (Wakade and Wakade 1984, Stjärne 1985, Stjärne and Åstrand 1985b). In the guinea-pig vas deferens a combination of TEA (20 mM) and 4AP (1 mM) strongly increased the secretion of ^{3}H-NA induced by field stimulation; the output per unit time reached a maximum on stimulation at 4 Hz (it was not further enhanced by a further increase in the stimulus frequency). The output per stimulus was inversely related to the frequency of stimulation and at 0.0625 Hz reached a level which may represent its theoretical maximum, namely one at which each nerve impulse released a quantum from every varicosity. Under these conditions the average varicosity released about 4 quanta per min; this output could be maintained during continuous stimulation for at least 30 min. Thus under steady state conditions the rate limiting step (recharging of 'preferred release sites'?) seems to require about 15 s (Stjärne 1985, Stjärne and Åstrand 1985b).

B. **The scope of the physiological control of transmitter secretion:** Transmitter secretion from sympathetic nerve varicosities is regulated by two opposing physiological control mechanisms: (1) It is increased by **'facilitation'** as a result of repetitive nerve impulses, particularly at high frequency, and (2) opposed by **'autoinhibition'** via activation of prejunctional α_2-adrenoceptors (Bevan et al 1984, Stjärne 1975, 1981, Langer 1977, 1981, Starke 1977, 1981, 1987, Westfall 1977, Vizi 1979, Alberts et al 1981). The two systems interact in a complex manner: The inhibitory effect of activation of autoreceptors is strongly restricted by trains of nerve impulses at high frequency (Stjärne 1987). Under resting conditions autoinhibition via the adrenoceptors does not depress the secretory mechanisms tonically; to become operative this control system requires 'priming' by a short train of nerve impulses. On the other hand, its inhibitory effect is increasingly counteracted, or overcome, when the train length or frequency of nerve impulses is increased (see Stjärne 1975, Langer 1977, 1980, Starke 1977, 1987).

There is good reason to assume that the two physiological control mechanisms operate by altering the **probability of monoquantal release** (Stjärne 1985; however, for the view that they control the ' **quantal content'**, i.e., the number of quanta the nerve impulse releases from a varicosity, see Blakeley et al 1984, 1986). Pharmacological block of prejunctional α_2-autorecptors increases the secretory response to nerve impulses at low frequency, by up to 5-fold. Hence this physiological control has a limited scope: In the absence of α-autoinhibition (i.e., when the prejunctional adrenoceptors are not exposed to a sufficiently high NA concentration) the probability that a nerve impulse will release a quantum from a varicosity may be as high as 0.05-0.1, but accumulation of endogenous NA 'normally' maintains the probability in the range 0.002-0.03. In other words: the degree of activation of prejunctional α-adrenoceptors may determine if a nerve impulse in the parent axon will activate the release mechanisms of only 200, or up to 1000 of the 20000 varicosities in the terminal arborization of individual sympathetic neurons (Stjärne 1985, Stjärne and Åstrand 1985b).

This is the **'digital'** aspect of the physiological control of sympathetic transmitter secretion, by facilitation and autoinhibition. However, the two systems in addition exert **'analogue'** control, determining the composition of the 'transmitter cocktail' released from a varicosity. As mentioned above, low frequency nerve stimulation and activation of α_2-autoreceptors promote release from SDVs, while block of α_2-autoreceptors, and/or bursts of nerve impulses at high frequency, preferentially increase secretion from LDVs, causing a sympathetic neuron to become in part 'peptidergic' (Lundberg and Hökfelt 1986).

HETEROGENEITY OF SYMPATHETIC NERVE VARICOSITIES: IMPLICATIONS

During recent years it has been debated whether or not the release sites in 'boutons-en-passant' terminals e.g., in the brain, establish a clear 'synaptic' relationship to effector cells. The possibility has been discussed that chemical neurotransmission may not be exclusively 'synaptic', but also 'para'- and even 'non-synaptic' (Schmitt 1984, Vizi 1985).

That this is the case in sympathetically innervated tissues is uncontroversial. Many tissues are innervated by two classes of sympathetic nerve fibres (Merrillees 1968, Bennett 1973, Burnstock and Costa 1975, Gabella 1981): (1) Fibres in small bundles, enclosed in a Schwann sheath, which shows discontinuities ('windows') at the varicosities.

These varicosities, which often are widely separated from the effector cells, are here termed **'nonjunctional'**. (2) Single naked (Schwann-free), varicose fibres; these varicosities which often are closely apposed with the effector cells, are here termed **'junctional'**.

In some tissues (e.g., the guinea-pig or mouse vas deferens) the roles of transmitter quanta from these two classes of varicosities can be distinguished by electrophysiological techniques: Spontaneous EJPs, and 'fast' stimulus-evoked EJPs are thought to be caused by (the ATP component of) single sympathetic 'SDV quanta' acting **'focally'** , i.e., on a small receptor patch in the narrow junctional gap. In contrast, 'slow' stimulus-evoked EJP are thought to be caused by (the ATP component of) many quanta from numerous 'nonjunctional' varicosities, acting **'diffusely'**, i.e., on receptors distributed all over the surface of the smooth muscle cells (Bennett 1973). Transmitters from the two classes of varicosities thus appear to have different and complementary roles in sympathetic neurotransmission.

TRANSMITTER ACTION IN 'INTRA'- AND 'EXTRAJUNCTIONAL' BIOPHASES

The close agreement between calculations of the secretory activity of individual sympathetic nerve varicosities, based on NA overflow (which reflects the secretory activity in all varicosities) and on electrophysiological analysis of 'discrete events' or EJCs (which reflects selectively the release of ATP quanta from 'junctional' varicosities (Bennett 1973, Stjärne and Åstrand 1984), suggests .that transmitter secretion from both classes of sympathetic nerve varicosities is highly intermittent and monoquantal (Stjärne 1985).

Transmitters in single quanta released by a nerve impulse from a random selection of about 1% of the numerous 'nonjunctional' varicosities act in the **'extrajunctional'** biophase. Due to the large volume of this space the effective transmitter concentration in this biophase is low, and the transmittere action **graded** according to the degree of accumulation of transmitters (Schipper et al 1980).

In contrast transmitters in single quanta released by a nerve impulse from about 1% of the few 'junctional' varicosities act in narrow neuro-effector gaps, i.e. in the truly **'intrajunctional'** biophase. Due to its small volume, the initial peak concentration of transmitter is extremely high (equal to that in the vesicles), and possibly always sufficient to saturate the local receptor patch; the effect may be **all-or-none**. Due to the extreme intermittency of release from the individual varicosity, transmitter presumably does not accumulate in individual neuro-effector gaps, i.e., 'intrajunctionally'.

In both biophases the effects of the transmitters is likely to be proportionate to their local concentration, which in turn depends on the balance between release from the nerves and physical and/or chemical elimination from the biophases, and the rate of unbinding from receptors. There may exist differences in this regard between the different sympathetic messengers: For each them the initial peak concentration at the site of release is rapidly lowered by diffusion; however the rates of diffusion of NA, ATP and NPY are different, due to differences in their molecular size as well as in various physico-chemical properties. The NA concentration is further reduced by neuronal and extraneuronal uptake, while ATP is rapidly eliminated due to degradation by ectoenzymes. Less is known about the fate of released NPY. However, there is no evidence that NPY is taken up either into nerves or effector cells, and since NPY is relatively stable (at least in the blood stream, Lundberg and Hökfelt 1986) and is only slowly released from its receptors (off-rate about 20 min,) NPY may exert more prolonged effects than either NA or ATP.

MESSENGERS ACT BOTH IN 'TRANSMITTER'- AND 'MODULATOR' MODE

Neuronal messengers are often classified either as 'classical transmitters' (for example, NA, acetylcholine etc) or as 'neuromodulators' (for example, neuropeptides) This distinction is misleading and semantically unwarranted. Perhaps one should rather speak of the 'direct' and 'indirect' effects a neuronal messenger may exert on an effector system.

This distinction may be illustrated by the actions of NA on the smooth muscle in the mouse vas deferens. In this tissue NA is the main mediator both of the rapid twitch and the prolonged secondary contractile response to sympathetic nerve stimulation (Stjärne and Åstrand 1985a).(1) The **direct** effect of NA, namely a rise in resting tension requires very high concentrations of exogenous NA (more than 10 μM). Presumably such concentrations occur exclusively 'intrajunctionally', as a result of secretion of endogenous NA. (2) The **indirect** effects of NA, namely a change in the pre-and postjunctional excitability of the tissue to other stimuli are exerted at much lower concentrations (in the range probably occurring 'extrajunctionally'). At these concentrations NA does not cause any change in the resting tension, but (a) depresses transmitter secretion, and hence neurotransmission, via prejunctional α_2-adrenoceptors, and (b) increases the responsiveness of the smooth muscle to the transmitters released by nerve stimulation, via postjunctional α_1-adrenoceptors. As a sympathetic messenger (**'transmitter'**) in this tissue NA thus both (at high concentrations) exerts direct effects as a phasic 'signal', and at low concentrations operates as a pre- and postjunctional **'modulator'** (Stjärne 1986a,b, 1987).

CONCLUSIONS: FEATURES OF A NEW PARADIGM

(1) Different sympathetic nerve fibres, even within the same tissue, contain different messengers and release them in different combinations, depending on the nerve impulse

pattern and on the degree of activation of prejunctional receptors. Effector responses to nerve impulses are modulated both pre- and postjunctionally via receptors to sympathetic or other neurotransmitters as well as to other signal substances, locally released from effector cells or reaching the area by diffusion from the blood stream (Schmitt 1984). Both pre- and postjunctionally each signal substance often addresses two or more classes of specific receptors which mediate effects either directly via ionic channels in the membrane or indirectly via different second messenger cascades. All these features, which greatly complicate the analysis of sympathetic neurotransmission, also imply that its information potential is much richer than earlier understood (Iversen 1986).

(2) Sympathetic nerve terminals have a high safety factor for longitudinal conduction, combined with severe restriction of the secretory responsiveness of the varicosities to depolarization by nerve impulses (to avoid rapid depletion of the highly limited store of quanta). The restrictive mechanisms are mainly 'intrinsic' and not subject to physiological regulation, but are to a marginal (but presumably physiologically highly significant) extent modulated by two opposing mechanisms, namely the facilitating effect of trains of nerve impulses, particularly at high frequencies, and the inhibitory effect of activation of prejunctional α_2-autoreceptors. The function of this control may be to locally fine-tune the secretory response to a nerve impulse, in part by influencing the selective secretion from SDVs or LDVs and hence the transmitter choice.

(3) Many tissues are innervated by two classes of sympathetic nerve varicosities, 'junctional' and 'nonjunctional'. Basically both operate in a 'digital' mode: Each nerve impulse 'normally' releases single quanta from a random (?) selection of about 200 of the 20000 varicosities in the terminals of each neuron. The individual varicosity is extremely intermittently activated, each time (?) releasing single quanta. Transmitters do not accumulate within the individual neuro-effector cleft, i.e., intrajunctionally, but in the extrajunctional biophase. Sympathetic neurotransmission is based on lateral interaction between quanta from the the two classes of varicosities: Quanta from the few 'junctional' varicosities act focally on the intrajunctional receptors, in an all-or-none fashion. In contrast transmitters in quanta from the numerous 'nonjunctional' varicosities diffuse widely within the extrajunctional biophase and act in a graded fashion.

(4) The working model of a new paradigm of sympathetic neurotransmission presented in this paper is tentative and will have to be tested in future experiments.

ACKNOWLEDGEMENTS

The research in this paper was supported by the Swedish Medical Research Council (project B87-14X-03027-18A) and Karolinska Institutets Fonder. I thank Mrs Eivor Stjärne and Mrs Ingmarie Eriksson for excellent technical assistance.

REFERENCES

Alberts P, Bartfai T and Stjärne L (1981) Site(s) and ionic basis of α-autoinhibition and facilitation of [^{3}H]noradrenaline secretion in guinea-pig vas deferens. *J Physiol (Lond)* **312**: 297-334.

Baines A.J. (1987) Synapsin I and cytoskeleton. *Nature.* **326**: 646.

Basbaum CB and Heuser JE (1979) Morphological studies of stimulated adrenergic axon varicosities in the mouse vas deferens *J Cell Biol* **80**: 310-325

Bennett MR (1973) Structure and electrical properties of the autonomic neuromuscular junction. *Phil Trans R Soc Lond B* **265**: 25-34.

Bevan JA, Chesher GB and Su C (1969) Release of adrenergic transmitter from termina lexus in artery. *Agents and Actions* **1**: 20-26.

Bevan JA, Tayo FM, Rowan RA and Bevan RD (1984) Presynaptic α-receptor control of adrenergic transmitter release in blood vessels. *Fed Proc* **43**: 1365-1370.

Blakeley AGH and Cunnane TC (1979) The packeted release of transmitter from the sympathetic nerves of the guinea-pig vas deferens: An electrophysiological study. *J Physiol (Lond)* **296**: 85-96.

Blakeley AGH, Mathie A and Petersen SA (1984) Facilitation at single release sites of a sympathetic neuroeffector junction in the mouse. *J Physiol (Lond)* **349**: 57-71.

Blakeley AGH, Mathie A and Petersen SA (1986) Interactions between the effects of yohimbine, clonidine and $[Ca]_0$ on the electrical response of the mouse vas deferens. *Br J Pharmacol* **88**: 807-814.

Bolton TB and Large WA (1986) Are junction potentials essential? Dual mechanism of smooth muscle cell activation by transmitter released from autonomic nerves. *Quart J Exp Physiol* **71**: 1-28.

Brock JA and Cunnane TC (1987) Relation between the nerve action potential and transmitter release from sympathetic postganglionic nerve terminals. *Nature* **326**: 605-607.

Burgoyne R.D. and Cheek T.R. (1987) Role of fodrin in secretion. *Nature,* **326**: 448.

Burnstock G (1976) Do some nerve cells release more than one transmitter? *Neuroscience* **1**: 239-248.

Burnstock G (1986) The changing face of autonomic neurotransmission. *Acta Physiol Scand* **126**: 67-91.

Burnstock G and Costa M (1975) *Adrenergic Neurons.* Chapman and Hall, London.

Burnstock G and Holman ME (1966) Junction potentials at adrenergic synapses. *Pharmacol Rev* **18**: 481-493.

Bähler M and Greengard P (1987) Synapsin I bundles F-actin in a phosphorylation dependent manner. *Nature* **326**: 704-707

Changeux J-P (1986) Coexistence of neuronal messengers and molecular selection. *Prog Brain Res* **68**: 373-403.

Cheung DW (1982) Two components in the cellular response of rat tail arteries to nerve stimulation. *J Physiol (Lond)* **328**: 461-468.

Costa M, Furness J.B. and Gibbins I.L. (1986) Chemical coding of enterie neurons. In: T. Hökfelt, K. Fuxe and B. Pernow. Progress in Brain Research. Vol. 68, 217-237.

Cunnane TC and Stjärne L (1984) Transmitter secretion from individual varicosities of guinea-pig and mouse vas deferens: Highly intermittent and monoquantal. *Neuroscience* **13**: 1-20.

Dahlström A and Häggendal J (1966) Some quantitative studies on the noradrenaline content in the cell bodies and terminals of a sympathetic adrenergic neuron system. *Acta physiol scand* **67**: 271-277.

Dahlström A, Häggendal J and Hökfelt T (1966) The noradrenaline content of the nerve terminal varicosities of sympathetic adre nergic neurons in the rat *Acta physiol scand* **67**: 289-294.

Dismukes RK (1979) New concepts of molecular communication among neurons. *Behav Brain Sci* **2**: 409-448.

Eccles JC (1986) Chemical transmission and Dale's principle. *Prog Brain Res* **68**: 3-13.

Euler US von (1956) Noradrenaline. Charles C. Thomas Publ., Springfield,Ill.

Fedan JS, Hogaboom GK, O'Donnell JP, Colby J and Westfall DP (1981) Contribution by purines to the neurogenic response of the vas deferens of the guinea pig *Eur J Pharmacol* **69**: 41-53

Fillenz M (1977) The factors which provide short-term and longterm control of transmitter release *Prog Neurobiol* **8**: 251-278

Finkel AS, Hirst GDS, van Helden DF (1984) Some properties of the excitatory junction current recorded from submucosal arterioles of the guinea-pig ileum. *J Physiol Lond* **351**: 87-98.

Fredholm BB, Fried G and Hedqvist P (1982) Origin of adenosine released from rat vas deferens by nerve stimulatin *Eur J Pharmac* **79**: 233-243.

Folkow B, Häggendal J and Lisander B (1967) Extent of release and elimination of noradrenaline at peripheral adrenergic nerve terminals. *Acta Physiol Scand Suppl* **307**: 1-38.

Folkow B and Häggendal J (1970) Some aspects of the quantal release of the adrenergic transmitter. *Bayer-Symposium II.* Springer, Berlin: 91-97.

Gabella G. (1981) Structure of smooth muscles. In; E. Bülbring, A.F. Brading, A.W. Jones and T. Tomita (eds), *Smooth muscle*, pp 1-46, Edward Arnold London.

Griffith SG, Crowe R, Haven AJ and Burnstock G (1982) Regional differences in the density of perivascular nerves and varicosities. Noradrenaline content and responses to nerve stimulation in the rabbit ear artery. *Blood Vessels* **19**: 41-53

Haefely W (1972) Electrphysiology of the adrenergic neuron. In: Blaschko H and Muscholl. *Catecholamines. Handbook of Experimental Pharmacology,* Vol. 33. Springer Berlin: 661-725.

Hirst GDS and Neild TO (1980) Some properties of spontaneous excitatory junction potentials recorded from arterioles of guinea-pig. *J Physiol Lond* **303**: 43-60.

Hökfelt T (1969) Distribution of noradrenaline storing particles in peripheral adrenergic neurons as revealed by electron microscopy. *Acta Physiol Scand* **76**: 427-440.

Hökfelt T, Holets VR, Staines W, Meister B, Melander T, Schalling M, Schultzberg M, Freedman J, Björklund H, Olson L, Lindh B, Elfvin L-G, Lundberg JM, Lindgren JÅ, Samuelsson B, Pernow B, Terenius L, Post C, Everitt B and Goldstein M (1986) Coexistence of neuronal messengers — an overview. *Prog Brain Res* **68**: 33-70.

Iversen L.L. (1986) Chemical signalling in the nervous system. In: T.Hökfelt, K. Fuxe and B. Pernow. Progress in Brain Research. Vol.68, 15-21.

Klein RL (1982) Chemical composition of the large noradrenergic vesicles. In: Klein RL, Lagercrantz H and Zimmermann H (eds). *Neurotransmitter Vesicles.* Academic Press, London: 133-150.

Korn H (1984) What central inhibitory pathways tell us about mechanisms of transmitter release. *Exp Brain Res Suppl* **9**: 201-224.

Kügelgen I v and Starke K (1985) Noradrenaline and adenosine triphosphate as co-transmitters of neurogenic vascontriction in rabbit mesenteric artery. *J. Physiol.*, **367**: 435-455.

Lagercrantz H and Fried G (1982) Chemical composition of the small noradrenergic vesicles. In: Klein RL, Lagercrantz H and Zimmermann H (eds). Neurotransmitter Vesicles. Academic Press London: 175-188.

Langer SZ (1977) Presynaptic receptors and their role in the regulation of transmitter release. *Br J Pharmacol* **60**: 481-497.

Langer SZ (1981) Presynaptic regulation of the release of catecholamines. *Pharmac Rev* **32**: 337-362.

Lundberg JM and Hökfelt T (1986) Multiple co-existence of peptides and classical

transmitters in peripheral autonomic and sensory neurons — functional and pharmacological implications. *Prog Brain Res* **68**: 241-262.
Meldrum L.A. and Burnstock G. (1983) Evidence that ATP acts as a co-transmitter with noradrenaline in sympathetic nerves supplying the guinea-pig vas deferens. *Europ. J. Pharmacol.*, **92**: 161-163.
Merrillees N.R.C. (1968) The nervous enviroment of individual smooth muscle cells of the guinea-pig vas deferens. *J. Cell. Biol.*, **37**: 794-817.
Muramatsu I. (1987) The effect of reserpine on sympathetic, purinergic neurotransmission in the isolated mesenteric artery of the dog: a pharmacological study. *Br. J. Pharmc* **91**: 467-474.
Neild TO (1987) Actions of neuropeptide Y on innervated and denervated rat tail arteries. *J Physiol (Lond)* **386**: 19-30.
Neild TO and Hirst GDS (1984) 'The γ-connection': a reply. *Trends Pharmac Sci* February: 56-57.
Pelletier G, Steinbusch HW and Verhofstad A (1981) Immunoreactive substance P and serotonin present in the same dense core vesicles. *Nature* **293**: 71-7
Potter DD, Matsumoto SG, Landis SC, Sah DWY and Furshpan EJ (1986) Transmitter status in cultured sympathetic principal neurons: plasticity, graded expression and diversity *Prog Brain Res* **68**: 103-120.
Ryan LJ, Tepper JM, Sawyer SF, Young SJ and Groves PM (1985) Autoreceptor activation in cental monoamine neurons: modulation of neurotransmitter release is not mediated by intermittent axonal conduction. *Neuroscience* **15**: 925-93
Schipper J, Tilders JH and Mulder H (1980) Extraneuronal catecholamine in the iris of the rat: A consequence of nonsynaptic neurotransmission? *Neuroscience* **5**: 745-751.
Schmitt F.O. (1984) Molecular regulations of brain function: A new view. *Neuroscience*, **13**: 991-1001.
Sneddon P and Burnstock G (1984) Inhibition of excitatory junction potentials in guinea-pig vas deferens by α, β-methylene-ATP: Further evidence for ATP and noradrenaline as cotransmitters. *Eur J Pharmacol* **100**: 85-90.
Sneddon P and Westfall DP (1984) Pharmacological evidence that adenosine triphosphate and noradrenaline are co-transmitters in the guinea-pig vas deferens. *J Physiol (Lond)* **347**: 561-580.
Starke K (1977) Regulation of noradrenaline release by presynaptic receptor systems. *Rev Physiol Biochem Pharmacol* **77**: 1-124.
Starke K (1981) Presynaptic receptors. *Ann Rev Phamacol Toxicol* **21**: 7-30.
Starke K (1987) Presynaptic α-autoreceptors. *Rev Physiol Biochem Pharmacol* **107**: 73-146.
Stjärne L (1975) Basic mechanisms and local feedback control of secretion of adrenergic and cholinergic neurotransmitters. In: Iversen LL, Iversen SD and Snyder SH (eds) *Handbook of Psychopharmacology. Vol. 6.* Plenum Press, New York: 179-233.
Stjärne L (1981) On sites and mechanisms of presynaptic control of noradrenaline secretion. In: Stjärne L, Hedqvist P, Lagercrantz H and Wennmalm Å (eds). *Chemical Neurotransmission 75 years.* Academic Press, London: 257-272.
Stjärne L (1985) Scope and mechanisms of control of stimulus-secretion coupling in single varicosities of sympathetic nerves. *Clinical Science* **68**: (Suppl. 10), 77s-81s.
Stjärne L (1986a) New paradigm: Sympathetic neurotransmission by lateral interaction between secretory units? *News in Physiological Sciences* **1**: 103-106.
Stjärne L (1986b) New paradigm: sympathetic transmission by multiple messengers and lateral interaction between monoquantal release sites? *Trends in Neurosci* **9**: 547-548.
Stjärne L (1987) New paradigm: A digital model of feedback regulation of sympathetic neurotransmitter secretion. In: Vanhoutte PM (ed) Proc Symp "*Mechanisms of Vasodilatation*", Rochester, Minnesota. Raven Press. In press.
Stjärne L and Lundberg JM (1986) On the possible roles of noradrenaline, adenosine 5'-triphosphate and neuropeptide Y as sympathetic cotransmitters in the mouse vas deferens. *Prog Brain Res* **68**: 263-278.
Stjärne L and Åstrand P (1984) Discrete events measure single quanta of adenosine

5'-triphosphate secreted from sympathetic nerves of guinea-pig and mouse vas deferens. *Neuroscience* **13** : 21-28.

Stjärne L and Åstrand P (1985a) Relative pre- and postjunctional roles of noradrenaline and adenosine 5'-triphosphate as neurotransmitters of the sympathetic nerves of guinea-pig and mouse vas deferens. *Neuroscience* **14**: 929-946.

Stjärne L and Åstrand P (1985b) Site of action of presynaptic inhibition mediated via adrenoceptors. In: Szabadi E, Bradshaw CM and Nahorski SR (eds). *Pharmacology of Adrenoceptors.* Macmillan: 157-166.

Stjärne L, Lundberg JM and Åstrand P (1986) Neuropeptide Y — A cotransmitter with noradrenaline and adenosine 5'-triphosphate in the sympathetic nerves of the mouse vas deferens? A biochemical, physiological and electropharmacological study. *Neuroscience* **18**: 151-166.

Suzuki H (1983) An electrophysiological study of excitatory neuromuscular transmission in the guinea-pig main pulmonary artery. *J Physiol (Lond)* **336**: 47-59.

Thuresson-Klein Å (1983) Exocytosis from large and small dense cored vesicles in noradrenergic nerve terminals. *Neuroscience* **10**: 245-252.

Vizi ES (1979) Presynaptic modulation of neurochemical transmission. *Prog Neurobiol* **12**: 181-290.

Vizi ES (1985) Non-synaptic Interactions Between Neurons: Modulation o Neurochemical Transmission. John Wiley & Sons. Chichester, New York, Brisbane, Toronto, Singapore.

Wakade AR and Wakade TD (1984) Do storage vesicle of peripheral sympathetic nerves have more than one life cycle? In: Catecholamines: Basic and Peripheral Mechanism. Pergamon Press: 89-103.

Westfall TC (1977) Local regulation of adrenergic neurotransmission. *Physiol Rev* **57**: 659-728.

Zhu PC, Thuresson-Klein Å and Klein RL (1986) Exocytosis from large dense cored vesicles outside the active synaptic zones of terminals within the trigeminal subnucleus caudalis: A possible mechanism for neuropeptide release. *Neuroscience* **19**: 43-54.

SYMPATHO-ADRENAL CO-STORAGE, RELEASE AND SYNTHESIS OF ENKEPHALINS AND CATECHOLAMINES INDUCED BY ACUTE CNS ISCHEMIA IN PIG

Richard L. Klein, Richard W. Duncan, Thomas J. Selva, Jae-Yang Kong, William E. Clayton, Yun-Long Liaw, Nimr F. Rezk and Åsa Thureson-Klein
University of Mississippi Medical Center, Department of Pharmacology,
Jackson, MS 39216 USA

INTRODUCTION

Proof of Co-storage and Co-release of Enkephalins and Catecholamines at High Concentrations from the Same Subcellular Particles

Earlier improvements in Ulf von Euler's original large dense cored vesicle (LDV) preparation from bovine splenic nerve (~98% sympathetic C fibers) were based on subcellular particle marker activity (Lagercrantz *et al* 1970; Yen *et al* 1973) and ultrastructural morphometry (Klein and Thureson-Klein 1971; Thureson-Klein *et al* 1973a,b; Klein and Thureson-Klein 1974). This LDV preparation produced a high vesicle yield at up to 80-90% purity which made it suitable for chemical composition study (Lagercrantz 1976; Klein *et al* 1982a; Klein and Thureson-Klein 1984) and comparable to that purity achieved to perform analogous studies on bovine adrenomedullary chromaffin granules (Winkler and Westhead 1980; Winkler *et al* 1986).

Collaborative studies proved that enkephalins (ENKs) were co-stored with norepinephrine (NE) at the highest concentration in the purest LDV fraction from bovine splenic nerve (Wilson *et al* 1980a; Klein and Lagercrantz 1981; Klein *et al* 1982b, 1984). This was similar both in concentration per vesicle and in ENK:catecholamine molar ratio to the co-storage calculated (Klein *et al* 1984) for purified bovine adrenal chromaffin granules reported a year earlier (Viveros *et al* 1979). **It was already evident from a perusal of the literature prior to 1981 that the probability was high for neuropeptide storage exclusively in large dense cored vesicles regardless of which co-stored neurotransmitter was present** (Klein *et al* 1982a). The original example may be that of substance P by Ulf von Euler (1963).

In studies on various species, it was also obvious that the ENK contents tend to increase with animal size in both adrenal medullae (Viveros *et al* 1979) and vasa deferentia (Douglas *et al* 1986), where the amount of ENKs paralleled the percentage

NATO ASI Series, Vol. H21
Cellular and Molecular Basis of Synaptic Transmission
Edited by H. Zimmermann

population of LDVs in the particular noradrenergic terminals. Thus, **exocytotic release of ENKs will be quantitatively greater in cow, pig and man than in dog , cat and rabbit, and least in small rodents like guinea pig, rat and mouse.**

Although the co-release phenomenon is inferred at sites where a classical neurotransmitter and neuropeptide are localized at the light microscopic level, actual proof of co-storage in a specific subcellular vesicle from a single cell type is limited to very few other systems; e.g., VIP in certain cholinergic LDVs (Johansson and Lundberg 1981; Lundberg *et al* 1981), substance P in certain serotonergic LDVs (Pelletier *et al* 1981) and, tentatively, NPY in certain noradrenergic LDVs (Fried *et al* 1985, 1986).

Significance of Noradrenergic Vesicle Distribution and Release of Neuropeptides

Prior to the era of discovery of neuropeptide-transmitter co-storage in noradrenergic varicose terminals, it was poorly appreciated that LDVs comprise up to 30-50% of all terminal vesicles equivalent to 80-95% of the total vesicle storage capacity in certain tissues and species, including man (Klein and Lagercrantz 1981; Thureson-Klein and Stjärne 1981; Klein *et al* 1982a; Klein and Thureson-Klein 1984).

In general, **the percentage of LDVs in noradrenergic terminals increases with animal size**; i.e., **axoplasmic transport distance.** This has been interpreted as a requirement to maintain a larger standing population of LDVs in larger animals (Klein *et al* 1982a; Klein and Thureson-Klein 1984; Douglas *et al* 1986), because the LDVs are synthesized and packaged only in the perikaryon. The LDVs provide the primary neurotransmitter synthesis capacity and contain the major store of protein and peptides releasable during stimulus-evoked exocytosis (Fig. 1). In fact, the ENK contents of purified LDVs remain constant during axoplasmic transport to the terminals, like dopamine *B*-hydroxylase, chromogranins (O'Connor and Klein, unpubl) and ATP, in contrast to NE which is continuously synthesized and accumulated (Dahlström 1984; Klein and Thureson-Klein 1984).

Careful ultrastructural examination has made it possible to accumulate micrographs showing literally 100's of formerly elusive exocytotic profiles in noradrenergic (and other) terminals; i.e., the core substance is caught in the act of extrusion. Examples come from many tissues and species, including man (Thureson-Klein and Stjärne 1981; Thureson-Klein 1983, 1984; Thureson-Klein *et al* 1986; Zhu *et al* 1986).

While it can be concluded with relative assurance that the SDVs are primarily responsible for NE release during steady, low frequency physiological stimulation, the **LDVs account for the major release of NE together with soluble core proteins and peptides during stressful stimulation** (Fig 1) (Thureson-Klein and Stjärne 1981; Klein and Thureson-Klein 1984; Thureson-Klein 1984).

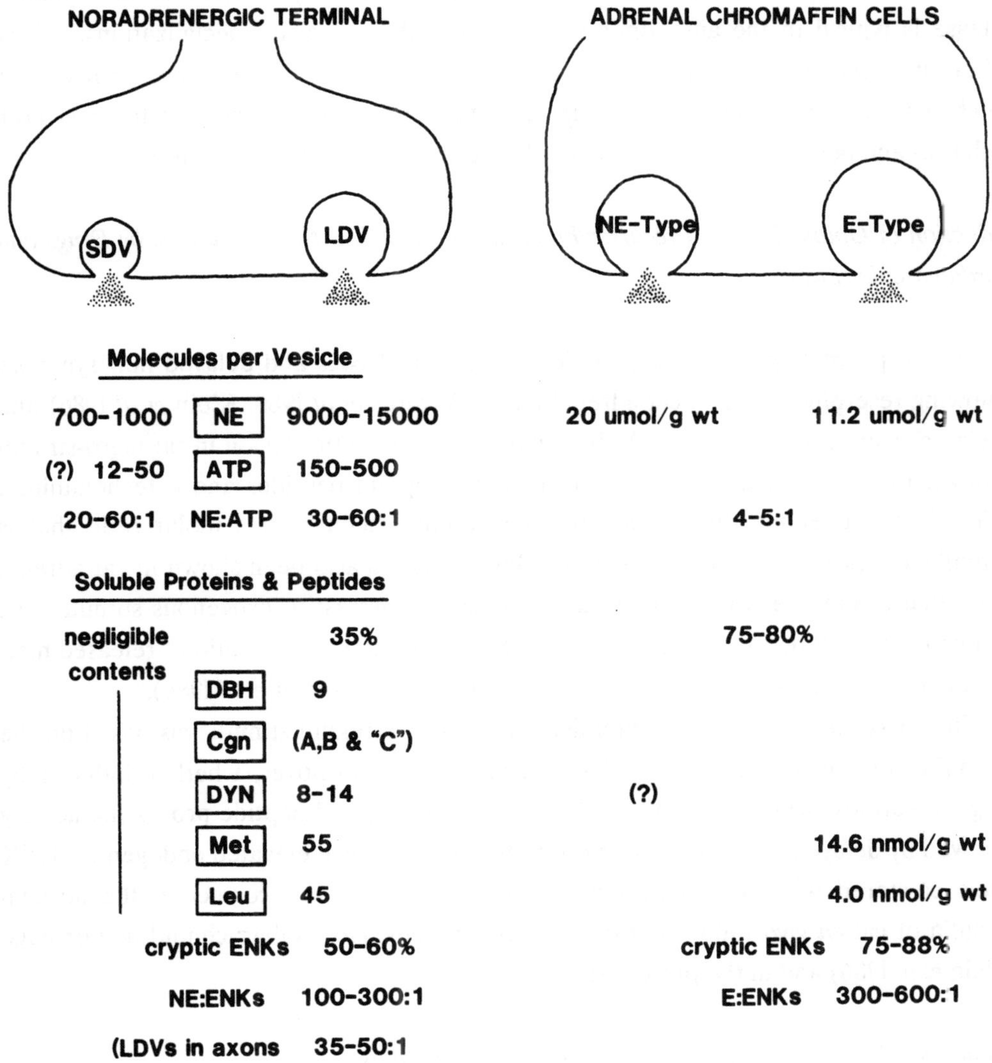

Fig. 1. Potential release of soluble contents from subcellular vesicles. The data compiled for noradrenergic terminals are based on the vasectomized rat vas deferens model for purified SDVs (Fried 1981; Fried *et al* 1984) and the bovine splenic nerve model for LDVs (Klein *et al* 1982a,b; Klein and Thureson-Klein 1984; Klein *et al* 1984; Douglas *et al* 1986; Hagn *et al* 1986). These are compared to partial data for the pig adrenal chromaffin tissue; also see Fig. 3a-d and refer to reviews on bovine chromaffin granules (see text). Abrev: D*B*H, dopamine *B*-hydroxylase; Cgn, chromogranin; Met, met-enkephalin; Leu, leu-enkephalin.

Morphological data from intact organisms subjected to stress strongly imply that

large dense cored vesicles *per se* **release their peptide and transmitter contents by exocytosis primarily at non-synaptic sites, even where synaptic specializations exist**, as in the CNS (Thureson-Klein *et al* 1986, 1987; Zhu *et al* 1986). Of course non-synaptic release is typical in the autonomic system where there are no specializations. Thus, different mechanisms of release can co-exist in the same terminal, providing a mechanism for wider dispersion of neuropeptides and neurotransmitter to react both with pre- and postsynaptic receptors (see Thureson-Klein *et al* this volume).

Induction of Opioid Peptide Precursor Processing and/or De Novo Synthesis by Drugs and Cardiovascular Stress

Acute processing of opioid peptide precursor to ENKs and delayed new synthesis following reserpine occur in cat adrenal gland (Viveros *et al* 1980; Klein *et al* 1986) and bovine chromaffin cell cultures (Wilson *et al* 1980b), and after lesion in rat nigro-striatal pathway (Thal *et al* 1983). Exocytotic release of opioid peptides and catecholamines, induced by exogenous stimuli, occurs from adrenal medulla in a molar ratio that is virtually identical to endogenous stores. Howver, it has also been shown in cat adrenal gland that a non-stressful physiological stimulus, in contrast to exogenous stimuli (e.g., KCl or ACh), will by itself induce acute processing to increase the ratio of released met-ENK:catecholamines relative to that in the gland at rest (Rossier *et al* 1984).

In this context, our original premise for studying stressful stimuli was based on the reasoning that catecholamine depletion and/or rapid turnover, whether induced by drugs or cardiovascular stress, will enhance precursor opioid peptide processing acutely followed by delayed induction of *de novo* synthesis to achieve elevated endogenous ENK levels co-stored with catecholamine. This reasoning proved correct in the adrenal medulla of cat *en vivo* following experimental "closed-space" subarachnoid hemorrhage (Klein *et al* 1986) and in the present studies on pig.

Physiological Roles for Endogenously Released Enkephalins

Potential roles for endogenous ENKs include a relatively generalized presynaptic receptor feedback inhibition of NE release (Pfeiffer and Illes 1984), possible postsynaptic vasodilatation in certain beds (Hanko and Hardebo 1978), negative chronotropism in the myocardium (Ruth and Eiden 1984), tachyphylaxis to NE (Sicuteri *et al* 1984), and protection against cardiovascular shock (Bernton *et al* 1985) and focal cerebral ischemia (Baskin *et al* 1984).

RESULTS AND DISCUSSION

Domestic male pigs, 30-40 kg, were pre-anesthetized with Ketamine (20 mg/kg, i.m.), induced with Pentothal Sodium (2% solution, i.v.) and maintained by Isoflurane inhalation (2% at 150 cc/min in O_2 at 2L/min). Blood was sampled from the vena cava via a femoral vein catheter and from the adrenal vein outflow via a lumbar vein catheter threaded into the renal vein to a point just past the insertion of the suprarenal vein. Acute CNS ischemia (CNS-I) was produced by first ligating the right common carotid, left and right subclavian and vertebral arteries and then initiating the response by clamping the left common carotid for 10 min. Tissues were removed 60-90 min following initiation of the response. Tissue catecholamines were quantitated by HPLC-EC after PCA extraction; blood plasma required prior purification (alumina). Tissue met- and leu-ENKs were quantitated by RIAs using specific antibodies after acetic acid extraction and lyopholization; blood plasma required prior purification (SepPak).

TABLE 1. CATECHOLAMINES AND ENKEPHALINS IN PERTINENT TISSUES OF THE DOMESTIC PIG

	Catecholamine (nmol/g±SEM)		Enkephalin (pmol/g±SEM)		
Tissue	Norepinephrine	Epinephrine	irMet-ENK	irLeu-ENK	Met-/Leu Molar Ratio
Adrenal medulla	20,300±1100 (n=32)	11.200±400 (n=30)	14,600±1500 (n=35)	4000±400 (n=37)	3.6
Vas Deferens	104±13 (n=19)	*	541±58 (n=25)	152±16 (n=26)	3.6
Mesenteric artery	39±6 (n=15)	*	145±30 (n=16)	20±5 (n=15)	9.3
vein	26±3 (n=13)	*	42±10 (n=16)	13±3 (n=16)	3.2
Coronary artery	47±15 (n=8)	*	88±15 (n=14)	23±4 (n=14)	3.8
Middle cerebral artery	128±86 (n=5)	*	496±119 (n=4)	66±14 (n=4)	7.5

Catecholamine and Enkephalin Contents of Tissues in Pig

The data in Table 1 summarize the catecholamine and enkephalin contents of tissues pertinent to this study. Tissues generally contain a 3-4:1 molar ratio of met-:leu-ENK like the adrenal medulla; however, much higher ratios can occur; e.g., mesenteric artery (9:1) and middle cerebral artery (7.5:1). The adrenal medulla NE and ENK values in pig rival those in the bovine species except that E-type chromaffin cells do not predominate.

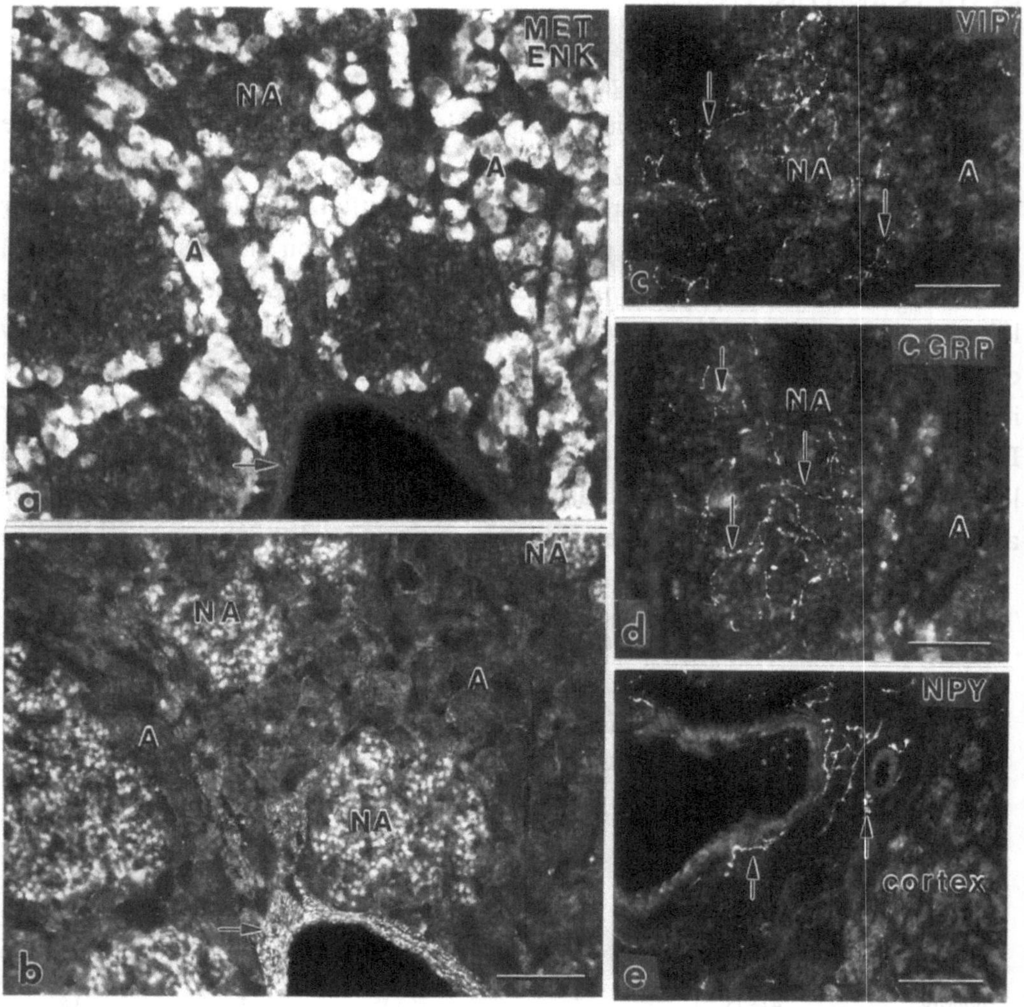

Fig. 2. Neuropeptide and catecholamine distribution in adrenal glands of domestic pig (*Sus scrofa*); bar 100 um. (a) Adrenaline (A) chromaffin cells with strong met-ENK immunoreactivity surround noradrenaline (NA) cells with little or no activity. A vessel (arrow) also lacks nerve fibers with met-ENK reactivity in the plane of section. (b) Islands with noradrenaline (NA) fluorescence among adrenaline (A) cells correspond to islands lacking met-ENK reactivity in 2a. The vessel wall (arrow) shows some noradrenergic fibers and elastic tissue autofluorescence. (c) Thin nerve fibers with VIP-immunoreactivity (arrows) among noradrenaline (NA) cells that are non-reactive, likely reflect cholinergic innervation. (d) CGRP-immunofluorescent fibers have a similar distribution among noradrenaline (NA) cells and a few fibers occur among adrenaline (A) cells. (e) NPY fibers occur mainly around blood vessels in pig adrenal medulla.

The NE and ENK contents of pig vas deferens also are similar to the bovine species (Douglas *et al* 1986), except for a higher met-:leu-ENK ratio. The vas deferens and middle cerebral artery in pig have exceptionally high met-ENK contents, which in part correspond to the dense noradrenergic fiber innervation and in part reflect the relatively high, 30-50%, population of LDVs (Douglas *et al* 1986).

Differential catecholamine fluorescence (formaldehyde at specific temperature) and met-ENK, VIP, CGRP and NPY immunofluorescence (FITC second antibody) distributions are shown in light micrographs of pig adrenal chromaffin tissue (Fig. 2a-e). Note by comparisons of adjacent sections that met-ENK distribution (Fig. 2a) corresponds precisely to epinephrine-type chromaffin cells and not to clusters of norepinephrine cells (Fig. 2b). Leu-ENK distribution (not shown) essentially mimics met-ENK and this morphological distribution is supported by the analytical results of acute CNS-I presented below. The pig chromaffin cells are unreactive to NPY, unlike the bovine species, but NPY fibers do innervate blood vessel walls (Fig. 2e).

Reserpine Treatment and Recovery: Catecholamine and Enkephalin Contents

Reserpine treatment of pigs at near antihypertensive doses (0.05 mg/kg, i.m., every other day); i.e., ~3 times the human loading dose recommended during the first 1-2 wks, were purposefully designed to cause partial adrenomedullary catecholamine depletion in order to induce ENK precursor peptide processing and *de novo* synthesis.

Reserpine caused an immediate and gradual depletion of E, while NE content was initially increased and then depleted to a lesser degree (Fig. 3a). After treatment with reserpine for 7 days, NE recovered to control within 3 days (Fig. 3b) while E was not fully recovered after 14 days cessation of treatment.

Conversely, both ENKs were markedly elevated already by 3 days (primarily processing?) and reached about 8 times controls (*de novo* synthesis) by 30 days reserpine (Fig. 3c). The elevated ENK values reached at 7 days treatment were maintained through at least 14 days recovery (Fig. 3d). Calculations of total ENK (x 10^3):E molar ratios changed from 1.7:1 in controls to 55:1 after 30 days reserpine.

Reserpine for 7 to 14 days caused a 70-85% depletion of NE in all of the noradrenergically innervated peripheral tissues studied (see control values, Table 1). A small increase or no change in ENK contents of most tissues was found. Mesenteric vein was a notable exception, where a time-dependent increase of both ENKs occurred to reach 3 times control at 7 days and 4.5 times controls at 14 days (data omitted). Following 7 days treatment with reserpine, NE contents of vas deferens had returned to control by 14 days, but those in blood vessels remained 20-40% depleted. Previously

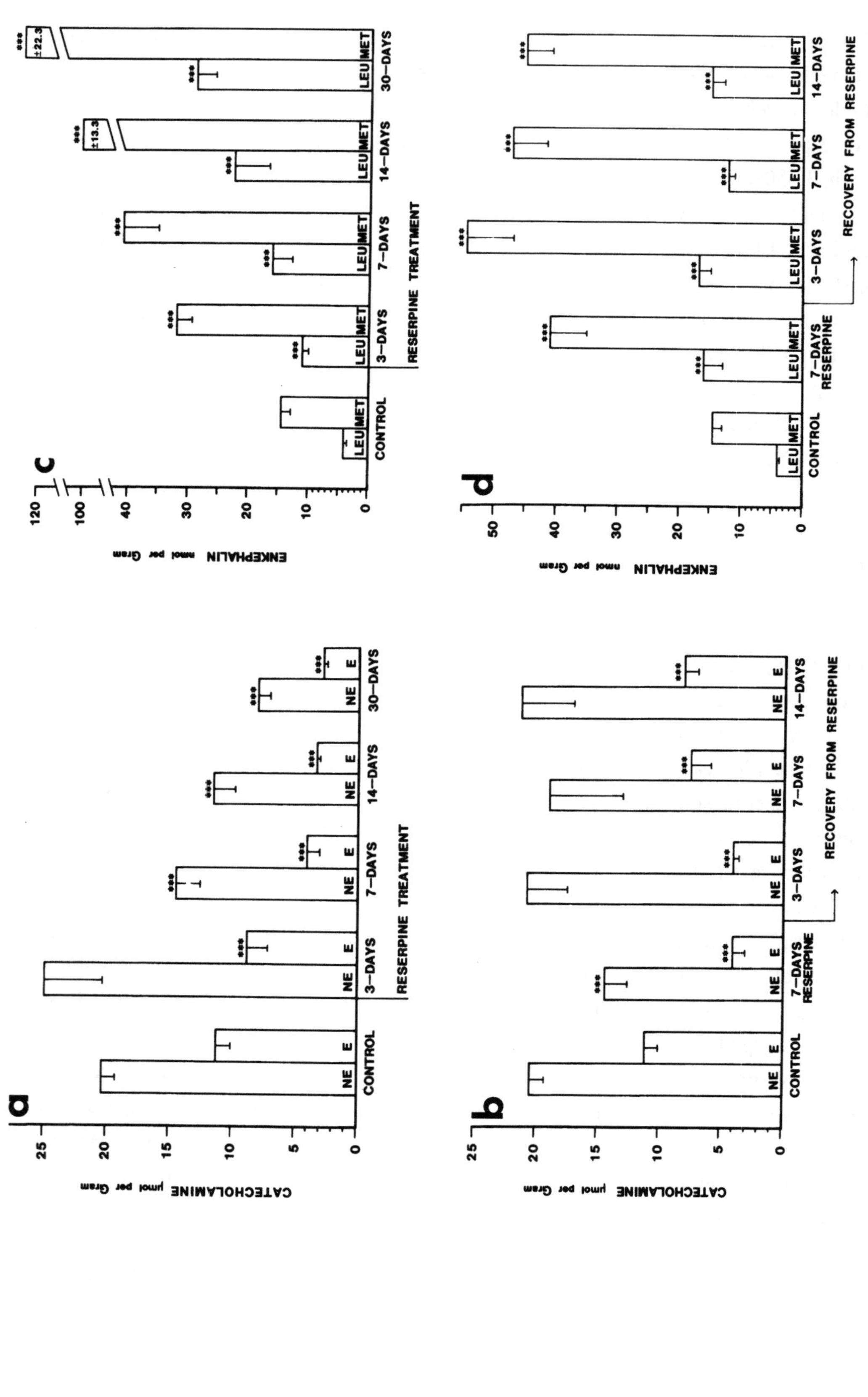
a
b
c
d
CATECHOLAMINE µmol per Gram
ENKEPHALIN nmol per Gram
NE
E
LEU
MET
CONTROL
3-DAYS
7-DAYS
14-DAYS
30-DAYS
RESERPINE TREATMENT
7-DAYS RESERPINE
RECOVERY FROM RESERPINE
±13.3
±22.3
***

elevated ENK contents (mesenteric vein) decreased to control and all other tissues remained in the control range at 7 and 14 days recovery from reserpine. Thus, **reserpine treatment has a tendency to induce acute opioid peptide precursor processing reflected by a small increase in ENK contents in the short term and/or *de novo* synthesis of ENKs in certain sympathetic nerves as well as in the adrenal medulla.** The amount of precursor susceptible to processing may vary in particular terminals.

Acute CNS Ischemia (CNS-I): Tissue Catecholamine and Enkephalin Contents and Circulating Plasma Levels

Control pigs:

Of all stimuli known to cause an increase in arterial pressure, acute CNS-I is by far the most powerful (Cushing 1901; Guyton 1948). In control pigs, the 10 min occlusion of cerebral blood supply (~85% reduction) produced an immediate rise in mean arterial pressure (MAP) of 75 ± 9 mm Hg above baseline (85-100 mm Hg) and a brisk tachycardia. This rise in MAP corresponds to a cerebral arterial pressure decrease to ~20 mm Hg (Sagawa *et al* 1961). Upon removal of the clamp, MAP returned to normal.

Adrenal NE content was significantly increased and E unchanged by acute CNS-I (Fig. 4a). This result is due to compensation by catecholamine synthesis induction which is evident from the markedly elevated circulating levels (Fig. 5). In the femoral vein (central venous sample), NE was increased 500% and E 650% above control at the peak sampling time, 5 min after the initiation of CNS-I (Fig. 5a). Adrenal vein outflow values are even more impressive with a 1450% increase in E at the peak in 1 min and a 650% increase in NE above control at its peak in 5 min (Fig. 5b).

Adrenal met-ENK was depleted about 50% by acute CNS-I and leu-ENK was unchanged (Fig. 4b). Induction of precursor peptide processing likely occurred, as circulating central venous levels of both ENKs are significantly increased, 60-75% at the 1 min peak sampling time (Fig. 5c) and even greater at the 1 min peak sampling time in the adrenal vein outflow (Fig. 5d).

The plasma disappearance halftimes from peak samples were estimated to be about 7-8 min for catecholamines and 10-12 min for ENKs.

In other tissues, the acute CNS-I caused 30-35% depletion of NE and 35-70% deple-

Fig 3 (opposite page): Pig adrenal medulla catecholamine contents (a) with reserpine treatment and (b) with recovery after 7 days reserpine; and enkephalin contents (c) with reserpine treatment and (d) with recovery after 7 days reserpine. Bars ± SEM; *** p 0.001 or less compared to the respective control.

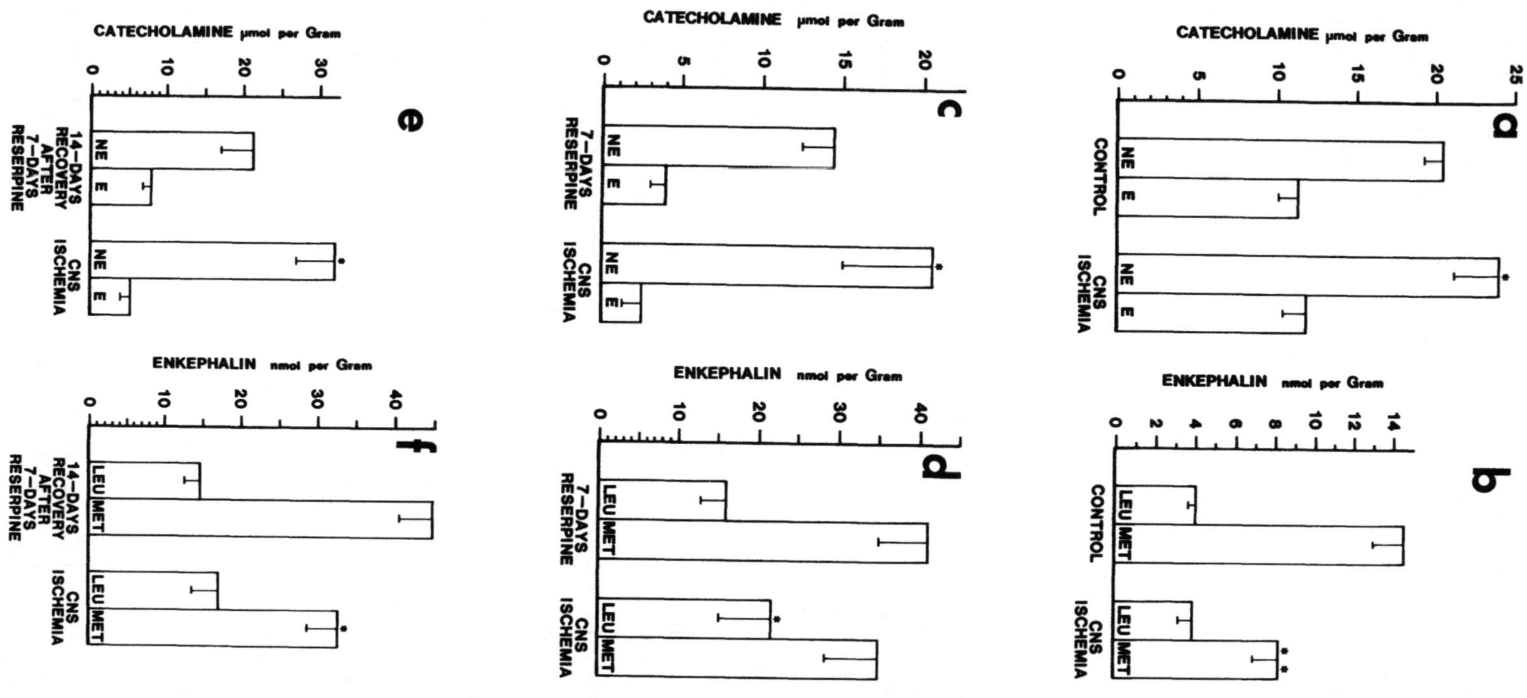

Fig 4. Effects of 10 min acute CNS ischemia on pig adrenal medulla: catecholamines in (a) control, (c) after 7 days reserpine and (e) at 14 days recovery from 7 days reserpine; and enkephalins in (b) control, (d) after 7 days reserpine and (f) at 14 days recovery from 7 days reserpine. Bars ± SEM; * p=0.05, ** p=0.01 or less compared to the respective control.

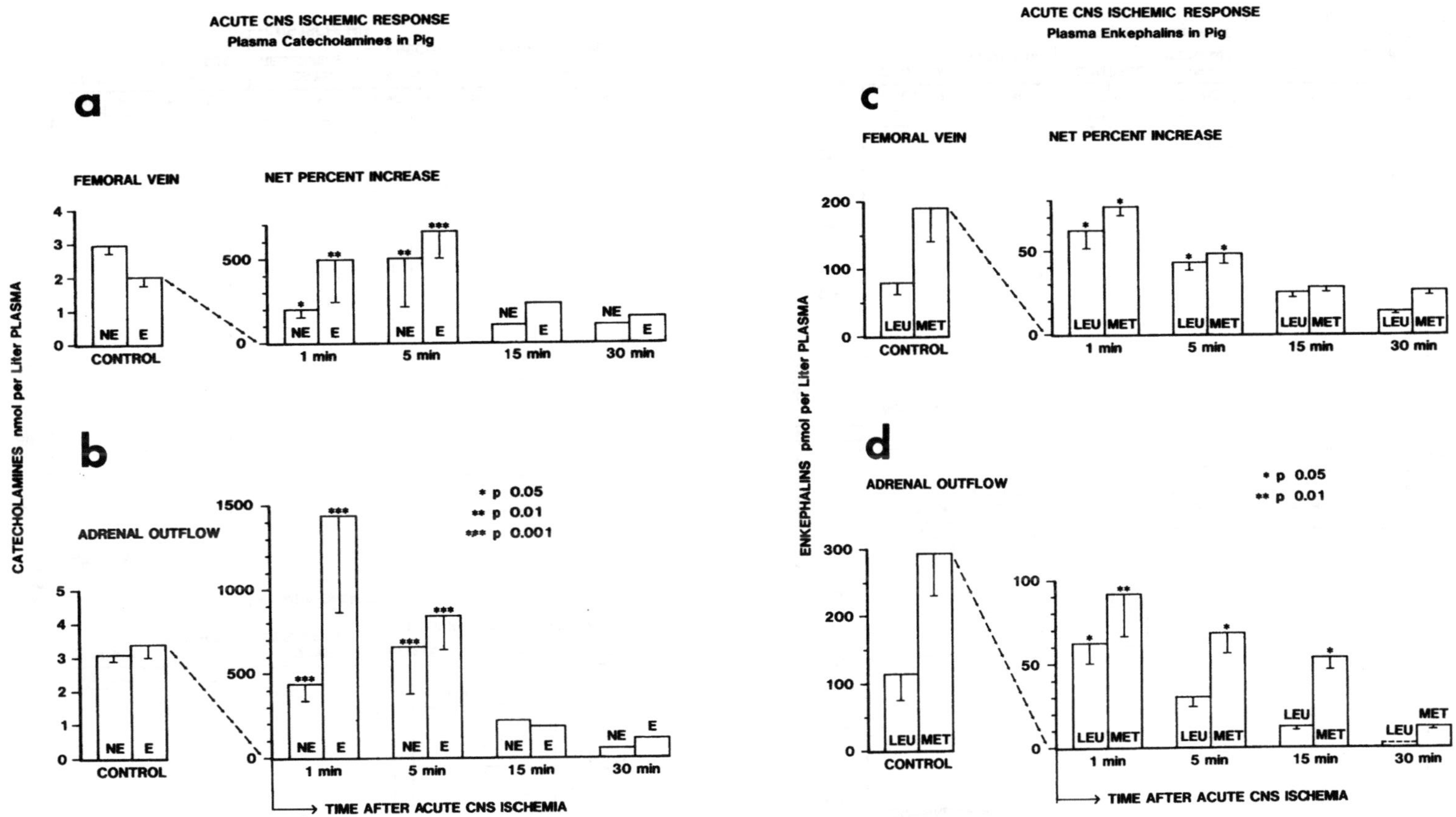

Fig 5. Plasma catecholamines in (a) femoral vein and (b) adrenal vein outflow, at various times after initiation of acute CNS ischemia; plasma enkephalins in (c) femoral vein and (d) adrenal vein outflow, at various times after initiation of acute CNS ischemia. Bars ± SEM; statistical significance of changes are compared to the respective controls.

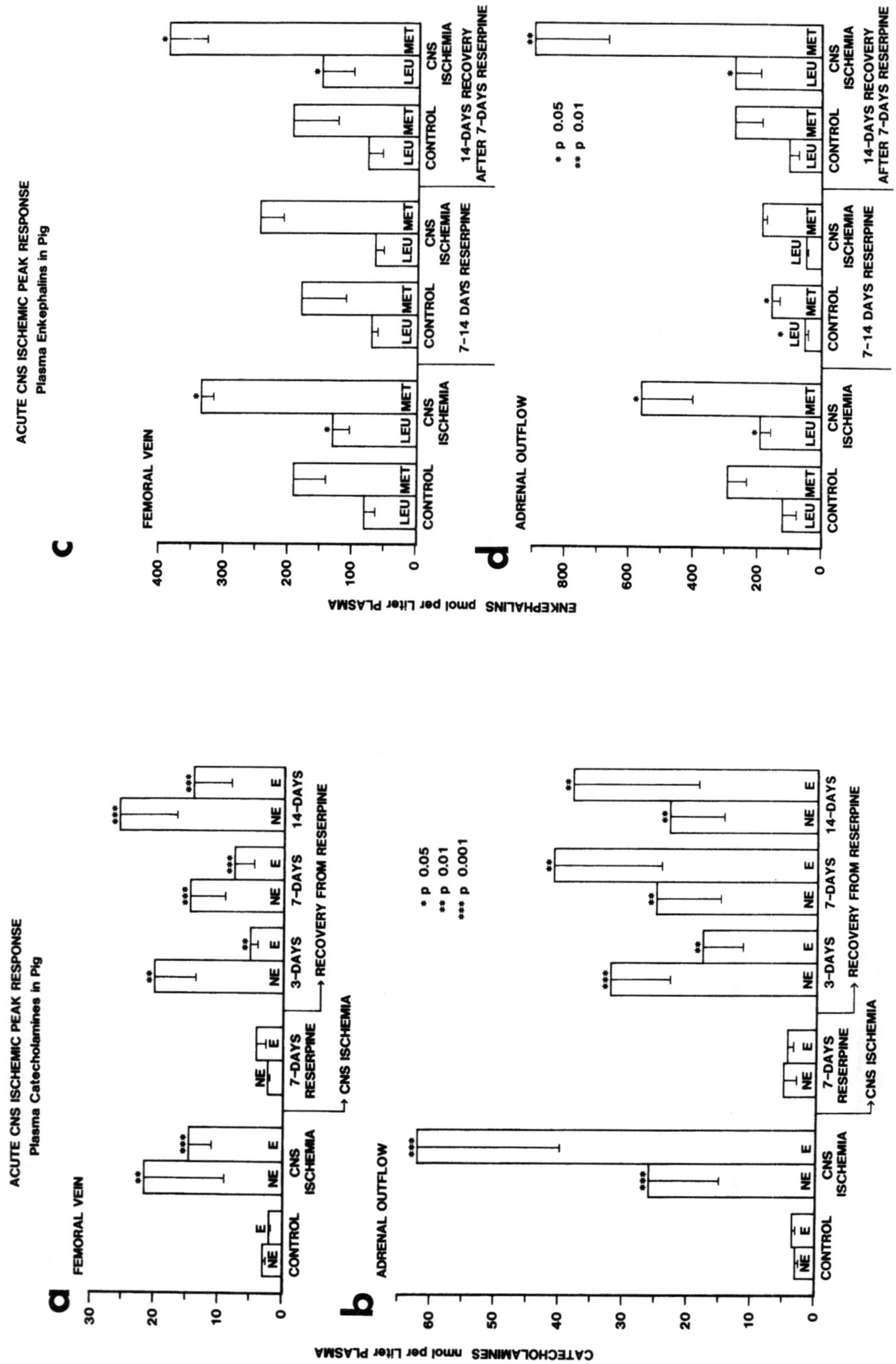
ACUTE CNS ISCHEMIC PEAK RESPONSE
Plasma Catecholamines in Pig
a
FEMORAL VEIN
b
ADRENAL OUTFLOW
CATECHOLAMINES nmol per Liter PLASMA
NE
E
CONTROL
CNS ISCHEMIA
7-DAYS RESERPINE
→ CNS ISCHEMIA
3-DAYS
7-DAYS
14-DAYS
→ RECOVERY FROM RESERPINE
* p 0.05
** p 0.01
*** p 0.001
ACUTE CNS ISCHEMIC PEAK RESPONSE
Plasma Enkephalins in Pig
c
FEMORAL VEIN
d
ADRENAL OUTFLOW
ENKEPHALINS pmol per Liter PLASMA
LEU
MET
CONTROL
CNS ISCHEMIA
7-14 DAYS RESERPINE
14-DAYS RECOVERY AFTER 7-DAYS RESERPINE
* p 0.05
** p 0.01

tion of both ENKs in all of the blood vessels studied (see control values, Table 1). Vas deferens was similarly depleted of ENKs, but NE actually increased 33%, suggesting that new synthesis predominated over release.

Reserpine treated pigs:

At the doses used, reserpine treatment for 7 or 14 days had no statistical effect on resting NE or E levels in the femoral or adrenal outflow plasma (Fig. 6a,b). At the doses used, femoral vein ENK levels were unchanged (Fig. 6c), but the higher adrenal vein outflow values typical of controls were significantly reduced (Fig. 6d) to a level not different from that in the central venous circulation.

The acute CNS-I response, so marked in control pigs, was nearly completely blocked by reserpine, including increases in MAP, peak increases in circulating NE and E levels (Fig. 6a,b) and peak increases in circulating ENKs (Fig. 6c,d).

In the adrenal medulla, partially depleted catecholamines responded to CNS-I like controls (Fig. 4c), but met-ENK depletion was not significant from the 3-fold, reserpine-induced, elevation and leu-ENK was actually increased (Fig. 4d). The latter is likely due to enhanced precursor processing in combination with blocked release.

In other tissues, acute CNS-I caused a further reduction (not significant) in NE contents. There was no change in ENK contents of vas deferens or middle cerebral artery by acute CNS-I after 7 days reserpine, which may reflect largely pre-processed peptide stores. However, there was an 80-450% elevation in met-ENK and a lesser increase in leu-ENK, especially in the coronary artery and mesenteric vein .

Recovery from reserpine:

After 7 and 14 days recovery from 7 days reserpine treatment, the increase in MAP due to acute CNS-I was completely recovered, however, a notable delay in the onset of a brisk tachycardia was evident.

The acute CNS-I effect on adrenal NE and E was similar to that seen in controls (Fig. 4e). Peak increases in circulating levels of NE were similar to controls already at 3 days recovery (Fig. 6a,b). Peak increases in circulating E were regained more gradually and approached control responses by 14 days recovery, but were ~33% below the net

Fig. 6 (opposite page): Plasma catecholamines in (a) femoral vein and (b) adrenal vein outflow showing peak responses (see Fig. 5a,b for sample times) to acute CNS ischemia; plasma enkephalins in (c) femoral vein and (d) adrenal vein outflow showing peak responses (see Fig. 5c,d for sample times) to acute CNS ischemia. Bars ± SEM; statistical significance of changes are compared to the respective controls.

1450% increase found in the control adrenal vein outflow after acute CNS-I (Fig. 6a,b)

Adrenal ENK content responses to acute CNS-I after 14 days recovery from reserpine (Fig. 4f) were similar to those in controls; i.e., no change in leu-ENK and a significant depletion of met-ENK, which now occurred from the reserpine-induced, 3-fold elevation in stores. The net increases at peak of ENKs due to CNS-I in the central venous circulation now were 35-50% above the control response (Fig. 6c), but were not quite significant at the p 0.05 level with the number of experiments performed. However, the absolute net increases of ENKs in the adrenal vein outflow, due to CNS-I after 14 days recovery from reserpine, were significantly elevated to 2.3 times that of the control response (Fig. 6d). The latter reflected reserpine elevated stores of ENKs which were now susceptible to exocytotic release after 14 days recovery.

In other tissues upon acute CNS-I there now was resistance to NE depletion even after 14 days recovery, except coronary artery in which a 33% decrease in NE occurred (data omitted). In contrast to the depleting effect in controls after CNS-I, ENKs also showed no change or an insignificant small decrease. **Thus, general resistance to enkephalin depletion as well as to norepinephrine depletion were evident in many peripheral tissues even after 14 days recovery from reserpine treatment.**

SUMMARY

1. Neuronal peptides are stored exclusively in LDVs, regardless of costored transmitter. Proof of co-storage in the same LDV population exists for: (a) ENK - NE; (b) NPY - NE (?); (c) VIP - ACh; (d) Sub P - 5HT.

2. Potential for exocytotic release of ENKs co-stored with NE in LDVs is: (a) greater in larger animals; (b) proportional to the percentage of LDVs in terminals - related to axoplasmic transport distance; (c) primarily at non-synaptic sites, even where specializations exist; (d) primarily during stress.

3. Acute processing of precursor opioid peptide and delayed *de novo* synthesis can result from cardiovascular stress (e.g., subarachnoid hemorrhage in cat adrenal; CNS-I in pig adrenal). The extent of induction in neuronal systems may depend in part on inherent amounts of stored precursor (e.g., LDVs in bovine splenic nerve contain only 50-60% of ENKs in unprocessed opioid peptide, whereas mesenteric vein in pig shows *de novo* synthesis and processing more like adrenal medulla.

Studies in pig

4. CONTROL: Resting plasma levels of NE reflect release primarily from noradrenergic SDVs together with E from adrenal chromaffin cells. Resting plasma

levels of leu- and especially met-ENK reflect release from E-type chromaffin cells more than from noradrenergic LDVs.

Acute CNS-I causes a 500-650% net increase in peak plasma NE and E and 60-75% net increase in ENK levels. NE originates from LDVs and SDVs more than from NE-type chromaffin cells; ENKs originate from E-type chromaffin cells slightly more than from LDVs, but both are greatly stimulated.

5. RESERPINE TREATED: Resting plasma levels of NE, E and ENKs are little changed (homeostasis); a decreased release of E and ENKs occurs from chromaffin cells. ENK precursor processing and *de novo* synthesis are greatly enhanced in E-type chromaffin cells, and to a lesser degree in certain peripheral sympathetic innervations with exceptions.

Acute CNS-I responses are nearly completely blunted including MAP increase, elevations in plasma NE, E and ENKs and tissue depletions. A centrally mediated reserpine inhibition of sympathetic outflow is apparent and immature stages of E-type chromaffin granules and noradrenergic LDVs may preclude release.

6. RECOVERY FROM RESERPINE: Resting plasma NE, E and ENKs are normal. Adrenal NE recovers rapidly prior to E (incomplete at 14 days); ENKs remain elevated from reserpine induction. Neuronal NE contents are also incompletely recovered at 14 days, while ENKs are normal.

Acute CNS-I peak plasma levels of NE and E resemble controls, but net increases in ENKs are 2-3 times the control response, reflecting elevated stores in E-type chromaffin cells. Usual tissue depletions of noradrenergic NE and ENKs are largely blunted at 14 days recovery. Homeostatic compensation by NE synthesis and acute peptide processing to ENKs may be compounded by residual reserpine effects to inhibit release.

ACKNOWLEDGEMENTS

Support from the American Heart Association, Mississippi Affiliate and BRSG 2 S07 RR05386-25 are greatfully acknowledged. The authors are indebted to Ulrika C. Matthiessen for skillful graphic illustrations and to Michael Moody for photography.

REFERENCES

Baskin DS, Hosobuchi Y, Loh HH and Lee NM (1984) Dynorphin (1-13) improves survival in cats with focal cerebral ischemia. Nature *312:* 551-552.

Bernton EW, Long JB and Holaday JW (1985) Opioids and neuropeptides: mechanisms in circulatory shock. Fed Proc FASEB *44*: 290-299.

Cushing H (1901) Concerning a definite regulatory mechanism of the vasomotor center which controls blood pressure during cerebral compression. Bull Johns Hopkins Hosp *12*: 290-292.

Dahlström A (1983) Presence, metabolism, and axonal transport of transmitters in peripheral mammalian axons. In: Lajtha A (ed). Handbook of Neurochemistry Vol 5. Plenum Publ, New York: 405-441.

Douglas BH II, Duff RB, Thureson-Klein ÅK and Klein RL (1986) Enkephalin contents reflect noradrenergic large dense cored vesicle populations in vasa deferentia. Regul Peptides *14* 193-210.

von Euler US (1963) Substance P in subcellular particles in peripheral nerves. Ann NY Acad Sci *104*: 449-461.

Fried G (1981) Small noradrenergic storage vesicles isolated from rat vas deferens - biochemical and morphological characterization. Acta Physiol Scand *111*, Supple 493: 1-28.

Fried G, Thureson-Klein Å and Lagercrantz H (1981) Noradrenaline content correlated to matrix density in small noradrenergic vesicles from the rat seminal ducts. Neuroscience *6*: 787-800.

Fried G, Lagercrantz H, Klein RL and Thureson-Klein Å (1984) Large and small dense cored vesicles - origin, contents and functional significance. In: Usdin E (ed). 5th Catecholamine Sympos, Neurology and Neurobiology Vol 8A. Alan R Liss, New York: 45-54.

Fried G, Lundberg JM and Theodorsson-Nordheim E (1985) Subcellulaar storage and axonal transport of neuropeptide Y (NPY) in relation to catecholamines in the cat. Acta Physiol Scand *125*: 145-154.

Fried G, Terenius L, Brodin E, Efendic S, Dockray G, Fahrenkrug J, Goldstein M and Hokfelt T (1986) Neuropeptide Y, enkephalin and noradrenaline coexist in sympathetic neurons innervating bovine spleen; biochemical and immunohistochemical evidence. Cell Tissue Res *243*: 495-508.

Guyton AC (1948) Acute hypertension in dogs with cerebral ischemia. Am J Physiol *154*: 45-54.

Hagn C, Klein RL, Fischer-Colbrie R, Douglas BH II and Winkler H (1986) An immunological characterization of five common antigens of chromaffin granules and of large dense-cored vesicles of sympathetic nerve. Neurosci Lett *67*: 295-300.

Hanko JH and Hardebo JE (1978) Enkephalin-induced dilation of pial arteries *in vitro* probably mediated by opiate receptors. Europ J Pharmacol *51*: 295-297.

Johansson O and Lundberg JM (1981) Ultrastructural localization of VIP-like immunoreactivity in large dense cored vesicles of "cholinergic type" terminals in cat exocrine glands. Neuroscience *6*: 847-862.

Klein RL and Lagercrantz H (1981) Noradrenergic vesicles: composition and function. In: Stjärne L, Hedqvist P, Lagercrantz H and Wennmalm Å (eds). Second Nobel Conf Stockholm 1980, Chemical neurotransmission - 75 years. Academic Press, London: 69-83.

Klein RL and Thureson-Klein Å (1971) An electron microscopic study of noradrenaline storage vesicles isolated from bovine splenic nerve trunk. J Ultrastruct Res *34*: 473-491.

ibid (1974) Pharmaco-morphological aspects of large dense cored adrenergic vesicles. Fed Proc FASEB *33*: 2195-2206.

ibid (1984) Noradrenergic vesicles: molecular organization and function. In: Lajtha A (ed). Handbook of Neurochemistry Vol 7. Plenum Publ, New York: 71-109.

Klein RL, Lagercrantz H and Zimmermann H (eds)(1982a) Neurotransmitter vesicles: composition, structure and function. Academic Press, London: 1- 384.

Klein RL, Wilson SP, Dzielak DJ, Yang WH and Viveros OH (1982b) Opioid peptides and noradrenaline co-exist in large dense-cored vesicles from sympathetic nerves. Neuroscience *7*: 2255-2261.

Klein RL, Lemaire S, Thureson-Klein Å and Day R (1984) Leu-enkephalin, dynorphin and bombesin contents of a highly purified large dense cored vesicle fraction from bovine splenic nerve. In: Fraioli F, Isadori A and Mazzetti M (eds). Opioid peptides in the periphery. Elsevier, Amsterdam: 205-212.

Klein RL, Yabuno N, Peeler DF, Thureson-Klein Å, Douglas BH II, Duff RB and Clayton WE (1986) Adrenal enkephalin and catecholamine contents following subarachnoid hemorrhage in cats. Neuropeptides *8*: 143-158.

Lagercrantz H (1976) On the composition and function of large dense cored vesicles in sympathetic nerves. Neuroscience *1*: 81-92.

Lagercrantz H, Klein RL and Stjärne L (1970) Improvements on the isolation of noradrenergic storage vesicles from bovine splenic nerves. Life Sci *9*: 639-650.

Lundberg JM, Fried G, Fahrenkrug J, Holmstedt B, Hökfelt T, Lagercrantz H, Lundgren G and Änggård Å (1981) Subcellular fractionation of cat submandibular gland: comparative studies on the distribution of acetylcholine and vasoactive intestinal polypeptide (VIP). Neuroscience *6*: 1001-1010.

Pelletier G, Steinbusch HWM and Verhofstad AAJ (1981) Immunoreactive substance P and serotonin present in the same dense-cored vesicles. Nature *293*: 71-72.

Pfeiffer N and Illes P (1984) Hypotensive effect of opioids: peripheral mechanisms. Trends Pharmacol Sci *5*: 414-415.

Rossier J, Foutz AS and Chaminade M (1984) Release of enkephalins and precursors from perfused cat adrenal glands. In: Fraioli F, Isidori A and Mazzetti M (eds). Opioid peptides in the periphery. Elsevier, Amsterdam: 39-46.

Ruth JA and Eiden LE (1984) Modulation of peripheral cardiovascular function by enkephalins. In: Fraioli F, Isidori A and Mazzetti M (eds). Opioid peptides in the periphery. Elsevier, Amsterdam: 103-110.

Sagawa K, Ross JM and Guyton AC (1961) Quantitation of the cerebral ischemic pressor response in dogs. Am J Physiol *200*: 1164-1168.

Sicuteri F, DelBianco PL and Michelacci S (1984) Main role of endogenous opioids in homeostatic tachyphylaxis of sympathetic vascular junction in man. In: Fraioli F, Isidori A and Mazzetti M (eds). Opioid peptides in the periphery. Elsevier, Amsterdam: 111-117.

Thal LJ, Sharpless NS, Hirschhorn ID, Horowitz SG and Makman MH (1983) Striatal met-enkephalin concentration increases following nigrostriatal denervation. Europ J Pharmacol *84*: 3297-3301.

Thureson-Klein Å (1983) Exocytosis from large and small dense cored vesicles in noradrenergic nerve terminals. Neuroscience *10*: 245-252.

ibid (1984) The roles of small and large noradrenergic vesicles in exocytosis. In: Usdin E (ed). 5th Catecholamine Sympos, Neurology and Neurobiology Vol 8. Alan R Liss, New York: 79-87.

Thureson-Klein Å and Stjärne L (1981) Dense-cored vesicles in actively secreting noradrenergic neurons. In: Stjärne L, Hedqvist P, Lagercrantz H and Wennmalm Å (eds). Second Nobel Conf Stockholm 1980, Chemical neurotransmission - 75 years. Academic Press, London: 153-164.

Thureson-Klein Å, Klein RL and Lagercrantz H (1973a) Highly purified splenic nerve vesicles: early post mortem effects on ultrastructure. J Neurocytol *2*: 13-27.

Thureson-Klein Å, Klein RL and Yen SS (1973b) Ultrastructure of highly purified sympathetic nerve vesicles: correlation between matrix density and norepinephrine content. J Ultrastruct Res *43*: 18-35.

Thureson-Klein Å, Klein RL and Zhu PC (1986) Exocytosis from large dense cored vesicles as a mechanism for neuropeptide release in the peripheral and central nervous system. Scan Electr Micr *I*: 179-187.

Thureson-Klein Å, Zhu PC and Klein RL (1987) Nonsynaptic exocytosis from large dense cored vesicles as a mechanism for non-directional neuropeptide release. In: Nobin A, Owman C and Arneklo-Nobin A (eds). Neuronal messengers in vascular function. Elsevier Sci Publ, Amsterdam: 211-217.

Viveros OH, Diliberto EJ Jr, Hazum E and Chang KJ (1979) Opiate-like materials in the adrenal medulla: evidence for storage and secretion with catecholamines. Molec Pharmacol. *16*: 1101-1108.

ibid (1980) Enkephalins as possible adrenomedullary hormones: storage, secretion and regulation of synthesis. In: Costa E and Trabucchi M (eds). Neural peptides and neural communication. Raven Press, New York: 191-204.

Wilson SP, Klein RL, Chang KJ, Gasparis MS, Viveros OH and Yang WH (1980a) Are opioid peptides co-transmitters in noradrenergic vesicles of sympathetic nerves? Nature *288*: 707-709.

Wilson SP, Chang KJ and Viveros OH (1980b) Synthesis of enkephalins by adrenal medullary chromaffin cells: reserpine increases incorporation of radiolabelled amino acids. Proc Nat Acad Sci, USA, *77*: 4364-4367.

Winkler H and Westhead E (1980) The molecular organization of adrenal chromaffin granules. Neuroscience *5*: 1803-1823.

Winkler H, Apps DK and Fischer-Colbrie R (1986) The molecular function of adrenal chromaffin granules: established facts and controversial results. Neuroscience *18*: 261-290.

Yen SS, Klein RL and Chen-Yen SH (1973) Highly purified splenic nerve vesicles: early postmortem effects on norepinephrine content and pools. J Neurocytol *2*: 1-12.

Zhu PC, Thureson-Klein Å and Klein RL (1986) Exocytosis from large dense cored vesicles outside the active synaptic zones of terminals within the trigeminal subnucleus caudalis: a possible mechanism for neuropeptide release. Neuroscience *19*: 43-54.

EXTRACELLULAR METABOLISM OF ATP AT THE CHOLINERGIC ELECTROMOTOR SYNAPSE: CHARACTERIZATION AND LOCATION OF ECTONUCLEOTIDASES

Ernst J.M. Grondal and Herbert Zimmermann
AK Neurochemie, Zoologisches Institut
der J.W. Goethe-Universität
6000 Frankfurt am Main
Federal Republic of Germany

INTRODUCTION

ATP is stored in cholinergic synaptic vesicles of the peripheral (Dowdall et al 1974; Volknandt and Zimmermann 1986) and central (Richardson and Brown 1987) nervous system, in adrenergic vesicles, in chromaffin granules and also in serotonin containing granules of blood platelets (ref. Zimmermann 1982). Both, for the peripheral and central cholinergic system (Silinsky 1975; Morel and Meunier 1981; Schweitzer 1987; Richardson and Brown 1987) as well as for the adrenergic nerve terminals (Stjärne and Åstrand 1984) there is evidence for a corelease of ATP and the classical neurotransmitter substance. This observation raises the two major questions: that of the physiological function and the metabolic fate of released ATP.

Whereas for the adrenergic sympathetic nervous system an excitatory postsynaptic action of ATP has been described, e.g. for the motor innervation of the rat vas deferens (Stjärne and Åstrand 1984) or the rabbit ear artery (Benham et al. 1987) and the action of ATP on smooth muscle may be via a receptor-operated Ca^{2+}-permeable ion channel (Benham and Tsien 1987), the situation is less clear for the cholinergic system. Cholinergic synaptic transmission at the neuromuscular junction is abolished by drugs which block recirculation of the neurotransmitter acetylcholine (ACh). This appears to exclude a direct role of ATP in cholinergic synaptic signal transmission. Several studies suggest, however, that ATP may potentiate the postsynaptic function of ACh and that it may increase cation influx into muscle cells

NATO ASI Series, Vol. H21
Cellular and Molecular Basis of Synaptic Transmission
Edited by H. Zimmermann

(ref. Häggblad and Heilbronn 1987). In chicken myotubes and isolated muscle cells (diaphragm) a P_2-receptor mediated accumulation of IP_3 has been observed (Häggblad 1987; Häggblad and Heilbronn 1987). Since IP_3 can liberate Ca^{2+} from cytosolic compartments the action of ATP may - in a modulatory manner - directly effect the excitation-contraction coupling in the postsynaptic muscle cell. If not a cotransmitter ATP may, at the cholinergic synapse, at least be a coreleased synaptic modulator.

In brain ATP has been observed to exert inhibitory functions. Other effects of extracellular ATP on neural membranes are the inhibition of Ca^{2+}-uptake into nerve terminals (Ribeiro et al 1979; Lindgren and Smith 1987) and also the phosphorylation of surface proteins (Ehrlich et al 1986).

EXTRACELLULAR METABOLISM OF ATP

Due to its high negative charge synaptically released ATP is unlikely to be recycled directly into adjacent cells. Are there extracellularly located enzymes - in analogy to ACh-esterase - to hydrolyze released nucleotides and thus terminate its physiological function? What is the final product of nucleotide hydrolysis? In a previous series of studies using fractions of intact cholinergic synaptosomes isolated from the Torpedo electric organ we could demonstrate that added ATP is hydrolyzed to adenosine (Zimmermann et al 1979). Extracellular adenosine in turn can be taken up via a high affinity uptake system (Km = 2 µM) into the nerve terminals where it is immediately phosphorylated to form AMP and eventually ATP. Nerve terminal ATP, derived from extracellularly administered adenosine, is taken up again into synaptic vesicles (Zimmermann 1978). Thus, ATP recycles at the cholinergic synapse (adenosine cycle) in a very similar way as ACh via its hydrolysis product choline (choline cycle) (Zimmermann 1982; Zimmermann et al 1986) (Fig. 1).

Our results suggest that synaptically released ATP may be inactivated by ectonucleotidases situated at the surface of synaptic membranes. However, whereas extracellular hydrolysis of ACh terminates its action, hydrolysis of ATP to adenosine

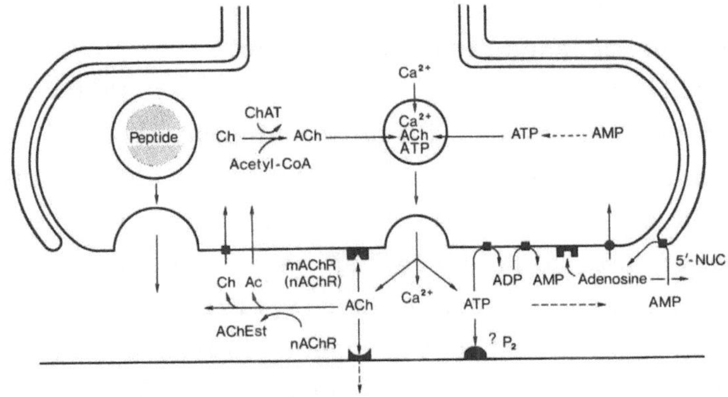

Fig. 1. Functional scheme of a peripheral cholinergic synapse. Both, the electron-lucent (representing the majority of vesicles) and the dense-cored synaptic vesicle type are included. Metabolic fate and possible functions of vesicular contents at the presynapse (autoreceptors) and the postsynapse are indicated by arrows.

may create a new physiological function: Adenosine has been demonstrated to reduce transmitter release via presynaptic autoreceptors (ref. Zimmermann 1982; Richardson et al 1987). A detailed understanding of the physiological role of ATP at the synapse thus requires both the biochemical characterization and exact cellular location of the enzymes involved in its extracellular hydrolysis.

LOCATION OF ECTONUCLEOTIDASES AS REVEALED BY SUBCELLULAR FRACTIONATION

Using a simple sucrose step gradient for the isolation of synaptosomes from the Torpedo electric organ enzyme activities for the hydrolysis of ATP, ADP and AMP (Mg^{2+}salts) are enriched in the synaptosome fraction on top of a layer of 0.55 M sucrose. Enzyme activity is assessed under isotonic conditions which should keep synaptosomes intact (Grondal and Zimmermann 1986). Similarly, when a continuous sucrose gradient is used to achieve a higher resolution of subcellular particles according to density, nucleotidase activities cosediment with synaptosomes. Synaptosomes are identified by their contents in choline-acetyltransferase, cholinesterases and binding activity for (^{3}H)l-quinuclidinylbenzilate (muscarinic receptors). As shown by the distribution of cytochrome-c-oxidase and binding activity for [^{125}I]-iodo-α-bungarotoxin mitochondria and postsynaptic membranes are denser and do not contaminate the synaptosome fraction.

In order to obtain evidence for an extracellular location of the catalytic site of nucleotidase activities intact fractions of synaptosomes were incubated under isotonic conditions with protease (pronase) or the low molecular weight, slowly permeating, covalent inhibitor of enzyme activity, diazotized sulfanilic acid (DSA) (Fig. 2). In either case, activity of nucleotidases

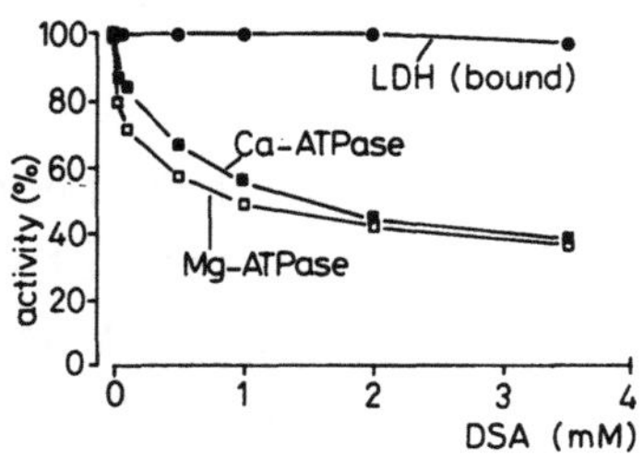

Fig. 2. Inhibition of activity of ATPase (4 mM Mg-ATP or Ca-ATP) and LDH in fractions of intact synaptosomes after application of DSA (5 min.) From Grondal and Zimmermann (1986).

is inhibited in a time or concentration-dependent manner, whereas activity of lactate dehydrogenase (LDH) which is occluded in the synaptosomal cytoplasm is not affected. If synaptosomes are made accessible to DSA by previous treatment with Triton X-100 (1%) LDH is equally affected as activity of nucleotidases (Keller and Zimmermann 1983; Grondal and Zimmermann 1986). These results suggest that fractions of isolated and intact synaptosomes from the Torpedo electric organ contain ecto-nucleotidase activity for the hydrolysis of ATP, ADP and AMP.

CHARACTERISTICS OF ECTO-ATPASE ACTIVITY IN SITU

The ecto-ATPase associated with sealed synaptosomes can be activated by either Ca^{2+} or Mg^{2+}-ATP to a similar extent (Km = 66 and 75 μM respectively, Vmax = 32.5 and 34.8 nmol/mg of protein respectively). At saturating concentrations of Mg^{2+} no further enzyme activation could be obtained by addition of Ca^{2+} and vice versa. Thus, both Ca^{2+}-ATP and Mg^{2+}-ATP are likely to act on the same enzyme which would qualify as a Ca^{2+} or Mg^{2+}-ATPase. ADP inhibits hydrolysis of ATP whereas AMP is without effect (Grondal and Zimmermann 1986).

An analysis of the substrate specificity (Fig. 3) reveals, that there is a broad specificity for the hydrolysis of both purine and pyrimidine-derived nucleotides. There is, however,

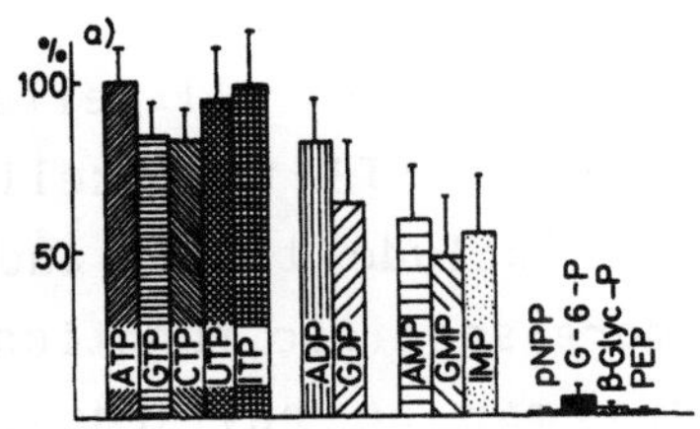

Fig. 3. Hydrolysis of phosphate-esters (1 mM) by fractions of intact synaptosomes in the presence of 4 mM Mg-ATP. Values of 100% correspond to 36.7 mU/mg of protein. From Grordal and Zimmermann (1986).

no hydrolysis of substances for unspecific phosphatases like p-nitrophenylphosphat (pNPP), glucose-6-phosphate (G-6-P), ß-glycerophosphate (ß-Glyc-P) or phospho(enol)pyruvate (PEP). The hydrolysis of GTP may be of functional significance, since cholinergic synaptic vesicles store GTP in addition to ATP (molar ratio GTP/ATP = 0.2) (Wagner et al 1978) and is likely to be co-released with ACh and ATP.

None of the known inhibitors of the mitochondrial Mg^{2+}-ATPase, Na^{+}, K^{+}-ATPase or the Mg^{2+}-dependent, Ca^{2+}-activated (Ca^{2+} + Mg^{2+})-ATPase of the sarcoplasmic (endoplasmic) reticulum had a significant inhibitory effect. However, a significant reduction of enzyme activity can be obtained with filipin (50 µg/ml) or queritin (100 µM) (Grondal and Zimmermann 1986).

ECTO-ATPASE ACTIVITY, A GENERAL PROPERTY OF ALL CELL SURFACES?

Enzyme activity for the hydrolysis of extracellularly applied nucleoside triphosphatases has been described at the surface of a variety of cell types. These include lymphocytes and granulocytes, vascular endothelial cells and smooth muscle cells in culture, neuroblastoma and glia cells in culture, skeletal muscle cells (ref. Grondal and Zimmermann 1986) and also nerve endings isolated from brain (Nagy et al 1986; Brown and Richardson 1987). Properties of these ecto-ATPases generally are: low substrate specificity for nucleoside triphosphates, full activation by either Ca^{2+} or Mg^{2+}-ATP, km-values for ATP in the range between 10 and 100 µM, insensitivity to inhibitors of other well described ATPases. In several cases it has been demonstrated that the hydrolysis of the nucleoside triphosphate proceeds to the nucleoside.

If activity of ectonucleotidases was a property of all cells one would expect also an ubiquitous release of ATP from cell surfaces by a number of stimuli. These may include trauma and hypoxia. However, in a number of cases a more specific function of extracellular hydrolysis of nucleotides may be envisaged. This relates to the function of ecto-ATPase activity of blood platelets in platelet aggregation, the hydrolysis of vascular ATP by endothelial cells (ref. Gordon 1986, Zimmermann et al 1986) and presumably also to synaptic transmission where a focal release of ATP together with the neurotransmitter can be expected. It is therefore essential to know the biochemical and molecular properties of the isolated enzymes involved in degradation of ATP as well as the exact subcellular localization in the synapse region.

ISOLATION AND MOLECULAR CHARACTERIZATION OF 5'-NUCLEOTIDASE FROM TORPEDO ELECTRIC ORGAN

The ectoenzyme involved in the final step of ATP hydrolysis, 5'-nucleotidase, has been purified from the Torpedo electric organ after solubilization in Triton X-100 and deoxycholate by affinity chromatography on concanavalin A-Sepharose and AMP-Sepharose (Grondal and Zimmermann 1987). The enzyme is a glycoprotein with an apparent Mr of 62,000 (Fig. 4). The molecular radius of the native deoxycholate-enzyme complex is 131,000 (gel filtration on Sephacryl 200) suggesting that the enzyme occurs as a homodimer.

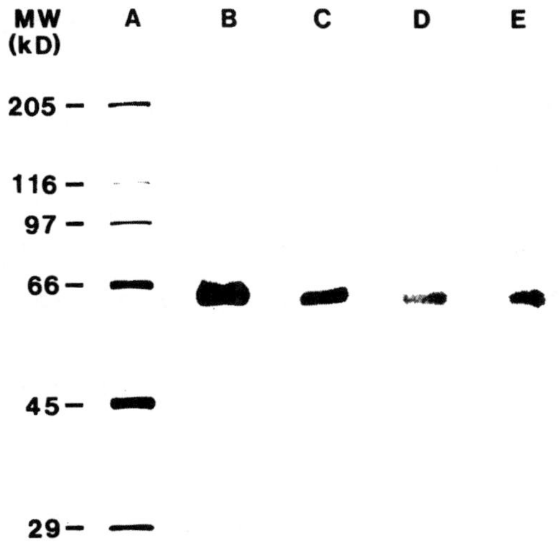

Fig. 4. Analysis of 5'-nucleotidase by SDS/polyacrylamide-gel electrophoresis and immunoblotting.
A) Standard proteins, B) purified 5'-nucleotidase, silver staining, C) purified 5'-nucleotidase after transfer to nitrocellulose and staining with concanavalin A-peroxidase, D) 5'-nucleotidase after transfer to nitrocellulose and staining with anti-5'-nucleotidase antiserum, F) synaptosome fraction from electric organ after transfer to nitrocellulose and staining with anti-5'-nucleotidase antiserum. From Grondal and Zimmermann (1987).

Of the purine and pyrimidine mononucleotides, AMP is hydrolyzed most effectively (Km for AMP = 38 µM, Vmax = 31 units/mg protein). ADP (Ki = 22 µM) and ATP (Ki = 48 µM) are competitive inhibitors of the enzyme. Glycerophosphate, phosphoenolpyruvate and p-nitrophenyl phosphate are not substrates for the enzyme.

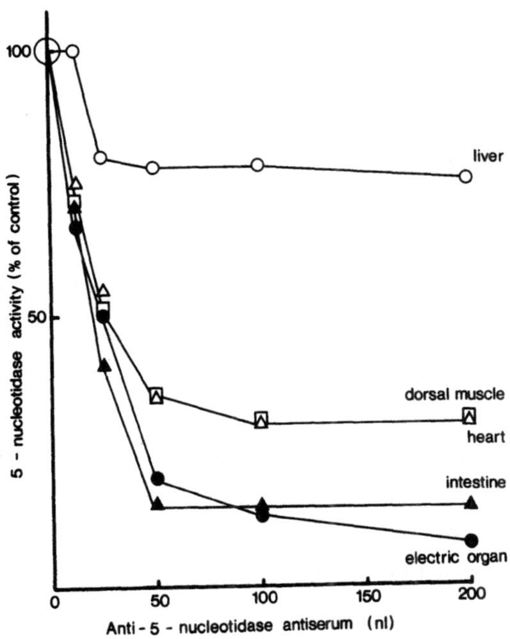

Fig. 5. Tissue specificity of 5'-nucleotidase as revealed by enzyme inhibition with monospecific antiserum. From Grondal and Zimmermann (1987).

A polyclonal antiserum raised against the purified native enzyme inhibits enzyme activity up to 100% and recognizes only a single band of Mr 62,000 on immunoblotting after transfer of a synaptosomal fraction to nitrocellulose (Fig. 4). The inhibitory effect of the antiserum raised against electric organ derived 5'-nucleotidase was tested against crude enzyme fractions derived from muscle, heart, liver and intestine (Fig. 5). The incomplete inhibition of enzyme activity in some tissues would be in accordance with the assumption of tissue-specific variants of the enzyme which are only partly recognized by the antiserum.

SUBCELLULAR DISTRIBUTION OF 5'-NUCLEOTIDASE AS REVEALED BY IMMUNOCYTOCHEMISTRY

Using indirect FITC-immunofluorescence 5'-nucleotidase can be allocated to the axon bundles and the numerous nerve terminal ramifications at the ventral side of the electroplaque cells in the electric organ (Fig. 6). The sites of innervation can also be marked by rhodamine-labelled α-bungarotoxin (postsynaptic nicotinic ACh-receptors (Grondal et al 1987). Significant fluorescence is also obeserved at the dorsal side of the electroplaque cells. Furthermore, the lumen and the endothelial cells of the small capillaries in the tissue and the connective tissue sheaths surrounding the columns of electroplaque cells fluoresce. No fluorescence occurs when the preimmune serum is used.

Using the colloidal gold postembedding technique and an affinity-purified anti-5'-nucleotidase antibody (Grondal et al 1987), the subcellular distribution of the enzyme could be revealed at the electron-microscopical level (Fig. 6c, d). The colloidal gold particles densely bind to the small Schwann cell lamella which covers the terminal axon ramifications up to the synaptic cleft. Binding of gold particles to other cellular structures in the synaptic region like the axonal cytoplasm, the presynaptic or postsynaptic plasma membrane and the basal lamina in the synaptic cleft is very low. Whereas the cytoplasm

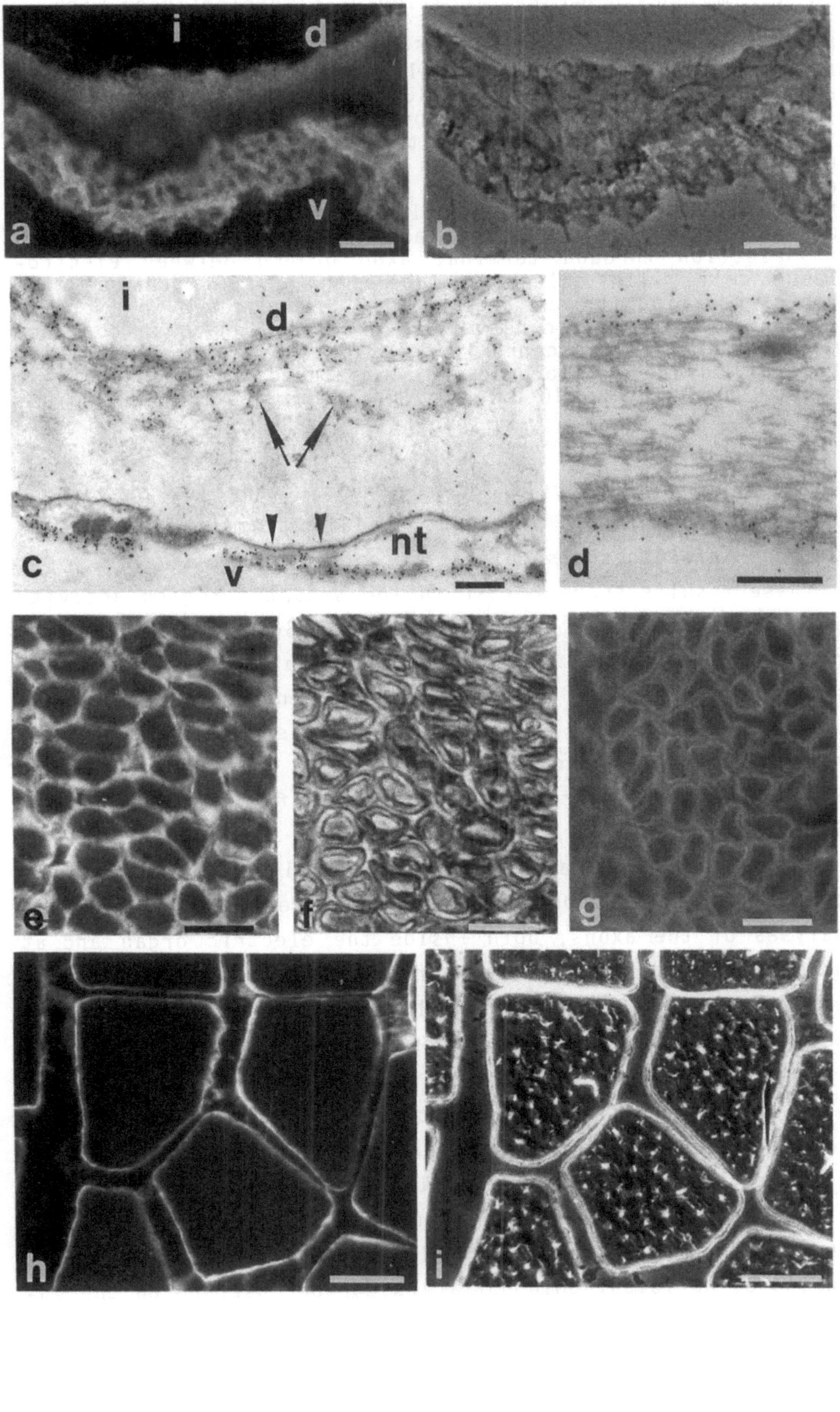
i
d
v
a
b
i
d
nt
c
v
d
e
f
g
h
i

Fig. 6. Immunocytochemical location of 5'-nucleotidase with FITC-fluorescence on colloidal-gold labelling. a) Cross-section through electroplaque cell with FITC-fluorescence of the terminal axon branches at the ventral side (v) and the non-innervated dorsal (d) side of the cell. i, interstitial space between electroplaque cells. b) Corresponding phase-contrast image. c) Labelling with antibody-coated colloidal-gold particles (post-embedding) of the Schwann cell processes covering the nerve terminals (nt) and of the infoldings of the dorsal membrane of the electroplaque cell (arrows). The postsynaptic membrane (arrow heads) is not labelled. d, i, v as for a). d) Cross-section through terminal axon branch between electroplaque cells with selective labelling of the Schwann cell process covering its surface. e) FITC-fluorescence in a cross-section of the electric nerve. f) Corresponding phase contrast image. g) FITC-fluorescence in cross-section of cross-striated skeletal muscle of Torpedo. i) Corresponding phase-contrast image. Bars: a, b = 10 µm; c, d = 500 nm; e, f, g = 20 µm; h, i = 40 µm.

of the electroplaque cell is - besides a very low background binding - free of gold particles, the dorsal membrane with its numerous infoldings also binds the gold particles.

These results suggest that in the synaptic region activity of 5'-nucleotidase is restricted to the Schwann cell process and that it is absent from (or contained only to a minor extent in) the axolemma. The observation that 5'-nucleotidase is considerably enriched in "higly purified" presynaptic plasma membrane preparations from Torpedo electric organ (Morel et al 1982) may, thus, represent an artifact of subcellular fractionation.

The notion of the allocation of 5'-nucleotidase to Schwann cells is further supported by light and electron-microscopical analyses of the axons, both inside the electric organ and at the level of the electromotor nerve (Fig. 6, d-g). Both, longitudinal and cross sections reveal bright immunofluorescence at the axonal surface. A comparison of the phase contrast and fluorescence images suggests that it is not the axolemma but the surrounding Schwann cell which binds the anti-serum. Whereas in untreated sections the fluorescence appears to be located rather at the surface of the myelin sheath, in sections treated with ethanol (up to 90%) the fluorescence is more clearly refined to the myelin sheath. On electron-microscopical analysis binding of colloidal gold particles can be observed on the entire surface of the Schwann cell but also in the myelin and at the perineurial sheath surrounding the axon when proceeding through the electric organ.

Interstingly, neither in the electric lobe which contains the pericarya of the electromotor neurons nor in any other part of the central nervous system investigated, any membrane of neural or glial origin show immuno-fluorescence. Only blood capillaries and the meningeal layers surrounding the brain tissue are labelled. The myelin derived from oligodendroglial cells which covers the intial part of the electromotor axon is unlabelled. Only when the axon leaves the brain and enters the root of the electric nerve immunofluorescence begins. These results suggest that 5'-nucleotidase is either absent from oligodendroglia cells or that our antiserum (which shows preference to specific tissue variants of 5'-nucleotidase) does not bind the oligodendroglial enzyme.

5'-nucleotidase is also associated with the surface of cross-striated muscle fibres taken from dorsal muscle of the fish (Fig. 6h, i).

LOCATION OF ECTO-ATPASE

The ecto-ATPase has not yet been isolated from any tissue discussed above. We performed a series of immunoprecipitation experiments using polyacrylamide beads and the anti-5'-nucleotidase antibody to determine whether ecto-ATPase was associated with the same subcellular particles as 5'-nucleotidase. After incubation of the synaptosome fraction with antiserum-coupled beads, and subsequent centrifugation, 5-nucleotidase was removed from the supernatant to a far greater extent that ecto-Mg^{2+}-ATPase or the cytoplasmic markers LDH and ATP (Table 1) (Grondal and Zimmermann 1987). Thus, the majority of ecto-ATPase is unlikely to be associated with the same membranes as 5'-nucleotidase and rather appears to be a constituent of the presynaptic plasma membrane. The cytochemical location of reaction product in the area of the Schwann cell neuronal plasma membrane interface demonstrated by Pappas et al (this volume) may indicate activity of ecto-ATPase in this part of the nerve terminal.

TABLE 1. Immunoprecipitation of particles contained in the synaptosome fraction using anti-5'-nucleotidase coupled polyacrylamide beads.

MARKER	% OF CONTENT OR ACTIVITY REMAINING IN THE SUPERNATANT
ATP	88.4 (6)
Acetylcholinesterase	67.1 (6)
Lactate dehydrogenase	89.1 (6)
5'-nucleotidase	39.4 (6)
Mg-ATPase	81.6 (4)

Subcellular particles were incubated with the beads and then centrifuged. Values are means of 4 - 6 experiments.

SUMMARY AND CONCLUSIONS

Our results suggest that ATP can be hydrolyzed extracellularly to adenosine at the cholinergic synapse. The adenosine formed may be salvaged by the nerve terminal via a high affinity uptake system and reused for the formation of ATP and reloading of synaptic vesicles. There are at least two enzymes responsible for this hydrolysis: a Ca^{2+} or Mg^{2+}-dependent ATPase which is likely to be located at the surface of the axonal plasma membrane as well as a 5'-nucleotidase located at the surface of the Schwann cell process surrounding the terminal axon ramifications. Whether ADP is hydrolyzed by the triphosphatase or a separate enzyme is not known. In contrast to intracellular ATPases, activity of ectonucleotidases would be activated by the availability of substrate ATP with Ca^{2+} and Mg^{2+} at saturating concentrations in the extracellular medium.

Both the location of the 5'-nucleotidase and preliminary estimates of the rate of hydrolysis of ATP suggest that ATP may be present in the synaptic region for a considerably longer period of time (several orders of magnitude) than ACh which becomes hydrolyzed very rapidly. This would support the notion of a more tonic effect of the nucleotide in cholinergic transmission than the phasic one of the transmitter ACh.

The function of the ectonucleotidases could be in the control of any effect which ATP may have pre- or postsynaptically and in the production of adenosine as an additional synaptic modulator. This final hydrolysis to adenosine is controlled by 5'-nucleotidase which is inhibited competitively both by ATP and ADP. Whether the energy of hydrolysis of extracellular ATP is made available to a cellular mechanism is not known.

An additional question raised by our immunocytochemical findings is the distribution of 5'-nucleotidase at the surface of the Schwann cells along the entire length of the axon, the perineurial layer and also the surface of muscle cells. As quoted above ecto-ATPases have been found at the surface of glial cells in culture and of skeletal muscle cells. This raises the question of a possible release of ATP from exitable tissues as a general phenomenon. Release of ATP has been implicated both from stimulated nerve truncs, skeletal muscle cells (Abood et al 1962; Boyed and Forrester 1968; Forrester 1972; Israël et al 1976). The observation that excited rabbit nerve fibres release into the medium non-phosphorylated metabolites rather than nucleotides (Maire et al 1982) may be due to the presence of ectonucleotidases.

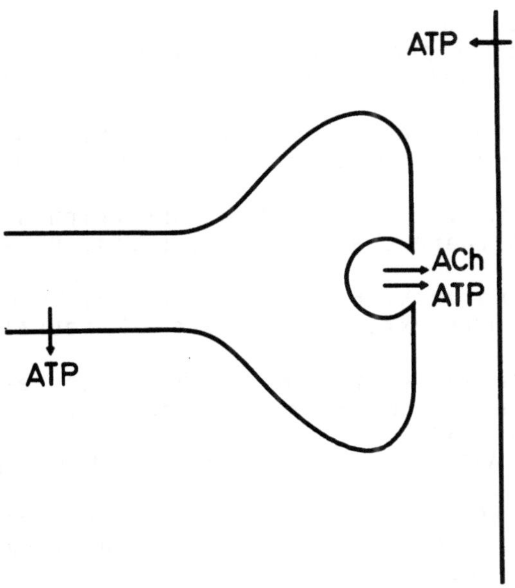

Fig. 7. Speculative scheme on the release of ACh and ATP from nerve terminal and muscle.

Our final model (Fig. 7) would, thus, envisage the possibility of exocytotic ATP release as a cotransmitter or coreleased modulator at the cholinergic nerve terminal but also its liberation from excitable tissue on excitation. By what mechanism ATP is released from these tissues and what its physiological function could be remains to be established. Excitable tissues appears to have build up several barriers for the excape of released nucleotide where it becomes hydrolyzed in order to salvage it in the form of adenosine. This relates not only to the surfaces of Schwann cells, but also to the perineurial sheaths of peripheral axons and nerves, the connective tissue sheaths surrounding individual columns of electroplaque cells in the electric organ or the meningeal tissue layers surrounding the brain which are all rich in 5'-nucleotidase-immunoreactivity.

REFERENCES

Abood LG, Koketsu K, Miyamoto S (1962) Outflow of various phosphates during membrane polarization of excitable tissues. Am J Physiol 202:469-475

Benham CD, Tsien RW (1987) A novel recptor operated Ca-2+-permeable channel activated by ATP in smooth muscle. Nature 328:275-278

Benham CD, Bolton TB, Byrne NG, Large WA (1987) Action of externally applied adenosine triphosphate on single smooth muscle cells dispersed from rabbit ear artery. J Physiol 387:473-488

Boyed IH, Forrester T (1968) The release of adenosine triphosphate from frog skeletal muscle in vitro. J Physiol 199: 115-135

Dowdall MJ, Boyne AF, Whittaker VP (1974) Adenosine triphosphate, a constituent of cholinergic synaptic vesicles. Biochem J 140:1-12

Ehrlich YH, Dawis TB, Bock E, Kornecki E, Lenox RH (1986) Ectoprotein kinase activity on the external surface of neural cells. Nature 230:67-70

Forrester T (1972) An estimate of adenosine triphosphate release into the venous effluent from excercising human forearm muscle. J Physiol 224:611-628

Gordon JL (1986) Extracellular ATP: effect sources fate. Biochem J 233:309-319

Grondal EJM, Zimmermann H (1986) Ectonucleotidase activities associated with cholinergic synaptosomes isolated from Torpedo electric organ. J Neurochem 47:871-881

Grondal EJM, Zimmermann H (1987) Purification, characterization and cellular localization of 5'-nucleotidase from Torpedo electric organ. Biochem J 245:805-810
Grondal EJM, Janetzko A, Zimmermann H (1987) Monospecific antiserum against 5'-nucleotidase from Torpedo electric organ: immunocytochemical distribution and its assocation with Schwann cell membranes. Neuroscience, in press
Häggblad J (1987) Neuromuscular junction revisited. Biochemical studies on mechanisms involved in tranmission events. Dissertation. Univ. Stockholm
Häggblad J, Heilbronn E (1987) Externally applied adenosine-5'-triphospate causes inositol triphosphate accumulation in cultured chick myotubes. Neurosci Lett 74:199-204
Heilbronn E, Häggblad J (1987) A case for ATP as a co-transmitter and trigger of postsynaptic Ca-2+-release at the neuromuscular junction? In: Dowdall MJ, Hawthorne JN (eds) Cellular and molecular basis of cholinergic function. Ellis Horwood, Chichester, and VCH, Weinheim, pp 444-453
Israël M, Lesbats B, Meunier FM, Stinnakre J (1976) Postsynaptic release of adenosine triphosphate induced by single impulse transmitter action. Proc R Soc Lond Biol 193:461-468
Keller F, Zimmermann H (1983) Ecto-adenosine triphosphatase activity at the cholinergic nerve ending of the Torpedo electric organ. Life Sci 33:2635-2641
Lindgren CA, Smith DO (1987) Extracellular ATP modulates calcium uptake and transmitter release at the neuromuscular junction. J Neurosci 7:1567-1573
Maire JC, Medilanski J, Straub RWC (1982) Uptake of adenosine and release of adenine derivatives in mammalian non-myelinated nerve fibres at rest and during activity. J Physiol 323:589-602
Morel N, Meunier FM (1981) Simultaneous release of acetylcholine and ATP from stimulated cholinergic synaptosomes. J Neurochem 36:1766-1773
Morel N, Manaranche R, Israël M, Gulik-Krzywicki T (1982) Isolation of presynaptic plasma membrane fraction from Torpedo cholinergic synaptosomes: evidence for a specific protein. J Cell Biol 93:349-356
Nagy AK, Shuster TA, Delgado-Esueta V (1986) Ecto-ATPase of mammalian synaptosomes: identification and enzyme characterization. J Neurochem 47:976-986
Ribeiro JA, Sa-Almeida AM, Namorado JM (1979) Adenosine and adenosine triphosphate decrease 45-Ca uptake by synaptosomes stimulated by potassium. Biochem Pharmacol 28:1297-1300
Richardson P, Brown SJ (1987) ATP release from affinity-purified rat cholinergic nerve terminals. J Neurochem 48:622-630
Richardson PJ, Brown SJ, Bailyes EM, Luzio JP (1987) Ectoenzymes control adenosine modulation of immunoisolated cholinergic synapses. Nature 327:232-234
Schweitzer E (1987) Coordinated release of ATP and ACh from cholinergic synaptosomes and its inhibition by calmodulin antagonists. J Neurosci 7:2948-2956
Silinsky EM (1975) On the association between transmitter secretion and the release of adenine nucleotides from mammalian motor nerve terminals. J Physiol 247:145-162

Stjärne L, Åstrand P (1984) Discrete events measure single quanta of adenosine 5'-triphosphate secreted from sympathetic nerves of guinea pig and mouse vas deferens. Neuroscience 13:21-28

Volknandt W, Zimmermann H (1986) Acetylcholine, ATP, and proteoglycan are common to synaptic vesicles isolated from the electric organs of electric eel and electric catfish as well as from rat diaphragm. J Neurochem 47:1449-1462

Zimmermann H (1978) Turnover of adenine nucleotides in cholinergic synaptic vesicles of the Torpedo electric organ. Neuroscience 3:827-836

Zimmermann H (1982) Coexistence of adenosine 5'-triphosphate and acetylcholine in the electromotor synapse. In: Cuello AC (ed) Co-Transmission. Macmillan, London, pp 243-259

Zimmermann H, Dowdall MJ, Lane DA (1979) Purine salvage at the cholinergic nerve endings of the Torpedo electric organ: the central role of adenosine. Neuroscience 4:979-993

Zimmermann H, Grondal EJM, Keller F (1986) Hydrolysis of ATP and formation of adenosine at the surface of cholinergic nerve endings. In: Kreutzberg WG, Reddington M, Zimmermann H (eds) Cellular Biology of Ectoenzymes. Springer, Berlin Heidelberg New York, pp 35-48

THE APPLICATION OF IMMUNOAFFINITY TECHNIQUES TO THE STUDY OF CHOLINERGIC NEUROBIOLOGY

Susan J. Brown, J. Paul Luzio and Peter J. Richardson
Department of Clinical Biochemistry,
University of Cambridge Clinical School,
Hills Road,
Cambridge CB2 2QR, UK

INTRODUCTION

Synaptosomes, i.e. functional nerve terminals, were first prepared from mammalian tissue by Gray and Whittaker (1962). Their ease of preparation (reviewed by Jones, 1975), and their metabolic competence, has led to their widespread use in the study of synaptic transmission. One major disadvantage of the synaptosome preparation is its heterogeneity, a consequence of the highly complex nature of the mammalian central nervous system. In order to study the molecular basis of transmission in cholinergic neurones, other non-mammalian systems have been used - in particular the purely cholinergic nerves which innervate the electric organ of *Torpedo* fish. Synaptosomes prepared from this source (Israel *et al*, 1976; Dowdall and Zimmermann, 1977) are homogeneous with respect to transmitter type, but there are still problems in extrapolating some data from *Torpedo* to mammalian systems; especially as it is now becoming apparent that there is heterogeneity in cholinergic nerve terminals derived from differing mammalian brain regions (Richardson *et al*, 1987).

To overcome some of these problems immunoaffinity techniques have been applied. This chapter describes the use of one immunoaffinity procedure to purify cholinergic nerve terminals from mammalian brain and the use of these purified terminals in studies of presynaptic modulation of acetylcholine release by purines.

NATO ASI Series, Vol. H21
Cellular and Molecular Basis of Synaptic Transmission
Edited by H. Zimmermann

IMMUNOAFFINITY PURIFICATION OF SYNAPTOSOMES

The advantage of using immunoaffinity techniques for the selective isolation of subcellular particles is that it relies on biological differences (i.e. in cell surface components), rather than physical parameters such as size, shape, density, charge or hydrophobicity. The indirect immunoadsorbent has proved to be practically the more successful (reviewed by Richardson and Luzio, 1986) and requires a specific, high-affinity anti-organelle antibody to a specific antigen, which in turn is picked up by a second antibody covalently bound to a solid support.

To produce specific antisera to cholinergic nerve terminals, synaptic membranes purified from the cholinergic electric organ of *Torpedo* were used as immunogens and injected into sheep. The antiserum was shown to bind to a ganglioside antigen (Chol-1) localised on cholinergic terminals in *Torpedo*, rat and guinea-pig (Jones *et al*, 1981; Richardson *et al*, 1982; Ferretti and Borroni, 1986). To complete the purification procedure a high capacity immunoadsorbent is required. A mouse monoclonal anti-(sheep IgG) antibody coupled to cellulose gave a capacity of 290 ug sheep IgG per mg of cellulose (Richardson *et al*, 1984). In brief, a crude mitochondrial rat brain fraction can be sensitized by the polyclonal sheep antiserum and the cholinergic terminals isolated rapidly via centrifugation after incubation with the immunoadsorbent. The purification obtained by this method is approximately 18-fold in rat cerebral cortex (Richardson *et al*, 1984) but can vary depending on the amount of cholinergic innervation in the brain region used.

Terminals separated in this manner show characteristic synaptosomal profiles, display osmotic sensitivity, metabolic activity, take up choline and synthesise and release ACh on stimulation (Richardson *et al*, 1984; Richardson, 1986). With this preparation studies into presynaptic cholinergic modulation could now be undertaken.

PURINE RELEASE FROM CHOLINERGIC NERVE TERMINALS

Purine nucleotides and nucleosides have potent extracellular actions on excitable membranes. Experimental evidence shows ATP and adenosine can be released from a variety of neuronal preparations, with possible involvement in physiological modulatory processes (as reviewed by Stone, 1981; Phillis and Wu, 1981).

ATP is released from Torpedo cholinergic synaptosomes, exhibiting similar kinetics to that of ACh release (Morel and Meunier, 1981). Indeed, cosedimentation of ACh and ATP was observed during the purification of cholinergic synaptic vesicles from the electric organ (Dowdall et al, 1974; Zimmermann, 1979). However, recent work on Torpedo synaptosomes showed that botulinum toxin Type A inhibited ACh release but not that of ATP (Marsal et al, 1987). This data suggests that there may be some differences in the release mechanisms of ACh and ATP.

In the mammalian neuromuscular junction, co-release of ACh and adenine nucleotides has also been demonstrated (Silinsky, 1975). Similarly, in the mammalian central nervous system Potter and White (1980) compared regions of ATP release with putative neurotransmitters with which it might be co-stored in rat brain synaptic vesicles. The K^+-evoked release of ATP correlated well with ACh distribution. Later work by the same authors on guinea-pig cortical synaptosomes, pretreated with botulinum toxin A, showed inhibition of K^+-induced ^{14}C -ACh release by over 50%. There was however no significant effect on ATP release (White et al, 1980). Since less than 10% of cortical synaptosomes are cholinergic (Richardson, 1981), the problem of synaptosomal heterogeneity again arises, since any corelease of ATP with ACh in this preparation could be masked by other terminals releasing ATP.

In order to determine whether ATP is released from mammalian central cholinergic nerve terminals, affinity purified rat striatal terminals were depolarized with either 22.6 mM KCl or 50 uM veratridine. Elevated KCl caused the release of 248 ± 41 pmol of ATP per mg of nerve terminal protein. This was the first direct demonstration of Ca^{2+}dependent ACh and ATP corelease from mammalian central nerve terminals (Richardson and Brown, 1987). Isolation of cholinergic synaptic vesicles from this preparation by continuous sucrose density gradient centrifugation showed a region of the gradient rich in both ACh and ATP. The coinciding peaks sedimented at a density of 0.45 M sucrose, characteristic for mammalian cholinergic vesicles (Whittaker and Sheridan, 1965). The molar ratio of ACh:ATP in the synaptic vesicles was 6.7 ± 1:1, whereas the ratio of ACh:ATP release varied between 9.2 ± 0.7:1 (25 mM KCl) to 11.2 ± 1.5:1 (veratridine). Therefore it is possible that ACh and ATP are released

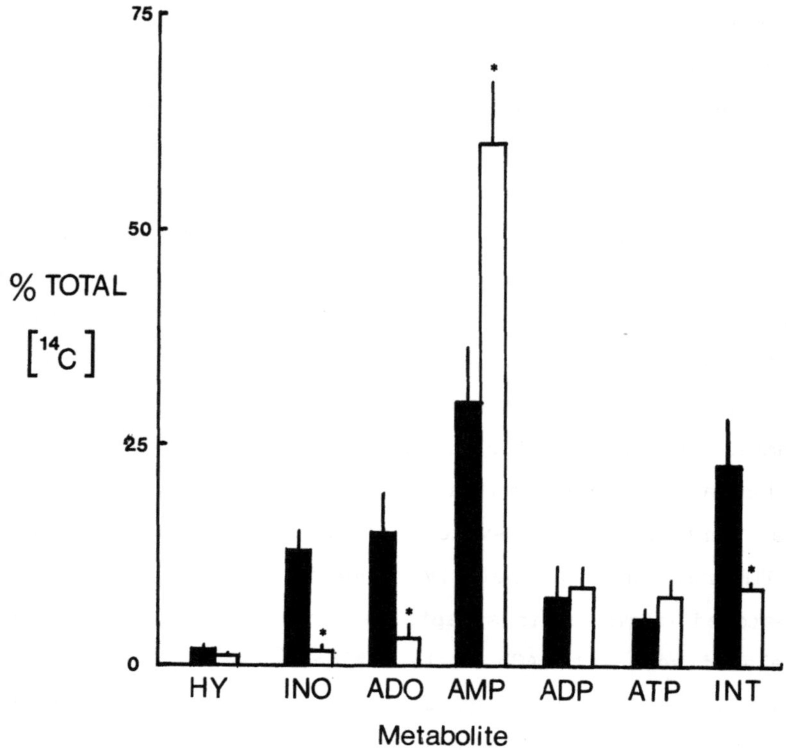

Fig. 1. Metabolism of extracellular ^{14}C ATP by affinity-purified cholinergic nerve terminals. ^{14}C ATP, 15 pmol, was added to cholinergic nerve terminals. After 2 min, the metabolites present in the supernatant were separated by HPLC, and their radioactivity was determined. The radioactivity taken up by the terminals (INT) was also measured. HY, hypoxanthine; INO, inosine; ADO, adenosine. Data are mean SEM (bars) values from three experiments in the presence (□) and absence (■) of a 1:50 dilution of anti-(5'-nucleotidase) serum. Values in the supernatants were compared between terminals with and without anti-(5'-nucleotidase): *p < 0.05.

via different mechanisms, which would comply with differences observed in kinetics of release (Stone, 1981) and differences in inhibition of release (Marsal et al, 1987).

Since ATP was released from the immunoaffinity purified terminals, its extracellular fate was investigated. ^{14}C -ATP was added to depolarised synaptosomes and the metabolites present after two minutes were separated and identified by HPLC. Data obtained confirmed a $>$ 90% hydrolysis of the nucleotide resulting in the accumulation of AMP, adenosine and inosine. Approximately 25% of the label was taken up by the terminals, while 30% accumulated as 5'-AMP (Fig.1). The uptake of label was thought to be via a high-affinity adenosine transporter, similar to the purine salvage system identified in the Torpedo (Zimmermann et al, 1979). Indeed, ^{3}H -adenosine uptake gave a classical Lineweaver-Burke plot with a Km of 16.6 ± 2.5 uM similar to that obtained for guinea-pig cortical synaptosomes (Barberis et al, 1981). The rapid degradation of the ATP demonstrated the presence of ecto-nucleotidases at these synapses (Nagy, 1986; Kreutzberg et al, 1986). In Torpedo synaptosomes, the rate-limiting step in ATP breakdown is between ATP and ADP (Zimmermann et al, 1986). However, as shown in Fig. 1 the rate limiting step in adenosine production at the striatal cholinergic nerve terminal is AMP hydrolysis. Indeed in the presence of an anti-(5'-nucleotidase) inhibitory antibody (Stanley et al, 1983) the production of adenosine and inosine are both significantly reduced, as is the uptake of label into the nerve terminal. This underlines the fact that uptake occurs after AMP hydrolysis i.e. via the adenosine transporter.

To examine the purine metabolites released upon stimulation, ^{3}H -adenosine was used to prelabel the intraterminal purine pool. After stimulation with 50 uM veratridine, 8% of the total label was released in a Ca-dependent manner. The metabolites released in the presence and absence of the inhibitory anti(5'-nucleotidase) antibody were analysed by HPLC. Inhibition of the enzyme caused a rise in extracellular AMP levels, but while having no significant effect on adenosine, increased extracellular inosine levels. This suggested that much of the extracellular adenosine and inosine label was not derived from released ATP. As shown in Fig. 2 the total release of label rose by 5% in the presence of the antibody, an enhancement also achieved in the presence of 10^{-5} M theophylline (an Al receptor antagonist) suggesting the possible involvement of an Al-receptor mediated feedback system at these terminals.

To quantify the relative amounts of ATP and nucleoside released, the nucleoside transporter inhibitor, dipyridamole, was used (Meghji et al, 1985). Anti-(5'-nucleotidase) antibody and dipyridamole together almost totally abolished the Ca^{2+} dependent accumulation of adenosine and inosine, while allowing the release of approximately 25% of the label as nucleotides (Richardson and Brown, 1987). This indicated that the

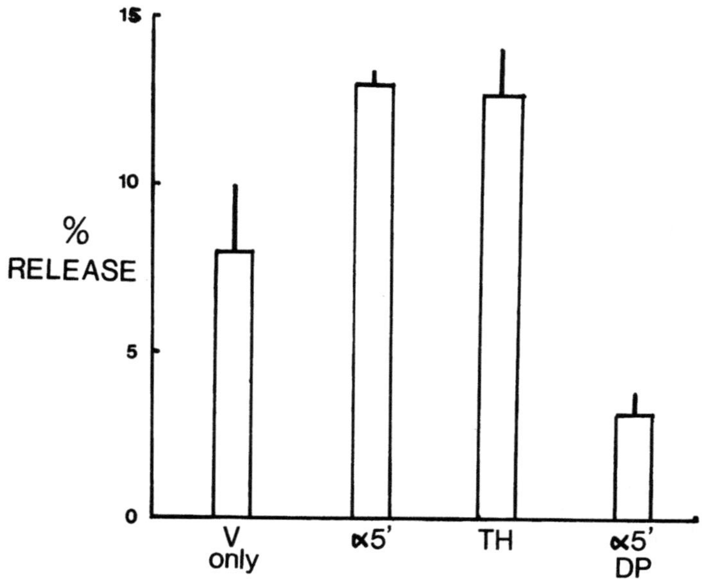

Fig. 2. Ca^{2+}-dependent release of ^3H after ^3H adenosine labelling of affinity-purified nerve terminals. After 5 min incubation with 1.1 μ M ^3H adenosine, affinity-purified terminals were washed and incubated for a further 20 min with inhibitor (i.e. anti-(5'-nucleotidase) serum (1:50 dilution), theophylline, or dipyridamole), washed, and resuspended with inhibitor with 50 μM veratridine. After 2 min, the terminals were spun down and the supernatant subjected to HPLC. Ca^{2+}-dependent release was determined after subtraction of the label released with veratridine in the absence of Ca^{2+} plus 0.5 mM EGTA. Data are mean ± SEM (bars) from three experiments and are percentages of the total label present. V, 50 μM Veratridine; α 5', 1:50 dilution of anti-(5'-nucleotidase) serum; TH, 10^{-5} M theophylline; DP, 50 μM dipyridamole.

Ca-dependent release of label consisted of 25% nucleotide and 75% nucleoside release. After allowing for the specific radio activities of the labelled nucleoside and ATP pools, 30-fold more ATP was found to be released than nucleosides.

PURINE MODULATION OF CHOLINERGIC NERVE TERMINALS

The existence of an extracellular purine receptor has been generally accepted for peripheral tissues. Burnstock (1983) distinguished two types of receptor: P1 for adenosine and P2 for adenine nucleotides. Purines have been shown to have a depressed action on central neurones

Phillis *et al*, 1975) and evidence points to P1 receptor involvement. It has now been recognised that P1 can be subdivided into at least two classes of extracellular adenosine receptor, termed A1 and A2 (Daly, 1982). Both are xanthine-sensitive, but A1 has a high affinity for adenosine (Km 10nM) and inhibits adenyl cyclase, whereas A2 has a lower affinity for adenosine (Km 10 uM) and stimulates adenyl cyclase. The concept of purinergic modulation of transmitter release in the mammalian nervous system has gained much support (reviewed by Stone, 1981; Phillis and Wu, 1981). ATP and adenosine have been shown to inhibit transmitter release from the neuromuscular junction (Ribeiro and Walker, 1975) and from peripheral sympathetic and parasympathetic nerve terminals (Vizi and Knoll, 1976). In the CNS, work on brain slices (Dunwiddie and Hoffer, 1980) and on heterogeneous synaptosome preparations (Pedata et al, 1986) have also confimed this modulatory action. However, only with the recent immunoaffinity purification techniques has it been possible to analyse the role of ATP and adenosine in feedback control of cholinergic terminals.

In order to study effects on ACh release, immunoaffinity purified synaptosomes were prelabelled with ^{3}H -choline, via the high-affinity choline uptake system. A continuous-flow superfusion system was used, fractions were collected every two minutes and the amount of radioactive transmitter present was determined. In brief, the Ca-dependent release of ACh was stimulated by 75 uM veratridine for 2 min, followed by a ten minute interval, in which modulators could be included in the medium, and then the stimulus was repeated. Such a dual stimulus experiment gave two periods of ACh release, termed S1 and S2 respectively, and changes in the S2/S1 ratio were used as indicators of modulation.

Initially, the effects of extraterminal adenosine, and its analogue 2-chloroadenosine, were examined. The terminals were pre-treated with adenosine deaminase (to remove endogenous extracellular adenosine) and dipyridamole was present to prevent adenosine inactivation by uptake. From dose-response studies, maximal inhibition of ACh release was obtained at 10^{-8}M adenosine (25±4%) and 10^{-10}M 2-chloroadenosine (33 ± 2%). These nM concentrations implicate the activation of A1 adenosine receptors, a conclusion further supported by the antagonism of their effects by both 10^{-5}M theophylline and 10^{-6}isobutylmethylxanthine (Richardson *et al*, 1987). Similar levels of adenosine mediated

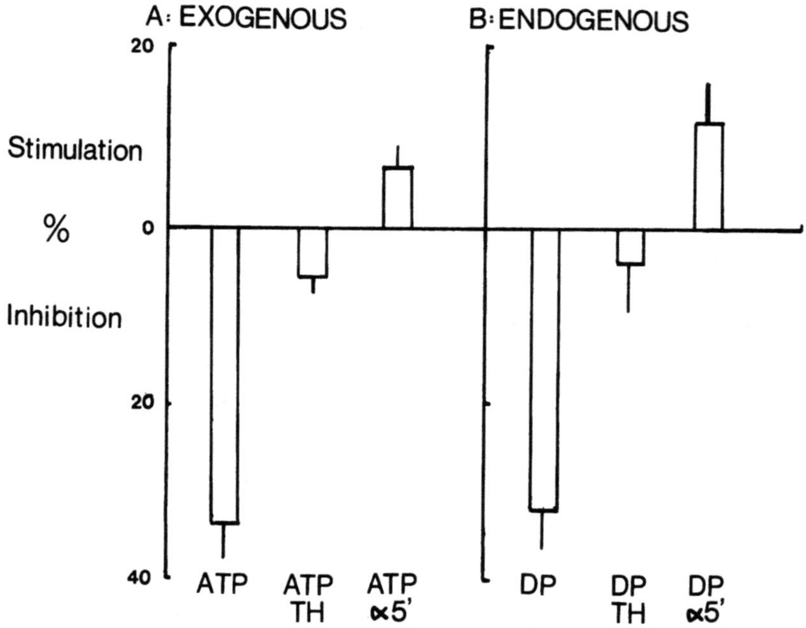

Fig. 3A. Exogenous ATP modulation of ACh release. Affinity-purified nerve terminals were labelled with ^3H-choline, placed in perfusion chambers and given two 2 min stimuli (S1 and S2) 10 min apart, with 75 μM veratridine. Adenosine deaminase treatment was stopped after S1 and the second stimulus occurred in the presence of modulator. Data are mean ± SEM (bars) percentages of inhibition or stimulation of the S2/S1 ratio compared with the control.
Fig. 3B. Endogenous feedback control of ACh release. Method and data expression as above but adenosine deaminase treatment was stopped prior to S1. ATP, 10^{-3} M extracellular ATP; TH-, 10^{-5} M theophylline; ∝5', 1:50 dilution of anti-(5'-nucleotidase) serum; DP, 50 μM dipyridamole.

inhibition of noradrenaline release have been observed during electrical stimulation of rat hippocampal slices (Jonson and Fredholm, 1984).

Knowing that ATP itself is released from the purified terminals (see above), a direct effect of this nucleotide on ACh release was sought. In the presence of 10^{-5}M theophylline, extraterminal ATP (100 uM) - a P2 agonist - or extraterminal αβ methylene ATP (1 uM) - a P2 antagonist - showed no significant inhibition of transmitter release. However, in the absence of theophylline, the rapid degradation of extracellular ATP resulted in the inhibition of ACh release by 33.5 ± 4%. Inhibition of ATP breakdown via an anti(5'-nucleotidase) antibody, caused total abolition of the ATP effect (Fig. 3).

To determine whether sufficient ATP was released to cause feedback inhibition after its hydrolysis to adenosine, adenosine deaminase treatment of the terminals was halted before S1. Thus, any adenosine accumulating outside the terminals during the first stimulation would remain active. Using this protocol, the addition of dipyridamole after S1 resulted in a 32 ± 6% inhhibition of ACh release. Under these conditions dipyridamole prevents any direct nucleoside release via the transporter system (see above), as well as preventing any uptake-mediated inactivation of adenosine that is formed extraterminally. This inhibition was relieved by adenosine deaminase or theophylline (Fig. 3), once more implicating an A1-receptor mediated effect. It therefore appeared that adenosine mediated feedback inhibition of these terminals was not a consequence of the direct release of adenosine, but may have been due to adenosine produced from the released ATP.

Control of adenosine modulation via the ectoenzymes concerned with ATP hydrolysis could also be examined in close detail with the immunoaffinity purified synaptosomes. Since ecto-5'-nucleotidase appeared to be the rate limiting step of this pathway at the striatal cholinergic nerve terminal (Fig. 1) regulation of this enzyme was attempted. Incubation of the terminals with the specific inhibitory anti-(5'nucleotidase) antibody led to the inhibition of ATP degradation, and the size of S2 increased compared to a control value (Fig. 4). Thus, ACh release was augmented. To examine the opposite effect, i.e. increased activity of the enzyme, it was necessary to affinity purify 5'-nucleotidase by immobilising it on a cellulose immunoadsorbent. This was done via a non-inhibitory mouse monoclonal antibody raised against the enzyme. Incubation with extra 5'-nucleotidase resulted in a decreased S2 peak i.e. a stronger inhibition of the transmitter release (Fig.4). These experiments demonstrated for the first time in the mammalian system a possible physiological role for enzymes in the ectophosphohydrolase pathway present in the synaptic cleft (Richardson et al, 1987). In the neuromuscular junction of the frog analogous results have been observed for (i) extracellular adenosine depressing neuromuscular transmission (ii) a dipyridamole-sensitive adenosine uptake system necessary for inactivation of the neuromodulator and (iii) released ATP contributing to the inhibitory action of extracellular adenosine, since inhibition of ecto-5'nucleotidase reduces the inhibition (Ribeiro and Sebastiao, 1987).

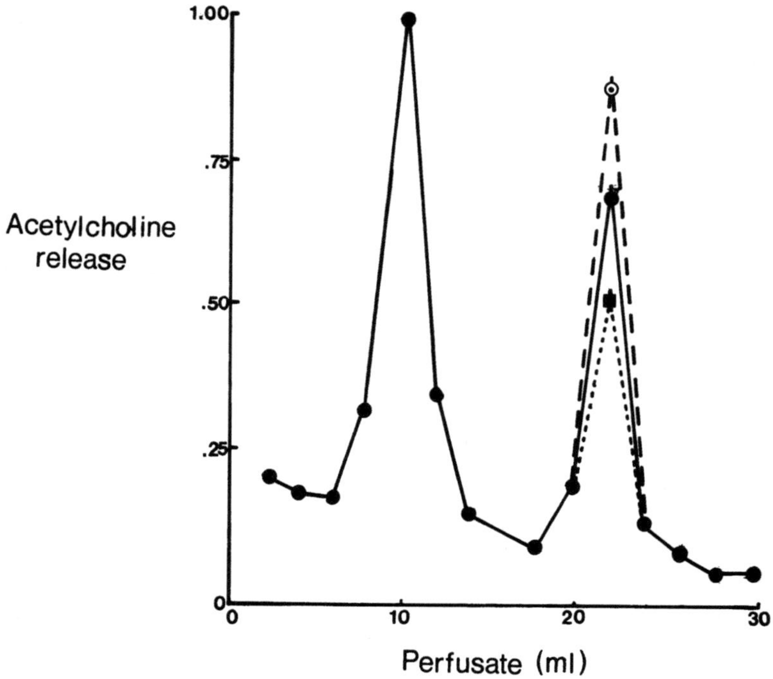

Fig. 4. Effect of ecto-5'-nucleotidase activity on ACh release from affinity-purified nerve terminals. The amount of ^{3}H -ACh released after each stimulation was calculated by subtracting the average basal release and the results normalized to a peak S1 release of 1.0. The control (●) ratio was 0.60 ± 0.02, whereas inhibition of 5'-nucleotidase (○) resulted in an increase in the S2/S1 ratio of 0.12 ± 0.03. An additional 20 mU of 5'-nucleotidase (■) reduced the S2/S1 ratio by 0.20 ± 0.03. Control and experimental perfusions were run on the same batch of terminals on the same day. Duration of stimulus by 75 μM veratridine is represented by the bars.

Thus it is possible that the degree of expression of synaptic ectophosphohydrolases may play a role in the autoinhibition of neurotransmission, and that other ectoenzymes e.g. ectoendopeptidases may serve similar purposes.

It is also of interest to note that the activities of these ectoenzymes can alter in different brain regions and could be used as an indicator for the type of autoinhibition found in the area. With immunoaffinity purified cholinergic terminals from the rat cerebral cortex the hydrolysis of extraterminal ^{14}C -ATP only produced 2% of the total label added in the form of nucleosides (as identified by HPLC).

This contrasts strongly with the 25% obtained from striatal synaptosomes. The activities of both the ecto-5'nucleotidase and the ecto-(ADP)ase were low in the cortical cholinergic terminals. Superfusion experiments with these terminals also showed no significant inhibition of ACh release by extracellular adenosine or ATP. Analogous results showing differing degrees of adenosine mediated inhibition of ACh release in distinct brain areas were also observed by Pedata et al (1986), with the cortex showing the least purinergic regulation. It is interesting that the rate limiting step in adenosine production at the cortical cholinergic nerve terminal (i.e. ATP hydrolysis to ADP), while differing from that in the striatum, is the same as that observed at the Torpedo electrocyte synapse.

The possibility therefore arises that there are two separate pathways of autoinhibition in cholinergic neurones. Presynaptic muscarine feedback has been demonstrated in both rat cortical and hippocampal nerve terminals (Marchi et al, 1981), but not on striatal terminals (Marchi et al, 1983). Similar experiments to those described by Marchi et al were performed with immunoaffinity purified striatal synaptosomes and no significant muscarine autoinhibition was detected. It remains to be determined why some cholinergic synapses are regulated via an ATP/adenosine feedback system and others via muscarinic or even nicotinic mechanisms.

Adenosine regulates adenylate cyclase in many tissues via A1 receptors which reduce its activity and A2 receptors which enhance it (as reviewed by Londos et al, 1981). In the rat brain it has been shown that A1 receptors are directly coupled with brain adenylate cyclase (Ebersolt et al, 1983). Using affinity purified cholinergic striatal membranes we have recently demonstrated that A1 agonists, e.g. 2-chloroadenosine, inhibit a GTP dependent adenylate cyclase (as measured by Solomon et al, 1974). This assay is also being used to investigate the presynaptic regulatory effects of other possible neuromodulators at this cholinergic terminal.

SUMMARY

With the advent of immunoaffinity techniques detailed studies can be made on the mammalian cholinergic system. The co-release of ATP with ACh has been demonstrated with this preparation, as has the resulting presynaptic modulation, via A1-Adenosine receptors, of striatal acetylcholine release. This preparation has enabled us to analyze the relative importance of ATP derived adenosine and directly released adenosine, in the control of the striatal cholinergic nerve terminal. It has also enabled us to demonstrate a role for ectonucleotidases at this synapse, and to show that the expression of one of these enzymes (ecto-5'-nucleotidase) may regulate ACh release. However, we have not as yet been able to determine why some cholinergic synapses utilize the ATP/adenosine feedback mechanism (e.g. in the striatum and at the neuromuscular junction) and some an ACh feedback (in the cortex, myenteric plexus and the heart). Indeed, in the hippocampus both feedback mechanisms appear to be present at cholinergic synapses - it remains to be seen whether this reflects the presence of different types of cholinergic terminal, or whether both mechanisms are active at the same synapses. In any case it will be of interest to analyze the mechanism of the inhibitory action of adenosine and to look for a link with the voltage-dependent Ca^{2+} channel (as hypothesised by Ribeiro and Sebastiao, 1986).

As more transmitter-specific surface antigens are identified, it is hoped that affinity purification will lead to the purification of other nerve terminals and neurones, so permitting more detailed analyses of their molecular mechanisms.

REFERENCES

Barberis C, Minn A and Gayet J (1981) Adenosine transport into guinea-pig synaptosomes. J Neurochem 36: 347-354

Burnstock G (1983) Recent concepts of chemical communication between excitable cells. In Dale's Principle and Communication Between Neurones (Osoborne N, ed): 7-37. Pergamon Press, Oxford

Daly JW (1982) Adenosine receptors: Targets for future drugs. J Med Chem 25: 197-207

Dowdall MJ, Zimmermann H (1977) The isolation of pure cholinergic nerve terminal sacs (T-sacs) from the electric organ of juvenile *Torpedo*. Neurosci 2: 405-421

Dowdall MJ, Boyne AF, Whittaker VP (1974) Adenosine triphosphate, a constituent of cholinergic synaptic vesicles. Biochem J 140: 1-12

Dunwiddie TV and Hoffer BJ (1980) Adenine nucleotides and synaptic transmission in the in vitro rat hippocampus. Br J Pharmacol 69: 59-68

Ebersolt C, Premont J, Prochiantz A, Perez M and Bockaert J (1983) Inhibition of brain adenylate cyclase by A1 adenosine receptors: pharmacological characteristics and locations. Brain Res 267: 123-129

Ferretti P and Borroni E (1986) Putative cholinergic-specific gangliosides in guinea-pig forebrain. J Neurochem 46: 1888-1894

Gray EG and Whittaker VP (1962) The isolation of nerve endings from brain: an electron-microscopic study of cell fragments derived by homogenization and centrifugation. J Anat (London) 96: 79-87

Israel M, Manaranche R, Mastour-Frachon P and Morel N (1976) Isolation of pure cholinergic nerve endings from the electric organ of Torpedo marmorata. Biochem J 160: 113-115

Israel M and Meunier FM (1978) The release of ATP triggered by transmitter action and its possible physiological significance; retrograde transmission. J Physiol (Paris) 74: 485-490

Jones DG (1975) Synapses and synaptosomes: 44-97, Chapman and Hall, London

Jones RT, Walker JH, Richardson PJ, Fox GQ and Whittaker VP (1981) Immunohistochemical localization of cholinergic nerve terminals. Cell Tissue Res 218: 355-373

Jonzon B and Fredholm BB (1984) Adenosine receptor mediated inhibition of noradrenaline release from slices of the rat hippocampus. Life Sci 35: 1971-1979

Kreutzberg GW, Heymann D and Reddington M (1986) 5'-nucleotidase in the nervous system. In Cellular Biology of Ectoenzymes (Kreutzberg GW, Reddington M and Zimmermann H, eds): 147-164. Springer Verlag, Berlin

Londos C, Wolff J and Cooper DMF (1981) Adenosine as a regulator of adenylate cyclase. In Purinergic Receptors (Burnstock G ed): 287-323

Marchi M, Paudice P and Raiteri M (1981) Autoregulation of ACh release in isolated hippocampal nerve endings. Eur J Pharmacol 73: 75-79

Marchi M, Paudice P, Caviglia A and Raiteri M (1983) Is ACh release from striatal nerve endings regulated by muscarinic autoreceptors? Eur J Pharmacol 91: 63-68

Marsal J, Solsona C, Rabasseda X, Blasi J and Casanova A (1987) Depolarization-induced release of ATP from cholinergic synaptosomes is not blocked by botulinum toxin type A. Neurochem Int 10: 295-302

Meghji P, Holmquist CA and Newby AC (1985) Adenosine formation and release from neonatal rat heart cells in culture. Biochem J 229: 799-805

Morrel N and Meunier FM (1981) Simultaneous release of ACh and ATP from stimulated cholinergic synaptosomes. J Neurochem 36: 1553-1557

Nagy A (1986) Enzymatic characteristics and possible role of synaptosomal ecto-ATPase from mammalian brain. In Cell Biology of Ectoenzymes (Kreutzberg GW, Reddington M and Zimmermann H, eds): 49-59. Springer Verlag, Berlin.

Pedata F. Giovannelli L, De Sarno P and Pepeu G (1986) Effect of adenosine, adenosine derivatives and caffeine on ACh release from brain synaptosomes: interaction with muscarinic autoregulatory mechanisms. J Neurochem 46: 1593-1598

Phillis JW, Kostopoulos GK and Limacher JJ (1975) A potent depressant action of adenine derivatives on cerebral cortical neurones. Eur J Pharmacol 30: 125-129

Phillis JW and Wu PH (1981) The role of adenosine and its nucleotides in central synaptic transmission. Prog Neurobiol 16: 187-239
Potter P and White TD (1980) Release of ATP from synaptosomes from different regions of rat brain. Neurosci 5: 1351-1356
Ribeiro JA and Walker J (1975) The effects of adenosine triphosphate and adenosine diphosphate on transmission at the rat and frog neuromuscular junctions. Br J Pharmacol 54: 213-218
Ribeiro JA and Sebastiao AM (1986) Adenosine receptors and calcium: basis for proposing a third (A3) adenosine receptor. Prog Neurobiol 26: 179-209
Ribeiro JA and Sebastiao AM (1987) On the role, inactivation and origin of endogenous adenosine at the frog neuromuscular junction. J Physiol 384, 571-585
Richardson PJ (1981) Quantitation of cholinergic synaptosomes from guinea-pig brain. J Neurochem 37: 258-260
Richardson PJ (1986) Choline uptake and metabolism in affinity purified cholinergic nerve terminals from rat brain. J Neurochem 46: 1251-1255
Richardson PJ and Luzio JP (1986) Immunoaffinity purification of subcellular particles and organelles. Appl Biochem Biotech 13: 133-145
Richardson PJ and Brown SJ (1987) ATP release from affinity-purified rat cholinergic nerve terminals. J Neurochem 48: 622-630
Richardson PJ, Siddle K and Luzio JP (1984) Immunoaffinity purification of intact, metabolically active, cholinergic nerve terminals from mammalian brain. Biochem J 210: 647-654
Richardson PJ, Brown SJ, Bailyes EM and Luzio JP (1987) Ectoenzymes control adenosine modulation of immunoisolated cholinergic synapses. Nature 327: 232-234
Salomon Y, Londos C and Rodbell M (1974) A highly sensitive adenylate cyclase assay. Anal Biochem 58: 541-548
Silinsky EM (1975) On the association between transmitter secretion and the release of adenine nucleotides from mammalian motor nerve terminals. J Physiol (London) 247: 145-162
Stanley KK, Burke P, Pitt P, Siddle K and Luzio JP (1983) Localisation of 5'-nucleotidase in a rat liver cell line using a monoclonal antibody and indirect immunofluorescent labelling. Exp Cell Res 144: 39-46
Stone TW (1981) Physiological roles for adenosine and ATP in the nervous system. Neurosci 6: 523-555
Vizi ES and Knoll J (1976) The inhibitory effects of adenosine and related mucleotides on the release of acetylcholine. Neurosci 1: 391-398
White T, Potter P and Wonnacott S (1980) Depolarisation-induced release of ATP from cortical synaptosomes is not associated with ACh release. J Neurochem 34: 1109-1112
Whittaker VP and Sheridan MN (1965) The morphology and ACh content of isolated cerebral cortical synaptic vesicles. J Neurochem 12: 363-372
Zimmermann H, Dowdall MJ and Lane DA (1979) Purine salvage at the cholinergic nerve endings of the Torpedo electric organ: the central role of adenosine. Neurosci 4: 979-993
Zimmermann H, Grondall EJM and Kellet F (1986) Hydrolysis of ATP and formation of adenosine at the surface of cholinergic nerve endings. In Cell Biology of Ectoenzymes (Kreutzberg GW, Reddington M and Zimmermann H, eds): 35-48. Springer Verlag, Berlin

TRANSMITTER/MODULATOR INDUCED EVENTS RELATED TO EXCITATION-CONTRACTION COUPLING IN SKELETAL MUSCLE OF VERTEBRATES. ACTION OF ATP.

Edith Heilbronn and Johan Häggblad
University of Stockholm, Unit of Neurochemistry and Neurotoxicology,
S-106 91 Stockholm,
Sweden

INTRODUCTION

Recently we have reported evidence that an adenosine-5'-triphosphate (ATP)-dependent purinergic cotransmission to, or modulation of the cholinergic neurotransmission exists at the vertebrate skeleto-neuromuscular junction (n-m j) (Häggblad et al, 1985b; Heilbronn and Häggblad, 1986; Häggblad and Heilbronn, 1987). ATP reaches the outside of the sarcolemma in two ways: It is released by the stimulated motor neuron together with acetylcholine (ACh) as the two compounds are costored in the synaptic vesicles. The ATP:ACh ratio in these vesicles is 1-4:10, depending on animal species. A few larger dense core vesicles are seen in motor neurons and probably contain peptides. It is not known if they also contain ATP, nor is it known if their content indeed is released at the n-m j, or elsewhere. Any nerve impulse triggered release or a leakage of cytosolic ATP from the nerve ending has, to our knowledge, not been described. Several papers describe, however, that depolarisation of the muscle cell (Abood et al, 1962) or the electrocyte (Israel et al, 1980) membrane releases intracellular ATP from these cells, in concentrations surpassing those derived from the synaptic vesicles. An ATP release is also seen from depolarized myotubes in culture (Häggblad, 1987). The precise sites for these depolarisation-triggered ATP releases are not known, but there is reason to suspect that they may occur

NATO ASI Series, Vol. H21
Cellular and Molecular Basis of Synaptic Transmission
Edited by H. Zimmermann

along the whole myotube or muscle fibre surface.

ATP that has been released into the synaptic cleft, probably survives as such for a longer period than released ACh. Its hydrolysis by ectophosphatases and 5'-ectonucleotidases, present in or near the endplate area (Grondahl and Zimmermann, 1986) and probably also outside this area is considerably slower than that of ACh. A continuous nerve stimulation could therefore result in an ATP accumulation at the endplate.

The purpose of the ATP releases is, however, far from clear. The presynaptic, vesicular release suggests a transmitter or modulator function for ATP, either at the presynapse, and then probably after adenosine formation and stimulation of an adenosine receptor as a step in the control of action potential triggered neurotransmitter release (Ribeiro and Sebastiao, 1986), or at the postsynapse, where it may be active either as adenosine or as nonhydrolyzed ATP. Its depolarisation-triggered postsynaptic release could be part of a transsynaptic regulatory system for transmitter release, also using the presynaptic adenosine receptor. The postsynaptic ATP release might also simply serve the purpose of reducing a high, intracellularly active, ATP-concentration through extrusion and extracellular hydrolysis. In the present paper, however, we will examine a third possibility, that of ATP and ACh cooperating via their respective receptors and second messenger systems. In this paper we will add some new results to those previously published from our laboratory and will discuss the implications of these results together with results obtained in some other laboratories.

SUMMARY OF RESULTS

An ATP-sensitive ion channel may exist in the sarcolemma.

Systems involved in ATP production are present in the developing muscle cell *in vivo* and *in vitro*. Results collected here are, however, mainly obtained with myotubes in culture. Kolb and Wakelam (1983) found, using patch clamp techniques on chick myotube in culture, two populations of ATP(1-100 M)-operated channels, which were unable to distinguish between Na^+ and K^+ and not significantly Ca^{2+} permeable. Two populations of channels were also found using ACh or carbachol, with conductance values and voltage-dependent mean channel lifetimes similar to those produced in response to ATP. Häggblad *et al* (1985a, 1985b) found an ATP-elicited, concentration dependent increased inward flux of the cation tracer $^{86}Rb^+$. The influx was biphasic, with an EC_{50} of the first phase of about 10 M, corresponding to the concentrations used by Kolb and Wakelam. This ion influx could not be blocked with α- bungarotoxin (α-bgt) or d-tubocurarine (d-tc) and was found to be additive to that produced by carbachol. Thus, the ATP- activated ion channel seems to be unrelated to the ion channel of the nicotinic acetylcholine receptor (nAChR). Breakdown products of ATP had little or no effect on the channel while slowly hydrolysing derivatives of ATP had an antagonistic action on the ATP-elicited influx. Myotube membrane depolarisation was observed only in some experiments. Ouabain did not block influx, which excludes involvement of the ouabain- sensitive Na/K ATPase, but still leaves open the possibility that other ATP-dependent ion-translocating enzymes are involved. Cell leakage caused by ATP was excluded by control experiments (Häggblad, 1987) with ^3H-deoxyglucose-loaded myotubes.

An ATP receptor mediated phosphoinositide turnover exists in myotubes and skeletal muscle. Occurrence of a P_2-purinoceptor and presumably of a G-protein.

Our results (Häggblad and Heilbronn, 1987) have demonstrated an association of ATP at the myotube membrane with the signalling system that uses phosphoinositide (PIns) degradation (Berridge, 1984; Berridge and Irvine, 1984), via a purinoceptor. Two polyphosphoinositide phosphatidyl inositol bisphosphate ($PInsP_2$) - derived products are known to act as second messengers, i.e. inositol 1,4,5-triphosphate ($InsP_3$) causing release of Ca^{2+} from intracellular pools, and diacylglycerol, activating a protein kinase C (above; Nishizuka, 1984; Majerus et al, 1985). We believe that these two compounds are involved in excitation - contraction coupling in skeletal muscle (see fig. 6).

We have pharmacologically characterized the ATP-receptor connected with $PInsP_2$-degradation as a P_2-purinoceptor, by the following criteria: Stimulation (chick, mouse and rat myotubes in culture) occurs with ATP (10^{-6}-10^{-3}M) or with β,γ-imido ATP (APPNP) but to a lesser degree with the breakdown products of ATP, while the photoaffinity ATP-analogue arylazido aminopropionyl-ATP (ANAPP3), which binds covalently to P_2-purinoceptors, largely blocks stimulation of InsP formation (Häggblad, 1987).

P_2-purinoceptors may be coupled to a phospholipase C, via a G-protein. The enzyme would serve to increase the turnover of $PInsP_2$, thus stimulating the formation of inositol phosphates. Some evidence for the presence of a G-protein in myotubes was obtained from the fact that GTP-γS, a non-hydrolysable GTP-analogue, stimulated the formation of $InsP_3$ (Häggblad and

Heilbronn, to be published. Fig. 1). It has been suggested that protein G_c, a cholera and pertussis toxin sensitive protein of about 40 kda, might be coupled to the PIns system. The characteristics of the G-protein found in the myotubes in culture only partly fit such a G-protein, as pertussis toxin only had a weak inhibitory action on ATP-produced InsP accumulation (Fig. 2) and cholera toxin was without effect. ADP-ribosylation experiments have to be performed.

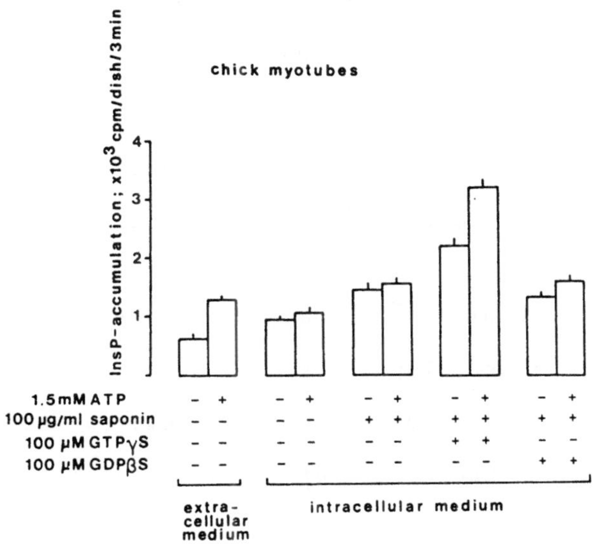

Fig. 1 Effect of GTPγS and GDPβS on myotube inositol phosphate levels after permeabilization with saponin. Myotubes were incubated for 3 min at 37°C. A high ATP-concentration was chosen in order to mimic intracellular conditions.

Purinoceptor-stimulated formation of $InsP_1$, $InsP_2$ and $InsP_3$ was shown as well in rat diaphragm with APPNP and other slowly hydrolyzing ATP-derivatives but not significantly with ATP which may have been removed too rapidly by hydrolysis.

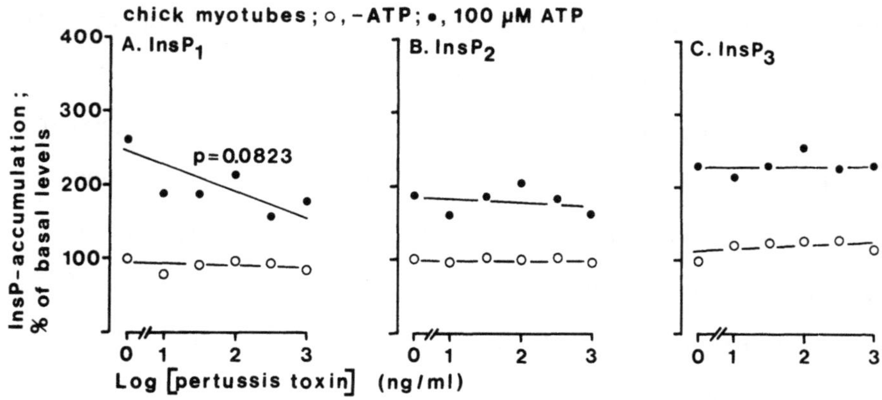

Fig. 2. Effect of pertussis toxin (4 h preincubation) on ATP-induced (3 min incubation) increases in InsP-accumulation. P-values were determined using F-test on regression analysis.

<u>ATP-stimulated purinoceptor-mediated formation of $InsP_3$ is related to release of Ca^{2+} from intracellular stores in myotubes and skeletal muscle?</u>

$InsP_3$ as a trigger of Ca^{2+} release from intracellular stores (review see e.g. Berridge, 1984) has been suggested for a variety of cell types, including smooth muscle (Suematsu <u>et al</u>, 1984; Somlyo <u>et al</u>, 1985; Carsten <u>et al</u>, 1985) and skeletal muscle (Vergara <u>et al</u>, 1985).

Häggblad and Heilbronn (unpublished) found that ATP or ATPγS-stimulated increase of $InsP_3$ in cultured chick myotubes results in increased concentrations of cytosolic Ca^{2+} (Fig. 3), as measured by increased quin-2 fluorescence. APPNP also stimulated, though to a lesser extent, both the formation of $InsP_3$ and increases in the concentration of cytosolic Ca^{2+}, as compared to controls. An exclusion of extracellular Ca^{2+}, by addition of EGTA, neither affected the ATP-stimulated InsP-accumulation nor the intracellular increases in cytoplasmic Ca^{2+}-concentrations. The time course for the ATP-triggered increase in cytosolic Ca^{2+} was rapid; maximal levels have been reached in less than the 30s the tests needed. Increases in $InsP_3$ and cytosolic Ca^{2+} concentration followed each other.

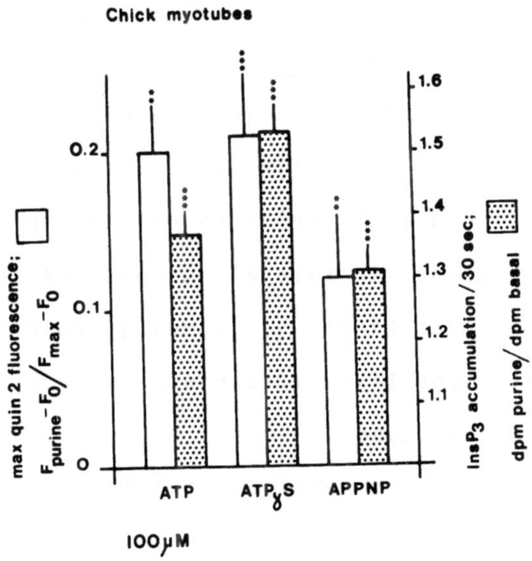

Fig. 3. Effect of ATP, ATPγS and APPNP on levels of $InsP_3$ and intracellular calcium in chick mytubes. **,$p<0.01$, ***, $p<0.001$ (Student´s t-test) as compared to basal levels. F=fluorescence

The experimental procedure used did not allow us to distinguish between different intracellular Ca^{2+}-storage organelles as the source of the $InsP_3$-stimulated Ca^{2+}-release. However, some recent experiments performed in our laboratory (Tassin <u>et al</u>, to be published) suggest that $InsP_3$ is not able to trigger Ca^{2+} release from intracellular stores of myotubes cultured from muscles of embryonic dysgenic mice. Such mice, in addition to numerous other defects, seem to have disturbancies in the Ca^{2+} storing and/or releasing system of the skeletal muscle fibers i.e. the sarcoplasmic reticulum (SR) (Beam <u>et al</u>, 1986; Rieger <u>et al</u>, 1987). The cultured dysgenic myotubes showed ATP-stimulated $InsP_3$ formation while no concomittant increase in cytosolic Ca^{2+} concentration was observed, suggesting that the normal source of the $InsP_3$-triggered increases in Ca^{2+} concentration seen in the cytosol indeed is the SR.

Does a nicotinic acetylcholine receptor-related $PInsP_2$ turnover exist in myotube and skeletal muscle?

Agonist-induced receptor-mediated stimulation of multiple second messenger systems is a wellknown phenomenon. It can certainly not be excluded that the nAChR belongs to the group of receptors which couples to several second messenger systems simultaneously. As determined by binding experiments no ^3H-quinuclidylbenzilate-sensitive muscarinic AChR was present on membranes of cultured myotubes.

Häggblad and Heilbronn recently tested the ability of cholinergic agonists to stimulate InsP formation. In myotubes (chick, mouse) carbachol (1mM) or ACh (3mM) produced a nonsignificant stimulation (Fig. 4). However, using a rat diaphragm preparation a significant stimulation was seen (carbachol, 1mM). This stimulation could, however, partially be blocked by atropine (10 μM) but not by d-tc (10 μM) (Fig. 5) and should therefore be attributed rather to a muscarinic than to a nicotinic receptor-mediated system. The location of muscarinic receptors (mAChR) in this preparation may have been on blood vessel membranes or, though still uncertain, in the presynaptic membrane (for a discussion on the presence/absence of mAChR at the motor nerve ending of vertebrates see Häggblad, 1987).

Adamo et al (1985) measuring InsP accumulation in the presence of Li^+ in (1,2 (n)-^3H) inositol - loaded chick myotubes in culture have suggested that ACh in a concentration- and time-dependent way stimulates phosphatidyl inositide turnover. After 20 min only stimulation of $InsP_1$ was found but no $InsP_2$ or $InsP_3$.

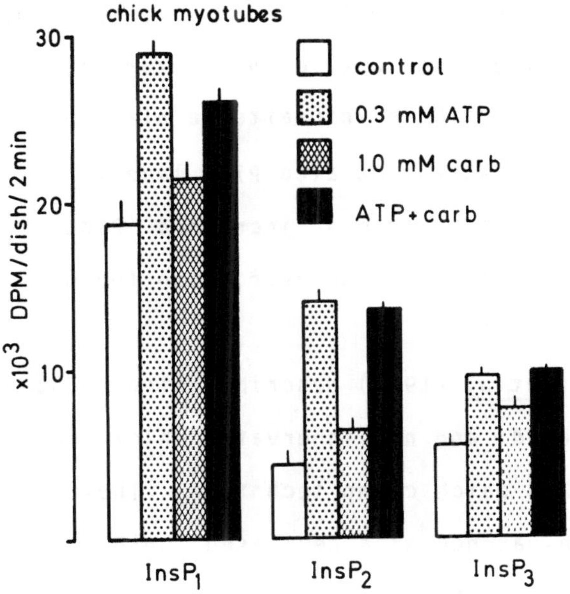

Fig. 4. Effect of ATP and carbachol on inositolphosphate levels in chick myotubes.

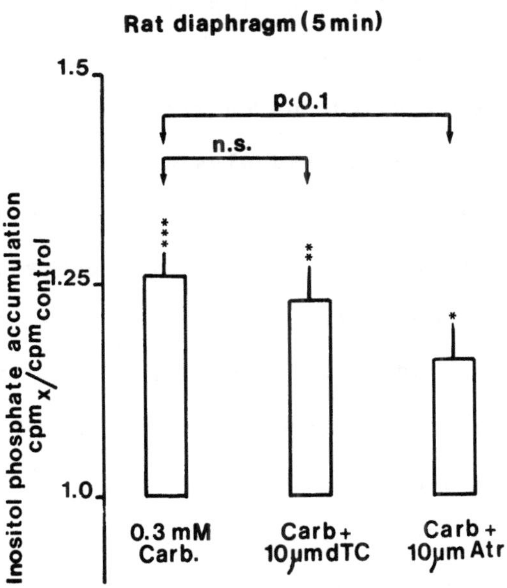

Fig. 5. Effect of carbachol on inositolphosphate levels in pieces of rat diaphragm (radially cut). ***, $p<0.001$, **, $p<0.01$, *, $p<0.05$ as compared to basal level (Student´s t-test).

The effect was both bgt (10^{-7}M) and d-tc($1,5x10^{-4}$M) sensitive. Significant ACh stimulation was found to start at 10^{-7}M, with a steep increase up to $2x10^{-5}$M whereafter a plateau was reached. The authors found an ACh-stimulated PIns labelling which they interpreted as resulting from an increased PIns-turnover. Increased labelling of PIns was, however, also found in the presence of Li^+ alone.

Recently Eusebi <u>et al</u> (1987) described single channel measurements on unstriated and non-innervated chick embryo myotubes using cell-attached patch clamp technique. They found that ACh applied to a non-patched area decreased both channel opening probability and conductance. This was also seen when the non-patched membrane was exposed to "Ca^{2+}-free" extracellular medium. The decrease in channel opening probability and conductance was absent when the membrane was treated with curare, an observation that suggested that they studied an nAChR channel. In whole-cell patch-clamp experiments ACh-induced currents decreased in time course and amplitude when the non-hydrolysable GTP analogue GTP-γS was loaded into the cells, but not when either GDP-βS or cAMP were used. Internal perfusion with GTP-γS also decreased channel opening probability and conductance. The authors interpreted these results as being indirect effects of ACh on the nAChR-channel by delivery of an intracellular messenger and by activation of a putative G-protein.

<u>Diacylglycerol, a second messenger to P_2-purinoceptor stimulation, may be a link between ATP and ACh cotransmission.</u>

Breakdown of $PInsP_2$ by phospholipase C results simultaneously in the formation of $InsP_3$ and diacylglycerol. Both Ca^{2+} and diacylglycerol stimulate protein kinase C (PKC), known to transfer phosphate groups to proteins.

In the myotube and muscle fiber obvious targets for the action of a diacylglycerol-stimulated PKC are intracellular parts of the nicotinic acetylcholine receptor such as the (open) ion channel. In model experiments Huganir _et al_ (1984) have found that nAChR is phosphorylated by at least three different protein kinases. Kinetic studies showed that phosphorylation by a cAMP-dependent protein kinase increased the desensitation rate coefficient (Huganir _et al_, 1986). A working hypothesis from our laboratory connecting stimulated $PInsP_2$ turnover with phosphorylation of the nAChR is seen in Fig. 6. Among other targets for protein phosphorylation in the myotube or muscle fiber the K^+ and Ca^{2+}-channels (Browning _et al_, 1985; Levitan, 1985) may be mentioned.

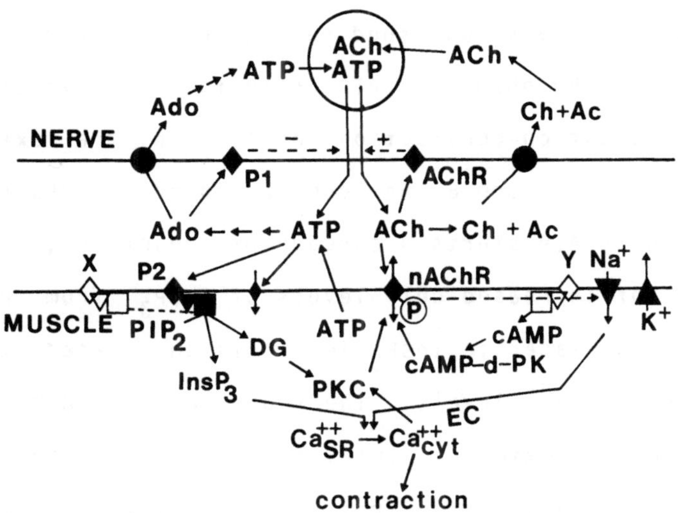

Fig. 6. Working hypothesis for the study of an ACh- and ATP-induced excitation and contraction of skeletal muscle, including one possible way to regulate nAChR. Following nerve stimulation, ACh and ATP are released. ACh-nAChR-interaction triggers depolarisation of the muscle membrane which leads to the formation of an action potential and to release of ATP from the muscle fibre into the extracellular space. Some Ca^{2+} may enter the muscle fibre

through the open nAChR channel. Through the excitation-contraction coupling mechanism, involving a chemical messenger between the tubular system and the SR, Ca^{2+} is released into the cytosol, thus triggering muscle contraction. The extracellular ATP stimulates P_2-purinoceptors, thus triggering intracellular formation of $InsP_3$. $InsP_3$ acts as a chemical messenger releasing Ca^{2+} which increases the ACh effects. Diacylglycerol (DG) formed simultaneously with $InsP_3$ together with Ca^{2+} activates a protein kinase C (PKC) which regulates nAChR function through phosphorylation. Presynaptically, ACh may enhance its own release through nicotinic autoreceptors. Adenosine (Ado), derived from released ATP, is known to decrease ACh release through P_1 - purinoceptors. Formed Ado and choline (Ch) are taken up by the nerve terminal.
Figure 6 also shows other modulators or hormones which may, through their receptors (designated X and Y, respectively), regulate $InsP_3$- and DG-formation as well as an adenylate cyclase activity. A cAMP-dependent protein kinase (cAMP-d-PK) may also regulate nAChR-function.

DISCUSSION

Externally applied ATP is known to potentiate ACh-nAChR induced muscle depolarisation (Ewald 1976; Akasu et al, 1986) and to stimulate skeletal muscle contraction (Buchtal and Folkow, 1948). ATP also induces myotube contraction (Häggblad, 1987). In this and previous papers we have published results which could provide an explanation for the observed ATP effects, an explanation involving the excitation-contraction coupling in the skeletal muscle, as outlined in figure 6. As seen, we found that P_2-purinoceptor stimulation by ATP starts a cascade of reactions, involving production of increased cell levels of $InsP_3$ from $PInsP_2$ breakdown. This increase in $InsP_3$ levels is correlated to an increase in cytosolic Ca^{2+}-levels. The source of this Ca^{2+} seems to be the sarcoplasmic reticulum. $InsP_3$ has previously been shown to release Ca^{2+} from SR isolated from rabbit fast twitch muscle (Volpe et al, 1985) and intracellular $InsP_3$ was shown to induce contraction in peeled skeletal muscle fibre from rabbit (Donaldson et al 1985; 1987).It is wellknown that a transient increase in cytosolic Ca^{2+}-levels is essential for muscle contraction and relaxation. Finally, the presence of enzymes necessary for the formation of $InsP_3$ in skeletal muscle has been verified (Hidalgo et al, 1986, Varsanyi et al 1986).

If, as it may seem, $InsP_3$ is in fact a chemical messenger at the triad of intact innervated skeletal muscle fiber and the only one, one would primarily assume that the location for $PInsP_2$ turnover and $InsP_3$ formation would be the tubular membrane where an activation of relevant enzymes may occur as a consequence of the arrival of an action potential triggered by the sequence of events following ACh-nAChR interaction and transient depolarisation of the sarcolemma. A P_2-purinoceptor may be present in the triadic membrane and be activated by changes in cytosolic ATP levels. Depolarisation-induced release of ATP into the extracellular space, followed by hydrolysis could then be a regulatory mechanism for this intracellular ATP function. We are currently studying these possibilities. An intracellular ATP and P_2-purinoceptor system may, however, not necessarily be the only possible mechanism for an $InsP_3$ formation, several other receptors activating a phospholipase C system are known.

In this paper, however, we have examined an external ATP-induced activation of a P_2-purinoceptor-mediated cascade which results in increased intracellular $InsP_3$ levels and which is followed by increased levels of cytosolic Ca^{2+}. We propose that the aim of this system may be reinforcement, possibly also prolongation of the mechanical responses of the skeletal muscle fiber to ACh-induced nAChR activation. ATP-induced contraction may be important in developing muscle, before the presence of a motor neuron which releases ACh. Such an ATP effect may play a role in the development of the properties of the muscle cell. We have recently found (to be published) that ATP applied to both myoblasts and myotubes stimulates increases in intracellular $InsP_3$ levels. Any directed source of external ATP before arrival of a functioning motor neuron is, however, not known unless mechanisms inducing depolarisation of the plasma membrane of the developing muscle fiber exist in noninnervated myotube and muscle and cause release of ATP from the interior of the cell. P_2-purinoceptors may well exist over the whole myotube membrane.

There are other occasions when the ATP-P_2-purinoceptor effect at the skeletal muscle may become evident. Facilitation, a process observed as a consequence of repetitive stimulation at high frequency of a motor neuron, is thought to be the result of increases in the number of releasable transmitter fractions. The process is observed especially after curarisation of skeletal muscle. Facilitation seems to change the nature of the muscle twitch. We suggest that potentiation of the muscle twitch during facilitation at least partly may be a consequence of an extensive ATP accumulation and action at a P_2-purinoceptor present in the sarcolemma of skeletal muscle.

ATP-stimulated increase of $InsP_3$ formation may be an "emergency" mechanism e.g. in a disease such as Myasthenia gravis, where reinforcement of the muscle twitch by ATP might improve muscle contraction possibilities. Perhaps slowly hydrolysed derivatives of ATP or even a locally applied stimulator of the putative G-protein linking the P_2-purinoceptor to phospholipase C might perhaps even be of therapeutic interest as the disease is characterized by a reduction in the quantity of functioning nAChR. Their use could probably only be considered in a limited form of the disease.

At least one other human disease is known where an alteration of free myoplasmic Ca^{2+} homeostasis in muscle fibre from patients exists, malignant hyperthermia (MH). An abnormal function of SR membrane may be involved in the pathophysiology of this disease. As shown by Condrescu et al (1987), the capacity for $^{45}Ca^{2+}$ uptake and the Ca^{2+}/Mg^{2+} - ATP ase activity in SR vesicles from intercostal muscle of MH-susceptible patients is lower than that of controls. An increase of resting free Ca^{2+} is seen in the myoplasm of skeletal muscle fibers of MH patients, (see also Endo et al 1983). The changes between SR Ca^{2+} uptake capacity were, however, not significantly different in normal and susceptable swine (Ohnishi et al, 1983), although the myoplasmic free Ca^{2+} concentration was about 4 times higher in MH susceptible animals than in controls (Lopez et al 1985). Whether or not an increased formation of $InsP_3$ would be of value cannot be decided before the defects present in the SR have been characterized.

Diacylglycerol, one of the second messengers produced by phospholipase C-induced $PInsP_2$ turnover is one of several mechanisms of controlling nAChR function, and its ability to stimulate protein kinase C is expected to represent one of the control mechanisms of nAChR function (Huganir and Guengard, 1983; Huganir et al, 1984). Experiments designed to clarify this point are in progress.

REFERENCES

Abood LG, Koketsu K and Miyamoto S (1962) Outflux of various phosphates during membrane depolarisation of excitable membranes. Am J Physiol 202:469-474

Adamo S, Zani BM, Nervi C, Senni MI, Molinaro M and Eusebi F (1985) Acetylcholine stimulates phosphatidylinositol turnover at nicotinic receptors of cultured myotubes. FEBS LETTERS 190: 161-164

Akasu T, Hirai K and Koketsu K (1981) Increase in acetylcholine-receptor sensitivity by adenosine triphosphate: a novel action of ATP on ACh-sensitivity. Br J Pharmac 74:505-507

Beam KG et al (1986) A lethal mutation in mice eliminates the slow calcium current in skeletal muscle cells. Nature 320:168-170

Berridge MJ (1984) Inositol triphosphate and diacylglycerol as second messengers. Biochem J 220:345-360

Berridge MJ and Irvine RF (1984) Inositol triphosphate, a novel second messenger in cellular signal transduction. Nature 312:315-321

Browning MD, Huganir R and Greengard P (1985) Protein phosphorylation and neuronal function. J Neurochem 45:11-23

Buchthal F and Folkow B (1948) Interaction between acetylcholine and adenosine triphosphate in normal, curarized and denervated muscle. Acta Physiol Scand 15:150-160

Carsten ME and Miller D Y (1985) Ca^{2+} release by inositol triphosphate from Ca^{2+} tansporting microsomes derived from uterine sarcoplasmic reticulum. Biochim Biophys Res Commun 130:1027-1031

Condrescu M, Lopez JR, Medina P and Alamo L (1987) Deficient function of the sarcoplasmic reticulum in patients susceptible to malignant hyperthermia. Muscle and Nerve 10:238-241

Donaldson SK (1985) Peeled skeletal muscle fibers. Possible stimulation of Ca^{2+} release via a transverse tubule sarcoplasmic reticulum mechanism. J gen Physiol 86:501-525

Donaldson SK, Goldberg ND, Walseth TF and Huetteman DA (1987) Inositol trisphosphate stimulates calcium release from peeled skeletal muscle fibers. Biochim Biophys Acta 927:92-99

Endo M, Yagi S, Ishizuka T, Horiuti K, Koga Y and Amaha K (1983) Changes in the Ca-induced Ca release mechanism in the sarcoplasmic reticulum of the muscle from a patient with malignant hyperthermia. Biom ed. Res 4:83-92

Eusebi F, Grassi F, Molinaro M and Zani BM (1987) Acetylcholine regulation of nicotinic receptor-channels through a putative G protein in the chick myotubes. Neuroscience 22 (Suppl):766

Ewald DA (1976) Potentiation of postjunctional sensitivity of rat diaphragm by high-energy-phosphate adenine nucleotides. J Membr Biol 29:47-65

Ewald DA, Williams A and Levitan IB (1985) Modulation of single Ca^{2+}-dependent K^{+}-channel activity by protein phosphorylation. Nature 315:503-506

Grondal EJM and Zimmermann H (1986) Ectonucleotidase activities associated with cholinergic synaptosomes isolated from Torpedo electric organ. J Neurochem 47:871-881

Heilbronn E and Häggblad J (1986) A case for ATP as a neurotransmitter and trigger of postsynaptic Ca release at the neuromuscular junction? In: Cellular and molecular basis of cholinergic function. (Dowdall M, ed). Ellis Horwood, Chichester:444-553

Hidalgo C, Carrasco MA, Magendo K and Jaimovich E, (1986) Phosphorylation of phoshpatidylinositol by transverse tubule vesicles and it possible role in excitation contraction coupling. FEBS Lett 202:69-73

Huganir RL, Milers K and Greengard P (1984) Phosphorylation of the nicotinic acetylcholine receptor by an endogenous tyrosine-specific protein kinase. Proc Natl Acad Sci (USA) 81:6968-6772

Huganir RL, Delcour AH, Greengard P and Hess GP (1986) Phosphorylation of the nicotinic acetylcholine receptor regulates its rate of desensitization. Nature 321:774-776

Häggblad J, Eriksson H and Heilbronn E (1985a) Effects of extracellular ATP on ^{86}Rb influx in chick myotubes; indications of a contransmitter role in neuromuscular transmission. In: Molecular basis of nerve activity (Changeux JP et al, eds) Walter de Gruyter and Co Berlin New York:185-193
Häggblad J, Eriksson H and Heilbronn E (1985b) ATP-induced cation influx in myotubes is additive to cholinergic agonist action. Acta Physiol Scand 125:389-393
Häggblad J (1987) Neuromuscular junction revisited. Biochemical studies on mèchanism involved in transmission events. Thesis, University of Stockholm, Sweden.
Häggblad J and Heilbronn E (1987) Externally applied adenosine-5´-triphosphate causes inositol triphosphate accumulation in cultured chick myotubes. Neurosci Lett 74:199-204
Israel M, Lesbats B, Manaranche R, Meunier FM and Frachon P (1980) Retrograde inhibition of transmitter release by ATP. J Neurochem 34:923-932
Kolb HA and Wakelam MJO (1983) Transmitter-like effects of ATP on patched membranes of cultured myoblasts and myotubes. Nature 303:621-623
Lopez JR, Alamo L, Jones D, Allen P, Papp I, Gergely J and Sieter FA (1985) Free myoplasmic calcium concentration in skeletel fibers from malignant hyperthermia susceptible swine, measured in vivo with Ca^{2+}-sensitive microelectrodes. Biophys J 47:313a
Majerus PW, Wilson DB, Connoly TM, Bross TE, Neufeld EJ (1985) Phosphoinositide turnover provides a link in stimulus-response coupling. Trends Biol Sci 19:168-171
Nishizuka Y (1984) Turnover of inositol phospholipids and signal transduction. Science 225:1365-1370
Ohnishi ST, Taylor S and Gronert GA (1983) Calcium-induced Ca^{2+} release from sarcoplasmic reticulum of pigs susceptible to malignant hyperthermia. The effects of halothane and dantrolene. FEBS Lett 161:103-107
Ribeiro JA and Sebastiao AM (1986) Adenosine receptors and calcium: Basis for proposing a third (A_3) adenosine receptor. Prog Neurobiol 26:179-209
Rieger F, Pincon-Raymond M, Tassin AM, Garcia L, Romey G, Fosset M and Lazdunski M (1987) Excitation-contraction uncoupling in the devloping skeletal muscle of the muscular dysgenesis mouse embryo. Biochimie 69: in press
Somlyo AV, Bond M, Somlyo AP and Scarpa A (1985) Inositol trisphosphate induced calcium release and contraction in vascular smooth muscle. Proc Natl Acad Sci USA 82:5231-5235
Suematsu E, Hirata M, Haskimoto T and Kuriyama H (1984) Inositol 1,4,5-trisphosphate releases Ca^{2+} from intracellular store in skinned single cells of porcine coronary artery. Biochem Biophys Res Commun 120:481-485
Varsanyi M, Messer M, Brandt N and Heilmeyer LMG Jr (1986) Phosphatidylinositol 4,5 biphosphate formation in rabbit skeletal and heart muscle membranes. Biochem Biophys Res Commun 138:1395-1404
Vergara J, Tsien RY and Delay M (1985) Inositol 1,4,5-triphosphate: A possible chemical link in excitation-contraction coupling in muscle. Proc Natl Acad Sci (USA) 82:6352-6356
Volpe P, Salviati G, Di Virgilio F and Pozzan T (1985) Inositol 1,4,5-triphosphate induces calcium release from sarcoplasmic reticulum of skeletal muscle. Nature 316:347-349

IDENTITY AND FUNCTIONAL ROLE OF NEUROPEPTIDES IN *TORPEDO* ELECTRIC ORGAN

Michael.J.Dowdall and David.L.Downie
Department of Zoology, University of Nottingham,
University Park, Nottingham. NG7 2RD. U.K.

INTRODUCTION

It is well known that the electromotor system of *Torpedinidae* rays is an excellent paradigm for elucidating the biochemical basis of chemical transmission at cholinergic synapses(for recent review see Whittaker, 1987). The system also has considerable potential in providing information on the broader aspects of chemical transmission such as the molecular basis of stimulus-secretion coupling, neuronal gene expression and synaptogenesis. One of the principle conundrums in contemporary neurobiology concerns the role(s) of neuropeptides which are, as a group, ubiquitous components of neurones. The discovery of the neuropeptides and their co-localisation with classical transmitters was the result of the application of immunochemical and particularly immunocytochemical techniques to the nervous system. Thus far observations on the functional effects of neuropeptides have been limited to a few well-defined systems, whereas those concerning the qualitative presence of immunopositive staining have been prolific. There is now a body of evidence implicating the presence of neuropeptides in the *Torpedo* electromotor system and this therefore may provide an additional test-bed system for exploring functional aspects of co-existence. This evidence will be reviewed in this chapter.

NEUROPEPTDES OF THE ELECTROMOTOR SYSTEM

Michaelson et al (1984 a, b), reported the presence of enkephalin-like peptide(s) in extracts of whole tissue and synaptosomes from the electric organ of *T.oscellata* using a radio-immuno assay (RIA), and antibodies against mammalian enkephalins. They also showed the inhibitory action of morphine on potassium-stimulated, calcium ion uptake into and

NATO ASI Series, Vol. H21
Cellular and Molecular Basis of Synaptic Transmission
Edited by H. Zimmermann

acetylcholine release from synaptosomes. These observations suggested a presynaptic localisation of enkephalins and a role as negative feedback inhibitors of synaptic secretion. More recently another neuropeptide, vasoactive intestinal peptide (VIP), has been reported to be present in the electromotor system of <u>T.marmorata</u>. (Agoston and Conlon, 1986). In this study VIP-like immunoreactivity was detected in the cell bodies, axons and terminals of the electromotor neurones using immunocytochemical and RIA techniques and an antiporcine VIP antisera. Chemical analysis of the endogenous peptide by HPLC revealed similarity but not complete identity with porcine VIP.

Following the earlier report of enkephalins in electric organ and concurrent observations in this laboratory on neuropeptidases in electric organ, (Turner and Dowdall, 1984) - see following section - the presence of opioid receptors was investigated using radioligand binding to membrane preparations from <u>T.marmorata</u> electric organ (Birchmoor et al). This receptor screening exercise employed various selective opioid antagonists in an attempt to sub-type any detectable specific binding. None of the four classical receptor binding sites was evident although some atypical d-sites were detected. These ´negative´ observations cast some doubt against the case for presynaptic enkephalin receptors in <u>Torpedo</u> electric organ - at least for <u>T.marmorata</u> and we therefore directed our attention towards VIP as a canditate co-existing neuropeptide. VIP has been isolated from a variety of vertebrates and as shown below in Table 1, there is some species variability in the amino acid sequence of this peptide.

TABLE 1. <u>Amino acid Sequences of Vertebrate VIP´s.</u>

SPECIES	AMINO ACID SEQUENCE[a]																												
G.Pig					L				T										M							V			
Pig	H	S	D	A	V	F	T	D	N	Y	T	R	L	R	K	Q	M	A	V	K	K	Y	L	N	S	I	L	N	NH_2
Chicken											S		F													V		T	
Dogfish											S		I										I			L		A	NH_2

a) Single letter code for amino acids with variants from the porcine sequence indicated. Dogfish sequence from Dimaline et al 1987.

In the view of these differences it seemed judicious to use an antisera least likely to give erroneous titres with extracts of Torpedo tissue.

In a collaborative study with Dr.R.Dimaline of the University of Liverpool, extracts prepared from the electromotor system and other tissues of T.marmorata have been assesed for VIP content by RIA, using an antisera directed against the non-variant, (compared to dogfish,) N-terminal portion of porcine VIP.

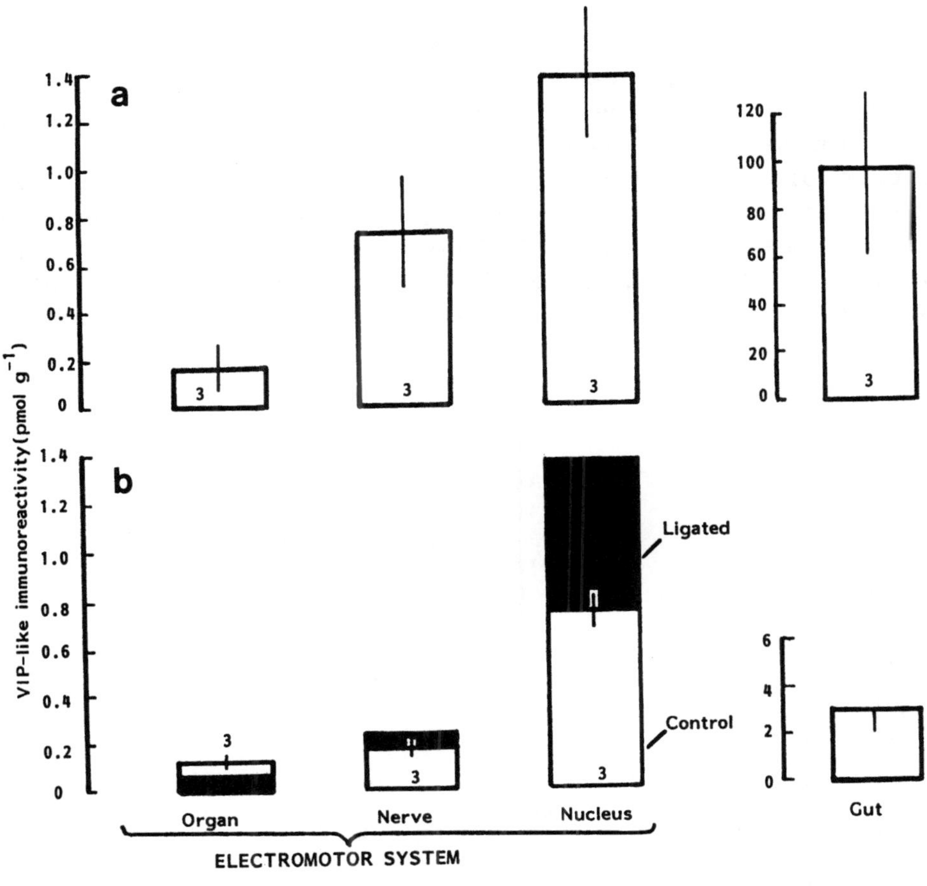

Figure 1. VIP-like immunoreactivity in T.marmorata tissues. Results in upper half (a) from present study using antisera against N-terminal portion of porcine VIP as described by Dimaline et al (1986), are compared to lower half (b) as reported by Agoston and Conlon (1986). The latter study also shows the effect of nerve ligation on VIP-like titres in the electromotor system.Means ±SD of 3 measurements are shown.

As shown in Fig 1 our study has essentially confirmed the observations of Agoston and Conlon (1984), both quantitatively and qualitatively. The most noticable difference between the two sets of results is the relatively low titre for Torpedo intestine reported by Agoston and Conlon, (1984). Our figure of 100 pmol/ g wet wt. for Torpedo gut is for small intestine and is very comparable to literature values (Dimaline et al 1986). We therefore feel some confidence in our estimates of about 1 pmol/ g wet wt. for the electromotor perikarya and axons. Compared to the acetylcholine content of these neurones (about 1 µmol/g wet wt.) this neuropeptide is therefore a very minor constituent.

Since seasonal differences in the physiological and biochemical properties of electric organ have been previously noted by others, (Babel-Guerin 1974 ; Pinchasi et al 1985), we nave measured VIP-like immunoreactivity in the electric organs of fish in different physiological states. Three parameters were tested - sex, body weight and starvation. As can be seen in Figure 2, VIP contents ranged from 0.1 - 0.5 pmol per g., but with no obvious sexual differences and no discernable variation with body weight or starvation period.

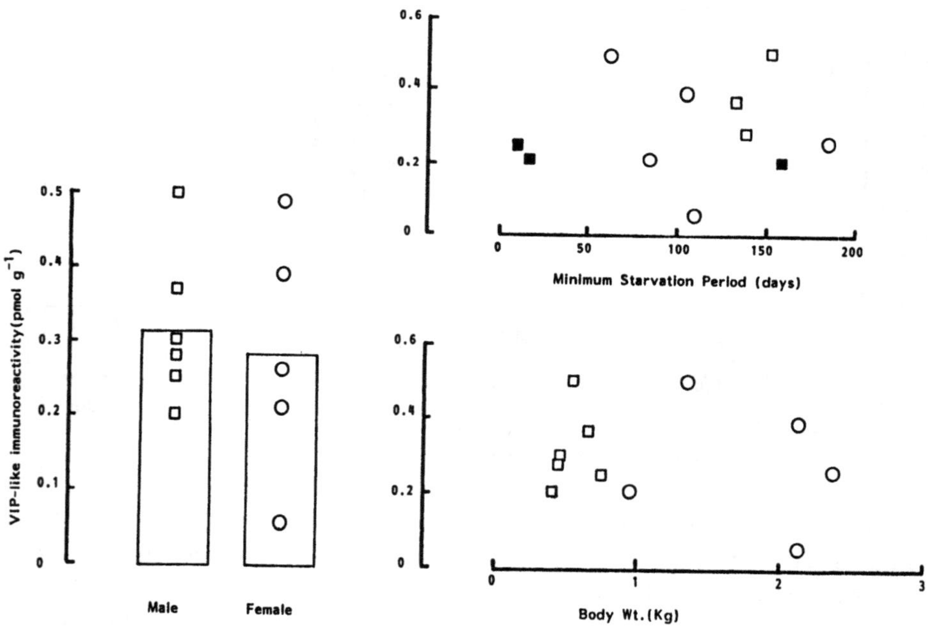

Figure 2. VIP levels in Torpedo electric organ from fish in different physiological states. VIP-like immunoreactivity was measured as described in Figure 1. Males are indicated by squares, females by circles; open symbols for tissue stored at -80°C before extraction, closed symbols for fresh tissue.

Pooling all the results for electric organ gives a mean concentration of 301 ± S.D. 131 f mols. VIP-like immunoreactivity per g (n=11). The ratio of ACh to VIP in this tissue is therefore approximately $3x10^6$:1 whereas the number of ACh molecules per vesicle has been estimated at about 10^5. It is difficult to escape the conclusion that not all vesicles can contain VIP. In fact even when generous allowance is made for the size of the cytoplasmic ACh pool , (say 70%), then only one vesicle in ten will contain a single copy of VIP! An alternative explanation is that the RIA technique seriously underestimates the concentration of VIP-like peptide which is present. However the technique gives high titres with extracts from Torpedo small intestine (Figure 1, a). Isolation and chemical characterisation of the Torpedo electric organ peptide would clearly resolve these uncertainties - the task however is a daunting one, particularly if its concentration is as low as the present results suggest.

NEUROPEPTIDASE INHIBITION AND SYNAPTIC TRANSMISSION.

Membrane fragments from T. marmorata, (Turner and Dowdall 1984), and T.californica, (Altstein et al 1984), have been shown to contain neuropeptidases. Two enzymes were found in T. marmorata - endopeptidase 24.11, (enkephalinase) and peptidyl dipeptidase A, (angiotensin-converting enzyme (ACE)). Since these peptidases are capable of hydrolysing a variety of peptides including Leu- and Met- enkephalins, (see Turner et al 1987), it has been speculated that endogenous neuropeptides may regulate cholinergic function in electric organ, (Turner and Dowdall 1984; Dowdall and Turner 1984; Altstein et al 1984). Specific inhibitors of the two enzyme activities are Phosphoramidon and Captopril, for the endopeptidase and peptidyl dipeptidase respectively. We have therefore tested the effect of these inhibitors on the electrical response of isolated prisms of electric organ to electrical stimulation. Our preliminary findings are given below.

The electrophysiological assay system on isolated prisms of electric organ was set up essentially as described by (Dunant et al 1980). Compound electroplax potentials, (epps), in response to single stimuli and trains of pulses were measured. Under appropriate conditions it can be shown that the epp amplitude is directly proportional to ACh secretion. The assay thus measures,indirectly, ACh output from the cholinergic terminals. Dunant and Walker (1982) used this system to demonstrate presynaptic regulation of ACh secretion via muscarinic receptors and we were able to confirm this atropine-senstive oxotremorine inhibition, (results not shown).

Results from Captopril-treated prisms are shown in Figure 3. Essentially this inhibitor, at a concentration which would totally inhibit the peptidyl dipeptidase enzyme, (see Turner and Dowdall 1984), had no effect on the pattern of synaptic transmission over a 3 h period.

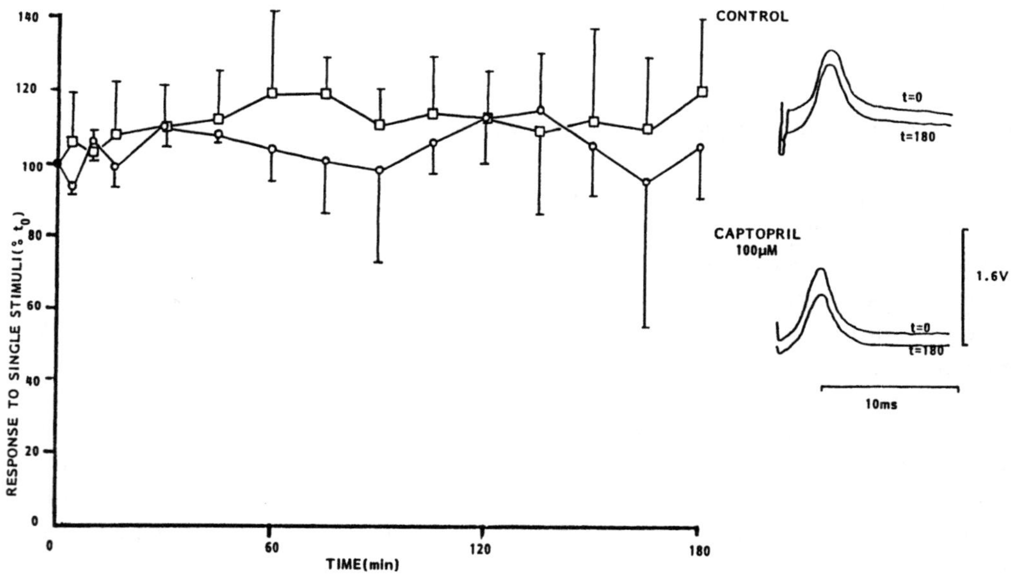

Figure 3. Responses of isolated prisms of electric organ to single stimuli in the presence and absence of Captopril. Superfused, excised prisms of electric organ were stimulated with single pulses at the times indicated and the electrical responses measured, (see inserts). Results are from three control prisms and three which included 100 μM Captopril added to the superfusion medium, - a Torpedo Ringer, (see Dunant et al 1980 for composition).

Lisinopril, an analogue of Captopril with a more potent

action towards peptidyl dipeptidase (Turner et al 1987) was also tested in a similar fashion and these results are shown in Figure 4. Again no effect of this inhibitor was detectable under these conditions.

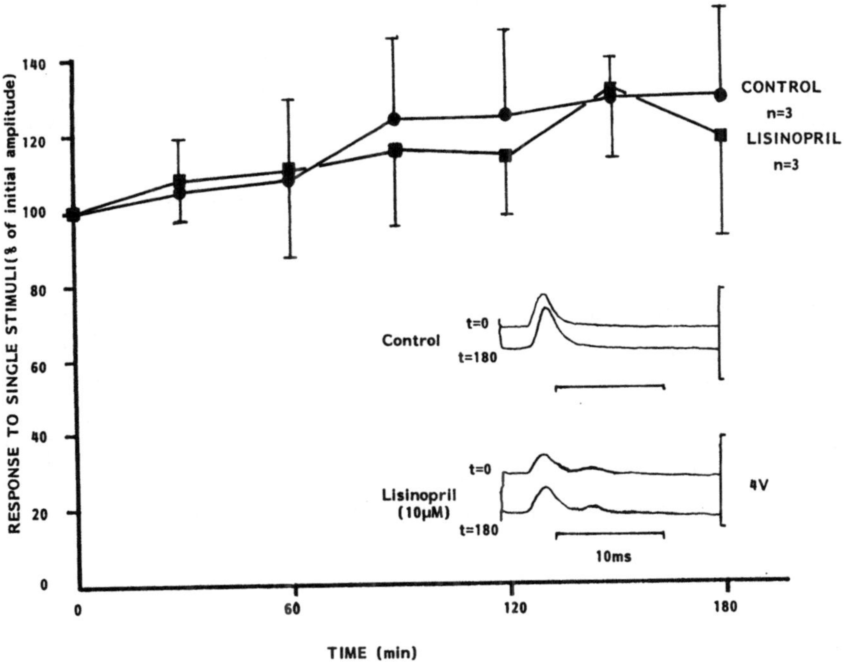

Figure 4. Responses of isolated prisms of electric organ to single stimuli in the presence and absence of Lisinopril. Experimental details are as for Figure 3., except that the drug concentration was 10 µM.

In other systems it has been observed that the release of neuropeptides from stimulated nerves is frequency - dependent (Lundberg and Hökfelt 1982) - with little or no release at low frequency stimulation. Our negative observations on peptidyl dipeptidase inhibition, after single stimuli, might be accounted for on this basis. In addition it seemed prudent in further experiments to inhibit both enzymes known to be present. We have therefore tested the combined effects of Captopril and Phosphoramidon on the response of prisms to trains of pulses at 20 Hz (Fig. 5) but as with the earlier experiments, using single stimuli, no inhibitor effects were evident.

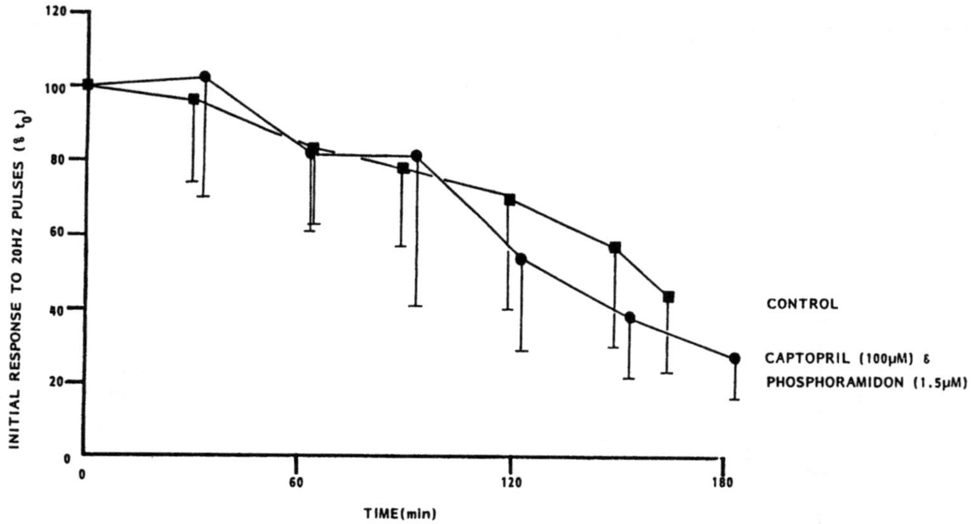

Figure 5. Responses of isolated prisms of electric organ to stimulation by impulse trains at 20 Hz in the presence of peptidase inhibitors. Details are as for Figure 3 except for stimulation conditions and the presence of the two peptidase inhibitors. Note that the initial response in all prisms declines with time. This should be compared to the steady responses for single stimuli (Figures 3 and 4).

DISCUSSION

Using T.marmorata we have confirmed the presence of a VIP-like peptide in the electromotor system and this therefore strengthens the general case for the co-existence of neuropeptides with ACh in these cholinergic neurones. Whether or not VIP-like peptides and those of the opioid class are both present in the electromotor neurones of all Torpedo species is an interesting question which arises from these and previous observations on T.marmorata (Agoston and Conlon, 1986) and T.oscellata (Michaelson et al, 1984). Conceivably the particular neuropeptide might be species specific.

Concerning the functional role of neuropeptides in this system the results reported here reveal little. The inhibitors used to block peptidase activity are certainly effective on isolated membranes from electric organ, but in superfused prisms of tissue these agents might not penetrate to the synaptic sites even at the elevated concentrations used. Also there may be additional enzymes present which degrade any released neuropeptide, but which are insensitive to the inhibitors so far used. In this connection it is interesting to note that porcine VIP is a poor substrate for the purified peptidyl dipeptidase A from electric organ (Turner & Dowdall unpublished observations). A specific investigation in to the VIP-degrading activity of Torpedo synaptic membranes should answer this question. The problem of drug penetration to synaptic sites in isolated blocks of tissue could be avoided by using synaptosomes from electric organ as a system for testing the effects of peptides and peptidase inhibitors on ACh secretion.

ACKNOWLEDGEMENTS

The authors thank the MRC for research support and Dr. R. Dimaline, Dept. Physiology, University of Liverpool, UK for measurements of VIP-like immunoreactivity.

REFERENCES

Agoston DV & Conlon JM (1986) presence of Vasoactive intestinal polypeptide-like immunoreactivity in the cholinergic electromotor system of Torpedo marmorata. J. Neurochem 47: 445-453

Altstein M, Dudai Y & Vogel Z (1984) Enkephalin-degrading enzymes are present in the electric organ of Torpedo californica. FEBS Lett 166: 183-188

Babel-Guerin E (1974) Calcium metabolism and acetylcholine release in the electric organ of Torpedo marmorata. J. Neurochem 23: 525-532

Birchmoor B, Clark CR & Dowdall MJ (unpublished observations)

Dimaline R, Thorndyke MC & Young J (1986) Isolation and partial sequence of elasmobranch VIP. Regulatory Peptides 14: 1-10

Dimaline R, Young J, Thwaite DT, Lee CM, Shuttlworth IJ & Thorndyke MC (1987) A novel vasoactive intestinal polypeptide (VIP) from elasmobranch intestine has full affinity for mammalian pancreatic VIP receptors. Biochem Biophys Acta 930: 97-100

Dowdall MJ & Turner AJ (1984) Neuropeptide metabolism in a model cholinergic system -Torpedo electric organ. In: Vizi ES & Magyar K (eds.). Regulation of Transmitter Function: Basic & Clinical Aspects. Akademiai Kiado, Budapest pp527-529

Dunant Y, Eder L & Servetiadis-Hirt L (1980) Acetylcholine release evoked by single or a few nerve impulses in the electric organ of Torpedo. J. Physiol Lond 298: 185-203

Dunant Y & Walker AI (1982) Cholinergic inhibition of acetylcholine release in the electric organ of Torpedo. Eur. J. Pharmacol 78: 201-212

Lundberg JM & Hokfelt T (1983) Co-existence of peptides and classical neurotransmitters. Trends Neurosci 6: 325-333

Michaelson DM, McDowell G & Sarne Y (1984a) The Torpedo electric organ is a model for opiate regulation of acetylcholine release. Brain Res 305: 173-176

Michaelson DM, McDowell G & Sarne Y (1984b) Opiates inhibit acetylcholine release from Torpedo nerve terminals by blocking Calcium influx. J. Neurochem 43: 614-618

Pinchasi I, Burstein M, Moldovan M & Michaelson DM (1985) Seasonal variations in the muscarinic regulation of acetylcholine release from Torpedo electric organ nerve terminals. Comp Biochem Physiol 81: 439-444

Turner AJ & Dowdall MJ (1984) The metabolism of neuropeptides: Both phosphoramidon-sensitve and captopril-sensitive metallopeptidases are present in the electric organ of Torpedo marmorata. Biochem J. 222: 255-259

Turner AJ, Hryszko J, Hooper NM & Dowdall MJ (1987) Purification and characterisation of a peptidyl dipeptidase resembling Angiotensin converting enzyme from the electric organ of Torpedo marmorata. J. Neurochem 48: 910-916

Whittaker VP (1987) Cholinergic transmission: Past adventures and future prospects. In: Dowdall MJ & Hawthorne JN (eds). Cellular and Molecular Basis of Cholinergic Function. Ellis Horwood, Chichester UK p495

MECHANISMS REGULATING THE EXPRESSION AND FUNCTION OF ACETYLCHOLINE RECEPTOR

Veit Witzemann* and Bert Sakmann#

*Abteilung Neurochemie und #Zellphysiologie, Max-Planck-Institut für biophysikalische Chemie, D-3400 Göttingen, FRG

INTRODUCTION

Acetylcholine receptors (AChR) are distributed throughout the developing skeletal muscle cell surface, but become highly concentrated at the neuromuscular junction in the innervated adult muscle. Concomitantly, AChRs become metabolically stable and their channel properties change (Edwards 1979; Schuetze and Role 1987). The synthesis of the AChR complex could be regulated on the level of transcription, translation or post-translation, affecting assembly, transport and insertion into the membrane. Regulation at these different levels during development may give rise to the expression of different AChR molecules in embryonic and adult muscle. In addition to AChR-associated changes, epigenetic factors may be involved in regulating or modulating AChR properties. Such factors could reside within the local environment and specific interactions may occur with lipids, proteins or enzymes. AChRs could furthermore communicate with components of the cytoplasmic phase as well as with components and factors originating from the presynaptic nerve terminal. With the introduction of recombinant DNA techniques it became evident that changing levels of the various AChR-subunit specific mRNAs could regulate the expression of different types of AChR (Mishina et al 1986; Evans et al 1987 and references therein). In bovine muscle, a developmentally regulated switch has been found in the expression of two types of AChR channels, which have different functional properties and which are composed of the α-, β-, γ- and δ-subunits or of the α-, β-, ϵ- and δ-subunits (Mishina et al

NATO ASI Series, Vol. H21
Cellular and Molecular Basis of Synaptic Transmission
Edited by H. Zimmermann

1986). Denervation experiments of skeletal muscle showed also pronounced effects on the biosynthesis of AChR and the changes are reminiscent of changes occuring during development: Denervation results in a large increase of newly synthesized AChRs, which become mostly incorporated into extrasynaptic regions. These AChRs have distinct biophysical properties and appear to be biochemically different from junctional AChRs. To understand the molecular basis of neurochemical transmission it is essential to establish a correlation between the structural organization of the synapse and the elements functionally involved in the transmission of specific signals. The electromotor system of Torpedo marmorata has been used as an ideal source for the purification and charactzerization of molecules associated with synapses. The high concentration of synaptic contacts in this purely cholinergic tissue enables one to obtain relatively homogeneous subcellular fractions of synaptic structures (Whittaker 1977). In addition these fish provide an excellent system to investigate developmental changes (Witzemann et al 1983).

Recently a protein was isolated and purified, the 43K protein, which may be necessary in stabilizing the mature postsynaptic membrane (Froehner 1986). Components localized in the junctional basal lamina may have a similar function. To date there is however no evidence that such components affect the gating properties of the AChR channels, and in Torpedo marmorata only one type of AChR complex appears to exist (Witzemann et al 1983a).

To investigate mechanisms which regulate the expression of different forms of AChR channels, rat muscle tissue has been used as a source material: Measuring the contents of the mRNAs encoding the γ- and ϵ-subunits of AChR in developing rat muscle as well as in innervated, denervated and reinnervated rat diaphragm it is found that the changes in the abundance of the two subunits mRNAs correlate with changes in the relative density of two different classes of AChR channels.

SYNAPSE-ASSOCIATED COMPONENTS

Considerable interest has been shown in studying proteins associated with the postsynaptic membrane. One such component, the 43 K protein, isolated from purified postsynaptic membranes of the electrocytes of the electric organ from Torpedo marmcrata has been found to be closely associated with the AChR (reviewed by Froehner 1986). Although a function has not been established, these polypeptides appear to be involved in anchoring the receptor molecules at the postsynaptic sites.

Following the appearance of these 43 K proteins in preparations of postsynaptic membranes isolated at various developmental stages of Torpedo marmorata it is found that both AChR as well as 43 K protein become increasingly enriched in these membranes (Witzemann et al 1983b). A more detailed analysis of the distribution of AChR and 43 K protein during development has been performed using thin sections of the electric organ stained with anti-43 K protein antibodies, raised and characterized by Froehner 1984, and rhodamine-labelled α-bungarotoxin. These studies show that incorporation of AChR into the developing postsynaptic membrane of the electrocytes preceedes the localization of 43 K protein at these sites (unpublished observations). With the formation of fully functional synaptic contacts increasing amounts of 43 K protein are co-localized with the AChRs supporting the view that these polypeptides serve as stabilizing components for AChR accumulated at sites of synaptic contacts.

A specialized structure, the junctional basal lamina,is although thought to contribute to the establishment of functional synapses. Of special significance appears to be the existence of an AChR-aggregating factor (Fallon et al 1985; Wallace et al 1985). Recently an antigen specifically associated with synaptic vesicles has been characterized (Stadler and Kiene 1987). It has been demonstrated that this antigen, or part of it, is released from the nerve terminal and deposited in the basal lamina. Using immunohistochemical methods, extracts from electric lobe (part of the brain which innervates the electric organ) and electric organ from Torpedo marmorata, obtained at various developmental stages, were tested for the presence of this antigen. The vesicular antigen is detected at early embryonic stages in electric lobe tissue only, but not in the electric organ. When axons start to invade the electric organ, the antigen is present in both tissues. During the functional maturation of the electromotor system the antigen levels increase strongly. At late embryonic stages and in the adult animal the antigen is much more abundant in the electric organ than in the electric lobe, supporting the view that this vesicular "core"- proteoglycan is deposited in the basal lamina and may be involved in specifying the correct intercellular interaction between neurons and their electric organ target. An interesting question is whether the AChR-aggregating factors and these vesicular proteoglycans are related.

Although the precise structural organization of the synapse is a requirement for neurochemical transmission it appears that receptor mediated iontranslocation is not affected by AChR associated components and changes in channel function observed during development may be the result of a change in the subunit composition (Mishina et al 1986). For these reasons it is necessary to investigate whether changes in subunit composition could be correlated with developmental changes, as described for calf muscle and in how far these changes are under the control of innervating motor nerves (Witzemann et al, in press).

CHANNEL TYPES IN RAT MUSCLE

In rat soleus muscle the fetal type of AChR channel with a conductance of about 40 pico Siemens (pS) and an open time of 5-8 ms converts into an adult type channel with a conductance of 60 pS and a significantly shorter mean open time of 1-2 ms (Brenner and Sakmann 1983). A similar observation was made analysing bovine muscle. A molecular explanation for this transition has been possible by using recombinant DNA cloning techniques: Knowing the NH_2-terminal sequences of the Torpedo-AChR subunits (Raftery et al 1980) it was possible to obtain all four cDNAs coding for the α, β, γ and δ subunits (for references see Mishina et al 1986). The next step was then, using these probes, to isolate the corresponding bovine muscle AChR subunits. Interestingly it was found that fetal AChR-complexes are composed of α, β, γ and δ subunits and that with development a new receptor type would be expressed. The AChR-complexes present in the adult muscle are compromised of α, β, ϵ and δ subunits (Mishina et al 1986). These findings indicate that low and high conductance channels differ in their subunit composition and that the exchange of the "fetal" γ-subunit by the "adult" ϵ-subunit could be the major molecular mechanism in the expression of different AChR channel types.

DIFFERENTIAL EXPRESSION OF γ- AND ϵ-mRNA IN RAT MUSCLE

To elucidate the role of the nerve in regulating the expression of two classes of AChRs the abundance of the γ- and ϵ-subunit specific mRNAs was measured in rat muscle (lower leg muscle and diaphragm).

A rat genomic DNA library was screened using probes carrying fragments of the bovine γ- and ϵ-subunit cDNAs (Takai et al 1984; Takai et al 1985). Hybridization-positive clones were isolated and characterized. On basis of the nucleotide and the deduced amino acid sequences, it was shown that specific probes for the rat γ- and ϵ-subunit were obtained, carrying sequences corresponding to the region of exons P 7 and P 8 as defined for the human γ-subunit gene (Shibahara et al 1985). Dot hybridization analysis showed in addition that these probes did not cross-hybridize with each other (Witzemann et al 1987).

Following the expression of these two subunits in the postnatal rat muscle the following results are obtained: The abundance of the γ-subunit specific mRNA, being high during the first few days, decreases rapidly to undetectable levels within the first 10 days of postnatal development. The ϵ-subunit specific mRNA on the other hand is very low or undetectable at day 1 and increases to reach a plateau level within 20 days of postnatal development. Thereafter the abundance is slowly reduced to reach levels, characteristic for the adult muscle. Concomitantly with the changing amounts of γ- and ϵ-subunit a switch in the gating properties of the AChR channels is observed (Witzemann, Barg and Sakmann, in preparation). The changes are qualitatively similar to the changes observed during the development of calf muscle, only that the time course of the observed changes are much faster in rat muscle than in calf muscle. These findings suggest that upon innervation and the establishment of functional synapses structural changes of the AChR-complex are induced. Denervation and reinnervation experiments were performed to investigate on the changes mediated by the motor nerves.

In innervated rat diaphragm, the ϵ-subunit mRNA is observed, whereas the γ-subunit mRNA is not detectable. Following denervation of the muscle, the content of the ϵ-subunit mRNA increases only slightly. At the same time a strong increase is observed in the contents of the γ-subunit mRNA. Measuring the amounts of the γ- and ϵ-subunit mRNAs in rat diaphragms denervated for 2-7 days it is found that in 7 days denervated diaphragm the γ-subunit mRNA increases to levels about one order of magnitude higher than those of the ϵ-subunit mRNA. The results indicate furthermore that the synthesis of the γ-subunit mRNA is switched on within the first two days after denervation.

To study reinnervation nerve crush experiments were performed. When the phrenic nerve is crushed 5-8 mm from its entry into the muscle, reinnervation takes place 7-9 days later and innervation is complete within 15-18 days following nerve crush. Seven days after nerve crush, the abundance of the γ-subunit mRNA is high, being comparable to that observed after denervation by cutting the nerve. The amount of the γ-subunit mRNA thereafter decreases as the diaphragm becomes reinnervated, and on day 30 after the crush the γ-subunit mRNA is undetectable. For the ϵ-subunit mRNA, elevated levels are measured being comparable to the levels observed after 7 days of denervation. With reinnervation the amount of the ϵ-subunit mRNA falls back to a level representative of normal innervated muscle. These changes are reminiscent of the changes observed during development and suggest that synaptic function leads to the repression of γ-subunit mRNA expression.

Merlie and Sanes (1985) have reported that the α- and δ-

subunit mRNAs are concentrated in the synaptic regions of innervated mouse diaphragm muscle. A similar spatial distribution of the γ- and ϵ-subunit mRNAs exists comparing end-plate-rich and end-plate-free muscle strips from denervated diaphragm in that both mRNAs are more abundant in the end-plate-rich regions. As a result of denervation, however, the γ-subunit mRNA levels are much higher than the ϵ-subunit mRNA levels in both regions. The uneven distribution of AChR subunit mRNAs in denervated muscle supports the notion that they are synthesized preferentially in synapse-associated nuclei.

CONCLUSION

The large increase in the rate of AChR synthesis upon denervation is well documented. Several reports confirmed this finding, demonstrating a significant increase in AChR subunit mRNA levels (e.g.Evans et al 1987; Merlie et al 1984).Denervation and reinnervation experiments show that the ϵ- and γ-subunit mRNAs are regulated differentially. The ϵ-subunit mRNA level is relatively independent of denervation or reinnervation of adult muscle, whereas the γ-subunit mRNA level is under tight control of the motor nerve. The changing levels of the γ- and ϵ-subunit mRNAs are closely correlated with changes of the functional properties of the AChR channels. In innervated muscle where only the ϵ-subunit mRNA is found, the high-conductance channel (~60 pS) is observed. In denervated muscle, both the γ- and ϵ-subunit mRNAs are present, but the γ-subunit mRNA is much more abundant. Concomitantly, the low-conductance channel (~40 pS) predominates (Witzemann et al 1987). These results support the view that important properties of the AChR channel depend on functional innervation. The differential expression of the various receptor-subunit mRNAs may be a major mechanism in regulating these properties.

REFERENCES

Brenner, H.R. and Sakmann, B. (1983). Neurotrophic control of channel properties at neuromuscular synapses of rat muscle. J. Physiol. London, 337, 159-171

Edwards, C. (1979). The effects of innervation on the properties of acetylcholine receptors in muscle. Neuroscience 4, 565-584

Evans, S., Goldmann, D., Heinemann, S. and Patrick, J. (1987) Muscle acetylcholine receptor biosynthesis. J. Biol. Chem. 262, 4911-4916

Fallon, J.R., Nitkin, R.M., Reist, N.E., Wallace, B.G. and McMahan, U.J. (1985). Acetylcholine receptor-aggregating factor is similar to molecules concentrated at neuromuscular junctions. Nature 315, 571-574

Froehner, S.C. (1984). Peripheral proteins of postsynaptic membranes from Torpedo electric organ identified with monoclonal antibodies. J. Cell. Biol. 99, 88-96

Froehner, S.C. (1986). The role of the postsynaptic cytoskeleton in acetylcholine receptor organization. Trends Neurosci. 9, 37-41

Merlie, J.P., Isenberg, K.E., Russel, S.D. and Sanes, J.R. (1984). Denervation supersensitivity in Skeletal muscle. Analysis with a cloned cDNA probe. J. Cell Biol. 99, 332-335.

Merlie, J.P. and Sanes, J.R. (1985). Concentration of acetylcholine receptor mRNA in synaptic regions of adult muscle fibre. Nature 317, 66-68.

Mishina, M., Takai, T., Imoto, K., Noda, M., Takahashi, T., Numa, S., Methfessel, C. and Sakmann, B. (1986). Molecular distinctions between fetal and adult forms of muscle acetylcholine receptor. Nature 321, 406-411

Raftery, M.A., Hunkapiller, M.W., Strader, C.D., Hood, L.E.(1980). Acetylcholine receptor: Complex of homologous subunits. Science 208, 1454-1457

Schuetze, S.M. and Role, L.W. (1987). Developmental regulation of nicotinic acetylcholine receptors. Ann. Rev. Neurosci. 10, 403-457

Shibahara, S., Kubo, T., Perski, H.J., Takahashi, H., Noda, M. and Numa, S. (1985). Cloning and sequence analysis of human genomic DNA encoding γ-subunit precursor of muscle acetylcholine receptor. Eur. J. Biochem. 146, 15-22

Stadler, H. and Kiene, M.-L. (1987). Synaptic vesicles in electromotor neurons. II. Heterogeneity of population is expressed in uptake properties. Exocytosis and insertion of a core proteoglycan into the extracellular matrix. Embo J. 6, 2217-2221

Takai, T., Noda, M., Furutani, Y., Takahashi, H., Notake, M., Shimizu, S., Kayano, T., Tanabe, T., Tanaka, K., Hirose, T. Inayama, S. and Numa, S. (1984). Primary structure of γ-subunit precursor of calf-muscle acetylcholine receptor deduced from the cDNA sequence. Eur. J. Biochem. 143, 109-115

Takai, T., Noda, M., Mishina, M., Shimizu, S., Furutani, Y., Kayano, T., Ikeda, T., Kubo, T., Takahashi, H., Takahashi, T., Kuno, M. and Numa, S. (1985). Cloning, sequencing and expression of cDNA for a novel subunit of acetylcholine receptor from calf muscle. Nature 316, 761-764

Wallace, B.G., Nitkin, R.M., Reist, N.E., Fallon, J.R., Moayeri, N.N. and McMahan, U.J. (1985). Aggregates of acetylcholinesterase induced by acetylcholine receptor aggregating factor. Nature 315, 574-577

Whittaker, V.P. (1977). The electromotor system of Torpedo a model cholinergic system. Naturwissenschaften 64, 606-611

Witzemann, V., Richardson, G. and Boustead, C. (1983a). Characterization and distributation of acetylcholine receptors and acetylcholinesterase during electric organ development in Torpedo marmorata. Neuroscience 8, 333-349

Witzemann, V., Schmid, D. and Boustead, C. (1983b). Differentiation-dependent changes of nicotinic synapse-associated proteins. Eur. J. Biochem. 131, 235-245

Witzemann, V., Barg, B. Nishikawa, Y., Sakmann, B. and Numa, S. (1987). Differential regulation of muscle acetylcholine receptor γ- and ϵ-subunit mRNAs. FEBS-Lett.in press

ANTIBODIES AS PROBES OF THE STRUCTURE OF THE NICOTINIC ACETYLCHOLINE RECEPTOR

Alfred Maelicke, Gregor Fels and Rita Plümer-Wilk
Max-Planck-Institut für Ernährungsphysiologie,
Rheinlanddamm 201, D-4600 Dortmund, West-Germany

ABSTRACT

Antibodies were raised against eight synthetic peptides matching preselected portions of the amino acid sequence of the nicotinic acetylcholine receptor (AChR) from *Torpedo marmorata*. To avoid ambiguity as to the exact sequence location of the antibody epitopes, synthetic peptides of only 5-7 amino acids in length were employed.

By means of solid phase binding assays we have probed the accessibility to antibody of the matching sequences within the receptor molecule. We find the sequence portions α81-85, α127-132, α137-142, α190-195 and α387-392 to be freely accessible both at the membrane-bound and purified receptor. To permit binding of anti-α137-142 immune serum, treatment of the receptor with endoglycosidase is required showing that Arg-141 indeed is glycosylated in the native AChR. The homologous sequence of the other subunits differing only in one sequence position from α137-142 is not accessible to antibody indicating clear differences in the folding of the receptor polypeptides. Accessibility of α387-392 suggests that this sequence portion is not part of a transmembrane domain as was previously suggested. In agreement with previous results, α161-166 is not accessible for antibody binding. We do not consider this finding sufficient evidence for a transmembrane domain in this region. We suggest a model with four α-helical transmembrane domains and extensive β-structure at the extracellular and cytoplasmic surfaces of the receptor.

NATO ASI Series, Vol. H21
Cellular and Molecular Basis of Synaptic Transmission
Edited by H. Zimmermann

INTRODUCTION

In the absence of x-ray data of sufficient resolution, epitope mapping (Lindstrom et al 1984; Fuchs and Safran 1986; Maelicke et al 1984) presently is the most successful approach to establish the transmembrane topography of nicotinic acetylcholine receptor (AChR) subunits already sequenced. Studies of this kind suggest that the N-termini of *Torpedo* receptor subunits are extracellular but in the native receptor are not accessible for antibody binding (Neumann et al 1984; Ratnam and Lindstom 1984) while the carboxy termini probably are cytoplasmic and accessible (Young et al 1985; Ratnam et al. 1986). This would suggest an uneven number of transmembrane crossings although hydrophilicity profiles predict an even number (four) of α-helical domains sufficient in length to span the bilayer (Noda et al 1982; Claudio et al 1983). Epitope mapping shows that at least parts of the proposed (fifth) amphipathic helix (Finer-Moore and Stroud 1984; Guy 1984) are at the surface and not membrane-spanning (Ratnam and Lindstom 1984; this publication). This poses a dilemma as to how to place the α-helical domain M4. Ratnam et al. (Ratnam et al 1986) have suggested this domain to be membrane-associated rather than membrane-spanning.

Most of the epitope mapping discussed above was performed with antibodies raised against rather long synthetic peptides. We have criticized this approach (Maelicke et al 1984; Plümer et al 1984) as it may lead to ambiguous results: Only under limiting conditions will the charge density pattern defining an antibody binding site be directly correlated to the sequence of the antigen. Thus, the more a synthetic peptide exceeds the size required to evoke an immune response, i.e. 4-6 amino acids, the less likely will the related antibodies recognize the *linear* sequence. As *any combination* of 4-6 amino acid residues of a longer peptide may form the recognition points of an epitope, the statistical probability that an anti-peptide antibody exclusively binds to sites formed by the matching sequence of the AChR is considerably lowered. The problem can be reduced by testing the sequence specificity of antibodies against shorter fragments of the original (long) peptide (Lindstrom 1986) but this approach has been applied (and works) in only a few selected cases.

We have avoided the above discussed ambiguity by raising antibodies against synthetic peptides only 5-7 amino acids in length (Maelicke et al 1984). These antibodies are likely to recognize at the receptor only epitopes of the same linear sequence as that of the synthetic peptide they were raised against. As their major disadvantage, most obtained antibcdies bind with rather low affinity to their antigens. This limits their range of application and, in particular, excludes in most cases electron microscopic studies with the colloidal gold double labelling technique.

MATERIALS AND METHODS

are presented in detail in the related full publications (Maelicke 1987; Fels et al 1986; Maelicke et al 1987).

RESULTS AND DISCUSSION

1. Synthetic peptides, their conjugates, production and properties of rabbit anti-peptide immune sera.

The synthetic peptides were selected according to the following criteria: (i) Their sequences match regions of the primary structure of the AChR which on the basis of hydrophilicity profiles are predicted to lie at the hydrophilic surface (Maelicke 1987). (ii) With the exception of peptide α81-85, the selected sequences are highly conserved between AChRs from different species. (iii) All peptides carry a negative net charge. These criteria are basic requirements for sequence regions to be involved in ligand binding and in channel gating. The sequences of the synthetic peptides and their positions in the primary sequence of the AChR from *Torpedo marmorata* are shown in table I.

Table I. SEQUENCES AND MATCHING POSITIONS IN THE ACHR PRIMARY STRUCTURE OF SYNTHETIC PEPTIDES

Peptide	Sequence	Matching AChR sequence
P1	PSDDV-Y	α 81- 85
P2	YCEIIV	α127-132
P3	FDQQNC-Y	α137-142
P4	FDWQNC-Y	β,γ,δ137-142
P5	ESDRPD	α161-166
P6	YTCCPD	α190-195
P7	KSDEES-Y	α387-392
P8	AAEEWKY	α395-400

The conventional one letter code for amino acids is used. Residues not contained in the matching receptor sequence (N-terminal tyrosines of some peptides) are separated by a dash from the original receptor sequence.

Bovine serum albumin (BSA) and keyhole limpet hemocyanin (KLH) were selected as carrier proteins as they apparently share only few, if any, immunogenic sites. Peptides were linked via their N-terminal amino groups or via terminal tyrosine residues to the above carriers as described in the original publication (Maelicke et al 1987). Antibodies were raised by conventional immunization of rabbits or by the murine hybridoma technique (Plümer et al 1984). To obtain antisera of sufficient titre, at least four booster injections of peptide-carrier conjugate were required. ELISA and RIA were employed to select antibody producing cell clones with the wanted epitope specificity and to characterize immune sera and monoclonal antibodies.

The epitope specificity of the obtained antisera was tested as follows:

(i) All antisera studied yielded positive ELISAs when complementary conjugates (same peptide cross-linked by the same method to a different carrier protein) or the free peptide were employed as antigen. Thus, the sera contained antibodies exclusively directed against the peptide portions of the conjugates.

(ii) With the exception of the immune sera raised against peptides P3 and P4, no cross-reactivity of immune sera with the other synthetic peptides was observed. The anti-P3 serum and the anti-P4 serum had to be used in approx. 50-100 fold higher concentration to yield the same ELISA readings with the respective other peptide. Thus, the immune sera clearly discriminated between different peptides.

(iii) All anti-peptide sera recognized the AChR as antigen. This was not the result of cross-reactivities between carrier protein and AChR, as antisera raised against BSA or KLH or against "non-sense" conjugates did not bind to receptor preparations. In some cases (see below), however, AChR binding required treatment by enzymes (endoglycosidases, proteases) and/or denaturation. AChR and free peptides competed for antibody binding supporting that identical epitopes existed in the oligopeptides and the macromolecule.

(iv) The AChR subunit specificity of the anti-peptide sera was determined by means of Western blots of *Torpedo* membrane fragments and ELISAs with purified AChR subunits as antigen. Because of the relatively low affinity of most anti-peptide antisera (Maelicke et al 1987), solid phase binding tests (ELISA) yielded more reliable results. Antisera raised against peptides P2, P6, P7 and P8 exclusively bound to α-subunit; P1 bound to α- and, more weakly, to ß-subunit. Anti-P3 serum recognized α-subunit of endoglycosidase-treated AChR. Anti-P4 serum was specific for the heavier subunits (δ-subunit) of the AChR. Proteolytic treatment of the AChR was required to expose the epitope for anti-P5 serum.

These data confirmed that the antisera contained antibodies specifically directed against the synthetic peptides employed as conjugates in rabbit immunization. Most likely the anti-peptide antibodies recognized three-dimensional rather than linear epitopes as the antisera against P3 and P4 did not completely discriminate against the respective other peptide. With the epitopes properly exposed, the anti-peptide antibodies also recognized the AChR as antigen. With the exception of anti-P4 antibodies, the anti-peptide antibodies were directed against AChR α-subunits.

2. Accessibility of peptide epitopes in the native AChR.

The immune sera raised against peptides P2 and P7 exhibited the best titres for AChR in its membrane-bound and purified form. Consequently, the regions defined by the respective peptide sequences probably are at the hydrophilic surface of the AChR. In the case of P2 (α127-132), this agrees with previous suggestions (Criado et al 1986; Barkas et al 1987). In the case of P7 (α387-392), it provides further evidence against the existence in this region of a membrane-spanning amphipathic helix M5 (Lindstrom et al 1984).

The immune sera raised against peptides P1 and P6 bind considerably less tightly to the native AChR. As both sera bind stronger to purified α-subunit, these sequences may have different conformations in the native AChR. This is supported by the fact that they both contain proline residues, and P6 in addition contains two cysteines suggested to form a disulfide bridge in the native AChR (Kao and Karlin 1986; Neumann et al 1986). Consistent with the latter suggestion, anti-P6 serum binds significantly more tightly to AChR treated with dithiothreitol. In summary, the regions α81-85 and α190-195 probably are at the hydrophilic surface of the AChR.

Anti-P8 serum does not bind to native *Torpedo* membranes but does quite well to detergent-solubilized AChR and to purified α-subunit. Thus, accessibility of the hydrophilic and charged region α395-400 sensitively depends on the conformation of the AChR, i.e. whether in membrane-bound or solubilized form.

As Asn-141 was suggested to be glycosylated in native AChR (Maelicke 1987), it is not surprising that antibodies raised against the (non-glycosylated) peptides P3 and P4 do not recognize glycosylated AChR as antigen. After treatment with endoglycosidase F, anti-P3 serum binds to AChR and to purified α-subunit. The same enzyme treatment does not suffice to render the AChR an antigen for anti-P4 serum. Consequently, the sequence α137-142 probably is at the hydrophilic surface of the AChR while the homologous sequence of the other three subunits is situated closer to the hydrophobic core of the receptor. This agrees with model considerations concerning the exchange of Trp for Gln in position 139 of this region (Smart et al 1984).

AChR must be denatured and partially digested to expose epitopes matching in structure the synthetic peptide P5. As all other tests of the epitope specificity of anti-P5 serum have been positive (see above), this suggests that the region α161-166 is situated within the hydrophobic core of the native AChR. This agrees with previous findings by Ratnam et al 1986.

DISCUSSION

1. Technical comments.

The results presented here suggest that the antigen binding sites of antibodies raised against *short* synthetic peptides indeed may be described as *linear* epitopes, i.e. that they are formed by the participation of *all* amino acids of the peptide. As the cross-reactivities of anti-P3 and anti-P4 immune sera with the respective other peptide already suggest, however, epitope discrimination is not complete: exchange of one amino acid in position three of the hexapeptide still permits recognition by the original immune serum albeit with lower affinity (by 1-2 orders of magnitude).

Obviously, the longer a peptide, the less probable will its immunogenic site(s) be linear. Thus, if peptides of 20 or more amino acids are employed (Lindstrom et al 1984; Lindstrom 1986), only antibodies directed against appropriately small parts of their sequence should be used for mapping the particular epitopes in proteins with matching sequence regions. Antibodies of the wanted specificity are best obtained by the hybridoma technique (Plümer et al 1984; Lindstrom 1986). If the conditions of a linear epitope are not fulfilled, the antibody or serum may be directed against *any* imaginable *combination* of 4-6 amino acid residues of the particular peptide. This conclusion is supported by the following findings with monoclonal antibodies raised against the whole receptor protein: The mAbs of our library which are directed against at least 12 different epitopes at the AChR (Watters and Maelicke 1983), did not bind to the free peptides or their conjugates described here. Furthermore, the large majority of them has been shown to bind conformation-dependend (Watters and Maelicke 1983; Fels et al 1986), i.e. these antibodies do not recognize denatured AChR. Consequently, antibodies recognizing linear sequences are rarely obtained from immunizations with larger antigens. Their specificity for linear epitopes must be properly established to avoid ambiguity in epitope mapping.

2. Transmembrane orientation of the Torpedo α-polypeptide.

The sequence regions α81-85 (P1), α127-132 (P2), α190-195 (P6), α387-392 (P7) all are more or less freely accessible for antibody binding to the native AChR. After endoglycosidase treatment, also α137-142 (P3) of the native AChR is accessible for antibody binding. Thus, the regions defined by these sequences lie at the hydrophilic surface of the receptor. In the case of peptide P7, this is further evidence against the existence of a membrane-spanning domain (Lindstrom et al 1984) in this region.

The sequence regions α161-166 (P5) and α395-400 (P8) are not accessible for antibody binding in the native AChR. Thus, they may be part of membrane-spanning domains as was suggested for parts of α161-166 (Ratnam et al 1986), or they may exist in the native AChR in a conformation not matched by the synthetic peptide. As anti-P8 serum bound to solubilized AChR and to purified α-subunits, the region α395-400 appears to be in a detergent-sensitive environment, i.e. associated to the hydrophobic domain M4 or the plasma membrane. This agrees with previous findings (Lindstrom et al 1984; Young et al 1985; Ratnam et al 1986) showing that the (cytoplasmic) C-terminus of the α-chain is not freely accessible to antibody binding in the native receptor.

Detergent treatment of the native AChR did not suffice to permit binding of anti-P5 serum. Similarly, treatment with urea or guanidinium or dithiothreitol was insufficient to expose this epitope. The sequence α161-166 therefore is part of the core region of the AChR. It could be part of a membrane-spanning domain (Ratnam et al 1986). As one of several other alternatives, it could also be part of core regions in the "entrance funnel" of the receptor-integral channel. As the putative α-helical region M2 is rather short for a membrane-spanning domain, it may be associated with other regions including α161-166 at the extracellular side of the membrane.

Based on its hydrophilicity profile and the available biochemical and immunological data (Maelicke 1987), we suggest the alternative models depicted in the Figure for the transmembrane organization of Torpedo α-subunit.

The region following the (cytoplasmic) phosphorylation sites presently is the most ambiguous in all models of the receptor's transmembrane topography. Our data suggest the matching sequence P7 to be freely accessible for antibody binding while sequence P8 is buried in the membrane-bound AChR. Anti-P7 serum efficiently recognized its epitope also in membrane preparations shown to be almost exclusively of the outside-out type and closed. This suggests the P7 epitope to possibly reside at the extracellular surface of membrane-bound AChR. Preliminary results with the electron microscopy double labelling technique with colloidal gold particles are ambiguous in this respect. Antibodies of higher affinity and/or a more sensitive test are required to unequivocally establish the location of the P7 epitope. It appears feasible, however, that this regions is part of a domain which according to functional requirements is more or less membrane associated and may even move through the membrane. Antibodies applied to the outside of closed membranes may also shift the equilibrium of receptor conformations (Fels et al 1986; Maelicke et al 1987) to ones less likely in their absence.

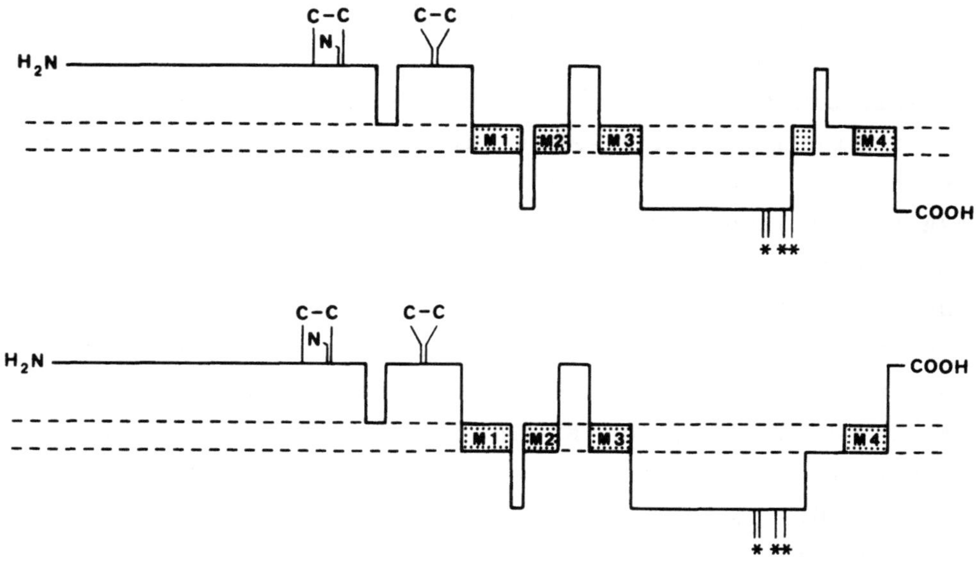

Figure legend: Models of the transmembrane topography of the α-subunit from *Torpedo* AChR compatible with the presented data.
In schemes a and b, M1-M4 are putative α-helical domains predicted by hydrophobicity plots of the α-subunit amino acid sequence (Maelicke 1987). Stars indicate the (cytoplasmic) phosphorylation sites.
Scheme a considers the C-terminus to be cytoplasmic (Young et al 1985; Ratnam et al 1986). This would require to assume an additional (5th) membrane-spanning domain located between the (cytoplasmic) phosphorylation sites and P7. As so far, there is no experimental evidence for this domain, scheme b considers the alternative of an extracellular (possibly membrane associated) C-terminus which would also agree with the epitope mapping data of Young et al 1985 and Ratnam et al 1986.

ACKNOWLEDGEMENTS

This work was supported in part by the Deutsche Forschungsgemeinschaft (SFB 168) and the Fonds der Chemischen Industrie.

REFERENCES

Barkas, T., Mauron, A., Roth, B., Alliod, C., Tzartos, S.J. & Ballivet, M. (1987) *Science* 235, 77-80.

Claudio, T., Ballivet, M., Patrick, J., & Heinemann, S. (1983) *Proc. Natl. Acad. Sci. USA* 80, 1111-1115.

Criado, M., Sarin, V., Fox, J.L. & Lindstrom, J. (1986) *Biochemistry* 25, 2839-2846.

Fels. G., Plümer-Wilk, R., Schreiber, M. & Maelicke, A. (1986) *J. Biol. Chem.* 261, 15746-15754.

Finer-Moore, J. & Stroud, R.M. (1984) *Proc. Natl. Acad. Sci. USA* 81, 155-159.

Fuchs, S. & Safran, A. (1986) in NATO ASI Series H: Cell Biology Vol. 3: *Structure and Function of the Nicotinic Acetylcholine Receptor* (ed. Maelicke, A.), pp. 49-59, Springer Verlag, Berlin.

Guy, H.R. (1984) *Biophys. J.* 45, 249-261.

Kao, P.N. & Karlin, A. (1986) *J. Biol. Chem.* 261, 8085-8088.

Lindstrom, J. (1986) *TINS* 9, 401-407.

Lindstrom, J., Criado, M., Hochschwender, S., Fox, J.L. & Sarin, V. (1984) *Nature* 311, 573-575

Maelicke, A. (1987) in *Handbook of Experimental Pharmacology* (Born, G.V.R., Farah, A., Herken, H. & Welch, A.D., eds), Volume: "The Cholinergic Synapse" (ed. V.P. Whittaker), Springer Verlag, Berlin, in press.

Maelicke, A., Plümer-Wilk, R., Fels, G., Engelhard, M. and Veltel, D. (1987) *Eur. J. Biochem.*, in press

Maelicke, A., Watters, D., Fels, G. & Plümer, R. (1984) *Journ. of Receptor Research* 4, 671-679.

Neumann, D., Barchan, D., Safran, A., Gershoni, J.M. & Fuchs, S. (1986) *Proc. Natl. Acad. Sci. USA* 83, 3008-3011.

Neumann, D., Fridkin, M. & Fuchs, S. (1984) *Biochem. Biophys. Res. Com.* 121, 673-679.

Noda, M., Takahashi, H., Tanabe, T, Toyosato, M., Furutani, Y., Hirose, T., Asai, M., Inayama, S., Miyata, T. & Numa, S. (1982) *Nature* 299, 793-797.

Plümer, R., Fels, G. & Maelicke, A. (1984) *FEBS Lett.* 178, 204-208.

Ratnam, M. & Lindstrom, J. (1984) *Biochem. Biophys. Res. Commun.* 122, 1225-1233.

Ratnam, M., Le Nguyen, D.L., Sargent, P.B. & Lindstrom, J. (1986) *Biochemistry* 25, 2633-2643.

Ratnam, M., Sargent, P.B., Sarin, V., Fox, J.L., Le Nguyen, D.L., Rivier, J., Criado, M. & Lindstrom, J. (1986) *Biochemistry* 25, 2621-2632.

Smart, L., Meyers, H.-W., Hilgenfeld, R., Saenger, W. & Maelicke, A. (1984) *FEBS Lett.* 178, 64-68.

Watters, D. & Maelicke, A. (1983) *Biochemistry* 22, 1811-1819.

Young, E.F., Ralston, E., Blake, J., Ramachandran, J., Hall, Z.W. & Stroud, R.M. (1985) *Proc. Natl. Acad. Sci. USA* 82, 626-630.

BRAIN α-NEUROTOXIN-BINDING PROTEINS AND NICOTINIC ACETYLCHOLINE RECEPTORS

Mark G. Darlison, Andrew A. Hicks, Victor B. Cockcroft, Michael D. Squire and Eric A. Barnard

MRC Molecular Neurobiology Unit, MRC Centre, Hills Road, Cambridge CB2 2QH, United Kingdom.

INTRODUCTION

The study of the nicotinic acetylcholine receptor (AChR) of skeletal muscle and fish electric organ has been greatly facilitated by the application of the α-neurotoxins (post-synaptic toxins) of elapid and hydrophid snake venoms, such as α-bungarotoxin (α-BTX), introduced by C. Y. Lee (Lee, 1973). These polypeptide toxins bind to the receptor with K_D values of between 10^{-9} and 10^{-12} M, causing blockade of function. It has been found that the peripheral (vertebrate muscle and electric organ) type of AChR invariably has two of these high-affinity α-toxin binding sites, one on each of the α subunits of its $\alpha_2\beta\gamma\delta$ pentameric structure (for reviews see Dolly and Barnard, 1984; Popot and Changeux, 1984). However, attempts to use α-bungarotoxin as a probe for the more poorly-characterised, neuronal nicotinic receptors have caused a great deal of confusion. Although a large number of $[^{125}I]$α-BTX-binding sites certainly exist in the vertebrate brain and in some autonomic ganglia, and a characteristic regional distribution for them in the brain has been consistently found (Marks and Collins, 1982; Clarke _et al_., 1985; Larsson and Nordberg, 1985), these often do not correlate with high-affinity $[^3H]$nicotine- or $[^3H]$acetyl-

NATO ASI Series, Vol. H21
Cellular and Molecular Basis of Synaptic Transmission
Edited by H. Zimmermann

choline-binding sites (Clarke *et al.*, 1985) or with the pattern of anti-AChR antibody staining (Swanson *et al.*, 1987). We will document here the results that have been obtained to date (using toxins, ligands, antibodies and cDNA clones) in an attempt to clarify the apparently confusing literature on this question.

PHARMACOLOGY OF LIGAND BINDING

The neurotoxin, α-BTX is obtained from the venom of the snake *Bungarus multicinctus*. Further purification from this venom yields a second toxin that appears in the literature under three names: kappa-bungarotoxin or kappa-toxin (Chiappinelli, 1983), bungarotoxin 3.1 (Ravdin and Berg, 1979) and toxin F (Loring *et al.*, 1984). While α-BTX binds to many sites in autonomic ganglia, for example, the chick ciliary ganglion, it fails to block nicotinic transmission in most of these (Oswald and Freeman, 1981). Kappa-toxin, however, does block nicotinic transmission in these ganglia, and has been shown to bind to two classes of sites, only one of which is able to bind α-BTX (Loring and Zigmond, 1987). Furthermore, α-BTX fails to prevent the inhibition of nicotinic transmission by kappa-toxin. It is, therefore, concluded that kappa-toxin binds to the functional AChR on these ganglia, as well as to a second entity which also binds α-BTX. The conclusion that kappa-toxin binds to the functional receptor is substantiated by the synaptic localisation of the α-BTX-insensitive kappa-toxin site (Loring and Zigmond, 1987) and by the fact that this same region is labelled by monoclonal antibodies that recognise the main immunogenic region of the *Torpedo* and *Electrophorus* AChR α subunits (Jacob *et al.*, 1984; Smith *et al.*, 1985). In contrast, the majority of α-BTX-sensitive kappa-toxin sites appear to localise to the extra-synaptic membrane and to cytoplasmic areas (Loring and Zigmond, 1987). It is also worth noting at this point that the ratio of α-BTX-sensitive to α-BTX-insensitive binding sites, in the chick ciliary ganglion, is estimated to be approximately 3 : 1 (Loring and Zigmond, 1987).

Comparison of the binding of [^{3}H]nicotine and [^{125}I]α-BTX on tissue sections reveals differences in the numbers and distribution of these sites (Clarke et al., 1985). An α-BTX-binding component has been isolated from chick optic lobe (Norman et al., 1982) and from rat brain (Wonnacott, 1986) using a toxin-affinity column, and the binding of α-BTX to this has been shown to be sensitive to high concentrations of (-)nicotine (K_i = 0.3 x 10^{-6} for chick and 6.3 x 10^{-6} M for rat). Further, nicotine-binding studies in rat brain (Wonnacott, 1986) reveal two classes of site; a high-affinity site with a K_D in the nanomolar range which correlates with the [^{3}H]acetylcholine-binding site (Martino-Barrows and Kellar, 1987) and a low-affinity nicotine site which is associated with an α-BTX-binding site (Wonnacott, 1986). Unlike the high-affinity site which displays an 80-fold preference for (-)nicotine, the low-affinity site displays little or no stereoselectivity. Similar experiments with the purified AChR from the electric organ of Torpedo marmorata (Wonnacott, 1986) show similar sensitivities to nicotine, and a comparable stereoselectivity to the brain low-affinity (α-BTX) sites. In summary, the α-BTX-binding component can be said to have an affinity for nicotine, to be enriched in nerve endings and to have a tissue localisation paralleling some, but not all, cholinergic pathways.

Further confusion as to the identity of the brain α-BTX-binding component derives from the use of monoclonal antibodies raised against the AChR isolated from electroplaque tissue. One such antibody, mAb35, has been used to localise AChRs in chick brain sections (Swanson et al., 1983). With the exception of certain layers within the optic tectum, the pattern of antibody staining does not correlate with the distribution of [^{125}I]α-BTX binding sites. Studies with mAb35 have also led to the intriguing discovery that up to two thirds of the AChRs associated with chick ciliary ganglion neurons in culture appear to be intracellular (Stollberg and

Berg, 1987), and that many of these may be associated with organelles that are known to comprise the biosynthetic pathway of integral plasma-membrane proteins (Jacob et al., 1986). It has also been suggested (Margiotta et al., 1987) that many of the surface receptors may be non-functional, as evidenced by the observation that electrophysiologically-functional AChRs may represent only 10% of the total number of receptors, as estimated by [^{125}I]mAb35 binding. Further, cAMP-dependent processes appear to increase the response of these chick ciliary ganglion neurons to acetylcholine by an unknown mechanism which does not involve either protein synthesis or alterations to the electrophysiological parameters of previously-functional receptors (Margiotta et al., 1987). It has, therefore, been proposed that large stores of intracellular receptor or non-functional surface receptor could be used as a means for controlling sensitivity to synaptic input, and some dependence upon second messenger systems may render this system responsive to environmental influences.

When the numbers of inactive receptors on the chick ciliary ganglion are correlated with the previous data regarding kappa-toxin binding, this leads to an interesting speculation that on these cultured neurons the α-BTX-sensitive kappa-toxin-binding component may represent some form of non-functional (i.e. electrophysiologically-inert) AChR. Since, the α-BTX-binding component has a low affinity for nicotine (Wonnacott, 1986), we suggest that, in brain, along with the "classical" ligand-gated ion channel AChR there may exist other proteins, currently of unidentified function, which may also modulate the effects of cholinergic ligands.

PROTEIN BIOCHEMISTRY

In an attempt to characterise the α-BTX-binding component from brain, several researchers have utilised the optic lobe of newborn chicks as an enriched source of this component

(Wang *et al*., 1978; Norman *et al*., 1982; Conti-Tronconi *et al*., 1985; Schneider *et al*., 1985). Further, in the optic tectum of the visual pathway of lower vertebrates there is electrophysiological evidence for the blockade of cholinergic transmission by micromolar concentrations of α-BTX (Oswald and Freeman, 1981). In the initial purification of this protein from the chick brain (Norman *et al*., 1982), an α-BTX-affinity column procedure yielded two forms, depending upon the conditions (including the type of detergent) employed: the smaller (9.1 S) had a simpler subunit pattern than the larger (12 S), and it seems likely that some subunits were lost from the smaller protein. This purification was subsequently improved (Conti-Tronconi *et al*., 1985); this confirmed that the α-BTX-binding protein of chick optic lobe *or* of the rest of the chick brain was comprised of four polypeptides (M_r 48 000, 56 000, 69 000 and 72 000). The M_r 48 000 subunit was found to be extremely sensitive to proteolytic degradation; when it was preserved and submitted to gas-phase micro-sequencing, it yielded a sequence that was homologous to the amino-termini of various AChR α subunits (Figure 1). None of the other subunits gave protein sequence information, presumably due to the presence of blocked amino-termini. Further, the M_r 56 000 subunit, and not the M_r 48 000 subunit, appears to carry the acetylcholine-binding site, since the former could be affinity-labelled by [^{3}H]bromoacetylcholine (Norman *et al*., 1982). The combined evidence suggests that the M_r 48 000 subunit is not a breakdown product of the larger molecular weight components. In addition, a monoclonal antibody (7B2), raised against the chick muscle AChR, was shown to immunoprecipitate the purified α-BTX-binding component. Taken together with the amino-terminal protein sequence homology, this strongly suggested that the α-BTX-binding component was a neuronal AChR.

In contrast, Schneider *et al*. (1985), using chick optic lobe extracts, separated a high-affinity acetylcholine-binding component from the α-BTX-binding component. Although these have similar subcellular distributions, hydrodynamic properties and similar rank order of affinities for nicotinic

AChR AMINO-TERMINAL SEQUENCES

	1	2	3	4	5	6	7	8	9	10	11	12	13	14	15	16	17	18	19	20	21	22	23	24	25
Chick brain 48K subunit	X	E	F	E	T	K	L	Y	K	E	L	L	K	N	Y	N	P	L	E	X	P	V	A	X	D
Torpedo electroplaque α subunit	S	E	H	E	T	R	L	V	A	N	L	L	E	N	Y	N	K	V	I	R	P	V	E	H	H
Chick muscle α subunit	Y	E	H	E	T	R	L	V	D	D	L	F	R	E	Y	S	K	V	V	R	P	V	E	N	H
Calf muscle α subunit	S	E	H	E	T	R	L	V	A	K	L	F	E	D	Y	N	S	V	V	R	P	V	E	D	H
Human muscle α subunit	S	E	H	E	T	R	L	V	A	K	L	F	K	D	Y	S	S	V	V	R	P	V	E	D	H

Figure 1. Alignment of the amino-terminal amino-acid sequences of the α-bungarotoxin-binding component M_r 48 000 subunit and peripheral AChR α subunits. The sequence of the chick brain M_r 48 000 subunit (48K subunit) was determined by gas-phase microsequencing (Conti-Tronconi et al., 1985). The sequences of the *Torpedo* electroplaque α subunit (Noda et al., 1982), chick muscle α subunit (Barnard et al., 1986) and the calf and human muscle α subunits (Noda et al., 1983) were derived from the corresponding cDNA or genomic clones. Boxes indicate that the same amino-acid residue is found (where determined) at the same position in all five sequences. X denotes an unassigned residue.

agonists, they differ in their affinities for nicotinic antagonists, in their regional distribution and in their sensitivity to thermal inactivation. It seems likely that the similarity in the rank order of agonist affinities of the acetylcholine-binding component and of the α-BTX-binding component reflect common structural features of these two nicotine-binding proteins.

An alternative approach to the purification of neuronal AChRs, which relies upon the use of monoclonal antibodies, has been adopted by Jon Lindstrom and co-workers. They obtained a monoclonal antibody (mAb35), raised against the AChR from electroplaque tissue, that was shown to bind, in chick, to the lateral spiriform nucleus, to axonal projections to the deeper layers of the optic tectum (Swanson _et al._, 1983) and to the synaptic membrane of chick ciliary ganglion neurons (Jacob _et al._, 1984). This antibody was used to immunopurify a protein from chick brain that, on SDS-PAGE, was found to be composed of two subunits of M_r 49 000 and 59 000 (Whiting and Lindstrom, 1986a). The M_r 49 000 subunit was found to be bound by both polyclonal antisera and monoclonal antibodies that recognised the α subunits of the AChR of electric organ and muscle. Monoclonal antibodies raised against the purified chick brain receptor identified a further chick brain subunit of M_r 75 000 (Whiting _et al._, submitted). It is proposed (Whiting _et al._, 1987b) that in chick brain two subtypes of AChR exist, each comprising two subunit types (αβ: M_r 49 000 and 59 000 respectively and αβ': M_r 49 000 and 75 000 respectively), with the α subunit common to both. Further, a monoclonal antibody (mAb270), which was raised against the purified chick brain receptor cross-reacts with a receptor from rat brain which appears homologous to the αβ' subtype of chick, that comprises subunits of M_r 51 000 and 79 000 respectively (Whiting and Lindstrom, 1987a). A surprising finding concerning these subunit types is that although antibodies raised against electric organ and vertebrate muscle AChR α subunits recognise the neuronal α subunits (Whiting and Lindstrom, 1986a), affinity reagents such as 4-(N-maleimide)-benzyltrimethylammonium react with the neuronal β subunits, M_r 59 000 and 75 000 in chick and M_r 79 000 in rat (Whiting and Lindstrom, 1987b). It is intriguing, therefore, that in the chick brain there appear to be two subtypes of AChR that have apparently identical α subunits exhibiting α-like immunogenicity. The similar affinities for (-)nicotine (Whiting _et al._, 1987b) are presumably dictated by the different molecular weight β subunits, suggesting an interesting structural conservation between these polypeptides. At least one of the two proposed subtypes (αβ) of AChR that have been immunopurified from chick brain fails to exhibit inhibition of nicotine binding by concentrations of α-BTX that approach micromolar (Whiting and Lindstrom, 1986b).

MOLECULAR BIOLOGY OF NEURONAL AChRs

To date, three laboratories have reported the isolation of cDNA and/or genomic clones that encode neuronal AChR subunits. Jim Patrick, Steve Heinemann and colleagues (Boulter *et al*., 1986) were the first to report the isolation of a clone that encoded a putative neuronal AChR polypeptide. This was obtained from a cDNA library that was constructed from polyA$^+$ RNA isolated from the rat phaeochromocytoma cell line PC12. The clone (now referred to as α-3) was isolated by cross-hybridisation under conditions of low stringency, using a cDNA probe from the mouse muscle AChR α subunit (now referred to as α-1). The polypeptide encoded by this cDNA possesses the pair of adjacent cysteine residues (positions 192-193, *Torpedo* α-subunit numbering) which are present in all electric organ and vertebrate muscle AChR α subunits. This was, therefore, designated as a neuronal α subunit. The mature, deglycosylated, polypeptide has a predicted molecular weight of 54 400 and exhibits an overall homology of 47%, at the amino-acid level, with the mouse muscle α subunit. Using the α-1 and the α-3 cDNA clones as probes, this same group was able to isolate another cDNA (designated α-4) from both a rat hypothalamic library and a rat hippocampal library (Goldman *et al*., 1987). The mature, deglycosylated, polypeptide encoded by this has a predicted size of 66 600 and also contains the vicinal cysteines (positions 192-193, *Torpedo* α-subunit numbering). The difference in size between the α-3 and α-4 subunits is largely due to an amino-acid insertion in the region between two proposed transmembrane helices (M3 and M4) that is generally considered to be intracellular (Guy and Hucho, 1987). Interestingly, the α-4 subunit possesses a potential tyrosine phosphorylation site (Taylor, 1987) within this region. *In situ* hybridisation experiments in rat brain have localised the α-3 mRNA to the medial habenula and the α-4 mRNA to the medial habenula, thalamus, hypothalamus and cortex (Goldman et al., 1987), indicating that they are expressed in different brain regions.

The α-3 clone was also used to isolate another cDNA clone, which the authors have designated β-2 since it lacks the pair of adjacent cysteines (Boulter et al., 1987). The full sequence of this clone remains to be reported. Expression experiments in the Xenopus oocyte, using RNA synthesised from the cDNA clones, reveals that only combinations of α-3 and β-2 or α-4 and β-2 can produce electrophysiologically-functional AChRs (Boulter et al., 1987). It should be noted, however, that electrophysiological responses can be obtained using a combination of subunit-specific RNAs that encode the Torpedo AChR even when one subunit RNA is missing (Kurosaki et al., 1987). Thus, more work is necessary in order to establish that neuronal receptor subtypes are composed of only two different subunits.

Independently of this, Marc Ballivet's group have reported the isolation of three genes that encode neuronal AChR subunits, which they designate α-2, α-3 and γ-2 (Nef et al., 1986). Their designation of α is again based upon the presence, in the corresponding polypeptides, of the pair of adjacent cysteines; in the γ-2 polypeptide this feature is absent. To date, however, the full sequences of these genes have not been reported. It is, therefore, not possible to compare their results with those of Patrick and Heinemann or with those from our own laboratory (discussed below).

Early molecular biological approaches in our laboratory were aimed at cloning the cDNA or gene that encoded the M_r 48 000 subunit that had previously been isolated from chick optic lobe and which had been subjected to gas-phase microsequencing (Conti-Tronconi et al., 1985). Thus, a combination of short (14-base and 17-base) degenerate oligonucleotides and longer "bestguess" (48-base and 75-base) oligonucleotides, designed on the basis of codon-usage tables (Grantham et al., 1981), were synthesised in order to correspond to the chemically-determined sequence. These were used to screen both a 1-dayold chick optic lobe cDNA library, constructed in λgt10, and a chicken genomic library, constructed in λEMBL3. To date, we have been unable to isolate a clone that encodes this aminoterminal peptide sequence. Recently, it has been shown (N. Ray, J. O. Dolly and E. A. Barnard, unpublished) that a polyclonal antibody, raised against a synthetic peptide that corresponds to a region of the M_r 48 000 amino-terminal sequence, that was disimilar to electric organ and muscle α subunit sequences, failed to immunoprecipitate the receptor and also failed to recognise the denatured protein on SDS gels. We tentatively suggest, therefore, that the chemicallydetermined sequence, which was obtained from very small amounts of labile polypeptide, may contain one or two inaccuracies.

Since the amino-terminal sequence of the M_r 48 000 subunit from chick optic lobe exhibited homology with electric organ and muscle AChR α subunits (see Figure 1), it was argued that neuronal and muscle AChR receptor subunits from chick would be related at both the amino-acid and DNA levels. Further, it is well established that the proposed transmembrane helical regions of AChR subunits are highly conserved, both between species and different subunit types. We therefore screened the 1-day-old chick optic lobe cDNA library using a 677-basepair BglII restriction fragment that encodes transmembrane helices M1 to M3 and part of the proposed extracellular domain of the chick muscle α subunit (Barnard et al., 1986). This strategy resulted in the isolation of two partial cDNA clones, λVC16 (308 basepairs) and λVC12 (938 basepairs). DNA sequencing of the cloned inserts indicated that both λVC16 and λVC12 derived from the same mRNA species, with the λVC16 sequence being contained wholly within that of λVC12. During attempts to clone the gene that corresponded to these cDNAs, a related, but different, sequence was isolated. This genomic clone has been designated λCG3.

Both λVC12 and λCG3 encode polypeptides that exhibit extensive amino-acid sequence homology to each other and to published muscle and neuronal sequences (Table 1). It is

Table 1: Homology of neuronal AChR subunits.

	Chick muscle α-1	Chick brain λVC12	Chick brain λCG3	Rat PC12 cell line α-3	Rat brain α-4
α-1	-	42	49	49	50
λVC12		-	48	41	52
λCG3			-	56	67
α-3				-	53
α-4					-

The percentage amino-acid identity between pairs of sequences has been calculated. The mature, neuronal, amino-acid sequences were deduced as follows: λVC12 from a partial chick optic lobe cDNA clone, λCG3 from a chick genomic clone, α-3 from a cDNA clone that was isolated from a rat PC12 cell line library (Boulter et al., 1986) and α-4 from a rat brain cDNA clone (Goldman et al., 1987). The chick muscle α-subunit sequence, α-1, was deduced from the corresponding cDNA (Barnard et al., 1986). Note that gaps, introduced to maximally align the sequences, are counted as one amino-acid substitution regardless of their length. For comparison, when the chick muscle α-subunit and the Torpedo α-subunit (Noda et al., 1982) amino-acid sequences are aligned, they are found to be 79% identical.

clear that these two polypeptides are not the chicken homologues of α-3 (Boulter et al., 1986) or α-4 (Goldman et al., 1987). The subunit encoded by λCG3 has a similar predicted transmembrane topology to known AChR subunits, e.g. the chick muscle α subunit (Barnard et al., 1986), as evidenced by the similarity of their hydropathy profiles (Figure 2). In addition, it contains the pair of adjacent

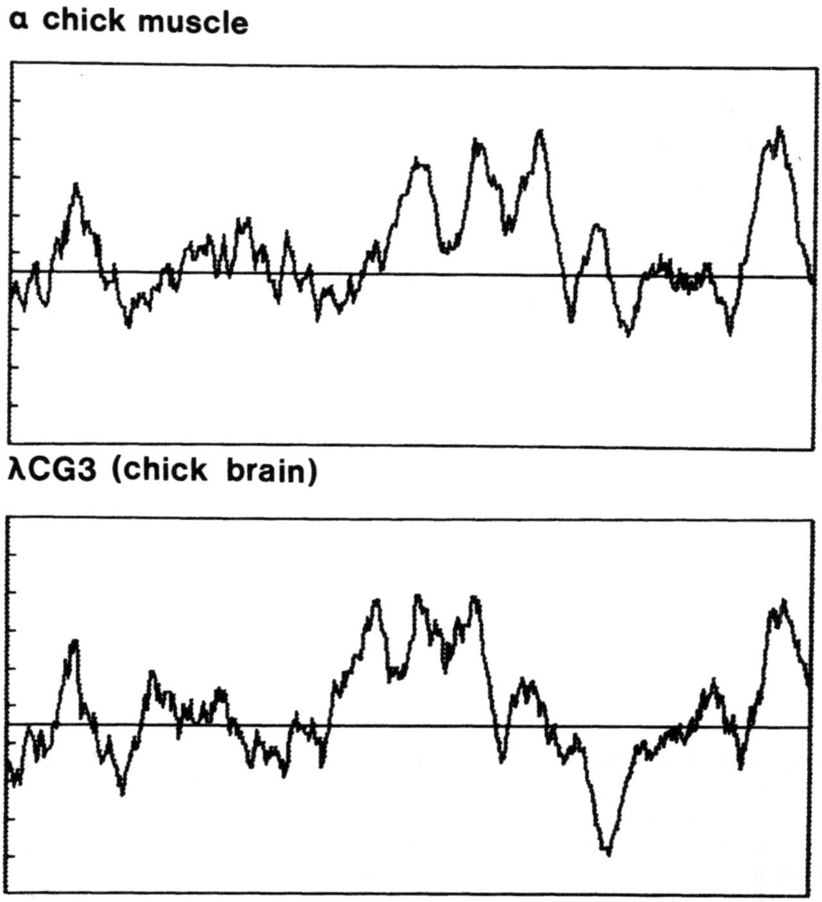

Figure 2. Hydropathy profiles of the chick muscle α subunit and a chick neuronal subunit. The amino-acid sequences of the mature chick muscle AChR α subunit and the mature chick brain AChR subunit were deduced from the corresponding cDNA (Barnard et al., 1986) and gene (λCG3) respectively. Hydropathy profiles were computed according to Kyte and Doolittle (1982), using a window size of 17 residues, and were plotted with a 1-residue interval. Regions of hydrophobic character appear as peaks above the horizontal line.

cysteine residues (positions 192-193, Torpedo α-subunit numbering) that have been proposed to be a characteristic feature of neuronal α subunits. Comparison of the very limited DNA and predicted protein sequence on the chicken neuronal subunit genes (Nef et al., 1986) leads us to speculate that λCG3 corresponds to the clone designated α-2 by Ballivet and co-workers. The predicted size of the mature, deglycosylated, polypeptide encoded by λCG3 is 58 100.

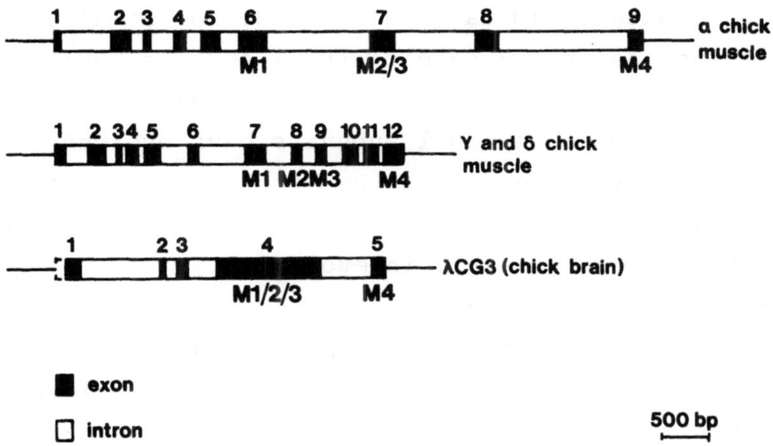

Figure 3. Schematic representation of the genomic organisation of chick muscle and brain AChR genes. The intron/exon structure was determined by DNA sequencing of clones encoding the chick muscle α-subunit gene (Nef et al., 1986), the chick muscle γ- and δ-subunit genes (Nef et al., 1984) and the chick brain gene (λCG3). Coding regions (exons) are represented by filled boxes and non-coding regions (introns) by open boxes. The exons are numbered above the schematic representation of each gene. The exons encoding the proposed membrane-spanning regions, M1 to M4, are indicated. The dashed open box at the 5' end of the chick brain gene indicates that the exact location of the signal peptide has not yet been established.

A notable feature of this amino-acid sequence is the presence of a stretch of 17 glutamic acid residues that are found between the proposed transmembrane helices M3 and M4. The function of this region, which is presumed to be intracellular, is currently unknown.

We have sequenced the entire λCG3 gene and much of its flanking sequences. Interestingly, the intron/exon organisation of this gene differs markedly (Figure 3) from the reported genomic structures of muscle AChR subunits (Noda et al., 1983; Nef et al., 1984, 1986; Shibahara et al., 1985). This neuronal gene contains fewer exons; the coding regions being separated by only five introns. Of note is the fact that the proposed transmembrane helices, M1 to M3, are encoded by a single exon. This is in contrast to muscle subunit genes in which M1 to M3 are encoded by either two exons, as in the case of the α subunit (Noda *et al*., 1983; Nef *et al*., 1986) or by three exons, as in the case of the γ and δ subunits (Nef *et al*., 1984; Shibahara *et al*., 1985). This would suggest that the genes that encode neuronal subunits have evolved independently of those encoding the muscle subunits and that the duplication event, separating muscle and neuronal receptor subunit genes, occurred early in evolution.

We do not yet know whether the polypeptide encoded by the partial cDNA, λVC12, contains the pair of adjacent cysteine residues that denote neuronal α subunits. It is clear, however, that λVC12 does not correspond to either α-2, α-3 or α-4. It is, therefore, tempting to speculate that the cysteine pair will be absent and that this polypeptide will correspond to the rat sequence designated β-2 by Patrick and Heinemann (Boulter *et al*., 1987). Attempts at cloning the full-length cDNA are now underway in order to address this question.

DISCUSSION

The expression experiments in the *Xenopus* oocyte (Boulter *et al*., 1987) using subunit-specific RNAs and the immuno-affinity purification studies from chick brain (Whiting and Lindstrom, 1986a; Whiting *et al*., 1987b) and rat brain (Whiting and Lindstrom, 1987a) suggest that the neuronal AChR is comprised of only two subunit types. This is in contrast to the AChR from electric organ and vertebrate muscle which is a pentamer that is composed of four homologous subunits (Dolly and Barnard, 1984; Popot and Changeux, 1984). Further, the expressed receptor/ion channels in the oocyte membrane are not blocked by 10^{-7}M α-BTX. However, they are sensitive to similar concentrations of kappa-toxin. Taken together, this evidence suggests that the cDNA clones α-3, α-4 and β-2 do, indeed, encode subunits of functional receptor/ion channel subtypes that may be activated by acetylcholine and nicotine.

In support of this, it has recently been reported (Whiting et al., 1987a) that the amino-terminal sequence of the M_r 79 000 subunit, that was immunoaffinity purified from rat brain, corresponds to that encoded by the α-4 cDNA. The difference in size, between that predicted from the DNA sequence (66 600) and that observed on SDS-PAGE gels (79 000), may be due to the presence of oligosaccharide residues or to the known aberrant mobility of such membrane-bound proteins in SDS gels. It remains to be seen whether the clones isolated by the group of Marc Ballivet and by ourselves also encode subunits that, in the appropriate combination, form other functional subtypes of neuronal AChR. The observation that α-BTX can block nicotinic transmission in some specialised regions of the central nervous system, e.g. the optic tectum of lower vertebrates (Oswald and Freeman, 1981), may be explained by the presence of an, as yet, unidentified AChR subtype found only there. Recent work (Bonner et al., 1987) on the cloning of the muscarinic receptor has demonstrated the fallibility of the classification of receptors purely on pharmacological evidence. The pharmacology of ligand binding has suggested the presence of two muscarinic receptor subtypes, M1 and M2 (Hammer et al., 1980). Cloning studies, however, have revealed the existence of a family of at least five subtypes that exhibit extensive amino-acid homology, yet which have different pharmacologies and which are expressed in different brain regions (Bonner et al., 1987; N. J. Buckley, personal communication).

The question remains, therefore, what is the function of the protein that is purified by α-BTX-affinity chromatography? This component presumably represents the α-BTX-binding sites that are found in brain and which do not co-localise with the high-affinity binding sites for nicotine or for acetylcholine (Clarke et al., 1985). It is not thought to be the subtype of neuronal AChR that can bind and be blocked by α-BTX as proposed above, since the α-BTX-binding protein can be purified from brain regions other than chick optic lobe (Conti-Tronconi et al., 1985) where α-BTX does not block nicotinic transmission. We believe that this toxin-binding component is structurally related to neuronal AChRs, as evidenced by the amino-terminal sequence homology (Conti-Tronconi et al., 1985) and by its ability to interact with nicotinic agonists (Schneider et al., 1985). In our laboratory we are purifying large amounts of this component for further peptide sequencing. This information will be used to construct specific oligonucleotide probes for use in screening libraries in order to isolate the corresponding cDNAs or genes. It will be of great interest to compare the sequences of these with those encoding the functional receptor/ion channel subtypes.

REFERENCES

Barnard, E. A., Beeson, D. M. W., Cockcroft, V. B., Darlison, M. G., Hicks, A. A., Lai, F. A., Moss, S. J. and Squire, M. D. (1986) in Nicotinic Acetylcholine Receptor: Structure and Function (Maelicke, A. ed.), NATO ASI Series H, Vol. 3, pp 389-415, Springer-Verlag, Berlin.

Bonner, T. I., Buckley, N. J., Young, A. C. and Brann, M. R. (1987) Science 237, 527-532.

Boulter, J., Evans, K., Goldman, D., Martin, G., Treco, D., Heinemann, S. and Patrick, J. (1986) Nature 319, 368-374.

Boulter, J., Connolly, J., Deneris, E., Goldman, D., Heinemann, S. and Patrick, J. (1987) Proc. Natl Acad. Sci. USA. 84, 7763-7767.

Chiappinelli, V. A. (1983) Brain Res. 277, 9-22.

Clarke, P. B. S., Schwartz, R. D., Paul, S. M., Pert, C. B. and Pert, A. (1985) J. Neurosci. 5, 1307-1315.

Conti-Tronconi, B. M., Dunn, S. M. J., Barnard, E. A., Dolly, J. O., Lai, F. A., Ray, N. and Raftery, M. A. (1985) Proc. Natl Acad. Sci. USA. 82, 5208-5212.

Dolly, J. O. and Barnard, E. A. (1984) Biochem. Pharmacol. 33, 841-858.

Goldman, D., Deneris, E., Luyten, W., Kochhar, A., Patrick, J. and Heinemann, S. (1987) Cell 48, 965-973.

Grantham, R., Gautier, C., Gouy, M., Jacobzone, M. and Mercier, R. (1981) Nucleic Acids Res. 9, r43-r74.

Guy, H. R. and Hucho, F. (1987) Trends Neurosci. 10, 318-321.

Hammer, R., Berrie, C. P., Birdsall, N. J. M., Burgen, A. S. V. and Hulme, E. C. (1980) Nature 283, 90-92.

Jacob, M. H., Berg, D. K. and Lindstrom, J. M. (1984) Proc. Natl Acad. Sci. USA. 81, 3223-3227.

Jacob, M. H., Lindstrom, J. M. and Berg, D. K. (1986) J. Cell Biol. 103, 205-214.

Kurosaki, T., Fukuda, K., Konno, T., Mori, Y., Tanaka, K., Mishina, M. and Numa, S. (1987) FEBS Letts 214, 253-258.

Kyte, J. and Doolittle, R. F. (1982) J. Mol. Biol. 157, 105-132.

Larsson, C. and Nordberg, A. (1985) J. Neurochem. 45, 24-31.

Lee, C. Y. (1973) in Proc. 5th Int. Congr. Pharmacol. pp 210-232, Karger, Basel.

Loring, R. H., Chiappinelli, V. A., Zigmond, R. E. and Cohen, J. B. (1984) Neuroscience 11, 989-999.

Loring, R. H. and Zigmond, R. E. (1987) J. Neurosci. 7, 2153-2162.

Margiotta, J. F., Berg, D. K. and Dionne, V. E. (1987) Proc. Natl Acad. Sci. USA. 84, 8155-8159.

Marks, M. J. and Collins, A. C. (1982) Mol. Pharmacol. 22, 554-564.

Martino-Barrows, A. M. and Kellar, K. J. (1987) Mol. Pharmacol. 31, 169-174.

Nef, P., Mauron, A., Stalder, R., Alliod, C. and Ballivet, M. (1984) Proc. Natl Acad. Sci. USA. 81, 7975-7979.

Nef, P., Oneyser, C., Barkas, T. and Ballivet, M. (1986) in Nicotinic Acetylcholine Receptor: Structure and Function (Maelicke, A. ed.), NATO ASI Series H, Vol. 3, pp 417-422, Springer-Verlag, Berlin.

Noda, M., Takahashi, H., Tanabe, T., Toyosato, M., Furutani, Y., Hirose, T., Asai, M., Inayama, S., Miyata, T. and Numa, S. (1982) Nature 299, 793-797.

Noda, M., Furutani, Y., Takahashi, H., Toyosato, M., Tanabe, T., Shimizu, S., Kikyotani, S., Kayano, T., Hirose, T., Inayama, S. and Numa, S. (1983) Nature 305, 818-823.

Norman, R. I., Mehraban, F., Barnard, E. A. and Dolly, J. O. (1982) Proc. Natl Acad. Sci. USA. 79, 1321-1325.

Oswald, R. E. and Freeman, J. A. (1981) Neuroscience 6, 1-14.

Popot, J.-L. and Changeux, J.-P. (1984) Physiol. Rev. 64, 1162-1239.

Ravdin, P. M. and Berg, D. K. (1979) Proc. Natl Acad. Sci. USA. 76, 2072-2076.

Schneider, M., Adee, C., Betz, H. and Schmidt, J. (1985) J. Biol. Chem. 260, 14505-14512.

Shibahara, S., Kubo, T., Perski, H. J., Takahashi, H., Noda, M. and Numa, S. (1985) Eur. J. Biochem. 146, 15-22.

Smith, M. A., Stollberg, J., Lindstrom, J. M. and Berg, D. K. (1985) J. Neurosci. 5, 2726-2731.

Stollberg, J. and Berg, D. K. (1987) J. Neurosci. 7, 1809-1815.

Swanson, L. W., Lindstrom, J., Tzartos, S., Schmued, L. C., O'Leary, D. D. M. and Cowan, W. M. (1983) Proc. Natl Acad. Sci. USA. 80, 4532-4536.

Swanson, L. W., Simmons, D. M., Whiting, P. J. and Lindstrom, J. (1987) J. Neurosci. 7, 3334-3342.

Taylor, S. S. (1987) BioEssays 7, 24-29.

Wang, G.-K., Molinaro, S. and Schmidt, J. (1978) J. Biol. Chem. 253, 8507-8512.

Whiting, P. J. and Lindstrom, J. M. (1986a) Biochemistry 25, 2082-2093.

Whiting, P. and Lindstrom, J. (1986b) J. Neurosci. 6, 3061-3069.

Whiting, P. J. and Lindstrom, J. (1987a) Proc. Natl Acad. Sci. USA. 84, 595-599.
Whiting, P. and Lindstrom, J. (1987b) FEBS Letts 213, 55-60.
Whiting, P., Esch, F., Shimasaki, S. and Lindstrom, J. (1987a) FEBS Letts 219, 459-463.
Whiting, P. J., Schoepfer, R., Swanson, L. W., Simmons, D.M. and Lindstrom, J. M. (1987b) Nature 327, 515-518.
Wonnacott, S. (1986) J. Neurochem. 47, 1706-1712.

HUMAN NEUROBLASTOMA CELLS: AN IN VITRO MODEL FOR THE STUDY OF MAMMALIAN NEURONAL NICOTINIC RECEPTORS.

Emanuele Sher, Cecilia Gotti, Diego Fornasari, Bice Chini, Azucena Esparis Ogando, Francesco Clementi, University of Milan, Department of Medical Pharmacology, CNR Center of Cytopharmacology, via Vanvitelli 32, 20129 Milan, Italy

1. INTRODUCTION

The nicotinic acetylcholine receptor (AChR) is a neurotransmitter receptor which, upon acetylcholine binding, undergoes a conformational change which triggers ion permeability by opening a self contained cation channel. AChR is present in muscle end plates, in ganglia and in CNS, with different structural and functional characteristics. The most complete information on structure and function are available for AChR of skeletal muscles and fish electric organs. In these organs AChR is a membrane complex of approximately 300 KDa composed by four protein subunits α, β, γ, δ, (Conti-Tronconi and Raftery, 1982 Popot and Changeux, 1984). In the functional receptor five subunits (α2, β, γ, δ) are arranged with pseudo-five-fold symmetry around the central cation conducting channel. Of the five subunits only the two α-subunits bind acetylcholine and cholinergic agents. Nicotinic acetylcholine receptors are found in many areas of the central nervous system (CNS) and pharmacological studies suggest that they could be divided in subtypes (Wonnacott, 1987). However, the number of AChR subtypes, their relationship to each other, the role they play in the various areas of the CNS are still controversial.

The presence of nicotinic receptors in the CNS and ganglia was indicated since a long time by pharmacological experiments. Intracerebral administration of low doses of nicotine stimulate the CNS,induces tremors and seizures, whereas high doses have a depressing action on spontaneous activity. Nicotinic effects include also hypertension,bradycardia,altera-

NATO ASI Series, Vol. H21
Cellular and Molecular Basis of Synaptic Transmission
Edited by H. Zimmermann

tions in body temperature, respiration, salivation and motor reflexes, release of vasopressin and corticosteroids, modification of sleep and control of arousal, nociception, behaviour and learning (Hall, 1984). Some of these changes are now thought to be mediate by postsynaptic effects, while some others by a presynaptic effect through a modulation of the release of all the major neurotransmitters in the brain (Balfour, 1982).

More recently data on biochemical and molecular characteristics of neuronal AChR have been put forward. We will briefly review here the latter findings and our recent data on this subject.

2. CHARACTERISTICS OF NEURONAL AChR

In ganglia and in the central nervous system the AChR is not a single entity, but it is represented by a family of molecules which bind cholinergic agents of nicotinic specificity but whose function and molecular structure are not yet completely resolved. The best way to characterize these molecules is on the basis of their pharmacological and immunological characteristics.

2.1 Tools for AChR characterization

The isolation and characterization of neuronal AChR has been made possible by the use of specific neurotoxins and monoclonal antibodies. Neurotoxins specific for AChR are present in the venom of several species (Chiung Chang, 1979), but we will mention here only those most used. α-Bungarotoxin (αBgtx), a toxin present as a major component in the venom of the snake Bungarus multicinctus, binds to muscle AChR and to some molecules in ganglia and CNS. P15 (Gotti et al. 1985) and K toxins (Chiappinelli, 1986), minor components of the venom of Bungarus multicinctus, both bind and block nicotinic functions in ganglia. K toxin, also referred to as Bgt 3.1 and F toxin, recognizes two classes of high affinity sites on chick ganglion cells in culture; binding to the first class is blocked by αBgtx while binding to the second class of sites is not blocked by αBgtx but is inhibited competitively by a variety of nicotinic cholinergic ligands (Chiappinelli, 1986). Another probe used for studying the neuronal AChR are monoclonal antibodies (mAbs). The most used is mAb35 directed against the main immunogenic region of the α subunit of AChRs of muscle and electric organs (Tzartos and Lindstrom, 1980). Using this mAb a membrane component of ciliary ganglion neurons and of chick and rat brain has been isolated whose properties are those expected for a neuronal AChR (Whiting et al., 1987a,b). The same component in chick ganglionic neurons is also recognized by the K toxin.

With the use of these tools at least three types of nicotinic receptors have been identified.

2.2 Types of AChRs

2.2.1 Nicotine binds in the CNS to high affinity sites and low affinity sites. The high affinity site binds acetylcholine and is blocked by chlorisondamine (Marks and Collins, 1982) by P15 and K toxins (Chiappinelli, 1986) but not by α-Bungarotoxin. It is localized in the thalamic nuclei, interpeduncular nucleus, superior colliculus medial habenula, hypothalamus, cortex and substantia nigra (Clarke et al., 1985). A synaptic localization of nicotinic receptors was indicated by the high concentration of ^{3}H nicotine binding sites in the synaptosomal fraction (Yoshida and Imune, 1979). A consistent presynaptic distribution of these receptors was suggested by lesion studies (Schwartz et al., 1984). For example, bilateral habenular distruction is correlated with a consistent decrease in nicotinic binding sites in nucleus interpeduncularis, the neuronal target of this cholinergic pathway (Clarke et al., 1986).

This receptor has been isolated from PC12 cells (Whiting and Lindstrom, 1986 and 1987a), chick ganglia (Smith et al., 1985), and rat brain (Whiting and Lindstrom, 1987b), both by affinity chromatography with specific monoclonal antibodies, or by cloning their genes (Boulter et al., 1986; Goldman et al., 1987). The data obtained suggest that the receptor is composed by two peptides (α and β) with a stechiometry of $\alpha_2\beta_2$ or $\alpha_3\beta_2$. The β-subunit binds the cholinergic agents. The β subunits although highly homologous, have a different Mw in rat and chick.

2.2.2 The other AChR is a ion channel operated by ACh and blocked by curare and αBgtx. It is present in chick optic lobe and probably in mammalian brain. In fact, it has been shown that some of the effects of light on circadian rythms appear to involve an αBgtx sensitive cholinerigc link in the hypothalamic suprachiasmatic nucleus (Zats et al., 1981). Furthermore, in the inferior colliculus, αBgtx blocks the effects of the cholinergic agonists on single neurons. This receptor has been isolated by affinity chromatography with αBgtx both from chick optic lobe (Norman et al., 1982;

Conti-Tronconi et al., 1985) and from the CNS of locust (Breer et al., 1985). Aminoacid sequence analysis of AChR α-subunit from optic lobe and muscle of chick shows that the two proteins are highly homologous, but different (Conti-Tronconi et al., 1985). Thus this receptor is probably similar in pharmacological properties and in structure to muscle AChR but maintains some peculiar differences. Probably similar to this type of receptor is the αBgtx binding site recently described in a human medulloblastoma cell line (Lukas, 1986). This receptor has pharmacological and functional properties very similar to those of muscle AChR.

2.2.3 The third molecule related to the AChR family is the receptor for αBgtx. This molecule is present on neurons (both in ganglia and CNS), binds nicotine with low affinity and αBgtx in reversible fashion in contrast to the end-plate receptor (Wonnacott, 1986). Toxin binding is displaced by nicotinic drugs: nicotine, d-tubocurarine and gallamine are potent competitors while carbachol, decamethonium, and particularly hexamethonium are very weak in this respect (Wonnacott, 1986). Thus this molecule has binding characteristics of a nicotinic receptor. Binding sites for ^{125}I-αBgtx, corresponding mainly to this third type of receptors, were found in hypothalamus, hippocampus, inferior colliculus, cerebral cortex and in some regions of pons and medulla including locus ceruleus, dorsal segmental nuclei, dorsal raphe nuclei, and in neuroblastoma cell lines (Clarke, 1985; Clementi et al., 1986). αBgtx binding sites are present in synaptosomal preparations (Leutz and Chester, 1977) and αBgtx conjugated with horseradish peroxidase labels synaptic junctions in many areas of CNS (Arimatzu et al., 1977; Vogel et al., 1977) and in sympathetic neurons (Marshall, 1981). Nevertheless an extrasynaptic distribution of αBgtx receptor has been reported in ganglionic cells (Jacob et al., 1983; Messig et al., 1983; Cocchia and Fumagalli, 1981).

Very little is known about the possible function of this molecule. Our data, reported in a later section of the review, indicate that this receptor is not an ACh operated ion channel (Gotti et al., 1976). One possibility is that other unknown second messenger system are activated by the interaction of nicotinic agents with this molecule; in agreement with this hypothesis is the described effect of nicotine on actin disassembly in chromaffin cells, an effect that does not seem to depend upon Na^{+} or Ca^{2+} fluxes (Cheek and Burgoyne, 1986). It is possible that ACh is not the physiological agonist of this receptor, but instead another molecule possibly with a structure more similar to αBgtx. It has been reported that in tissues where αBgtxR is present exists a soluble compound of a Mw higher than 1000 daltons which inhibits the binding of the toxin to its receptor (Quick, 1982).

In vivo studies suggest that this receptor may have a physiological significance. The antagonistic effect of mecamylamine on nicotine-induced seizures, for example, is exerted by the interaction with a low affinity nicotinic binding site, possible the αBgtxR. Miner and coll. (Miner et al., 1985) have reported a high correlation between seizure sensitivity and the number of hippocampal αBgtxRs in an inbred strain of mice.

3. THE FAMILY OF AChR

From the above mentioned results it appears that nicotinic receptors in the CNS are not an homogeneous entity but on the contrary, a family of nicotinic receptors. We have at least three type of molecules that bind specifically nicotine; all of them have a sovramolecular structure of 5 subunits, although the subunits may be different from one form to another; all the subunits have a large homology in aminoacid sequence among them and also with the subunits of the muscular AChR.

The generation of a family of central nicotinic receptors can be explained a) by the existence of different related genes codifying for different related protein, or b) by differential splicing. In fact, the nucleotide sequence analysis of two clones of α-subunits of two AChRs cloned from rat brain, shows that these two molecules are the products of the same gene by differential splicing mechanisms (Goldman et al., 1987).

Recent data on the GABA and Glycine receptors (Barnard et al., 1987) suggest that different receptors with a self contained ion channel have a very similar sovramolecular structure and in some parts an highly homologous aminoacid sequence. It is possible to postulate that all the ionic receptors are members of a unique large family of molecules.

4. THE αBgtxR IN IMR32 NEUROBLASTOMA CELLS

The αBgtxR is the less known molecule among the central AChRs and we have attempted to study it in some detail.

The first problem we have faced was to select a proper cellular model

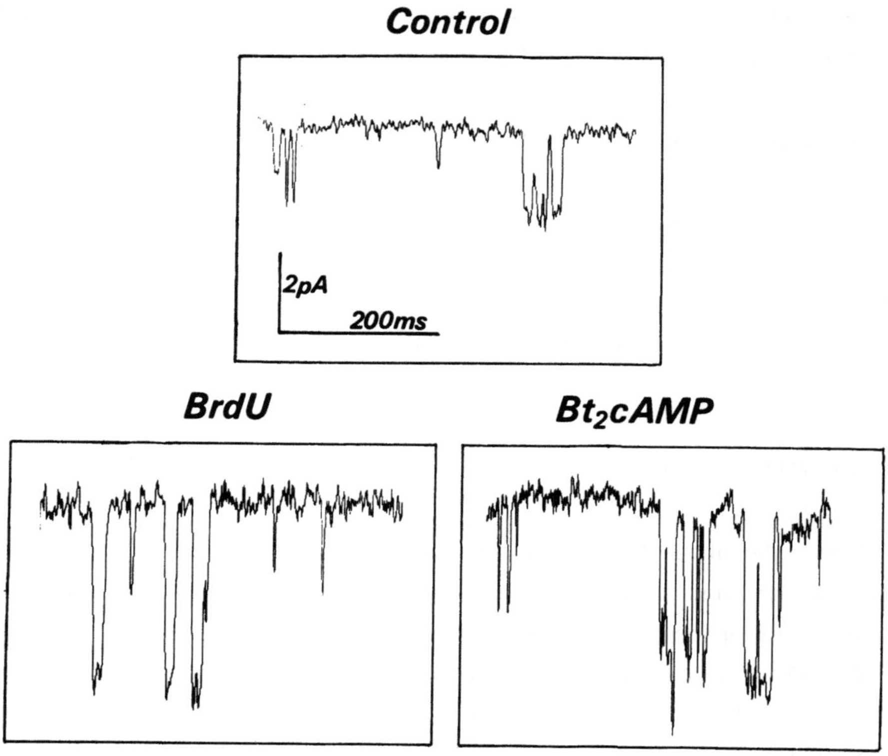

Fig.1 - Recordings showing examples of nicotinic channel openings in control and differentiated cells. The poor signal-to-noise ratio is due to the whole-cell patch-clamp configuration. Measurements of channels properties gave the following results: 18±3 and 30±4 and 58±6 pS (n=5) for Bt2cAMP-treated cells. Furthermore, the channels of differentiated cells also showed a shorter lifetime.

The ^{125}I-αBgtx had a PI of 5.3 and when run in a 5-20% sucrose gradient had a sedimentation coefficient of 9.5S.

Table 1 - Representative purification of αBgtxR from IMR32 cell line

	Receptor (pmol)	Specific activity (pmol/mg protein)	Purification fold
Membrane	36	0.13	1
Triton extract	70	-	-
Column flowthrough	6	-	-
Column eluate by CCh	26	650	5000

in which the study of the function and structure of αBgtxR was feasible. We have recently shown that IMR32, a human neuroblastoma cell line, have numerous binding sites for αBgtx (Clementi et al., 1986). The binding site was similar as far as its pharmacological characteristics to that described in CNS. In fact, αBgtx binding on these cells was displaceable by nicotine, carbachol, decamethonium and d-tubocurarine. Hexamethonium was not active as atropine or pirenzepine. We have shown, both by autoradiography and immunolocalization, that αBgtx binding sites are not localized in particular spots on the cell surface but are randomly distributed (Gotti et al., 1987) on the plasma membrane in both control and differentiated cells.

The next step was to investigate if these binding sites were connected with ion channels driven by ACh. We have measured membrane currents via a patch pipette in whole cell configuration in the presence and absence of ACh at different holding potentials, so that we could measure the single-channel events. We found that only 1 or 2 ACh operated Na^+ channels were expressed in a single cell (fig.1) in large contrast with the 3000 αBgtx binding sites found in each cell (Gotti et al., 1986).

The channels were hexamethonium but not αBgtx sensitive. We concluded that these cells were a good model for studies on the biochemical and functional properties of the αBgtxR, being the presence of AChR numerically irrelevant.

Furthermore IMR32 cells have other advantages, for example they can be induced to differentiate by several pharmacological stimuli and this could be a useful tool to study the regulation of the expression of the αBgtxR.

5. αBgtxR PURIFICATION AND CHARACTERIZATION

At least $100x10^7$ cells were used for each purification of αBgtx. Cells were first incubated with 1% Triton X-100 and the extract passed through an affinity column with bound αBgtx. A representative purification is shown in Table 1. The αBgtxR was eluted from the affinity column with 1M carbamylcholine and after extensive dyalisis radioiodinated with ^{125}I by a modification of the Chloramine T method (Greenwood et al., 1963). When run on SDS-PAGE αBgtxR, radioiodinated with ^{125}I or stained with silver staining gave, consistently four major bands with apparent MW of 76.3±2.3, 65±2.8, 57±2.7 47±2 Kd (fig.2).

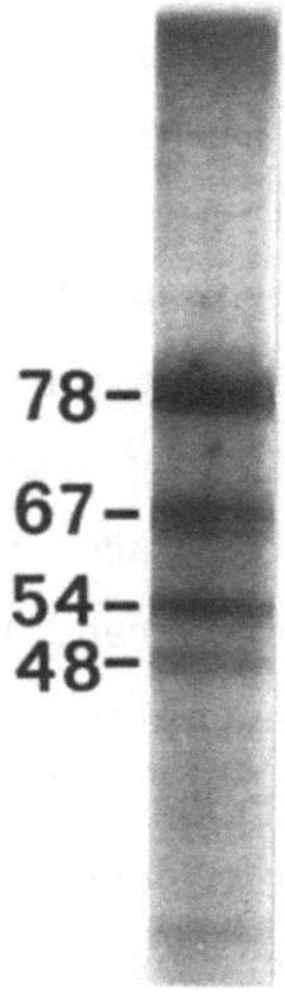

Fig.2 - 9% SDS-PAGE of ^{125}I-αBgtx of IMR32 cell line

This result is consistent with a molecular organization of this receptor similar to that of peripheral AChR and not with that of CNS AChR formed by only two peptides (Whiting and Lindstrom, 1987).

However, the preliminary data obtained with molecular biology techniques indicate that IMR32 mRNA hybridizes at low stingency with the probe pCA48 which recognizes an α-subunit of PC12 cells AChR. We are screening a cDNA library of IMR32 cells with the same probe and at the moment we have identified 10 positive clones. Immunological data are in line with the above reported findings, indicating similarities with both the muscle and ganglionic AChR. Policlonal antibodies and some monoclonal antibodies raised against veal muscle AChR precipitates αBgtxR but with low efficiency. Antibody against Torpedo AChR do not crossreact with IMR32 αBgtxR.

6. MODULATION OF αBgtxR DURING DIFFERENTIATION

IMR32 cells can be differentiated by a treatment with dibutyrril cAMP (Bt2cAMP) or with 5-bromodeoxyuridine (BrdU). During differentiation the cells acquire the ability of extending long neurites, increase the synthesis of neurotransmitters (Gotti et al. 1987), modulate the expression of muscarinic (Gotti et al., 1987) and adenosine (Abbracchio et al., unpublished results) receptors, and acquire a neurosecretory function. In this cell model of neuronal differentiation we studied the modulation of the expression of αBgtxR. The level of ^{125}I-αBgtx binding sites in control cells remained constant during the time of culture. The addition of Bt2cAMP (1mM) or BrdU (2.5 μM) to the culture medium significantly increased the number of ^{125}I-αBgtx binding sites in a time-dependent manner (fig.3). In Bt2cAMP- and BrdU-treated cells, the B_{max} values for ^{125}I-αBgtx binding increased two fold compared to control cells. No significant change in the Kd value of ^{125}I-αBgtx binding was found in Bt2cAMP- and BrdU-treated cells as compared with that observed in control cells.

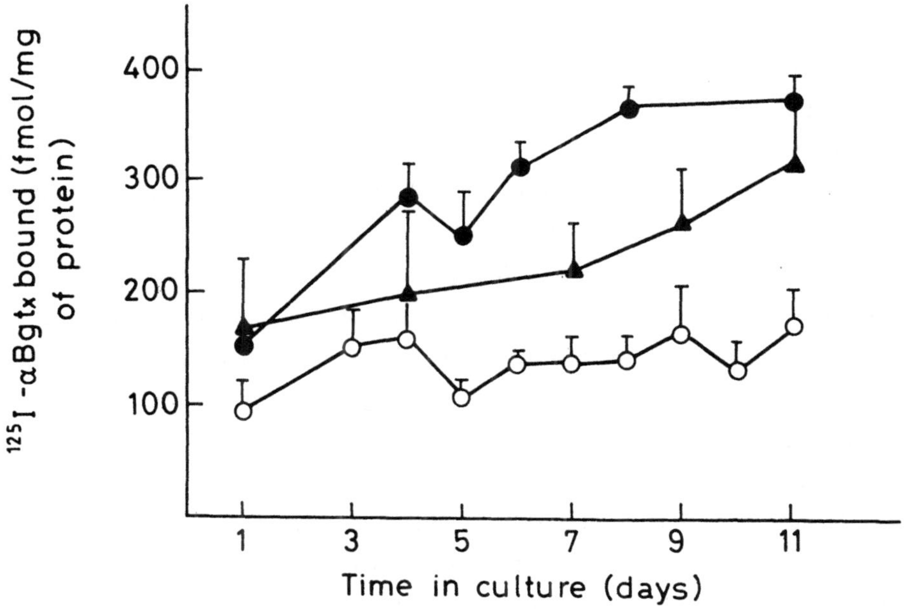

Fig.3 Time course of the expression of ^{125}-αBgtxR in IMR32 cells during treatment with Bt2cAMP (O---O) or BrdU (_---_) and in control (O---O) cells. Binding experiments were performed using 25 nM ^{125}I-αBgtx for total binding and 10 μM αBgtx for unspecific binding. The data (means ± SE) are expressed as femtomoles per mg of membrane protein.

Not only the number of αBgtx doubled, but also the few nicotinic channels increased their number (fig.1). This indicates that during differentiation induced by pharmacological agents the two molecules are similarly regulated.

7. ROLE OF αBgtxR IN NEUROTRANSMITTER SECRETION AND ON Ca^{2+} HOMEOSTASIS IN IMR32 CELLS

We have shown that non differentiated IMR32 cells express voltage dependent Ca^{2} channels (Sher et al., 1987) but that they are not able to secrete neurotransmitters (Clementi et al., 1986). After differentiation they acquire neurosecretory activity and dopamine is released by Ca^{2+} ionophores and α-Latrotoxin (E. Sher, personal communication). In differentiated cells amines are stored in granules, since they are reserpine sensible, and we have shown that secretory granules are present in neurites and enlarged nerve terminals of these cells (Gotti et al. 1987).

We investigated if carbamylcholine was able to induce neurotransmitter secretion from IMR32 cells as it does in adrenal medullary cells or in PC12 cells (Knight and Kesteren, 1983). We were unable to measure any dopamine secretion induced by carbachol. One possibility is that our assay is not sufficiently sensitive to detect low levels of secretion. Another possibility is a failure at the level of carbachol induced Ca^{2+} channels opening and Ca^{2+} mobilization. We investigated this point in cells that have been loaded with Quin2, an established fluorescent probe for measuring intracellular Ca^{2+} levels (Tsien et al., 1982).

We found that carbamylcholine induced an increase of intracellular Ca^{2+} ($[Ca^{2+}]_i$) in IMR32 cells. However, preincubation with d-tubocurarine (fig.4a) or αBgtx (fig.4b), failed to prevent the effects of carbachol; furthermore nicotine itself was without effects on $[Ca^{2+}]_i$ and failed to modify subsequent KCl- (fig.4C) or carbachol- (fig.4d) induced $[Ca^{2+}]_i$ rises. When nicotine and αBgtx were tested on differentiated cells (where compared to controls the nicotinic binding sites are increased two fold (Gotti et al., 1987)), the same lack of effect was found. This indicate that nor ACh driven ion channels nor αBgtx receptors are involved in carbachol control of calcium homeostasis in IMR32 cells. The effects of carbachol on Ca^{2+} homeostasis are mediated in these cells by the activation of muscarinic receptors (Sher et al., 1987).

It is thus interesting that the αBgtx binding sites (which are highly expressed in IMR32 cells and not coupled to known second messengers systems), do not influence calcium homeostasis nor neurotransmitter secretion. We have mentioned that some physiological effects of neuronal nicotinic receptor not due to Na^{+} ion and Ca^{2+} fluxes have been described (Check and Burgoyne, 1986), and further investigations on this direction are necessary before the function of this molecule could be clarified. For the moment the αBgtxR still remains a "receptor in search of a function and of a second messenger".

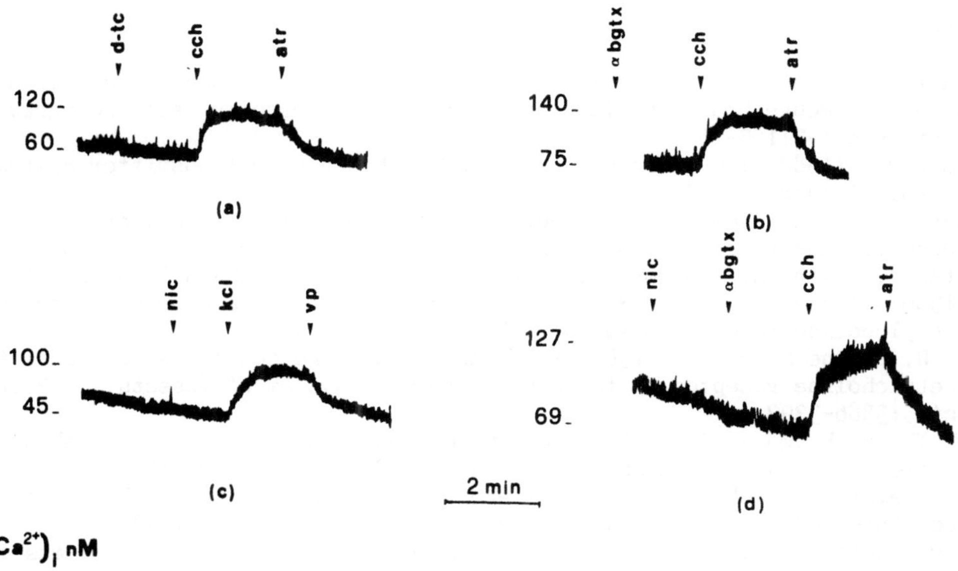

Fig.4 - Nicotinic receptors and $[Ca^{2+}]_i$. Neither d-Tubocurarine (d-Tc, 10µM) (a) nor αBgtx (0.2 µM) (b) were able to antagonize carbachol-induced $[Ca^{2+}]_i$-rises in IMR32 cells. Nicotine (Nic, 100µM) was without effects on resting calcium levels and did not influenced KCl- (50 mM) (c) or Carbachol- (d) induced $[Ca^{2+}]_i$ rises. Calibration is indicated on the left of each single trace.

8. CONCLUSIONS

In conclusion, we know that in the nervous system are present at least three types of molecules related, for pharmacological and structural characteristics, to the acetylcholine nicotinic receptor. We have studied in detail the structure and function of αBgtxR in a cellular model, a human neuroblastoma, that expresses practically only this type of molecules. αBgtxR has a four peptide organization, as the muscular AChR but at the molecular level it seems to have only a low degree of similarity with both AChR from CNS and muscle.

The αBgtxR is not involved in controlling calcium homeostasis and probably neurotransmitter secretion. Data obtained in vivo indicate that this receptor could be responsible for some actions of nicotine on CNS, i.e. convulsions. However the data accumulated up to now are too scanty to suggest any certain neurotransmitter and any possible function for this receptor.

REFERENCES

Arimatzu Y, Sebo A and Aaamo T (1977) Localization of α-Bungarotoxin binding site in mouse brain by light and electron microscopic autoradiography. Brain Res 147:165-169.

Balfour DJK (1982) The effects of nicotine on brain neurotransmitter systems. Pharmacol Ther 16:269-282.

Barnard EA, Darlison MG and Berburg P (1987) Molecular biology of the GABA receptor: the receptor/channel superfamily. Trend Neurol. Sci. 10:502-509.

Boulter J, Evans K, Goldman D, Martin G, Treco D, Heinemann S and Patrick J (1986) Isolation of a cDNA clone coding for a possible neuronal nicotinic acetylcholine receptor α-subunit. Nature 319:368-374.

Breer H, Kleene R, Hinz G (1985) Molecular formas and subunit strcture of the acetylcholine receptor in the central nervous system of insects. J Neurosci 5:3386-3392.

Cheek TR and Burgoyne RD (1986) Nicotine evoked disassembly of cortical actin filaments in adrenal chromaffin cells. FEBS Lett 207: 110-114.

Chiappinelli V (1986) Actions of snake venom toxins on neuronal nicotinic receptors and other neuronal receptors. Pharmacol Ther 31:1-32.

Chiung Chang C (1979) The action of snake venoms on nerve and muscle. Handbook of experimental Pharmacology, Springer Verlag, Berlin, 52:309-376.

Clarke P., Schwartz D, Paul SM, Pert CB, Pert A (1985) Nicotinic binding in rat brain: autoradiographic comparison of ^{3}H-acetylcholine, ^{3}H-Nicotine and ^{125}I-αBungarotoxin. J Neurosci 5:1307-1315.

Clarke PBS, Hamill GS, Nadi NS, Jakobowitz DM and Pert A (1986) ^{3}H-Nicotine and αBungarotoxin-labeled nicotinic receptors in the interpeduncular nucleus of rats. II. Effects of habenular deafferentation. J Comp Neurol 251:407-483.

Clementi F, Cabrini D, Gotti C and Sher E (1986) Pharmacological characterization of cholinergic receptors ina human neuroblastoma cell line. J Neurochem 47:291-297.

Cocchia D and Fumagalli L (1981) Immunocytochemical localization of α-Bungarotoxin receptors in the chick ciliary ganglion: synaptic and extrasynaptic sites. Neurochem Int 3:123-128.

Conti Tronconi B, Raftery MA (1982) The nicotinic cholinergic receptor: correlation of molecular structure with functional properties. Ann Rev Biochem 5:491-530.

Conti Tronconi BM, Dunn SMJ, Barnard EA, Dolly JO, Lai FA, Ray N and Raftery MA (1985) Brain and muscle nicotinic acetylcholine receptors are different but homologous proteins. Proc Natl Acad Sci USA 82:5208-5212.

Goldman D,Deneris E, Luyten W, Kochnar A, Patrick J and Heinemann S (1987) Members of a nicotine acetylcholine receptor gene family are expressed in different regions of the mammalian central neurons system. Cell 48:965-973.

Gotti C,Omini C, Berti F, Clementi F (1985) Isolation of a polypeptide from the venom of Bungarus multicinctus that binds to ganglia and blocks ganglionic transmission in mammals. Neuroscience 15:563-576.

Gotti C, Wanke E, Sher E, Fornasari D, Cabrini D and Clementi F (1986) Acetylcholine operated ion channel and α-Bungarotoxin binding site in a human neuroblastoma cell line beside on different molecule. Biochem Biophys Res Comm 137:1141-1147.

Gotti C, Sher E, Cabrini D, Bondiolotti G, Wanke E, Mancinelli E and Clementi F (1987) Cholinergic receptors ion channels, neurotransmitter synthesis, and neurite outgrowth are independently regulated during the in vitro differentiation of a human neuroblastoma cell line. Differentiation

34:144-155.
Greenwood FE, Hunter W and Glover JG (1963) The preparation of ^{125}I-labeled human growth hormone of high specific radioactivity. Biochem J 89:114-117
Hall GIT (1984) Pharmacological responses to the intracerebral administration of nicotine. In: Nicotine and tobacco smoking habit (Balfour DJK, ed) International Enciclopedia of Pharmacology and Therapy, Pergamon Press, New York, pp 45-60.
Jacob M and Berg DK (1983) The ultrastructural localization of α-Bungarotoxin binding sites in relation to synapses on chick ciliary ganglia neurons. J Neurosci 3:260-271.
Knight DE and Kesteren NT (1983) Evoked transient intracellular free Ca^{2+} changes and secretion in isolated bovine adrenal medullary cells. Proc R Soc Lond B Biol Sci 218:177-199.
Leutz TL and Chester J (1977) Localization of acetylcholine receptors in central synapses. J Cell Biol 75:258-267.
Lukas RJ (1986) Characterization of curaremimetic neurotoxin binding sites on membrane fractions derived from the human medulloblastoma clonal line, TE 671. J Neurochem 46:1936-1941.
Marks MJ and Collins AC (1982) Characterization of nicotine binding in mouse brain and comparison with the binding of αBungarotoxin and Quinidyl Benzylate. Mol Pharmacol 22:554-565.
Marshall LM (1981) Synaptic localization of α-Bungarotoxin binding which blocks nicotinic transmission of frog sympathetic neurons. Proc Natl Acad Sci USA 78:1948-1952.
Messing A and Gonatas NK (1983) Extra-synaptic localization of α-bungarotoxin receptors in cultured chick ciliary ganglion neurons. Brain Res 269:172-176.
Miner LL, Marks MJ and Collins AC (1985) Relationship between nicotine-induced seizures and hippocampal nicotinic receptors. Life Sci 37:75-83.
Norman N, Mehraban F, Barnard EA and Dolly JO (1982) Nicotinic acetylcholine receptor from chick optic lobe. Proc Natl Acad Sci USA 71:1321-1325.
Popot GL, Changeux JP (1984) Nicotinic receptor of acetylcholine: structure of an oligomeric integral membrane protein. Physiol Rev 64:1162-1239.
Quick M (1982) Presence of an endogenous factor which inhibits binding of α-Bungarotoxin 2.2 to its receptor. Brain Res 245:57-67.
Schwartz RD, Lehman J and Kellar KJ (1984) Presynaptic nicotinic cholinergic receptors labeled by ^{3}H-acetylcholine on catecholamine and serotonin axons in brain. J Neurochem 42:1495-1498.
Sher E Gotti C, Pandiella A, Madeddu L and Clementi F (1988) Intracellular calcium homeostasis in a human neuroblastoma cell line: modulation by depolarization, cholinergic receptors and α-latrotoxin. J Neurochem, in press.
Smith MA, Stallberg J, Berg DK and Lindstron J (1985) Characterization of a component in chick ciliary ganglia that cross-reacts with monoclonal antibodies to muscle and electric organ. J Neurosci 5:2726-2731.
Tsien RY, Pozzan T and Rink TJ (1982) Calcium homeostasis in intact lymphocytes: Cytoplasmic free calcium monitored with a new intracellularly trapped fluorescent indicator. J Cell Biol 94:325-334.
Tzartos S, Lindstrom JM (1980) Monoclonal antibodies used to probe acetylcholine receptor structure, localization of the main immunogenic region and detection of similarities between subunits. Proc Natl Acad Sci USA 77:755-759.
Vogel Z, Moloney GJ, Luig A and Daniels MP (1977) Identification of synaptic acetylcholine receptor sites in retina with peroxidase-labeled α-Bungarotoxin. Proc Natl Acad Sci USA 74:3268-3772.
Whiting PJ, Lindstrom JM (1986) Purification and characterization of a nicotinic acetylcholine receptor from chick brain. Biochemistry 25:2082-

2093.
Whiting PJ, Shoepfer R, Swanson LW, Simmons DM, Lindstrom JM (1987a) Functional acetylcholine receptor in PC12 cells reacts with a monoclonal antibody to brain nicotinic receptors. Nature 327:515-518.
Whiting PJ and Lindstrom JM (1987b) Purification and characterization of a nicotinic acetylcholine receptor from rat brain. Proc Natl Acad Sci USA 84:595-599.
Wonnacott S (1986) α-Bungarotoxin binds to low affinity nicotine binding sites in rat brain. J Neurochem 47:1706-1712.
Wonnacott S (1987) Brain nicotine binding sites. Human Toxicol 6:343-353.
Yoshida K and Imune H (1979) Nicotinic cholinergic receptors in brain synaptosomes. Brain Res 172:453-459.
Zats M and Brownstein MJ (1981) Injection of α-Bungarotoxin near the suprachiasmatic nucleus blocks the effects of light on nocturnal pineal enzyme activity. Brain Res 213:438-442.

MOLECULAR BIOLOGY APPROACHES TO THE FUNCTION AND DEVELOPMENT OF CNS SYNAPSES.

Dieter Zopf, Gabriele Grenningloh, Irm Hermans-Borgmeyer, Axel Rienitz, Cord-Michael Becker, Bertram Schmitt, Eckart D. Gundelfinger and Heinrich Betz.

Zentrum für Molekulare Biologie, Universität Heidelberg, Im Neuenheimer Feld 282, D-6900 Heidelberg, FRG.

Introduction

Synapses are the structural elements of the neuronal network through which cellular communication occurs. A prerequisite for understanding synaptic information transfer is the identification of the various components of the synaptic machinery. Recently, different molecular genetic techniques have been developed for the isolation of neuron-specific gene products, such as differential screening, gene transfer, deletion mutant analysis, and screening of expression libraries with monoclonal antibodies or selective ligands. Here, we describe two cloning approaches which have been successfully used in our laboratory. The first illustrates how neural gene products can be isolated and identified by selectively cloning mRNAs which appear in the avian optic lobe during the major period of synaptogenesis.

NATO ASI Series, Vol. H21
Cellular and Molecular Basis of Synaptic Transmission
Edited by H. Zimmermann

The second describes the cloning of a previously characterized protein, the ligand-binding subunit of the glycine receptor (GlyR) from rat spinal cord.

A. Cloning of the Chick Middle-Molecular-Weight Neurofilament (NF-M) Protein

Strategy for the selection of developmentally regulated mRNAs from chick optic lobe

In an attempt to isolate gene products expressed in target neurons upon synapse formation, we used a selective cloning procedure (Fig. 1) similar to that developed by Davis et al. (1984). Basically, cDNA synthesized from poly(A)$^+$RNA of innervated nervous tissue was enriched for "differentiation-specific" mRNAs by subtractive hybridization with an excess of poly(A)$^+$RNA of the same non-innervated brain region. We have chosen the optic lobe of the chick visual system because of its well-defined morphology, its extensive electrophysical characterization and its accessibility throughout development (Hunt and Brecha, 1984).

The first axons of the chick optic nerve arrive at the contralateral optic lobe at embyonic days 6-7. Therefore, poly(A)$^+$RNA was prepared from the "non-innervated" lobes at embryonic day 7 (OL E7). A 30-50fold excess of this RNA was hybridized in 0.4 M sodium phosphate buffer, pH 6.8, to radiolabelled ss cDNA reverse-transcribed from poly(A)$^+$RNA

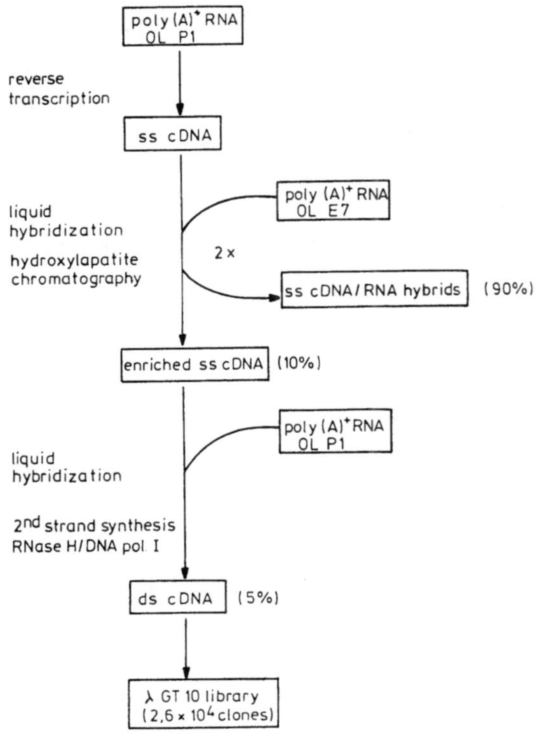

Figure 1: Strategy for construction of the "selective cDNA-library" (for details, see text).

of post-hatching day 1 optic lobes (OL P1). At this time, the chick optic system is completely innervated and functionally mature. After hybridization, ss cDNA strands were separated from ss cDNA/RNA hybrids on a hydroxylapatite column. At 60°C, the ss cDNA was eluted with 120 mM sodium phosphate buffer, pH 6.8, whereas the hybrid molecules remained bound. After a further round of hybridization with excess OL E7 poly(A)$^+$RNA, ≈90% of the ss cDNA, which represents poly(A)$^+$RNAs occuring in both RNA populations, had been removed. The residual ss cDNA (10%) was re-hybridized to homologous (OL P1) poly(A)$^+$RNA, converted into ds cDNA using the enzymes RNase H and DNA polymerase I (Gubler and Hoffmann, 1983), and, after addition of EcoRIlinkers, cloned into the EcoRI site of the phage vector lgt10. By this procedure a library of $2,6x10^4$ recombinants was established ("selective library").

Isolation and characterization of "differentiation-specific" cDNA clones

Screening with ss cDNA probes prepared from poly(A)$^+$RNA of OL P1, or liver (P1), provided a first qualitative analysis of the selective library. Fourty percent of the phage plaques hybridized with both probes, indicating incomplete subtraction. A further 40% gave no clear signals above background; it remains to be resolved whether these

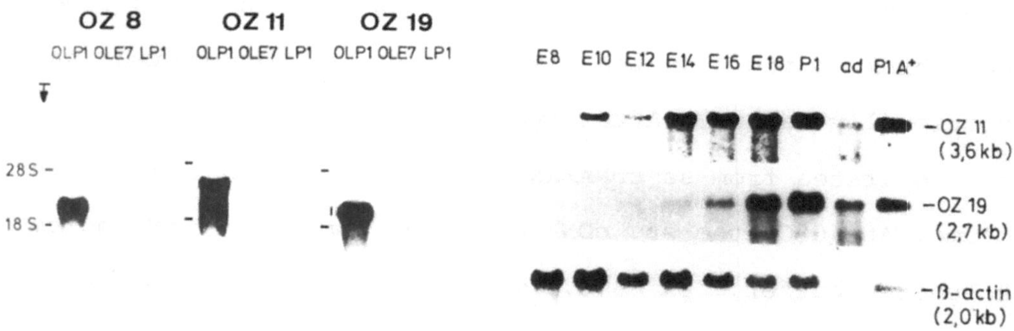

Figure 2: Tissue specificity and developmental regulation of OZ 8, OZ 11 and OZ 19 transcripts. a) Northern blots with poly(A$^+$)RNA (1µg/lane) isolated from optic lobe at days P1 (OL P1) and E7 (OL E7), and from liver at day P1 (L P1) were hyridized with radiolabelled OZ 8, OZ 11 and OZ 19 cDNAs as described (Zopf et al., 1987). b) Total RNA of chick optic lobe (10µg/lane), isolated at various developmental stages, (ad, adult) was probed with OZ 11, OZ 19 and β-actin (Cleveland et al., 1980) cDNAs. One µg of poly(A$^+$)RNA from day P1 (P1A$^+$) served as a control.

recombinants lack inserts, or correspond to mRNAs of rather low abundancy. The residual 20% hybridized only with the OL P1 probe. Northern blot analysis of three randomly selected clones of the latter class (OZ 8, OZ 11, OZ 19) confirmed their tissue-specific and developmentally regulated expression. The corresponding mRNAs were abundant in OL P1, but undetectable in liver (L P1) and OL E7 (Fig. 2A). During optic lobe development, OZ 11 mRNA synthesis starts at day E8, i.e. shortly after arrival of the optic nerve, increases until day P1, and then drops to a lower adult level (Fig. 2B). In contrast, OZ 19 mRNA appears later, around day E14, and rapidly increases to its highest level between days E18 and P1, before declining to its adult level. Control hybridizations with a chick β-actin cDNA probe revealed a different pattern of expression for actin mRNA. A high level was found early in development (day E8) which gradually decreased to adulthood.

Since the cDNA inserts of clones OZ 8, OZ 11 and OZ 19 were incomplete, appropriate libraries were screened for full length cDNAs and genomic DNAs, respectively. Furthermore, *in situ* hybridization experiments were performed in order to localize the corresponding trancripts.

One example of *in situ* hybridization analysis is shown in Figure 3. Transversal frozen tissue sections of day P1 chick brain (presenting the optic lobes, the midbrain and the

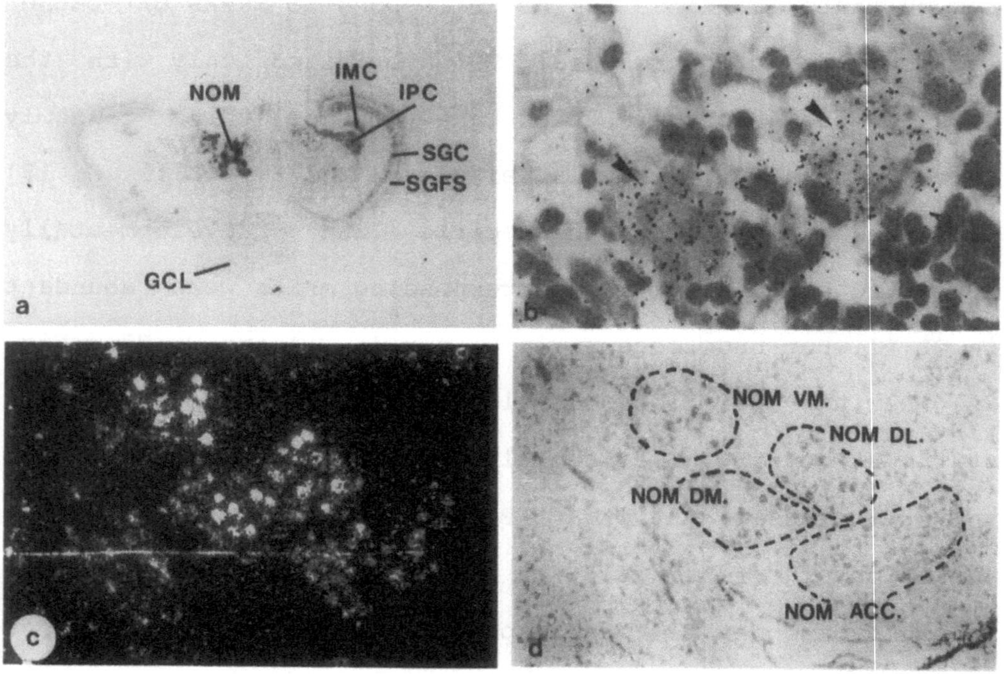

Figure 3: Localization of OZ 11 mRNA in the day P1 chick CNS. a) In situ hybridization of ^{35}S-radiolabelled OZ 11 probe to a transversal brain section showing the midbrain, optic lobes and cerebellar folds. b) High-magnification photomicrograph of Purkinje cells expressing OZ 11 mRNA (250x). c,d) Darkfield/brightfield photomicrographs of the nucleus oculomotorius (NOM) taken at a magnification of 50x. Tissue sections were prepared, hybridized and washed as described (Shivers et al., 1986). (VM.: ventromedialis; DL.: dorsolateralis; DM.: dorsomedialis; ACC.: accessorius).

cerebellar folds) were hybridized with ^{35}S-radiolabelled OZ 11 cDNA. In the optic lobe, specific hybridization was found in the stratum griseum centrale (SGC) and the lamina i stratum griseum et fibrosum superficiale (SGFS) of the optic tectum, and in the isthmic nuclei (nucleus pars parvocellularis (IPC) and nucleus pars magnocellularis (IMC). In the midbrain, intense labelling occurred in the nucleus oculomotorius (NOM, Fig. 2C,D) and in the nucleus trochlearis (not shown). In the cerebellum, the granular cell layer (GCL) showed a weak homogeneous labelling and punctated hybridization signals at its margin, which originated from the Purkinje cells (Fig. 2B). In conclusion, OZ 11 mRNA was found at high levels in brain regions which contain large multipolar and/or long projection neurons.

OZ 11 encodes the chick NF-M protein

Based on DNA sequence information derived from overlapping cDNA and genomic clones, we were able to identify OZ 11 as a middle-molecular-weight neurofilament (NF-M) cDNA. Neurofilaments are neuron-specific intermediate filaments and constitute part of the neuronal cytoskeleton. They contain three homologous polypeptides which are distinguished by their size (Weber et al., 1983). The identification of OZ 11 as a NF-M cDNA also explains the in situ hybridization data. Obviously, the rate of NF-M mRNA expression of a particular neuron correlates with the number and length of its neurites. Far-projecting neurons, e.g. the motoneurons of the NOM (Fig. 2C,D), are particularly rich in NF-M mRNA.

Recently, the human and mouse NF-M genes have been sequenced (Myers et al., 1987; Julien et al., 1986). In Figure 4, the deduced proteins are compared to chicken NF-M. All three NF-M proteins are highly homologous (>80%) in their head and rod domains, but only partly in their tail regions. The major difference between the mouse and the human protein is a stretch of 13 amino acids in the central part of the tail domain which is present once in mouse, but repeated six times in human. In contrast, a different sequence of six amino acids repeated 17 times is found in the chick NF-M tail piece. Even though these repeat regions are located at the same relative position, the chicken repeats are surrounded by an amino acid sequence with no obvious homology to human or mouse proteins. At the carboxyterminal end, a high degree of homology is resumed. Interestingly,

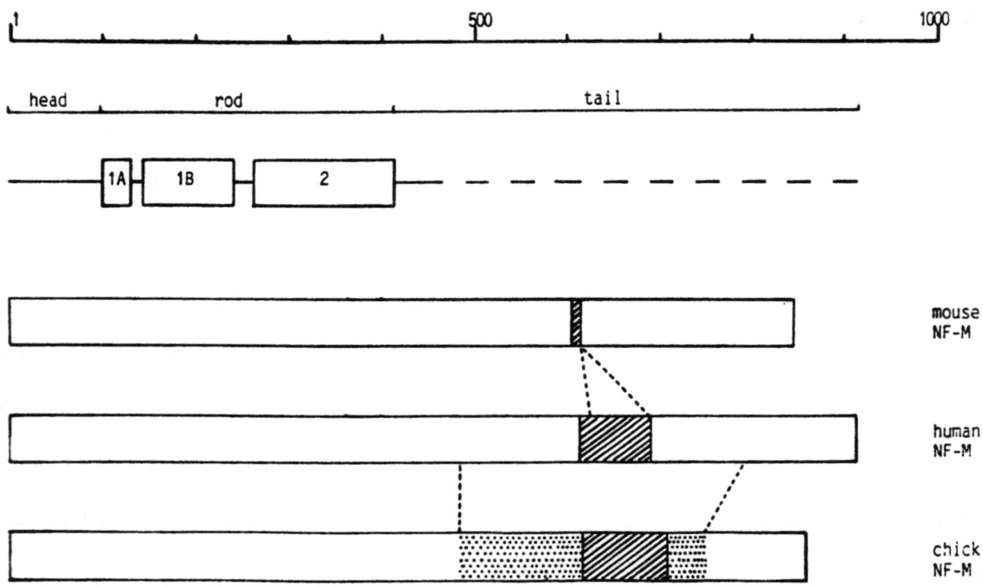

Figure 4: Structural comparison of the mouse, human and chick NF-M proteins. A common structural model of all intermediate filament proteins is shown above the alignment; 1A, 1B and 2 indicate the α-helical regions of the rod domain. Open boxes refer to regions of high homology (>80%). Repeat sequences are hatched, and the nonhomologous region of the chick protein is stippled. (Scale, amino acid position).

axonal NF-M proteins are highly phosphorylated (Julien and Mushynski, 1982), and for various arguments the tail repeat domains have been implicated as phosphorylation sites (Zopf et al., 1987). The variability of these repeats between species may indicate a divergent evolution of NF-M protein regulatory functions.

B. Cloning and Structure of the 48K Subunit of the Glycine Receptor (GlyR)

Neurotransmitter receptors and ion channel proteins are crucial for signal transmission at chemical synapses. A broad spectrum of such proteins has been characterized by electrophysiological and/or pharmacological techniques. However, primary sequence information on most of these proteins is lacking. Our strategy for cloning cDNAs of the 48K subunit of the rat GlyR was to screen cDNA libraries by hybridization with oligonucleotide mixtures synthesized on the basis of partial amino acid sequence data derived from the affinity-purified protein.

Purification, subunit composition and microsequencing of the GlyR

The convulsive alkaloid strychnine has been shown to selectively antagonize the hyperpolarizing effect of glycine, the major inhibitory neurotransmitter in brain stem and spinal cord (Aprison and Daly, 1978). Coupling of

2-aminostrychnine, a biologically active derivative of strychnine, to agarose beads allowed the affinity-purification of the GlyR from solubilized synaptic membranes of mammalian spinal cord (Pfeiffer et al., 1982; Graham et al., 1985; Becker et al., 1987). Biochemical and immunological analysis of the purified GlyR resulted in a structural model as shown in Figure 5 (Betz, 1987). The receptor "core" contains two glycosylated integral membrane proteins of Mr = 48K, and 58K. These polypeptides are thought to form, in a yet unknown stoichiometry, the chloride channel of the GlyR. Photoaffinity-labelling data indicate that the antagonist binding site is localized on

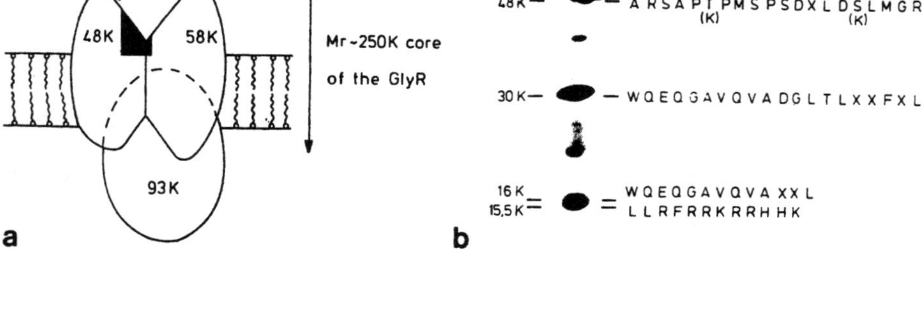

Figure 5: Structure and microsequencing of the GlyR. a) Model of the postsynaptic GlyR complex. The strychnine-binding core of the GlyR is assumed to be a 250 K heterooligomer of the 48 K and 58 K polypeptides; only one copy of each subunit is shown here. The strychnine binding region is indicated in black. symbolize carbohydrate side chains. b) <u>Staphylococcus aureus</u> V8 protease cleavage of 48K subunit and corresponding N-terminal polypeptide sequences identified by HPLC detection of phenylthiohydantoin (PTH) amino acids. X, unidentified amino-acid position; residues in parentheses, additionally detected amino acids.

the 48K subunit (Graham et al., 1983). A third peripheral membrane protein of Mr=93K is associated with the cytoplasmic domain of the postsynaptic GlyR complex (Schmitt et al., 1987). Its function is presently unknown, but may be related to the localization and/or anchoring of the receptor in the postsynaptic membrane.

For designing oligonucleotide probes, partial amino acid sequences were obtained by microsequencing the 48K subunit and three of its major cleavage products from a limited *Staphylococcus aureus* V8 protease digestion (Fig. 5B). The 30K and 16K fragments were found to have identical N-terminal sequences, and oligonucleotide mixtures corresponding to amino acids 1-7 (20mer) and 1-12 (35mer) of this sequence were synthesized (Grenningloh et al., 1987).

Verification of oligonucleotide probes in the frog oocyte translation system

Sumikawa et al. (1984) have shown that *Xenopus* oocytes express functional GlyR after injection of poly(A)$^+$RNA from mammalian brain. We have used this translation system for monitoring the appearence of GlyR mRNA in developing rat spinal cord and for verifying our synthetic oligonucleotide probes.

The number of strychnine binding sites in rat spinal cord has been shown to increase rapidly within the first 2-3 weeks after birth (Benavides et al., 1981). Accordingly,

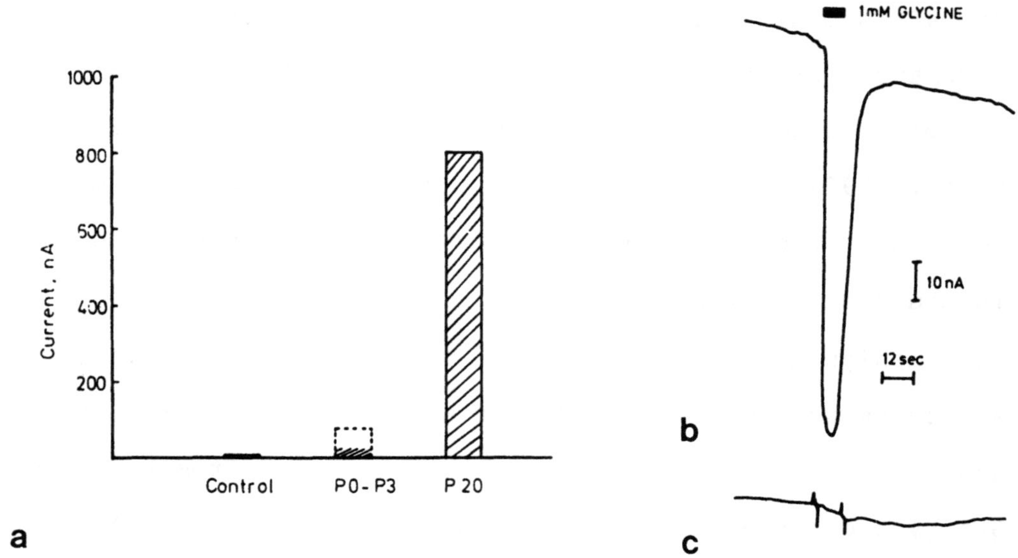

Figure 6: GlyR expression in frog oocytes. a) Histogram of oocyte responses after injection of 30-50 nl of poly(A)+RNA (1 µg/µl) from newborn (P0-P3) and 3-weeks old rats (P20). Control oocytes did not receive RNA. b) Representative recordings from oocytes injected with day P20 poly(A)+RNA together with either `sense' (b) or `antisense' (c) 20mer oligonucleotide mixtures.

poly(A)+RNA from newborn (P0-P3), or day P20, rat spinal cord was injected into frog oocytes. Oocytes which had received day P20 poly(A)+RNA developed a strong strychnine-sensitive response to glycine, whereas oocytes injected with day P0-P3 poly(A)+RNA were much less responsive (Fig. 6A). Consequently, day P20 poly(A)+RNA was used for the construction of cDNA libraries and for the verification of synthetic oligonucleotides in a translation arrest assay.

To this end, an excess of a 20mer, or 35mer, oligonucleotide mixture complementary to the mRNA (`antisense') was injected into oocytes one hour after priming with day P20 poly(A)$^+$RNA. Upon further incubation these oocytes did not develop glycine sensitivity (Fig. 6C). Co-injection of 20mer 'sense' or other unrelated oligonucleotides did not block the oocyte response to glycine (Fig. 6B). The inhibitory oligonucleotides were then used to isolate cDNA clones of the 48K polypeptide from previously established phage libraries.

Structure of the 48K GlyR subunit, a member of the ligand-gated ion channel protein family

Sequencing of several overlapping cDNA clones allowed the determination of the complete primary structure of the mature 48K subunit of the GlyR (Grenningloh et al., 1987). Analysis of the predicted amino acid sequence revealed four hydrophobic segments (M1 to M4) long enough to form transmembrane α-helices. Their arrangement is similar to that of nicotinic acetylcholine and the recently cloned $GABA_A$ receptor subunits (Schofield et al., 1987). Furthermore, significant amino-acid sequence homology exists between all these receptor proteins. An alignment of the 48K GlyR subunit sequence with those of the $GABA_A$ receptor polypeptides is shown in Figure 7. In addition to a relatively high homology of the transmembrane regions M1 to M3 which are thought to form the ion channel, considerable sequence conservation is found in the extracellular region, in particular between two precisely conserved

```
  1 QPSLQDELKDNTTVFTRILDRLL---DGYDNRLRPGLGERVTEVKTDIFVTSFGPVSDHDMEYTIDVFFRQSWKDERLKFKG-PMTVLRLNNLMASKIWT
  1 -----ARSAPKPMSPSDFLDKLMGRTSGYDARIRPNFKGPPVNVSCNIFINSFGSIAETTMDYRVNIFLRQQWNDPRLAYNEYPDDSLDLDPSMLDSIWK
  1 ---HSANEPSNMSYVKETVDRLL---KGYDIRLRPDFGGPPVDVGMRIDVASIDMVSEVNMDYTLTMYFQQSWKDKRLSYSGIPLN-LTLDNRVADQLWV
      .    .          .*.*.    *** *.** ......  *     *..  *..  ...  *.*.     ...*.*.* **  .  .  *    * *..  ...  .*
      A    N          LDRLL    GYD RLRP FGGPPV  V     IFV  SFG  VSE  MDYT     FFRQSWKD RL  Y  G  P    L LDN  MAD  IW
                                                ▼               ▼
 97 PDTFFHNGKKSVAHNMTMPNKLLRITEDGTLLYTMRLTVRAECPMHLEDFPMDAHACPLKFGSYAYTRAEVVYEWTREPARSVVVAEDGSRLNQYDLLGQ
 96 PDLFFANEKGAHFHEITTDNKLLRISRNGNVLYSIRITLTLACPMDLKNFPMDVQTCIMQLESFGYTMNDLIFEWQEQGAVQVADGLTLPQFILKE--EK
 94 PDTYFLNDKKSFVHGVTVKNRMIRLHPDGTVLYGLRITTTAACMMDLRRYPLDEQNCTLEIESYGYTTDDIEFYW--NGGEGAVTGVNKIELPQFSIVDY
    **..* * *..  *  *  *...*.  .*..**  *.*  ...*.*.*  .*.*  .  *  .  .*..**   .   ..*    ..   .. .      . .
    PDTFF N KKS  H  T  NKLLRI  DGTVLY  RIT  TAACPMDL  FPMD  Q  C  L  ESYGYT   D   FEW    GA   VV G      L Q
                               ————————M1 —————————      ———————M2 ————————          ————————M3 -
197 TVDSGIVQSSTGEYVVMTTHFHLKRKIGYFVIQTYLPCIMTVILSQVSFWLNRESVPARTVFGVTTTVLTMTTLSISARNSLPKVAYATAMDWFIAVCYAF
194 DLRYCTKHYNTGKFTCIEARFHLERQMGYYLIQMYIPSLLIVILSWISFWINMDAAPARVGLGITTVLTMTTQSSGSRASLPKVSYVKAIDIWMAVCLLF
192 KMVSKKVEFTTGAYPRLSLSFRLKRNIGYFILQTYMPSTLITILSWVSFWINYDASAARVALGITTVLTMTTISTHLRETLPKIPYVKAIDIYLMGCFVF
     .   .    ** .        *.*.* .**. .*.* *. ...***..***.* .. .**. .*.******** *   * .***. *..*.*.   ..* *
     S   V    TG Y        FHLKR IGYF IQTY PS LIVILSWVSFWIN DA PARV LGITTVLTMTT S   R SLPKV YVKAIDI   AVC F
    ———————————
297 VFSALIEFATVNYF------------------TKRGYAWDGKSVVPEKPKKVKDPLIKKNNTYAPTATSYTPNLARGD--PGLATIAKSATIEPKEVKPE
294 VFSALLEYAAVNFV-------------------SRQHKELLRFRRKRRHHKDDEGGEGRFNFSAYGMGPACLQAKDGISVKGANNNNTTNPAPAPSKSPE
292 VFLALLEYAFVNYIFFGKGPQKKGAGKQDQSANEKNKLEMNKVQVDAHGNILLSTLEIRNETSGSEVLTGVGDPKTTMYSYDSASIQYRKPMSSREGYGR
    **.**.*.* **.                   .   .  .  .    .    .. ......          . .   . . .    .     .  ..
    VFSALLEYA VNY                   R   E  K  V    K    LE RNNTSA          K G   G A I    P     E  PE
                                        ————————M4 ——————
377 --------------------PEPKKTFNSVSKIDRLSRIAFPLLFGIFNLVYWATYLNREPQLKAPTPHQ      GABA α
375 --------------------EMRKLFIQRAKKIDKISRIGFPMAFLIFNMFYWIIYKIVRREDVHNK---      48K
392 ALDRHGAHSKGRIRRRASQLKVKIPDLTDVNSIDKWSRMFFPITFSLFNVVYWLYYVH------------      GABA β
                           .     . .**. **. **  * .**  .**  *
                           K     V KIDK SRI FP  F IFN  VYW  Y
```

Figure 7: Alignment of sequences of the 48K GlyR polypeptide (Grenningloh et al., 1987) with the α and β subunits of the $GABA_A$ receptor (Schofield et al., 1987). Putative membrane spanning regions (M1-M4) are overlined. The conserved extracellular cysteines are marked by ▼ . Gaps are introduced for maximal homology. Numbering starts with the first N-terminal amino acid of the mature protein. ((*)=3, (•)=2 identical amino acids)

cysteine residues. The function of this region is unknown; the cysteines have been proposed to form a disulfide bridge essential for receptor tertiary structure.

Conclusions and perspectives

Using different approaches, we have cloned gene products which are important determinants of neuronal shape and function.

i) The NF-M protein is one of the brain-specific cytoskeletal subunits of the chick NF protein triplet. Neurofilaments are thought to determine neuronal geometry and axonal diameter. NF-M mRNA was found to be highly expressed in multipolar and long-projection neurons. The isolation and identification of the NF-M cDNA confirms the selectivity of the cloning procedure chosen here. Its specificity and sensitivity may be improved by narrowing the interval between developmental stages, and by using enriched cDNA probes generated by subtractive hybridization of different optic lobe mRNA populations.

ii) The primary sequence of the 48K subunit of the glycine receptor, a neurotransmitter-gated chloride channel in the CNS, has been determined. This polypeptide shows significant structural and amino acid sequence homology to $GABA_A$ and nicotinic acetylcholine receptor subunits, a finding which reflects the close functional similarities of these membrane proteins. The conservation of structural features between the subunits of different neurotransmitter-gated ion channels strongly suggests that all ion-conducting receptors of excitable membranes belong to a large protein super-family. DNA probes corresponding to conserved receptor domains may allow the isolation and analysis of yet unaccessible neurotransmitter-gated channels in the brain, in particular the excitatory amino acid receptors. Site-directed mutagenesis and reconstitution experiments in frog oocytes should further facilitate the identification of the amino acid residues and protein domains which determine agonist specificity and ion selectivity of the different ligand-gated channels.

Acknowledgements:

This work was supported by the Bundesministerium für Forschung und Technologie (BCT 365/1), Deutsche Forschungsgemeinschaft (SFB 317) and the Fonds der Chemischen Industrie. We thank Drs. K. Beyreuther, C. Methfessel, B. Sakmann and M. Zensen for experimental help at various stages of this work, Dr. P. Schofield for providing the alignment shown in Figure 7, C. Schröder, C. Udri and H. Krischke for expert technical assistance, and I. Nonnenmacher for help with the preparation of the manuscript.

References

Aprison, M.H. and E.C. Daly. 1978. Biochemical aspects of transmission at inhibitory synapses: the role of glycine. Advances in Neurochemistry 3: 203-294.

Becker, C.-M., I. Hermans-Borgmeyer, B. Schmitt and H. Betz. 1986. The glycine receptor deficiency of the mutant mouse spastic: evidence for normal glycine receptor structure and localization. J. Neurosci. 6: 1358-1364.

Benavides, J., J. Lopez-Lahoya, F. Valdivieso and M. Ugarte. 1981. Postnatal development of synaptic glycine receptors in normal and hyperglycinemic rats. J. Neurosci. 37: 315-320.

Betz, H. 1987. Biology and structure of the mammalian glycine receptor. Trends in Neurosci. 10: 113-117.

Cleveland, D.W., M.A. Lopata, R.J. MacDonald, N.J. Cowan, W.J. Rutter and M.W. Kirschner. 1980. Number and evolutionary conservation of α- and β-Tubulin and cytoplasmic β- and g-Actin genes using specific cloned cDNA probes. Cell 20: 95-105.

Davis, M.M., D.I. Cohen, E.A. Nielsen, M. Steinmetz, W.E. Paul and L. Hood. 1984. Cell-type specific cDNA probes and the murine I region: The localization and orientation of A^d. Proc. Natl. Acad. Sci. USA 81: 2194-2198.

Graham, D., F. Pfeiffer and H. Betz. 1983. Photoaffinity-labelling of the glycine receptor of rat spinal cord. Eur. J. Biochem. 131: 519-525.

Graham, D., F. Pfeiffer, R. Simler and H. Betz. 1985. Purification of the glycine receptor of pig spinal cord. Biochemistry 24: 990-994.

Grenningloh, G., A. Rienitz, B. Schmitt, C. Methfessel, M. Zensen, K. Beyreuther, E.D. Gundelfinger and H. Betz. 1987. The strychnine-binding subunit of the glycine receptor shows homology with nicotinic acetylcholine receptors. Nature 328: 215-220

Gubler, K. and B.J. Hoffmann. 1983. A simple and very efficient method for generating cDNA libraries. Gene 25: 263-269.

Hunt, P.H. and N. Brecha. 1984. The avian tectum: A synthesis of morphology and biochemistry. In Comparative neurology of the optic tectum (ed. H. Venegas) pp. 619-648. Plenum, New York and London.

Julien, J.P. and W.E. Mushynski. 1982. Multiple phosphorylation sites in mammalian neurofilament polypeptides. J. Biol. Chem. 257: 10467-10470.

Julien, J.-P., D. Meyer, D. Flavell, J. Hurst, and F. Grosveld. 1986. Cloning and developmental expression of the murine neurofilament gene family. Mol. Brain Res. 1: 243-250.

Myers, M.W., R.A. Lazzarini, V. M.-Y. Lee, W.W. Schlaepfer and D.L. Nelson. 1987. The human mid-size neurofilament subunit: a repeated protein sequence and the relatioship of its gene to the intermediate filament gene family. EMBO J. 6: 1617-1626.

Pfeiffer, F., D. Graham and H. Betz. 1982. Purification by affinity chromatography of the glycine receptor of rat spinal cord. J. Biol. Chem. 257: 9389-9393.

Schmitt, B., P. Knaus, C.-M. Becker and H. Betz. 1987. The M_r 93000 polypeptide of the postsynaptic glycine receptor complex is a peripheral membrane protein. Biochemistry 26: 805-811.

Schofield, P.R., M.G. Darlison, N. Fujita, D.R. Burt, F.A. Stephenson, H. Rodriguez, L.M. Rhee, J. Ramachandran, V. Reale, T.A. Glencorse, P.H. Seeburg and E.A. Barnard. 1987. Sequence and functional expression of the $GABA_A$ receptor shows a ligand-gated receptor super-family. Nature 328: 221-227.

Shivers, B., B.S. Schachter and D.W. Pfaff. 1986. In situ hybridization for the study of gene expression in the brain. Methods Enzymol. 124: 497-510.

Sumikawa, K., I. Parker and R. Miledi. 1984. Partial purification and functional expression of brain mRNAs coding for neurotransmitter receptors and voltage-operated channels. Proc. Natl. Acad. Sci. 81: 7994-7998.

Weber, K., G. Shaw, M. Osborn, E. Debus, and N. Geisler. 1983. Neurofilaments, a subclass of intermediate filaments: structure and expression. Cold Spring Harbour Symp. Quant. Biol. 46: 717-729.

Zopf, D., I. Hermanns-Borgmeyer, E.D. Gundelfinger and H. Betz. 1987. Identification of gene products expressed in the developing chick visual system: characerization of a middle-molecular-weight neurofilament cDNA. Genes and Development 1: 699-708.

RECENT STUDIES ON OPIOID RECEPTORS

Eric J. Simon
Departments of Psychiatry and Pharmacology
New York University Medical Center
New York, NY 10016
U.S.A.

INTRODUCTION

It is a great honor and pleasure to be invited to give a lecture at this NATO Symposium in honor of my good friend Victor Whittaker, who seems to me to be far too active and dynamic to be retiring.

Victor's contributions to Neurobiology have been numerous, as have been the scientists who greatly benefitted by their contacts with him, among whom I am happy to be included. The summers at Woods Hole during which I was privileged to share a laboratory with Victor and Mike Dowdall and collaborate with them were very happy and for me educational ones. We devised methods for isolating synaptosomes from teleost fish and, though this is not widely known, discovered high affinity uptake of choline into squid synaptosomes.

Our understanding of the neuronal synapse has been increased enormously since the isolation of synaptosomes was made possible by the pioneering work of Dr. Whittaker and his collaborators. I wish him and Margaret the best of health and happiness in the numerous activities they will surely engage in following Victor's official retirement.

The story of the discovery of receptors for opiate narcotic analgesics also began with the isolation of the crude synaptosomal (P_2) fraction from animal brain. It was here that the stereospecific binding of opiates was first discovered in our laboratory (Simon et al 1973) and simultaneously in two other laboratories (Terenius 1973; Pert and Snyder 1973). It has since been found that the concentration of binding sites is as high in the P_3 fraction and it is now customary to spin down a high speed fraction that contains P_2 as well as P_3 fractions (the P_1 fraction containing mostly nuclei is removed in some laboratories prior to the high speed spin). A great deal of progress has been made in this area

NATO ASI Series, Vol. H21
Cellular and Molecular Basis of Synaptic Transmission
Edited by H. Zimmermann

and I will give a summary of the state of the science before proceeding to describe some of our own more recent studies.

HISTORICAL REVIEW

Considerable evidence has been accumulated by a number of laboratories which strongly suggests that the stereospecific binding sites found in animal CNS, and later in some peripheral tissues as well, are pharmacologically relevant opiate receptors. They were found to be present in all vertebrate species examined, including man (Hiller et al 1973), as well as in a number of invertebrates (Stefano 1980). Detailed studies of the distribution of opioid binding sites within the CNS have been carried out in a number of species including human autopsy material, monkeys, cows, guinea pigs and rats. The early experiments were done by dissecting and homogenizing discrete regions of the brain and spinal cord and binding labeled opiates to the homogenates (Hiller et al 1973; Kuhar et al 1973). More recent studies have used autoradiography of brain and spinal cord slices to achieve detailed mapping of receptors even in sub-nuclei of various regions. Two methods have been used successfully, injection of animals with labeled opiates and isolation of tissue slices at various times (Atweh and Kuhar 1977a,1977b,1977c; Pearson et al 1980) and the incubation of tissue slices from various CNS regions with labeled opiates in vitro (Young and Kuhar 1979). The results from many different laboratories can be summarized by stating that large differences exist in the levels of opioid binding sites between different CNS regions. Moreover, the areas rich in opioid receptors tend to be in, or associated with, the limbic system and in all areas that have been implicated in pain perception and modulation and some of the other actions of opiate drugs. Opiate receptors have also been found in a number of peripheral tissues such as the gastrointestinal tract and some endocrine glands, such as the pituitary and adrenal glands.

The conservation of these receptors in so many species throughout evolution led to the suggestion that they must serve ligands endogenous to the organism. This resulted in the discovery of the endogenous opioid peptides, the putative natural ligands for these receptors. The first such peptides were discovered and characterized by Hughes and Kosterlitz (Hughes et al 1975). They were two pentapeptides, Tyr-Gly-Gly-Phe-Met and

Tyr-Gly-Gly-Phe-Leu, which the discoverers named Met-and Leu-enkephalin, respectively. Since that time a number of other peptides with opioid activity has been discovered, including the endorphins (a term coinec by the author of this paper, a contraction of endogenous morphine) which are constituents of the pituitary hormone ß-lipotropin (Cox et al 1976), dynorphin (Goldstein et al 1979; 1981) and several others. There are some 15 or so peptides with opioid activity currently known. Many investigators now prefer the term opioid receptors to the earlier term opiate receptors to indicate that we believe that their major function is to bind endogenous opioids and that their ability to bind opiate alkaloids is incidental. This term will be used hereafter in this paper.

A lot of information has been obtained about the distribution, metabolism and biosynthesis of opioid peptides, a detailed discussion of which is beyond the scope of this paper. In spite of an enormous amount of work already done in many laboratories the physiological functions of these peptides and their receptors are at present still unknown, though evidence suggests that they may play roles in a large variety of moods, emotions and behaviors. A role for opioidergic neurons in pain modulation seems reasonably firmly established.

On the cellular level the role of these neuropeptides is thought to be in synaptic transmission. In those instances where opioid receptors are located postsynaptically the peptides may alter trans-synaptic potentials, much like the classical neurotransmitters. However, most opioid receptors appear to be presynaptic. In this case the peptides are thought to act as neuromodulators, regulating the release at the synapse of classical neurotransmitters or neuropeptides acting as transmitters. This idea is supported by the knowledge that the opioid peptides inhibit the release of a wide variety of transmitters, a fact known to be true for opiate alkaloids long before the opioid peptides were discovered.

The biosynthesis and molecular biology of the opioid peptides can be summarized by stating that all of them are made via one of three large precursor proteins called proenkephalin, proopiomelanocortin and prodynorphin, which are coded in three separate genes. The complete structures of the precursors as well as of all three genes are known for several species (for reviews of this important work see Numa 1984, Gubler 1986). It should be noted that proopiomelanocortin was the first precursor found to give rise to several peptides with different biological activities. This molecule was first discovered in the laboratories of

Herbert (Roberts and Herbert 1977) and of Mains and Eipper (Mains et al 1977) to be a common precursor of ACTH and ß-endorphin. When the total sequence of the precursor became known (Nakanishi et al 1979) it was also found to contain several melanocyte-stimulating hormones (α,ß, and γ-MSH); hence the name proopiomelanocortin first suggested by Sidney Udenfriend.

In the first few years after the discovery of opioid receptors it was thought that they represent a single homogeneous class. However, the discovery of numerous peptides with opiate-like activity cast doubt on this idea. Martin and collaborators (Martin et al 1976, Gilbert and Martin 1976) provided the first clearcut pharmacological evidence for multiple opioid receptors. They postulated the existence of three separate classes of opioid receptors based on the differing pharmacological profiles of morphine and some of its benzomorphan congeners in chronic spinal dogs. These drugs were also unable to substitute for each other in the suppression of withdrawal symptoms in dogs made dependent on one of them. Based on the drugs used, Martin named the receptors mu (μ) for morphine, kappa (κ) for ketocyclazocine and sigma (σ) for SKF 10,047 (N-allylnormetazocine). After the discovery of the enkephalins, Kosterlitz and his group (Lord et al 1977) provided evidence for yet another type of opioid receptor which shows preference for enkephalins and some of their congeners. This receptor was named delta (δ) for deferens because the mouse vas deferens, an in vitro bioassay system, discovered and used very effectively by the Kosterlitz team, has a predominance of this type of receptor, whereas the myenteric plexus of the guinea pig ileum seems to have primarily μ receptors.

The existence of these 4 major types of opioid receptors has been confirmed by a large variety of in vivo and in vitro experiments. A number of other types (as well as subtypes, such as μ_1 and μ_2, κ_1 and κ_2) have been postulated, but the evidence for their existence is still relatively preliminary. There is now also some doubt whether the σ receptor is really an opioid receptor. Effects mediated via this receptor are not reversed by the opioid antagonist naloxone. Moreover, the σ receptor may be the receptor for another abused drug, phencyclidine (Zukin and Zukin 1979).

A word is in order about the current state of our knowledge of the steps triggered by the binding of opioids to their receptors. There is evidence that the major types of opioid receptors use cyclic AMP as their second messenger. Experiments done mainly in NG108-15 cells, (Sharma et al 1975) but also in brain homogenates, (Collier and Roy 1974) indicate that opioid receptors are coupled to adenylate cyclase in a negative way, i.e. opioids inhibit the enzyme via the G_i protein (the GTP binding protein used for inhibition of the cyclase). There has also been progress by the electrophysiologists in delineating the ionic channels involved in the effects mediated by the three major opioid receptor types. Briefly, the evidence suggests that binding of opioids to μ and δ receptors leads to an opening of synaptic potassium channels, whereas binding to κ sites results in the opening of calcium channels.

CHARACTERIZATION OF THE THREE MAJOR TYPES OF OPIOID RECEPTORS

The present paper will be restricted to a discussion of the three major and best established opioid receptor types, μ, δ, and k. As mentioned, their existence seems to be clearly established. However, by the use of in vivo pharmacological and in vitro membrane binding experiments, it is not possible to determine whether the different types of receptors represent different molecular species or different conformers of the same molecule. The more recent experiments to be discussed in this paper were done on solubilized receptor molecules in an attempt to address this question. The definitive answer will only become available when all seemingly different receptors have been purified and sequenced. The purification to homogeneity of the μ receptor by our laboratory will therefore be included in this discussion.

Before proceeding, there are several topics that need to be discussed briefly. The study of receptor heterogeneity has been greatly helped by the recent availability of highly selective ligands. The most selective ligand for μ receptors is D-Ala^2-MePhe^4-Gly-ol^5-enkephalin (DAGO), for δ sites it is D-Pen^2-D-Pen^5-enkephalin (DPDPE). The most selective κ ligands are two compounds synthesized by scientists at Upjohn, but not yet available in labeled form. Binding to κ sites is commonly measured by using a benzomorphan, such as tritiated bremazocine or ethylketocyclazocine, and blocking the μ and δ sites with saturating concentrations of μ and δ ligands.

The question concerning the endogenous ligands of the various receptor types has not been completely resolved. At present it is felt that κ receptors may serve the various peptides derived from the precursor prodynorphin, i.e. dynorphin A and B and α and ß-neoendorphin. The δ receptor is still thought to be the enkephalin-preferring receptor. The μ receptor has not been clearly assigned a putative endogenous ligand. Candidates include ß-endorphin, the enkephalins, which have lower but still respectable affinity for this receptor (μ and δ may be "isoreceptors"), or the recently discovered authentic morphine that seems to be present in some animal tissues (Oka et al 1985), including mammalian brain (Goldstein et al 1985; Donnerer et al 1986) and human cerebrospinal fluid (Cardinale et al 1987).

The functions of the various opioid receptor types and their endogenous ligands are still unknown and represent a high priority in current research. There is evidence suggesting that analgesia can be produced via all three of the major receptor types, μ, δ and k. The finding that tolerance and physical dependence seem to be largely properties of μ receptors has stimulated interest in the synthesis and study of drugs and peptides that produce analgesia via δ or κ receptors. Matching of the many actions of the opioids with appropriate receptors will require considerable additional work.

Since much of the work to be summarized was carried out on solubilized receptors, a word is in order concerning the way in which receptors, able to bind opioids in solution, are extracted from mammalian brain membranes.

We reported in 1975 (Simon et al 1975) the solubilization of opioid binding sites prebound to ^{3}H-etorphine. This was a significant advance, but the solubilized receptor was neither able to exchange bound ligand for free nor to bind labeled opioids after dissociation of the bound ^{3}H-etorphine. It was only in 1980 that several laboratories were able to achieve solubilization of opioid receptors capable of binding opiates in solution. The first such successful solubilization in our laboratory was carried out on the brain of the toad, Bufo marinus (Ruegg et al 1980, 1981). A good yield of opioid receptors (50-65%) was obtained using digitonin (0.5-1.0%). It was subsequently found by R. Howells in our laboratory (Howells et al 1982) that a modification of this technique was applicable to mammalian brain. The modification involved the addition of high concentrations of sodium chloride (0.5-1.0M) to the extraction medium. It is of interest that this is not a non-specific effect of high ionic strength, but a specific property of sodium salts. This method, which yields in active form in solution 30-50% of binding sites present in the membranes, is still the method of choice in our laboratory. Other techniques have been found useful in other laboratories (Bidlack and Abood 1980, Simonds et al 1980, Cho et al 1981, Demoliou-Mason and Barnard 1984).

The soluble receptor has been used for a variety of studies. Most notably it allowed us to establish the glycoprotein nature of opioid receptors (Gioannini et al 1982). This was done by the use of a variety of immobilized lectins. Of 9 lectins used only wheat germ agglutinin (WGA)-agarose was found to retain the receptors, which could be specifically eluted in ca. 50% yield with N-acetylglucosamine, the sugar bound specifically by this lectin. This lectin is also known to bind sialic acid. A role of this group in opioid receptor binding to WGA-agarose is therefore possible. All of the three major types of opioid receptors are retained on immobilized WGA (Valette et al 1986).

SEPARATION OF κ SITES

Using the opioid binding sites solubilized from guinea pig brain, Dr Itzhak in our laboratory (Itzhak et al 1984) tried to separate different receptor types. Due to the presence of high salt and digitonin the crude extracts, though they bind antagonists with high affinity, do not bind agonists very well. Dr Itzhak achieved high affinity agonist binding by centrifugation of the extracts on sucrose density gradients containing no NaCl and a low concentration of digitonin. Centrifugation into such a gradient and determination of binding in the various fractions showed two peaks of binding activity. The first peak was able to bind only ligands of the κ type, benzomorphans, such as bremazocine and ethylketocyclazocine, while the second peak exhibited high affinity for ligands selective for μ sites, such as morphine and DAGO and for δ sites, such as D-Ala^2-Leu$_5$-enkephalin (DADLE). Considerable evidence was accumulated supporting the hypothesis that κ sites had been separated from μ and δ sites. To cite one of the strongest pieces of evidence, extracts of guinea pig cerebellum, previously shown by Robson et al (1984) to contain mainly κ receptors, when placed on the same sucrose gradient, gave only one major peak identical to the first peak obtained with receptors solubilized from guinea pig brain.

Separation of κ sites from other opioid receptor types has also been reported by Chow and Zukin (1983), who used gel filtration columns on

CHAPS extracts of rat brain membranes.

BINDING SUBUNITS OF μ AND δ SITES

Since we were unable to achieve separation of active μ and δ sites from each other we resorted to a different approach to the question of whether these are separate molecules or different forms of a single molecule. We set out to determine whether the two receptors have the same or different binding subunits.

Early in his graduate research Andrew Howard became aware of the commercial availability of human ^{125}I-ß-endorphin. Iodination of the tyrosine in position 1 has long been known to destroy virtually all binding affinity of ß-endorphin. The peptide isolated from human tissues is the only one that contains a tyrosine residue in position 27. Derek Smyth in London devised ways of iodinating human ß-endorphin with ^{125}I and separating the I_{27} from the I_1-labeled peptide, a not insignificant feat, since the two peptides have the same size and charge. This labeled ligand (^{125}I-ß-EndH) is marketed by Amersham. Dr. Andrew Howard used this material to do covalent crosslinking experiments which will now be described (Howard et al 1985, 1986).

Membrane binding studies with ^{125}I-ß-End demonstrated a high degree of specific binding (ca.90%) and high affinity for both μ and δ sites but very low affinity for κ sites, in agreement with previous reports. Incubation of membranes, prebound with ^{125}I-ß-End_H and washed thoroughly, with a crosslinking reagent, such as bis[2-(succinimido-oxycarbonyloxy)ethyl]sulfone (BSCOES) resulted in the covalent linking of about 40-50% of the prebound ligand to the μ and δ receptors. The crosslinked receptors were extracted with SDS and run on SDS-polyacrylamide gel electrophoresis (SDS-PAGE). Autoradiography of the gels resulted in the visualization of several labeled bands, the pattern varying somewhat with the tissue used. Four labeled bands have been observed in extracts of brain tissue, of molecular sizes 65K, 53K, 41K and 38K daltons. The significance of the two smaller bands is as yet not established, though their labeling is suppressed by excess opiates, suggesting that they are constituents of binding sites. The ratio of 65K to 53K varied roughly in proportion to the μ:δ ratio in the tissue. In rat thalamus, a tissue rich in μ sites, the 53K band is undetectable, while in bovine frontal cortex (μ:δ=1:3), both bands are heavily labeled. When β-End_H is crosslinked to neuroblastoma x glioma hybrid cells (NG108-15), known to contain only δ sites, the 53K band is heavily labeled while the 65K band is absent. (There is an additional band at 25K, not seen in brain regions). These results support the notion that μ and δ receptors contain binding components that differ in molecular size. However, further proof is needed since other explanations, such as proteolysis of the 65K polypeptide to the 53K during isolation, could explain the data.

For additional evidence we did competition studies using highly selective μ and δ ligands. When the binding of ^{125}I-β-End_H was carried out in the presence of the μ-selective ligand DAGO, followed by crosslinking, the autoradiogram of the SDS-PAGE showed a drastic reduction in the labeling of the 65K band while the 53K band was only minimally reduced in intensity, if at all. Similarly, binding in the presence of the δ-selective peptide DPDPE led to a virtual disappearance of the 53K protein, while little effect was seen on the labeling of the 65K band in the same extract. These results were seen in guinea pig brain and in several bovine brain region. They have also now been extended to a cell line, SK-N-SH, which has a preponderance of μ sites and which was compared to NG 108-15, which has only δ sites (Keren et al 1987). Reduction in the labeling of the appropriate band is observed at concentrations of the selective ligands as low as 2nM.

All of these results provide strong evidence in favor of binding components of different molecular size for μ and δ receptors and suggest that these two types of opioid receptors may be distinct molecular species. It must be stressed that we do not yet know whether the differences reside in the polypeptide chains or in post-translational modifications. We cannot therefore state whether or not separate genes code for these two receptor types.

The two other labeled bands, one at 41K Da which was present only in guinea pig brain and bovine caudate, and one at 38K Da which was seen in all brain tissues studied are sensitive to both μ and δ ligands. Their significance is currently not clear. Several possibilities present themselves. They could be separate receptors (e.g. epsilon) with some affinity for μ and δ ligands, they could be common subunits of the μ and δ receptors, or they could be proteolytic breakdown products of both sites that happen to migrate to the same spot, unlikely but possible.

PURIFICATION OF OPIOID RECEPTORS

The ultimate goal of this research is the purification to homogeneity of all types of opioid receptors and the determination of the full amino acid sequence and sugar composition of these glycoproteins. We will describe the purification to apparent homogeneity of an opioid receptor from bovine striatal membranes and will present evidence that it is of the μ type.

It should be noted the other laboratories have also reported success in receptor purification. Simonds et al (1985) have purified the δ receptor from NG108-15 cells, affinity-labeled with a derivative of fentanyl ("superFIT"). Active μ receptors have been purified in the laboratories of Bidlack and Abood (1981), Zukin (Maneckjee et al 1985), Loh et al (Cho et al 1986), and Barnard (personal communication).

The purification scheme used in our laboratory (Gioannini et al 1985) involves two major steps. The most powerful step is passage through a ligand affinity column and elution with high concentrations of a ligand. The column consists of a derivative of naltrexone, in which the carbonyl group in position 6 is replaced by an ethylenediamine moiety. This molecule was synthesized by T.L. Gioannini and has been called naltiesylethylonedidmine (NED) (Gioannini et al 1984). It was coupled to CH-Sepharose beads (Biorad) using a water soluble carbodiimide. In one pass through such a column the receptor is purified 3000-5000-fold and recovered in a yield of 20-30%, based on opioid binding. This step is followed by chromatography on a column of wheat germ agglutinin, a lectin to which, as we showed earlier, opioid receptors bind quite efficiently by virtue of their N-acetylglucosamine (and perhaps sialic acid) residues. Elution in about 50% yield is achieved with high concentrations of N-acetylglucosamine. This step gives an additional 15-20-fold purification, yielding an overall purification of ca. 60000-70000 at an overall yield of 5-10%. Assuming a single binding site per molecular weight of 65K the theoretical purification needed to achieve homogeneity is 77000 and the theoretical specific binding activity is 15400 pmol/mg protein, which is approximated by our purified receptors (13000-14000 pmol/mg protein). These results indicate that we have achieved virtual homogeneity, which is further supported by the finding that a single band of molecular weight 65K is seen when radioiodinated, purified material is run on SDS-PAGE. Only one major spot is seen on two-dimensional gels and the protein was found to have an isoelectric point of 6.0-6.2 by isoelectric focusing.

The evidence that the purified protein is a component of the μ receptor is as follows: 1. The NED-agarose column has a preference for μ sites. 2. There are very few δ sites in bovine striatum. 3. The molecular weight of 65K is in agreement with that obtained by crosslinking μ receptors to ß-endorphin (see above), and 4. Most importantly, it has been possible to crosslink ^{125}I-ß-EndH to the purified binding sites, which cannot be done with κ sites.

WORK CURRENTLY IN PROGRESS TOWARD RECEPTOR SEQUENCING

We have recently succeeded in scaling up the purification procedure so that we can now purify several hundred picomoles of receptor in a few weeks. We have found that the amino terminal of the 65K protein is blocked, which significantly increases the amount of pure receptor required. In collaboration with Dr. Sidney Udenfriend and my former student Richard Howells, who is working in his laboratory at the Roche Institute of Molecular Biology, we are exploring methods for fragmenting the protein and purifying the resulting peptides for microsequencing. We are also actively trying to produce both monoclonal and polyclonal antibodies to the receptor, which would be extremely useful for isolation and identification and expression cloning. cDNA libraries have been prepared and are awaiting suitable oligonucleotide probes for screening for cDNA complementary to opioid receptor message. A knowledge of the total amino acid sequence of the major types of opioid receptors, the goal of this research, should bring significant advances in our understanding of the function of these receptors and their ligands. This has certainly been the case for the few receptors for which the amino acid sequence is known.

ACKNOWLEDGEMENTS

The research carried out in the author's laboratory was supported by grant No. DA-00017 from the National Institute on Drug Abuse.

REFERENCES

Atweh SF, Kuhar MJ (1977a) Autoradiographic localization of opiate receptors in rat brain I. Spinal cord and lower medulla. Brain Res 124:53-67

Atweh SF, Kuhar MJ (1977b) Autoradiographic localization of opiate receptors in rat brain II. The brainstem. Brain Res 129: 1-12

Atweh SF, Kuhar MJ (1977c) Autoradiographic localization of opiate receptors in rat brain III. The telencephalon. Brain Res 134:393-406

Bidlack JM, Abood LG (1980) Solubilization of the opiate receptor. Life Sci 27:331-340

Bidlack JM, Abood LG, Osei-Gyimah P, Archer S (1981) Purification of the opiate receptor from rat brain. Proc Natl Acad Sci USA 78: 636-639

Cardinale GJ, Donnerer J, Finck AD, Katrowitz JD, Oka K, Spector S (1987) Morphine and codeine are endogenous components of human cerebrospinal fluid. Life Sci 40:301-306

Cho TM, Hasegawa JI, Ge BL, Loh HH (1986) Purification to apparent homogeneity ot a u-type opioid receptor from rat brain. Proc Natl Acad Sci USA 83:4138-4142

Cho TM Yamato C Cho JS, Loh HH (1981) Solubilization of membrane bound opiate receptors from rat brain. Life Sci 28:2651-2657

Chow T, Zukin RS (1983) Solubilization and preliminary characterization of mu and kappa opiate receptor subtypes from rat brain. Mol Pharmacol 24:203-212

Collier HOJ, Roy AC (1974) Hypothesis: Inhibition of E-prostaglandin-sensitive adenyl cyclase as the mechanism of morphine analgesia. Prostaglandins 7:361-376

Cox BM, Goldstein F, Li CH (1976) Opioid activity of a peptide ß-lipotropin-(61-91) derived from ß-lipotropin. Proc Natl Acad Sci USA 73:1821-1823

Demoliou-Mason C, Barnard EA (1984) Solubilization in high yield of opioid receptors retaining high affinity delta, mu and kappa binding sites. FEBS Lett 170:378-382

Donnerer J, Oka K, Brossi A, Rice KC, Spector S (1986) Presence and formation of codeine and morphine in the rat. Proc Natl Acad Sci USA 83:4566-4567

Gilbert PE, Martin WR (1976) The effects of morphine- and nalorphine-like drugs in the nondependent morphine-dependent and cyclazocine-dependent chronic spinal dog. J Pharmacol Exp Therap 198:66-82

Gioannini TL, Foucaud B, Hiller JM, Hatten ME, Simon EJ (1982) Lectin binding of solubilized opiate receptors: Evidence for the glycoprotein nature. Biochem Biophys Res Commun 105:1128-1134

Gioannini TL, Howard A, Hiller JM, Simon EJ (1984) Affinity chromatography of solubilized opioid binding sites using CH-Sepharose modified with a new naltrexone derivative. Biochem Biophys Res Commu 119:624-629

Gioannini TL, Howard AD, Hiller JM, Simon EJ (1985) Purification of an active opioid-binding protein from bovine striatum. J Biol Chem 260:15117-115121

Goldstein A, Barrett RW, James IF, Lowney LI, Weitz CJ, Knipmeyer LL, Rapoport H (1985) Morphine and other opiates from beef brain and adrenal. Proc Natl Acad Sci USA 82:5203-5207

Goldstein A, Tachibana S, Lowney LI, Hunkapiller M, Hood L (1979) Dynorphin (1-13) an extraordinarily potent opioid peptide. Proc Natl Acad Sci USA 76:6666-6670

Goldstein A, Fischli W, Lowney LI, Hunkapiller M, Hood L (1981) Porcine pituitary dynorphin: complete amino acid sequence of a biologically active heptadecapeptide. Proc Natl Acad Sci USA 78:7219-7223

Gubler U (1986) Enkephalin genes In:Habener JF (ed) Molecular cloning of hormone genes. The Humana Press, Clifton, NJ pp 229-276

Hiller JM, Pearson J, Simon EJ (1973) Distribution of stereospecific binding of the potent narcotic analgesic etorphine in the human brain: predominance in the limbic system. Res Commu Chem Pathol Pharmacol 6:1052-1062

Howard AD, de la Baume S, Gioannini TL, Hiller JM, Simon EJ (1985) Covalent labeling of opioid receptors with radioiodinated human ß-endorphin. J Biol Chem 260:10833-10839

Howard AD, Sarne Y, Gioannini TL, Hiller JM, Simon EJ (1986) Identification if distinct binding site subunits of μ and δ opioid receptors. Biochemistry 25:357-360

Howells RD, Gioannini TL, Hiller JM, Simon EJ (1982) Solubilization and characterization of active opiate binding sites from mammalian brain. J Pharmacol Exp Ther 222:629-634

Hughes J, Smith TW, Kosterlitz HW, Fothergill LA, Morgan BA, Morris HR (1975) Identification of two related pentapeptides from the brain with potent opiate agonist activity. Nature 258:577-579

Itzhak Y, Hiller JM, Simon EJ (1984) Solubilization and characterization of μ δ and κ opioid binding sites from guinea pig brain: physical separation of κ receptors. Proc Natl Acad Sci USA 81:4217-4221

Keren O, Gioannini TL, Hiller JM, Simon EJ (1986) Affinity crosslinking of ^{125}I-human B-endorphin to cell lines possessing either μ or δ type opioid binding sites. Brain Res in press

Kuhar MJ, Pert CB, Snyder SH (1973) Regional distribution of opiate receptor binding in monkey and human brain. Nature 245: 447-450

Lord JAH, Waterfield AA, Hughes J, Kosterlitz HW (1977) Endogenous opioid peptides: multiple agonists and receptors. Nature 267:495-499

Mains RE, Eipper BA, Ling N (1977) Common precursor to corticotropins and endorphins. Proc Natl Acad Sci USA 74:3014-3018

Maneckjee R, Zukin RS, Archer S, Michael J, Osei-Gyimah P (1985) Purification and characterization of the μ opiate receptor from rat brain using affinity chromatography. Proc Natl Acad Sci USA 82:594-598

Martin WR, Eades CG, Thompson JA, Huppler RE, Gilbert PE (1976) The effects of morphine- and nalorphine-like drugs in the nondependent and morphine-dependent chronic spinal dog. J Fharmacol Exptl Ther 197:517-532

Nakanishi S, Inoue A, Kita T, Nakamura M, Chan ACY, Cohen SN, Numa S (1979) Nucleotide sequence of cloned cDNA for bovine corticotropin-B-lipoprotein precursor. Nature 278:423-427

Numa S (1984) Opioid peptide precursors and their genes. In: Udenfriend S, Meienhofer J (eds) The peptides - analysis synthesis biology, Volume 6 opioid peptides: biology chemistry and genetics. Academic Press, New York, pp 1-23

Oka K, Kantrowitz JD, Spector S (1985) Isolation of morphine from toad skin. Proc Natl Acad Sci USA 82:1852-1854

Pearson J, Brandeis L, Simon E, Hiller J (1980) Radioautography of binding of tritiated diprenorphin to opiate receptors in the rat. Life Sci 26:1047-1052

Pert CB, Snyder SH (1973) Opiate receptor:demonstration in nervous tissue. Science 182:1359-1361

Roberts JL, Herbert E (1977) Characterization of a mommon precursor to corticotropin and B-lipotropin: Cell-free synthesis of the precursor and identification of corticotropin peptides in the molecule. Proc Natl Acad Sci USA 74:4826-4830

Robson LE, Foote RW, Maurer R, Kosterlitz HW (1984) Opioid binding sites of the kappa- type in guinea pig cerebellum. Neuroscience 12:621-627

Ruegg UT, Cuenod S, Hiller JM, Gioannini TL, Howells RD, Simon EJ (1981) Characterization and partial purification of solubilized active opiate receptors from toad brain. Proc Natl Acad Sci USA 78:4635-4638

Ruegg UT, Hiller JM, Simon EJ (1980) Solubilization of an active opiate receptor from Bufo marinus. Eur J Pharmacol 64:367-368

Sharma SK, Klee WA, Nirenberg M (1975) Dual regulation of adenylate cyclase accounts for narcotic dependence and tolerance. Proc Natl Acad Sci USA 72:3092-3096

Simon EJ, Hiller JM, Edelman I (1973) Stereospecific binding of the potent narcotic analgesic ^{3}H-etorphine to rat brain homogenate. Proc Natl Acad Sci USA 70:1947-1949

Simon EJ, Hiller JM, Edelman I (1975) Solubilization of a stereospecific macromolecular complex from rat brain. Science 190:389-390

Simonds WF, Burke TR Jr, Rice KC, Jacobson AE, Klee WA (1985) Purification of the opiate receptor of NG108-15 neuroblastoma-glioma hybrid cells. Proc Natl Acad Sci USA 82:4974-4978

Simonds WF, Koski G, Streaty RA, Hjelmeland LM, Klee WA (1980) Solubilization of active opiate receptors. Proc Natl Acad Sci USA 77:4623-4627

Stefano GB, Kream RM, Zukin RS (1980) Demonstration of stereospecific opiate binding in the nervous tissue of the marine mollusc Mytilus edulis. Brain Res 181:440-445

Terenius L (1973) Stereospecific interaction between narcotic analgesics and a synaptic plasma membrane fraction of rat cerebral cortex. Acta Pharmacol Toxicol 32:317-329

Young WS III, Kuhar MJ (1979) A new method for receptor autoradiography: [^{3}H] opioid receptors in rat brain. Brain Res 179:255-270

Zukin SR, Zukin RS (1979) Specific [^{3}H] phencyclidine binding in rat central nervous system. Proc Natl Acad Sci USA 76:5372-5376

Valette A, Rouge P, Coulais E, Potonnier G, Gros J, Simon EJ (1987) Interaction with lectin of kappa opiod binding sites solubilized from human placenta. Life Sci 40:535-540

AUTHOR INDEX

SUBJECT INDEX

NATO ASI Series H

Vol. 1: **Biology and Molecular Biology of Plant-Pathogen Interactions.** Edited by J.A. Bailey. 415 pages. 1986.

Vol. 2: **Glial-Neuronal Communication in Development and Regeneration.** Edited by H.H. Althaus and W. Seifert. 865 pages. 1987.

Vol. 3: **Nicotinic Acetylcholine Receptor: Structure and Function.** Edited by A. Maelicke. 489 pages. 1986.

Vol. 4: **Recognition in Microbe-Plant Symbiotic and Pathogenic Interactions.** Edited by B. Lugtenberg. 449 pages. 1986.

Vol. 5: **Mesenchymal-Epithelial Interactions in Neural Development.** Edited by J.R. Wolff, J. Sievers, and M. Berry. 428 pages. 1987.

Vol. 6: **Molecular Mechanisms of Desensitization to Signal Molecules.** Edited by T.M. Konijn, P.J.M. Van Haastert, H. Van der Starre, H. Van der Wel, and M.D. Houslay. 336 pages. 1987.

Vol. 7: **Gangliosides and Modulation of Neuronal Functions.** Edited by H. Rahmann. 647 pages. 1987.

Vol. 8: **Molecular and Cellular Aspects of Erythropoietin and Erythropoiesis.** Edited by I.N. Rich. 460 pages. 1987.

Vol. 9: **Modification of Cell to Cell Signals During Normal and Pathological Aging.** Edited by S. Govoni and F. Battaini. 297 pages. 1987.

Vol. 10: **Plant Hormone Receptors.** Edited by D. Klämbt. 319 pages. 1987.

Vol. 11: **Host-Parasite Cellular and Molecular Interactions in Protozoal Infections.** Edited by K.-P. Chang and D. Snary. 425 pages. 1987.

Vol. 12: **The Cell Surface in Signal Transduction.** Edited by E. Wagner, H. Greppin, and B. Millet. 243 pages. 1987.

Vol. 13: **Toxicology of Pesticides: Experimental, Clinical and Regulatory Perspectives.** Edited by L.G. Costa, C.L. Galli, and S.D. Murphy. 320 pages. 1987.

Vol. 14: **Genetics of Translation. New Approaches.** Edited by M.F. Tuite, M. Picard, and M. Bolotin-Fukuhara. 524 pages. 1988.

Vol. 15: **Photosensitisation. Molecular, Cellular and Medical Aspects.** Edited by G. Moreno, R.H. Pottier, and T.G. Truscott. 521 pages. 1988.

Vol. 16: **Membrane Biogenesis.** Edited by J.A.F. Op den Kamp. 477 pages. 1988.

Vol. 17: **Cell to Cell Signals in Plant, Animal and Microbial Symbiosis.** Edited by S. Scannerini, D. Smith, P. Bonfante-Fasolo, and V. Gianinazzi-Pearson. 414 pages. 1988.

Vol. 18: **Plant Cell Biotechnology.** Edited by M.S.S. Pais, F. Mavituna, and J.M. Novais. 540 pages. 1988.

Vol. 19: **Modulation of Synaptic Transmission and Plasticity in Nervous Systems.** Edited by G. Hertting and H.-C. Spatz. 457 pages. 1988.

Vol. 20: **Amino Acid Availability and Brain Function in Health and Disease.** Edited by G. Huether. 487 pages. 1988.

NATO ASI Series H

MIX
Papier aus verantwortungsvollen Quellen
Paper from responsible sources
FSC® C105338

If you have any concerns about our products,
you can contact us on
ProductSafety@springernature.com

In case Publisher is established outside the EU,
the EU authorized representative is:
Springer Nature Customer Service Center GmbH
Europaplatz 3, 69115 Heidelberg, Germany

Printed by Libri Plureos GmbH
in Hamburg, Germany